Development of Laser Welding and Surface Treatment of Metals

Development of Laser Welding and Surface Treatment of Metals

Editor

Aleksander Lisiecki

MDPI • Basel • Beijing • Wuhan • Barcelona • Belgrade • Manchester • Tokyo • Cluj • Tianjin

Editor
Aleksander Lisiecki
Department of Welding
Engineering
Silesian University of
Technology
Gliwice
Poland

Editorial Office
MDPI
St. Alban-Anlage 66
4052 Basel, Switzerland

This is a reprint of articles from the Special Issue published online in the open access journal *Materials* (ISSN 1996-1944) (available at: www.mdpi.com/journal/materials/special_issues/welding_metal).

For citation purposes, cite each article independently as indicated on the article page online and as indicated below:

LastName, A.A.; LastName, B.B.; LastName, C.C. Article Title. *Journal Name* **Year**, *Volume Number*, Page Range.

ISBN 978-3-0365-6701-3 (Hbk)
ISBN 978-3-0365-6700-6 (PDF)

Cover image courtesy of Aleksander Lisiecki

Contents

About the Editor

Aleksander Lisiecki

Aleksander Lisiecki is a Professor at Silesian University of Technology in Gliwice, Poland, where he has been working since graduating in 1998. He is also a lecturer at the WSB University in Poznań. He completed internships at the University of Warwick, Warwick Manufacturing Centre in Coventry, United Kingdom. He obtained a Ph.D. in the discipline of materials science in 2001, and then a higher doctorate in 2017. He is a Vice-Head of the Department of Welding Engineering. He is involved in teaching and research. He executed many scientific projects and was involved in many research and development projects as R&D manager in the field of manufacturing technologies, especially welding, laser processing and coatings. He participated in many international conferences, and he is the author or co-author of over 250 publications in Polish and English, 15 patents and 17 patent applications. His main areas of research and interests include advanced methods of material manufacturing and processing, advanced structural materials and materials with special properties, design and manufacturing of structures, assurance and quality control, and automation and robotization of production processes.

He is an expert in many organizations and a technical advisor to many manufacturing and service enterprises. Author of many expert opinions, including court opinions. The member of many associations and organizations, including the Polish Welding Society, the Polish Production Management Association, and the Polish Innovation Management Association.

Preface to "Development of Laser Welding and Surface Treatment of Metals"

Welding is a strong industry in various regions of the world, regardless of the level of economic development. However, in countries with a high level of economic development, the most advanced welding technologies and materials processing technologies are dynamically developed. One of the most advanced technologies of welding and materials processing are based on high power density heat sources, such as the focused laser beam.

The aim of the book was to present the latest achievements in the field of research results and the state of the art in the field of laser welding and surface treatment by experienced and also young scientists who are dynamically active in this field.

The main scope of the book deals with important issues related to the development of science and technology in the field of application the laser beam for joining, surface treatment and coatings. However, the thematic scope is not limited only to the above issues. The scope of the book covers also issue related with advances in computational modelling of laser and arc heat sources, very unique techniques of underwater welding, advanced and unique techniques of localised forced cooling the weld metal during welding, hybrid process of laser deposition of metal matrix composite coatings under cryogenic conditions, microstructural and mechanical characterisation of coatings and joints produced by different welding technologies.

The motivations for writing the book was collecting the valuable information, both theoretical and practical research results in the field of advanced technologies of joining, surface treatment and coatings, quality control and assessment, as well as management of the technological processes.

The book is addressed to young scientists, engineers, and students in the field of welding and surface engineering, materials science and manufacturing engineering. The book contains 22 chapters presented results of interesting and recent researches.

The wide range of experimental, numerical or analytical study conducted and presented by 58 authors, representing 19 academic or research centers and 3 industrial centers, from 5 countries (Poland, China, Slovakia, Lithuania, and Ukraine) and 16 cities, proved the topicality, and importance of this Special Issue.

Concluding, I thank to all the authors, editors, and technical assistants who have made invaluable contributions to make this edition of the book a success.

Aleksander Lisiecki
Editor

materials

Editorial

Development of Laser Welding and Surface Treatment of Metals

Aleksander Lisiecki

Department of Welding Engineering, Faculty of Mechanical Engineering, Silesian University of Technology, Konarskiego 18A Street, 44-100 Gliwice, Poland; aleksander.lisiecki@polsl.pl

Abstract: This Special Issue on Development of Laser Welding and Surface Treatment of Metals contains as many as twenty-two research articles mainly related to the application of lasers, but also on other welding processes that may be competitive to laser technologies under specific conditions. Despite the introduction of lasers for material processing in the 1960s, the continuous development of laser devices also leads to the development and expansion of laser technology applications. This Special Issue is a compendium of knowledge in the field of fusion welding, the manufacturing of surface layers and coatings with increased wear resistance and tribological characteristics, as well as corrosion resistance and the characterization of coatings and surface layers. The topics of the presented research articles include aspects related to laser welding (eight articles), especially technological conditions, the properties of different types of joints, and analytical and numerical aspects of modelling the laser heat sources. The second dominant issue concerns laser cladding and laser surface treatment of different ferrous and nonferrous metallic and composite materials (six articles). In addition, there are interesting results of the study of fusion welding under forced cooling of the deposit or underwater conditions (four articles), results on the characterization of wear resistance coating produced by different technologies that can be competitive for laser cladding (three articles), and an original study on local strengthening of the thin-walled structure by laser treatment (one article). This Special Issue provides very wide and valuable knowledge based on theoretical and empirical study in the field of laser and fusion welding, laser and related coating technologies, characterization of coatings, and wear phenomena.

Keywords: welding; laser processing; joining; tribology; cladding; properties of surface layers

Citation: Lisiecki, A. Development of Laser Welding and Surface Treatment of Metals. *Materials* **2022**, *15*, 1765. https://doi.org/10.3390/ma15051765

Received: 3 February 2022
Accepted: 20 February 2022
Published: 26 February 2022

Publisher's Note: MDPI stays neutral with regard to jurisdictional claims in published maps and institutional affiliations.

1. Introduction

The manufacturing and processing of materials is the basis of the modern economy and of civilization progress. Striving for sustainable development as well as limiting the harmful impact on the environment, including the reduction in CO_2 emissions and the reduction in energy consumption, leads to a continuous increase in the effectiveness and efficiency of technological processes, machines, devices, vehicles, tools, and building structures [1–3].

Therefore, new materials are being developed with higher mechanical properties, higher wear resistance and thus higher operating parameters. The conventional welding methods and technologies usually do not provide satisfactory results in applications such as joining and cladding of modern materials, e.g., high strength steel, nonferrous metals and alloys, composite materials, and nanostructured and hybrid materials [1,4–8].

The main disadvantages of typical methods of welding and cladding such as conventional arc or plasma arc methods are excessive heat input and the relatively large volume of the molten pool. This leads to overheating of the material, internal stresses and deformations, and unfavorable grain growth or dissolution of the reinforcing phase particles, e.g., carbides or nitrides. Therefore, the attention of researchers and industry is focused on the search for methods of joining and processing materials that ensure a minimum thermal effect on the material and the ability to precisely control the amount of heat and the thermal cycle of the process [9–13].

One of the dynamically developed areas of material processing technology in recent years is laser material processing technology. In the area of joining materials, laser welding shows several advantages because it provides high-power densities and low diameter of the laser beam spot, and thus high penetration depth, high welding speed, and low, precise, and controllable heat input. The above features are particularly important for the joining of modern and advanced alloys such as ultra-high-strength steels (UHSS), advanced high-strength steels (AHSS), modern stainless duplex and super duplex steel, nonferrous metals, and light metals, e.g., titanium and aluminum alloys [6,11,13–17].

Another area of laser beam application that is currently being intensely developed is the shaping of the properties of surface layers and the production of coatings for enhanced wear characteristics such as corrosion resistance, tribological properties, abrasive resistance, thermal and mechanical fatigue resistance, and resistance to impact load [1–8,18].

In the area of surface treatment and cladding, the laser beam as a heat source is also advantageous. The most significant advantages include low and controllable heat input, and thus limited thermal impact on the substrate, reduced internal stresses, minimized distortions, high solidification and cooling rates, low penetration depth, and low dilution of the clad layer by the substrate material. Low heat inputs and related high cooling rates during laser processing are beneficial in many applications of surface treatment and cladding as they provide high quality clads and surface layers. Usually, surface layers are characterized by superior metallurgical bonding and a fine-grained and refined microstructure, which are decisive for providing superior tribological characteristics, high abrasive or erosion wear resistance, and corrosion or fatigue resistance [1,8,18].

It is worth noting that laser-made clads and surface layers show higher functional properties compared to surface layers made by plasma arc (PTA) or conventional arc cladding processes, even if the same additional materials are applied. This is due to intense grain refinement caused by very fast cooling and solidification of the molten pool, typically an order of magnitude higher than plasma (PTA) and conventional arc cladding (e.g., GTA–gas tungsten arc) [1,12].

Thanks to the advantages mentioned above, lasers are used in different processes of surface treatment such as laser surface hardening (LSH), melting (LSM), shocking (LSS), texturing (LST), alloying (LSA), cladding (LSC), remelting (LSR), and surface deposition (LSD), and in additive manufacturing processes such laser metal deposition (LMD) or selective laser sintering and melting (SLS/SLM) [1].

Continuous development of laser generators and related improvement of laser radiation and laser beam characteristics enable further development in the field of laser material processing, including laser welding and laser surface treatment. In addition to the gaseous CO_2 and solid-state rod type Nd:YAG laser generators used for the last few decades, new generations of solid-state lasers are currently available such as high-power diode lasers (HPDL), disk lasers, and fiber lasers [1,14–17]. Modern generations of laser devices offer new or expanded technological possibilities. For example, as shown in the literature, high-power diode lasers with a square or rectangular beam spot and uniform energy distribution (multimode beam) are favorable for surface treatment applications. The design of diode lasers has been developed to the extent that today there are direct diode lasers (DDL) which emit single-mode radiation with power up to 8 kW, suitable for cutting applications. Moreover, the single-mode DDL lasers are even competitive with CO_2 gas lasers and disk lasers, as well as fiber lasers [1,14,17,18].

There has also been significant progress in the design and construction of fiber generators in the last decade. Modern fiber lasers are also beneficial due to their compact design, high wall plug efficiency, low beam divergence, and possibility to focus the beam into a very small spot, even at relatively high power [18]. Moreover, these fiber lasers can provide a single-mode beam up to 20 kW, while the multimode industrial applications reach a power level of 100 kW. As a result, lasers are currently used for one-sided or two-sided welding, even for thick steel plates with a thickness up to approx. 30 mm.

Another interesting example of dynamic development in the field of material processing technology is hybrid processes. According to the definition, the hybrid process combines at least two different machining processes which must be carried out simultaneously and in the same processing area (e.g., one weld pool created during fusion welding). In the case of welding, the laser beam is often combined with the arc, e.g., laser + GMA (gas metal arc), laser + GTA (gas tungsten arc) or laser + PTA (plasma tungsten arc). Such combinations of two different heat sources allow taking the advantages of both welding methods. The laser beam is responsible for deep penetration of the material, while the arc processes provide additional heat and material to fill the gap. Consequently, it is possible to weld at high speed and ensure the correct shape of the weld at high depth. Hybrid processes are also being developed in the field of surface treatment and coating applications, ensuring higher functional parameters or the possibility of producing coatings with special properties [1,18].

2. Short Characterization of the Special Issue

The Materials and Methods should be described with sufficient details to allow others to replicate and build on the published results. Please note that the publication of your manuscript implicates that you must make all materials, data, computer code, and protocols associated with the publication available to readers. Please disclose at the submission stage any restrictions on the availability of materials or information. New methods and protocols should be described in detail, while well-established methods can be briefly described and appropriately cited.

This Special Issue, entitled "Development of Laser Welding and Surface Treatment of Metals", is a complementary and valuable resource of knowledge in the following fields:

- Study of technological conditions of laser welding, investigation of test joints properties [19–26];
- Analytical and numerical models of heat sources of laser beam and arc welding processes [19,20,22,23];
- Study of laser cladding and investigations of the coating's properties and tribological characteristics [27–33];
- Study and development of fusion welding processes, investigation of properties of the deposit produced under forced cooling or underwater conditions [34–36];
- Study of cladding, surfacing, and thermal spraying processes of wear-resistant coatings, characterization of the coatings [37–39];
- Study of application of local strengthening of the thin-walled structure by laser treatment [40].

In the work presented by Wang et al. [19], plates of X100 pipeline steel with a thickness of 12.8 mm were welded by a high-power laser on a robotized cell. The authors provided a quantitative correlation between thermal cycling and the microstructure of the welded joint by means of empirical and numerical study. They also showed that the effects of austenitization are more significant than those of the cooling rate on the final microstructures of the laser-welded joint of the investigated steel grade. The results of the presented work can provide scientific guidance and a reference for the simulation of the temperature field during laser welding, especially in the case of laser welding of X100 pipeline steel grade.

Kik [20] presented different computational techniques for comparison in the case of single and multipass arc and laser beam welding processes. The numerical analyses were conducted by the SYSWELD software package and the author showed differences between the applied computational techniques, as well as the benefits and disadvantages of the applied computational techniques. The presented results of the numerical simulations may be used for preparing an efficient and fast optimization of the laser and related fusion welding processes, according to the criteria of minimizing deformations or identification of potential defects of structures.

In the second article, Kik [23] presented new possibilities for modifying heat source models in numerical simulations of laser welding processes conducted using VisualWeld

(SYSWELD) software. He proposed a modification of heat source models and methods of defining the heat introduced into welded materials using solid-state and high-power diode lasers. He provided calibration and validation of the proposed models of heat sources based on metallographic tests and thermal cycles; therefore, the presented results may be helpful for further numerical analysis of laser processing by different beam shape and energy distribution.

The study presented by Górka [21] was focused on investigation of the influence of the basic parameters of autogenous laser welding of 700MC 10.0 mm thick plates on their microstructure and mechanical properties. He provided a comprehensive analysis of the microstructure and a detailed analysis of the phase constituents. The quality of the test joint was estimated as meeting level B, in accordance with 13919-1 standard. He concluded that in cases of S700MC steel, the analysis of the phase transformation of austenite exposed to welding thermal cycles and the value of carbon equivalent cannot be the only factors taken into consideration when assessing weldability.

The laser welding process was also investigated by Danielewski and Skrzypczyk [22]. They presented results of numerical and experimental analyses of the gaseous CO_2 laser welding of lap joints of low-carbon structural steel. They determined hardness profiles and weld geometry for the welds produced at different parameters. They also provided detailed analysis of the metallographic structure of fusion zone (FZ) and heat-affected zone (HAZ), as well as a quality assessment of the tested joints.

The study presented by Landowski et al. [24] concerned fiber laser welding of dissimilar butt joints between 316 L austenitic and 2304 lean duplex stainless steel plates of 8.0 mm thickness. Based on nondestructive tests and metallographic examinations, they showed that the tested joints meet the acceptance criteria for level B, in accordance with EN ISO 13919–1 standard. Finally, they proved that the elaborated procedure can be applied for welding of 316 L–2304 stainless steel plates with a thickness of 8.0 mm without ceramic backing. The results also provide valuable guidance for further research in this field.

The results of the study presented by Pańcikiewicz et al. [25] also concerned laser welding of dissimilar joints of AISI 430F (X12CrMoS17) martensitic stainless steel and AISI 304 (X5CrNi18-10) austenitic stainless steel. They provided detailed analysis of the microstructure of the base metals and the dissimilar test joint. However, the test joint meets the quality level C, in accordance with the ISO 13919-1 standard, due to gas pores in the weld metal.

Another study on laser welding was presented by Tofil et al. [26]. They conducted a test of laser welding of thin-walled AISI 316 L steel pipes, with diameters of 1.5 and 2.0 mm, used in medical equipment. They determined the range of optimal parameters of laser welding with a SISMA LM-D210 Nd:YAG laser. The results of the study showed good joint properties with a strength of more than 75% in the thinner pipe, uniform distribution of alloying elements, and a complex dendritic structure characteristic of pulse welding. Therefore, the presented results may also be guidance for further research in this field.

The topic of laser surface treatment and cladding begins with Lisiecki's article [27] on the study of the optical properties of surface layers produced on the titanium alloy substrate by surface melting and nitriding. The aim of this study was to determine the optical properties of the surface of titanium alloy Ti6Al4V. The specular, diffuse, and total reflection was determined for different surface conditions (roughness and oxides layer) and after laser surface treatment. The experiments were conducted for 808 nm wavelength, typical for diode lasers. The main conclusion of the study was that the distinct differences in absorption affected the heat transfer and thermal conditions of laser heating, and thereby the penetration depth during laser melting and nitriding of the titanium alloy.

A highly interesting study on laser surface modification of an aluminum alloy by boron carbides was carried out by Sroka et al. [28]. The fiber laser YLS-4000 with a maximum output power of 4.0 kW was used in this study. The effect of basic laser-alloying parameters of the foundry AlMg9 alloy with the B_4C particle on the microstructure and chosen properties was investigated and determined.

Another interesting study on laser cladding was performed by Li et al. [29]. In their study, the high aluminum and chromium Fe-B-C coating was produced by laser cladding on a 2Cr13 steel substrate. They investigated the microstructure and tribological characteristics. The results showed that the coating exhibited excellent wear resistance compared to the substrate. They concluded that the results could provide technical support in improving the properties of the Fe-based laser-cladded coating.

The next study, presented by Majkowska-Marzec et al. [30], is related to investigation of the mechanical and corrosion properties of a laser surface-treated titanium alloy with carbon nanotube coating. The coating was initially formed by electrophoretic deposition (EPD) and then subjected to laser treatment. The authors pointed to some advantages of such a coating. The main advantages were increased corrosion resistance, a lack of surface cracks, increased strength, and favorable contact angles between 46° and 82° (resulting in hydrophilic surfaces suitable for cell adhesion).

A comparative study of laser and plasma cladding was presented by Czupryński et al. [31]. The coatings of Inconel 625 on the 16Mo3 steel pipes were produced by plasma (PTA) cladding and high-power direct diode laser (HPDDL) cladding. The authors determined the quality of the coating by NDT testing, metallographic examinations, and microstructure analysis. They concluded that both tested and compared methods can provide high-quality protective coatings that may be operated at elevated temperatures up to 625 °C.

Another study on the effect of laser treatment on tribological properties was presented by Lubas et al. [32]. They analyzed the influence of different engine oils on the tribological parameters of sliding couples with a laser-borided surface layer produced on specimens of AISI 5045 steel, by laser remelting of a surface layer coated with amorphous boron. The tribological and wear characteristics were investigated on the pairs of AISI 5045 steel and SAE-48 bearing alloy, lubricated with 5W-40 and 15W-40-type engine oils. The detailed analysis of wear mechanisms was identified and described.

The thematic series on laser surfacing and cladding ends with the article presented by Lisiecki and Ślizak [33] on hybrid laser deposition of composite coatings at cryogenic conditions. This study demonstrated the effect of forced and localized cooling by a nitrogen vapor stream under cryogenic conditions during laser deposition of WC-Ni powder on the geometry and microstructure of clad layers and the dry sliding wear resistance of the coatings. Additionally, the quantitative relationship between heat input, cooling conditions, and carbide grain size distribution, as well as carbide share in relation to the matrix, was determined. It was shown that the novel demonstrated technique of localized forced cooling during laser cladding has advantageous effects. The forced localized cooling provides approximately 20% lower penetration depth and dilution, decreases tendency for tungsten carbide decomposition, and provides more uniform distribution and a higher share of massive eutectic W_2C-WC carbides across the coating.

Another very novel and interesting subject related to arc welding processes was presented by Szczucka-Lasota and Szymczak et al. [34,35]. The study presented in two articles was related to the application of a novel and patented technique of localized cooling of the weld by micro-jet streams. The first presented article is focused on the influence of processing conditions, especially localized cooling on the properties of the weld of Docol 1200 M steel. They found that the novel technique of cooling can provide a high fatigue limit at the level of at least 480 MPa. The second article presented by Szymczak et al. [35] is related to a similar issue but a different type of steel: S960MC sheet of 2.0 mm thickness. In this case, the Wöhler S–N curve of the weld was determined, indicating that the value of the fatigue limit of the tested weld was 100 MPa. The weld at the Union NiMoCr welding wire was indicated as the joint with the highest resistance on static and fatigue loadings.

Some manuscripts dealt with other arc processes and technologies, but the presented solutions were very innovative. An example of a highly interesting subject was presented by Tomków et al. [36]. They investigated underwater welding in terms of the role of the bead sequence during welding under such conditions. The test beads were produced by covered electrodes with the use of normalized S355G10 + N steel. They reported that welding in the underwater environment carries many problems related to arc stability, and thus the properties of the welds. Moreover, they revealed the effects of refining and tempering the microstructure in heat-affected zones of the previous beads, which resulted in a reduction in hardness.

The study presented by Mician et al. [37] was related to the influence of bead sequence on the deposit and heat-affected zone properties during arc surfacing. The process applied to surfacing was gas metal arc welding (GMAW) and the substrate plate was an S355 steel plate. The authors provided a comparative analysis of structure and hardness, considering the thermal impact of the bead sequence. As a result of the study, the most favorable sequence in terms of structure and hardness distribution, maximum hardness, and range of hardness has been established.

The second study presented by Czupryński [38] was focused on flame spraying of aluminum coatings reinforced by carbon nanotubes as an alternative for laser cladding technology. The microstructure of the coatings was analyzed, as well as abrasion and erosion resistance. He concluded that the flame spraying of aluminum coatings reinforced with the carbon nanotubes can be an effective alternative for laser cladding technology. Moreover, such coatings can be implemented in the automotive industry for the production of components characterized by high strength, wear resistance, good thermal conductivity, and low density.

The next article provided by Czupryński [39] presented results of the investigations of abrasive wear characteristics of wear-resistant plates made by cladding with innovative tubular electrodes with a metallic core and an experimental chemical composition. Detailed analyses of microstructure and wear tests of different reference plates were conducted. The author showed a high metal–mineral abrasive wear resistance of the deposit produced by the experimental tubular electrode.

Another original issue was presented in the study by Kapustynskyj et al. [40]. They focused on the numerical and analytical study of the effects of laser surface processing of thin-plate low-carbon structural steel. The authors developed the original analytical methodology of the estimation of the cross-section area of the laser-processed metal, that can be applied to the choosing of a reasonable distance between the centers of the laser-processed tracks. The results of the experimental and finite element numerical and analytical analyses showed that the laser treatments of the surface of the steel plate increase the yield point of the laser-processed region, as well as the axial and flexural stiffness of the plate.

3. Concluding Remarks

This Special Issue was very successful due to the valuable articles that were submitted, the wide variety of topics and research problems that were undertaken, and in-depth analysis of the state of the art in different fields of laser and related processes. This Special Issue contains a significant number of research articles (22 in total); however, the entire number of manuscripts submitted to this Special Issue was half that higher. Unfortunately, some of the manuscripts have not gone through a very rigorous review process. Such a large interest and quantity of articles shows the importance of the issue and themes. It should also be emphasized that each research article in the field of materials processing is the result of tedious, long-term, and interdisciplinary research, usually conducted by a team of scientists.

The topics of the articles were mainly focused on laser welding, laser cladding, and laser surface treatment, as well as related processes of fusion welding and manufacturing of coatings.

The wide range of experimental, numerical, or analytical study conducted and presented by 58 authors, representing 19 academic or research centers and 3 industrial centers from 5 countries (Poland, China, Slovakia, Lithuania, and Ukraine) and 16 cities, proved the topicality and importance of this Special Issue.

Funding: This publication received no external funding.

Acknowledgments: As the Editor of the Special Issue, I would like to thank all the authors of the submitted articles, as well as the reviewers, editors, and everyone who contributed to publishing the Special Issue.

Conflicts of Interest: The author declares no conflict of interest.

References

1. Lisiecki, A. Tribology and Surface Engineering. *Coatings* **2019**, *9*, 663. [CrossRef]
2. Kusinski, J.; Kąc, S.; Kopia, A.; Radziszewska, A.; Rozmus-Górnikowska, M.; Major, B.; Major, L.; Marczak, J.; Lisiecki, A. Laser modification of the materials surface layer—A review paper. *Bull. Pol. Acad. Sci. Tech. Sci.* **2012**, *60*, 711–728. [CrossRef]
3. Janicki, D.; Górka, J.; Czupryński, A.; Kwasny, W.; Żuk, M. Diode laser cladding of Co-based composite coatings reinforced by spherical WC particles. In Proceedings of the SPIE 10159, Laser Technology 2016: Progress and Applications of Lasers, Jastarnia, Poland, 8 December 2016. [CrossRef]
4. Soffel, F.; Lin, Y.; Keller, D.; Egorov, S.; Wegener, K. Laser Remelting Process Simulation and Optimization for Additive Manufacturing of Nickel-Based Super Alloys. *Materials* **2022**, *15*, 177. [CrossRef] [PubMed]
5. Lisiecki, A. Titanium matrix composite Ti/TiN produced by diode laser gas nitriding. *Metals* **2015**, *5*, 54–69. [CrossRef]
6. Dobrzanski, L.A.; Bonek, M.; Hajduczek, E.; Klimpel, A.; Lisiecki, A. Application of high power diode laser (HPDL) for alloying of X40CrMoV5-1 steel surface layer by tungsten carbides. *J. Mater. Process. Technol.* **2004**, *155*, 1956–1963. [CrossRef]
7. Lisiecki, A.; Kurc-Lisiecka, A. Erosion wear resistance of titanium-matrix composite Ti/TiN produced by diode-laser gas nitriding. *Mater. Tehnol.* **2017**, *51*, 29–34. [CrossRef]
8. Grabowski, A.; Lisiecki, A.; Dyzia, M.; Łabaj, J.; Stano, S. The effect of laser wavelength on surface layer melting of the AlSi/SiC composite. *J. Manuf. Process.* **2022**, *75*, 627–636. [CrossRef]
9. Górka, J. Assessment of the Weldability of T-Welded Joints in 10 mm Thick TMCP Steel Using Laser Beam. *Materials* **2018**, *11*, 1192. [CrossRef]
10. Sommer, N.; Warres, C.; Lutz, T.; Kahlmeyer, M.; Böhm, S. Transmission Electron Microscopy Study on the Precipitation Behaviors of Laser-Welded Ferritic Stainless Steels and Their Implications on Intergranular Corrosion Resistance. *Metals* **2022**, *12*, 86. [CrossRef]
11. Górka, J.; Ozgowicz, A. SStructure and properties of laser-beam-welded joints of low-alloy high-strength steel docol 1200M with a martensitic structure. *Mater. Tehnol.* **2018**, *52*, 189–193. [CrossRef]
12. Tomków, J.; Rogalski, G.; Fydrych, D.; Łabanowski, J. Advantages of the application of the temper bead welding technique during wet welding. *Materials* **2019**, *12*, 915. [CrossRef]
13. Kurc-Lisiecka, A.; Lisiecki, A. Laser welding of new grade of advanced high strength steel Domex 960. *Mater. Technol.* **2017**, *51*, 199–204.
14. Zhang, G.; Yu, F. Ultra-Narrow Gap Fiber Laser Conduction Welding Technology for 304 Stainless Steel Thick Plates and the Mechanical Properties of Welding Joints. *Coatings* **2022**, *12*, 59. [CrossRef]
15. Wang, L.; Li, H.; Huang, Y.; Wang, K.; Zhou, M. Effect of Preheating on Martensitic Transformation in the Laser Beam Welded AH36 Steel Joint: A Numerical Study. *Metals* **2022**, *12*, 127. [CrossRef]
16. Ramaphoko, N.A.; Skhosane, S.; Maledi, N. Microstructural Characterization and Mechanical Properties of Laser Beam-Welded Dissimilar Joints between A6000 Aluminum Alloy and Galvanized Steel. *Materials* **2022**, *15*, 543. [CrossRef]
17. Zhang, H.; Jiang, M.; Chen, X.; Wei, L.; Wang, S.; Jiang, Y.; Jiang, N.; Wang, Z.; Lei, Z.; Chen, Y. Investigation of Weld Root Defects in High-Power Full-Penetration Laser Welding of High-Strength Steel. *Materials* **2022**, *15*, 1095. [CrossRef] [PubMed]
18. Lisiecki, A.; Ślizak, D. Hybrid Laser Deposition of Fe-Based Metallic Powder under Cryogenic Conditions. *Metals* **2020**, *10*, 190. [CrossRef]
19. Wang, G.; Wang, J.; Yin, L.; Hu, H.; Yao, Z. Quantitative Correlation between Thermal Cycling and the Microstructures of X100 Pipeline Steel Laser-Welded Joints. *Materials* **2020**, *13*, 121. [CrossRef] [PubMed]
20. Kik, T. Computational Techniques in Numerical Simulations of Arc and Laser Welding Processes. *Materials* **2020**, *13*, 608. [CrossRef]
21. Górka, J. Assessment of the Effect of Laser Welding on the Properties and Structure of TMCP Steel Butt Joints. *Materials* **2020**, *13*, 1312. [CrossRef]
22. Danielewski, H.; Skrzypczyk, A. Steel Sheets Laser Lap Joint Welding—Process Analysis. *Materials* **2020**, *13*, 2258. [CrossRef]
23. Kik, T. Heat Source Models in Numerical Simulations of Laser Welding. *Materials* **2020**, *13*, 2653. [CrossRef] [PubMed]
24. Landowski, M.; Świerczyńska, A.; Rogalski, G.; Fydrych, D. Autogenous Fiber Laser Welding of 316L Austenitic and 2304 Lean Duplex Stainless Steels. *Materials* **2020**, *13*, 2930. [CrossRef] [PubMed]
25. Pańcikiewicz, K.; Świerczyńska, A.; Hućko, P.; Tumidajewicz. M. Laser Dissimilar Welding of AISI 430F and AISI 304 Stainless Steels. *Materials* **2020**, *13*, 4540. [CrossRef] [PubMed]
26. Tofil, S.; Danielewski, H.; Witkowski, G.; Mulczyk, K.; Antoszewski, B. Technology and Properties of Peripheral Laser-Welded Micro-Joints. *Materials* **2021**, *14*, 3213. [CrossRef]

27. Lisiecki, A. Study of Optical Properties of Surface Layers Produced by Laser Surface Melting and Laser Surface Nitriding of Titanium Alloy. *Materials* **2019**, *12*, 3112. [CrossRef]
28. Sroka, M.; Jonda, E.; Pakieła, W. Laser Surface Modification of Aluminium Alloy AlMg9 with B4C Powder. *Materials* **2020**, *13*, 402. [CrossRef]
29. Li, J.; Ju, J.; Chang, W.; Yang, C.; Wang, J. Investigation on the Microstructure and Wear Behavior of Laser-Cladded High Aluminum and Chromium Fe-B-C Coating. *Materials* **2020**, *13*, 2443. [CrossRef]
30. Majkowska-Marzec, B.; Tęczar, P.; Bartmański, M.; Bartosewicz, B.; Jankiewicz, B.J. Mechanical and Corrosion Properties of Laser Surface-Treated Ti13Nb13Zr Alloy with MWCNTs Coatings. *Materials* **2020**, *13*, 3991. [CrossRef]
31. Czupryński, A.; Wyględacz, B. Comparative Analysis of Laser and Plasma Surfacing by Nickel-Based Superalloy of Heat Resistant Steel. *Materials* **2020**, *13*, 2367. [CrossRef]
32. Lubas, J.; Szczypiński-Sala, W.; Woś, P.; Zielińska, E.; Miernik, K. Experimental Analysis of Tribological Processes in Friction Pairs with Laser Borided Elements Lubricated with Engine Oils. *Materials* **2020**, *13*, 5810. [CrossRef]
33. Lisiecki, A.; Ślizak, D. Hybrid Laser Deposition of Composite WC-Ni Layers with Forced Local Cryogenic Cooling. *Materials* **2021**, *14*, 4312. [CrossRef] [PubMed]
34. Szymczak, T.; Szczucka-Lasota, B.; Węgrzyn, T.; Łazarz, B.; Jurek, A. Behavior of Weld to S960MC High Strength Steel from Joining Process at Micro-Jet Cooling with Critical Parameters under Static and Fatigue Loading. *Materials* **2021**, *14*, 2707. [CrossRef] [PubMed]
35. Szczucka-Lasota, B.; Węgrzyn, T.; Szymczak, T.; Jurek, A. High Martensitic Steel after Welding with Micro-Jet Cooling in Microstructural and Mechanical Investigations. *Materials* **2021**, *14*, 936. [CrossRef] [PubMed]
36. Tomków, J.; Fydrych, D.; Rogalski, G. Role of Bead Sequence in Underwater Welding. *Materials* **2019**, *12*, 3372. [CrossRef]
37. Mičian, M.; Winczek, J.; Gucwa, M.; Koňár, R.; Málek, M.; Postawa, P. Investigation of Welds and Heat Affected Zones in Weld Surfacing Steel Plates Taking into Account the Bead Sequence. *Materials* **2020**, *13*, 5666. [CrossRef]
38. Czupryński, A. Comparison of Properties of Hardfaced Layers Made by a Metal-Core-Covered Tubular Electrode with a Special Chemical Composition. *Materials* **2020**, *13*, 5445. [CrossRef]
39. Czupryński, A. Flame Spraying of Aluminum Coatings Reinforced with Particles of Carbonaceous Materials as an Alternative for Laser Cladding Technologies. *Materials* **2019**, *12*, 3467. [CrossRef]
40. Kapustynskyi, O.; Višniakov, N.; Zabulionis, D.; Piščalov, A. Feasibility Evaluation of Local Laser Treatment for Strengthening of Thin-Walled Structures from Low-Carbon Steel Subjected to Bending. *Materials* **2020**, *13*, 3085. [CrossRef]

materials

Article

Hybrid Laser Deposition of Composite WC-Ni Layers with Forced Local Cryogenic Cooling

Aleksander Lisiecki [1,*] and Dawid Ślizak [2]

1 Department of Welding Engineering, Faculty of Mechanical Engineering, Silesian University of Technology, Konarskiego 18A Str., 44-100 Gliwice, Poland

2 Additive Manufacturing Laboratory, PROGRESJA S.A., Żelazna 9 Str., 40-851 Katowice, Poland; dslizak@progresja.co

* Correspondence: aleksander.lisiecki@polsl.pl

Abstract: The purpose of this study was to demonstrate the effect of forced and localized cooling by nitrogen vapours stream under cryogenic conditions during laser deposition of WC-Ni powder on the geometry, microstructure of clad layers and dry sliding wear resistance of the coatings. For this purpose, comparative tests were performed by conventional laser cladding at free cooling conditions in ambient air and by the developed novel process of laser deposition with additional localized cooling of the solidifying deposit by nitrogen vapours stream. Due to presence of gaseous nitrogen in the region of the melt pool and solidifying deposit, the process was considered as combining laser cladding and laser gas nitriding (performed simultaneously), thus the hybrid process. The influence of the heat input and cooling conditions on the geometrical features, dilution rate, share of carbides relative to the matrix, and the fraction share of carbides, as well as hardness profiles on cross sections of single stringer beads was analysed and presented. The XRD, EDS analysis and the sieve test of the experimental powder were used to characterize the composite WC-Ni type powder. The OM, SEM, EDS and XRD test methods were used to study the microstructure, chemical and phase composition of clad layers. Additionally, ball-on-disc tests were performed to determine the wear resistance of representative coatings under dry sliding conditions. The results indicate that the novel demonstrated technique of localized forced cooling of the solidifying deposit has advantageous effect, because it provides approximately 20% lower penetration depth and dilution, decreases tendency for tungsten carbides decomposition, provides more uniform distribution and higher share of massive eutectic W_2C-WC carbides across the coating. While the conventionally laser cladded layers show tendency for decomposition of carbide particles and resolidifying dendritic complex carbides mainly M_2C, M_3C and M_7C_3 containing iron, nickel, and tungsten, and with Ni/Ni_3B matrix. The quantitative relationship between heat input, cooling conditions and the carbides grain size distribution as well as carbides share in relation to the matrix was determined.

Keywords: laser cladding; laser deposition; hybrid process; cryogenic conditions; composite coatings; WC-Ni coatings; fiber laser

Citation: Lisiecki, A.; Ślizak, D. Hybrid Laser Deposition of Composite WC-Ni Layers with Forced Local Cryogenic Cooling. *Materials* **2021**, *14*, 4312. https://doi.org/10.3390/ma14154312

Academic Editor: Antonio Santagata

Received: 31 May 2021
Accepted: 26 July 2021
Published: 2 August 2021

1. Introduction

Laser cladding offers some advantages over the other methods of cladding or coatings [1–10]. The most significant advantages are high power density, localized and precise heating, high scanning speed, thus low heat input, low penetration depth and low dilution. Laser cladding is often used to produce wear-resistant layers based on metallic and composite materials. One group of composite materials are metal matrix composites (MMC) based on nickel matrix reinforced with tungsten carbides. Such composites are characterized by good wear resistance, corrosion resistance and also satisfactory dynamic load resistance. The high abrasive wear resistance is provided by hard particles of tungsten carbides distributed in the ductile nickel matrix. In turn, the ductile matrix provides the ability to withstand high loads, especially compressive stresses. Moreover, such composites show satisfactory cohesion between hard WC phases and the matrix of Ni solution, due to good wettability of WC particles by the Ni-based alloys. The process of laser cladding is considered as advantageous for manufacturing the composite clad layers due to low heat input, thus limited reaction between the ceramic phases and the metallic matrix [11–14]. However, even in the case of laser cladding, especially in the output power range of several kilowatts, the heat input may be too high producing unfavourable microstructure due to partial and even complete melting of WC particles. Moreover, under specific conditions the WC particles tend to fall off in the melt pool, causes uneven distribution of the carbides of various size across the surface layer. Zhang et al. indicate that the accumulation of WC particles at the bottom of the clad coating is detrimental for the properties of the coating [12]. They also point that it is the biggest obstacle to engineering applications of such composite coatings. Various methods for providing the uniform distribution of ceramic particles in the composite coatings, such mechanical vibration, ultrasonic, magnetic fields were studied and described in the literature [13]. Li et al. demonstrated an original technique of laser cladding of WC-Ni composite coatings assisted by micro-vibrations generated by a vibration exciter system made of magneto strictive materials [14,15]. They pointed the beneficial effect of the micro-vibrations on microstructure, including uniform distribution of tungsten carbides [15]. In turn Huang et al. demonstrated a cladding technique involving a pulsed Nd:YAG laser for reducing the heat input [16]. They successfully produced dense and crack-free WC-Ni composite clad layers on H13 substrate with the thickness up to 1.0 mm.

Another technique that allows to reduce the effect of the heat on the material and to control the solidification rate is a forced cooling. An original technique of micro-jet cooling by compressed gas streams of the deposit during arc cladding was elaborated and demonstrated by Węgrzyn et al. [17]. They proved the beneficial effect of the forced cooling of the deposit at solidification stage resulted in refinement of microstructure and enhanced wear resistance.

Liquid nitrogen bath providing cryogenic conditions of cooling was also tested by some researchers during laser surface melting or heat treatment, mainly for nonferrous alloys [18–23]. Such technique of liquid nitrogen bath cooling of the substrate was adopted and developed by the authors for laser powder deposition of metallic and composite coatings, as demonstrated in several previous publications [1,24–31]. It is worth to note, that complexity and difficulty of the laser powder deposition under such cryogenic conditions with coaxial powder delivery into the melt pool is significantly higher than the laser surface melting of the substrate supercooled by liquid nitrogen bath. Moreover, the results obtained so far indicate that due to intensive evaporating of liquid nitrogen, the gaseous nitrogen present in the region of powder deposition and melt pool provides an active atmosphere typical for laser gas nitriding (LGN) or laser alloying (LA). Therefore, the proposed new method of laser coating is considered as a hybrid process combining laser powder deposition and laser gas nitriding, additionally conducted under cryogenic conditions.

The preliminary tests of laser cladding of composite WC-Ni coatings, conducted by the authors, have shown that the forced cooling of the substrate by liquid nitrogen bath can provide some beneficial effects. The most significant are limited penetration depth, limited dilution, limited tendency for tungsten carbide melting, uniform and dense distribution of carbides, favourable hardness distribution and high values of hardness. However, some disadvantages and limitations of the liquid nitrogen bath (volume cooling) applied for laser powder deposition were identified. Due to high extent of supercooling the substrate and intensive dissipation of the heat during laser heating, the tendency for incomplete or lack of penetration of the clad was observed, especially at the lowest heat input range. Therefore, another technique of forced cooling by means of liquid nitrogen based on stream local cooling of the deposition region, that allows controlled and less intensive cooling of the substrate was developed and investigated by the authors.

According to the authors' knowledge and experience, the presented results are fully original and unique, and the results of similar studies have not been available nor published so far.

2. Materials and Methods

2.1. Materials

The non-alloy structural steel S235JR (according to EN 10025-2) was chosen as the substrate for technological tests. This grade of steel with a low content of alloying elements, good plasticity and excellent weldability was chosen to minimize the influence of the substrate material on the cladded layer, especially minimize the internal stresses and tendency for cracking of clad layers, Table 1. Specimens for laser cladding tests were cut from a steel plate 5.0 mm thick into coupons with dimensions 100 × 100 mm^2.

Table 1. Chemical composition of non-alloy structural steel S235JR (EN 10025-2) based on the supplier's (Thyssenkrupp, Dąbrowa Górnicza, Poland)) certificate (wt %).

C	Mn	Si	P	S	N	Cu	Al	Fe
0.05–0.14	0.2–0.8	0.1	0.025	0.015	0.01	0.2	0.015–0.08	Bal.

In turn, the powder for cladding was experimentally composed in such a way as to ensure high abrasion resistance, and corrosion resistance of deposit even at elevated temperatures. Therefore, tungsten carbides, as the reinforcing phase, were mixed with nickel-based powder providing the ductile metal matrix characterized by high resistance for corrosion and high temperature. The nickel-based metal powder was gas atomized in argon atmosphere and the spherical particles size range was 63 ÷ 160 µm. While the tungsten carbide irregular shaped (crushed) particles size range was 100 ÷ 160 µm. The sieve test of the mixed experimental powder was performed for statistical analysis according to the standard PN-EN 24497/ISO 4497. Sieve shaker LPzE-2e (Multiserw, Brzeźnica, Poland) and the moisture analyser with the laboratory scales BTS110 (AXIS, Gdańsk, Poland) were applied in the sieve test. The sieve test was repeated three times for different batches of powder weighing approx. 100 g each, for statistical reasons. The mass fraction share and cumulative particle size distribution curve determined for the experimentally composed powder are presented in Figure 1. The morphology, chemical and phase composition of the experimental powder are presented in Figures 2 and 3.

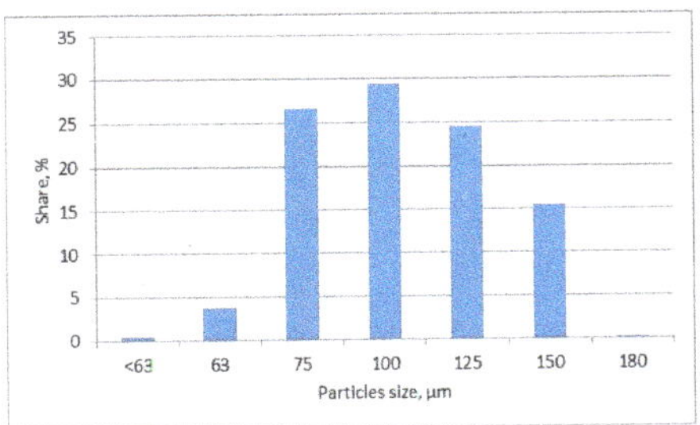

(a)

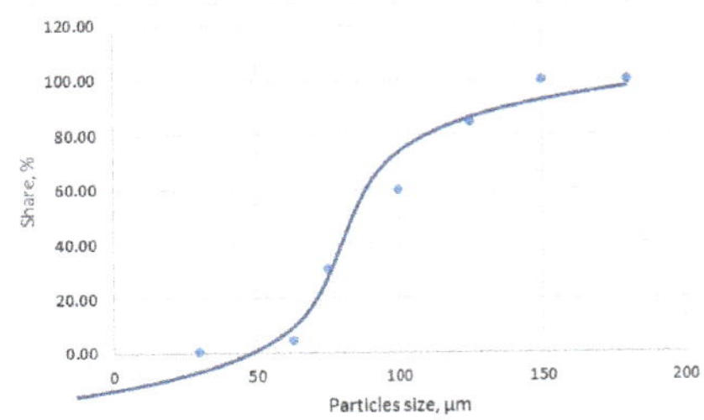

(b)

Figure 1. The mass fraction share (**a**) and cumulative particle size distribution curve (**b**) determined by sieve test (PN-EN 24497/ISO 4497) for the experimental composite WC-Ni powder, Table 2.

Table 2. Nominal composition of experimental powder as a mixture of tungsten carbides with nickel-based metal matrix (wt %), Figures 2 and 3.

WC	Fe	Si	B	C	Ni
60	2.0	3.0	3.0	0.02	Bal.

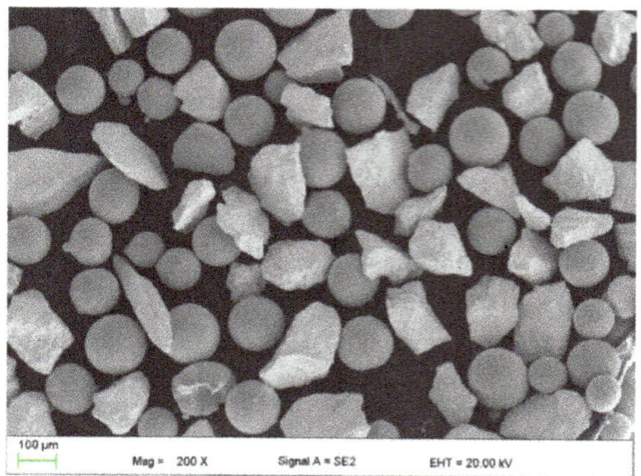

Figure 2. Morphology of the experimental composite WC-Ni powder; mixture of irregular WC particles and spherical metallic Ni-based particles, Table 2.

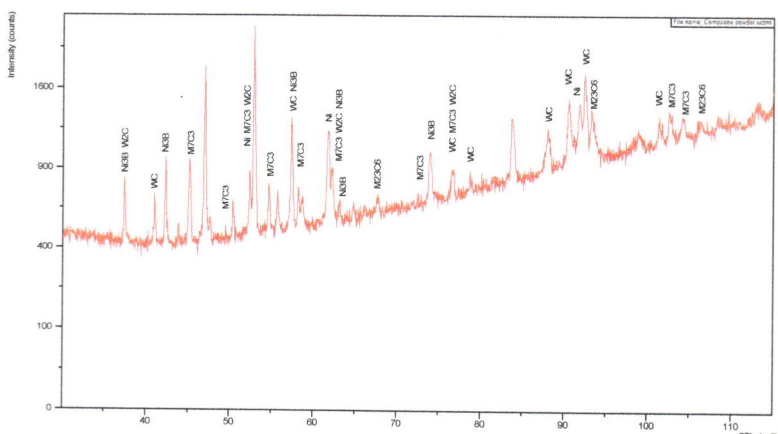

Figure 3. XRD pattern of the experimental WC-Ni composite powder, Table 2, Figures 1 and 2.

2.2. Laser Deposition Tests

The tests of laser deposition were performed by means of a prototype robotized stand. The stand was equipped with a six-axis robot Panasonic GII TL-190 (Industrial Solutions Company, Panasonic Corporation, Osaka, Japan) with maximum load capacity 6.0 kg, high power fiber laser (HPFL) IPG YLS-3000-CT-Y15 (IPG Photonics, Oxford, MS, USA) with maximum output power 3.0 kW, emitting at a wavelength 1.07 µm, and characterized by a single mode energy distribution across the spot (Gaussian intensity distribution), custom made powder delivery system, and a specially designed system for localized delivery of liquid nitrogen or nitrogen vapours stream. The laser beam was transmitted from the laser generator to the focussing head IPG FLW by 20 m long HLC-8 fiber with a core diameter 200 µm. The focal length of the applied FLW head was 200 mm, while the collimator focal length was 100 mm. The beam parameter product (BPP) of the laser beam determined for the applied optics configuration was 5.8 mm·mrad. The BPP was measured by Primes Focus Monitor FM-120 (PRIMES GmbH, Pfungstadt, Germany).

The stream localized cooling system was designed based on the experiences of previous preliminary study of laser cladding the steel substrate immersed partially in liquid nitrogen bath, to overcome the encountered limitations of such technique of volume cryogenic cooling [1]. During the previous preliminary tests, a wide range of processing parameters was investigated, at different scanning speed and output laser power. In this study the parameters were narrowed and optimized in order to provide the width of a single bead approx. 2.0 ÷ 3.0 mm, and proper fusion to the steel substrate, Table 3, Figures 4 and 5.

Table 3. Parameters of laser deposition of experimental composite WC-Ni powder at free cooling conditions (conventional laser cladding) and forced localized cooling by nitrogen vapours stream under cryogenic conditions (hybrid laser deposition process), Figure 6.

No.	Surface Layer Indication Free Cooling/Hybrid	Scanning Speed (mm/min)	Laser Power (W)	Energy Input * (J/mm)	Remarks Free Cooling/Hybrid
1	LC1/HC1	500	500	60	SP/HQ
2	LC2/HC2	500	1000	120	SP, HQ/HQ
3	LC3/HC3	500	1500	180	SP, HQ/HQ
4	LC4/HC4	500	2000	240	SP/SP, HQ

Remarks: UB—uneven bead, SP—single pore, IF—incomplete fusion, V—voids, HQ—high quality, LF—lack of fusion. * energy input is calculated by simply dividing the laser power by scanning speed, while the heat input should also include the heat transfer efficiency. Other process parameters; powder feeding rate: 8.5 g/min, diameter of the nozzle tip for nitrogen vapours stream: 5.0 mm, average consumption of liquid nitrogen: 80 ÷ 100 g/min.

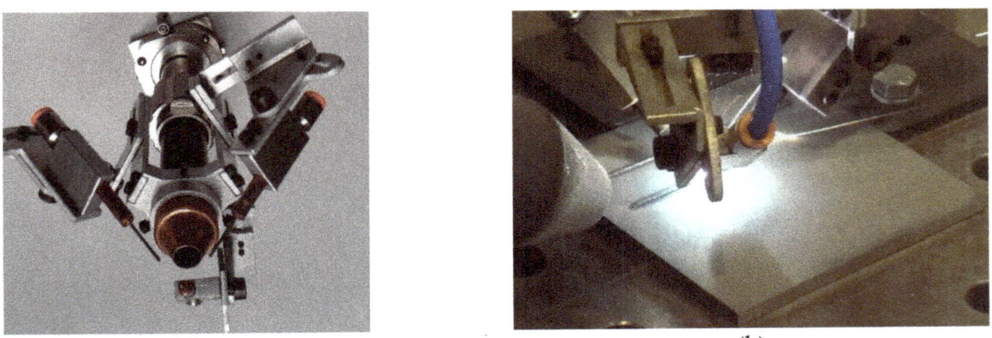

(a) (b)

Figure 4. (**a**) A view of the mount of three coaxial nozzles for powder delivery and injection into the melt pool on the laser focusing head; and (**b**) a view of the hybrid laser deposition of composite WC-Ni powder at cryogenic conditions by the localized cooling with the nitrogen vapours stream and the head with three coaxial nozzles.

Prior to the laser cladding tests, the specimens were sandblasted (surface roughness Ra 25 ÷ 60 µm) and degreased with acetone to remove surface contamination and provide stable and repeatable surface conditions for absorption of laser radiation. Surface roughness was determined by a portable surface roughness tester SJ-210 Surftest (Mitutoyo Corporation, Kanagawa, Japan). The test cladded layers were produced as single stringer beads by laser deposition of the experimentally composed powder at different heat input and different conditions of cooling. The length of the individual bead was 80 mm, while the shift between beads was 20 mm, Figure 6. One set of test clad layers was produced during conventional laser cladding with natural cooling (so called free cooling) of the specimen in the ambient air. The second set of test clad layers was produced at the same processing parameters but additionally with the application of forced cooling by localized stream of liquid nitrogen. The temperature of the liquid nitrogen in the container was below −196 °C, while the temperature of the nitrogen vapours stream measured at the tip of the nozzle by a thermocouple was approx. −160 ÷ 164 °C. In such a way deep cryogenic treatment (DCT) was provided in the localized area of forced cooling, Table 3, Figure 4b.

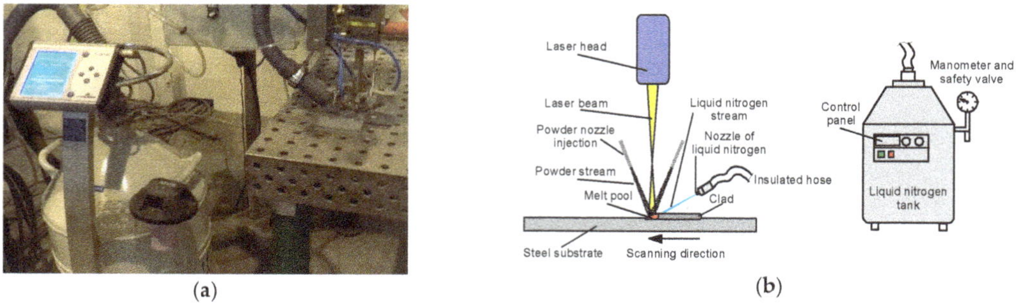

Figure 5. (**a**) A view of the experimental setup for laser powder deposition of composite layers with forced local cryogenic cooling by the nitrogen vapours stream (from left: cryogenic dewar with automatic control unit, elastic hose for nitrogen delivers, nitrogen stream nozzle and laser head with coaxial nozzles for powder injection); (**b**) a scheme of the laser powder deposition with forced cooling of the substrate by localized nitrogen vapours stream.

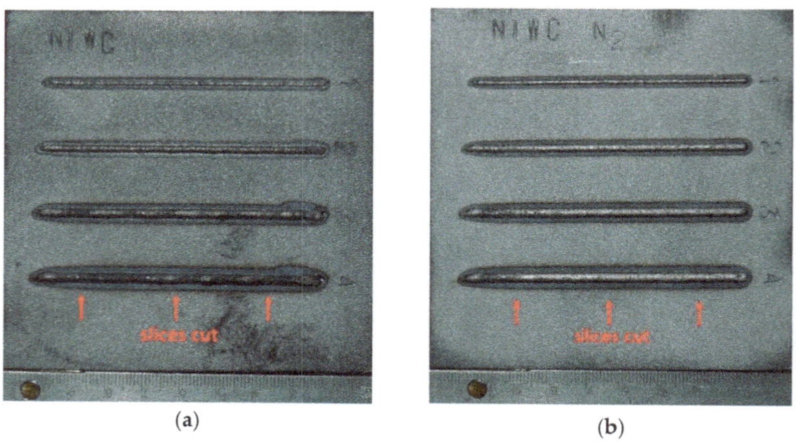

Figure 6. A view of stringer beads produced by laser deposition of experimental composite WC-Ni powder at free cooling conditions (conventional laser cladding); (**a**) and forced localized cooling by nitrogen vapours stream under cryogenic conditions (hybrid laser deposition process); (**b**), Table 3.

The scanning speed was kept constant at 500 mm/min, while the laser output power was varied from 500 W, 1000 W, 1500 W, and 2000 W. The feeding rate of the composite WC-Ni powder was maintained constant at 8.5 g/min. The powder was fed into the melt pool by three coaxial nozzles with a diameter of 0.8 mm each. The nozzles were attached to head body providing the ability to change the angle of inclination and vertical shift, as can be seen in Figure 4a. The individual streams from the nozzles were focused on the melt pool surface in the region of laser beam interaction, Figure 4. The powder feeding rate was controlled by a rotary disc feeder PFU4 (Durum, Willich, Germany), with the chamber filled in by argon at the pressure 1.5 bar, used also as the carrier gas. The flow rate of the carrier argon gas was kept at 8.0 L/min.

The laser beam was set perpendicularly to the surface of the substrate, and it was transmitted through a 10.0 mm diameter cylindrical nozzle in argon atmosphere, at the flow rate 20.0 L/min. The argon flow was used mainly for protection the optics inside the laser head against spatter. While the task of shielding the melt pool was negligible due to long distance (160 mm) between the nozzle and the substrate surface, as a result of defocusing the laser beam spot, Figure 4b. The laser beam spot diameter was 300 μm at the applied configurations of optics. In order to provide wider area of laser beam interaction on the substrate material, the beam was defocused by lifting the laser head and thus focal plane over the top surface of the substrate. In this way, as a result of lifting the focal plane to the distance of 160 mm from the substrate, the diameter of the laser beam was approximately 3.0 mm on the top surface, Figures 4b and 6.

Nitrogen vapours from a pressurized container (cryogenic dewar) were transported via a flexible hose into the jet nozzle. The nozzle tip was designed to be replaceable. Thanks to that different diameter tip nozzles could be applied. In this study a nozzle tip with a diameter of 5.0 mm was used, providing localized cooling of the clad area just behind the melt pool. The jet nozzle was fixed on the laser processing head, and the distance between the nozzle tip and the target surface was maintained constant at 40.0 mm, Figures 4b and 5a. The outflow rate of the nitrogen vapours stream from the nozzle depended on the pressure inside the container, which was maintained constant by an automatic pressure control unit KRIOSAN (KrioSystem, Wrocław, Poland), Figure 5. The average consumption of liquid nitrogen was ranged in 80 ÷ 100 g/min.

2.3. Macro and Microstructure Examinations

The specimens with test clad layers produced both by conventional laser cladding at free cooling conditions in ambient air, and the clad layers produced at forced cooling under cryogenic conditions were examined by visual inspection first. All the clad layers fulfilled the main criterion of complete penetration of the substrate. Therefore, all the test layers were cut perpendicularly and samples for metallographic study were prepared. Three sliced sections were taken from each test clad layer. One was taken from the middle region, and next two at a distance 15 mm from the beginning and the end of the clad layer, Figure 6. The sliced sections were first mounted in thermosetting phenol resin with graphite filler Electro-WEM (Metalogis, Warsaw, Poland), and next the samples were wet grinded by water papers with grit 120 to 2500 using an automatic grinding/polishing machine Struers Labopol-2 (Struers, Rodovre, Denmark). Next the cross-sections were polished with 1 μm diamond suspension Metkon Diapat-M (Metkon Instruments Inc., Bursa, Turkey). After polishing, the cross-sections were etched by the HNO_3 + 3HCL reagent to disclose the microstructure.

Observations of macrostructure at low magnifications (up to 25×) were carried out by means of a stereoscopic microscope OLYMPUS SZX9 (Olumpus Corporation, Tokyo, Japan), while the microstructure observations were done by the inverted metallographic microscope NIKON Eclipse MA100 (Nikon Corporation, Tokyo, Japan). The microstructure was additionally examined by scanning electron microscopy SEM (Carl Zeiss, Oberkochen, Germany), equipped with the Energy Dispersive Spectrometer EDS (Oxford Instruments, Abingdon, GB, USA) and Phenom Pro-X SEM equipped with EDS and BSD detectors (Thermo Fisher Scientific, Eindhoven, The Netherlands). The phase composition was determined by X-Ray diffraction (Panalitycal, Almelo, The Netherlands) with CuKα source of radiation, with the scanning range of the diffraction angle 2θ from 0 to 140°.

2.4. Hardness Measerements

The hardness was measured on the cross-section of the test clad layers after microstructural studies by Vickers test, at the load 5 N and the dwell time 10 s. The hardness tester WILSON WOLPERT 401 MVD (Wolpert Wilson Instruments, Aachen, Germany) was applied in the study. Measurements were taken along the vertical axis of symmetry, starting from the under surface region of the clad layer (face of the clad). The first measuring point was 0.15 mm under the top surface, while the distance between the subsequent points was constant at 0.2 mm. In such a way, the hardness distribution from the under surface region, through the clad, fusion zone, heat affected zone, till to the base metal was determined.

2.5. Tribological Test

The tribological tests of coatings produced by laser deposition of experimental composite powder at free cooling and under cryogenic conditions were conducted by a ball-on-disc tribometer T-01M under room temperature of 22 °C, according to the ASTM G99 standard. The relative humidity was about 45% ± 5%. The specimens were prepared in a form of discs with diameter of 45.0 mm balls made of bearing steel (EN 100Cr6, AISI 5210) with a diameter of 10.0 mm were used as the counter face material. The normal load was set as 20 N. The number of revolutions was 1500, while radius of the track was 15 mm. Therefore, the sliding distance was 141.3 m, while the sliding speed was 0.157 m/s. The tangential force of friction and displacement value were continuously measured and recorded during tests using a data acquisition system with PC computer. While the coefficient of friction μ was calculated by dividing the value of tangential force of friction by the value of normal load used:

$$\mu = T/F_n, \quad (1)$$

where:

F_n—normal load (30 N), T—tangential force of friction.

3. Results and Discussion

3.1. Macrostructure and Single Bead Geometry

At the first stage of the study the cross-sections of the clad layers produced as single stringer beads by conventional laser cladding at free cooling conditions were compared with the single stringer beads produced by the novel technique considered as a hybrid laser deposition under cryogenic conditions, Figure 7 Since the clad layers were produced at constant scanning speed, the influence of energy input on the bead geometry can be also considered. The energy input is defined as follows:

$$Ev = P/v \text{ (J/mm)}, \quad (2)$$

where P is the output power of laser beam (W), and v is the scanning speed (mm/s).

It is worth to note, when the heat transfer efficiency into the material is known and considered in the calculation, this parameter is called heat input. In laser processing the heat transfer efficiency is related with the absorption of laser energy, and under certain conditions it can take the value 1. Therefore, for simplicity the term "heat input" will be used in the remainder of the manuscript text. As seen in Figure 7, clear interface lines between the clads and the substrate can be distinguished, as well as the heat affected zone regions. A rough comparison of the geometry of clad layers produced at the same heat input but different cooling conditions, indicates that the forced localized cooling by nitrogen vapours stream leads to a clear decrease of penetration depth and thus dilution.

The value of dilution "D" was calculated by the following formula:

$$D = \frac{A_{FZ}}{A_{FZ} + A_{CL}} \cdot 100\%, \quad (3)$$

where: A_{FZ} is the cross-section area of the fusion zone, and the A_{CL} is the cross-section area of the clad layer (Figure 7).

The increase in laser output power, and therefore increase in heat input, leads to an increase in the curvature of the interface line, penetration depth, height and width of the clad, Figures 7 and 8.

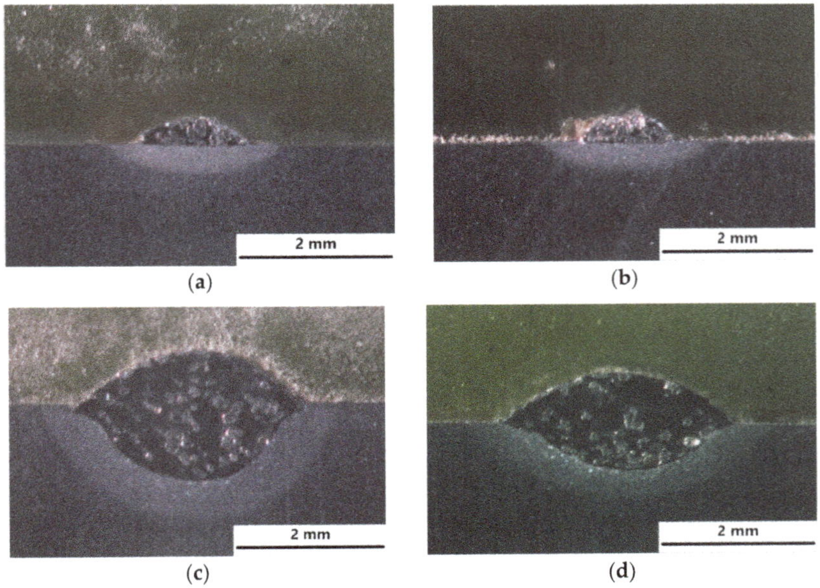

Figure 7. Macrostructure and the single bead geometry of the clad layers produced by laser cladding of experimental composite WC-Ni powder at constant scanning speed 500 mm/min, Table 3: (**a**) output power 500 W at free cooling; (**b**) output power 500 W at forced localized cooling; (**c**) output power 2000 W at free cooling; (**d**) 2000 W at forced localized cooling.

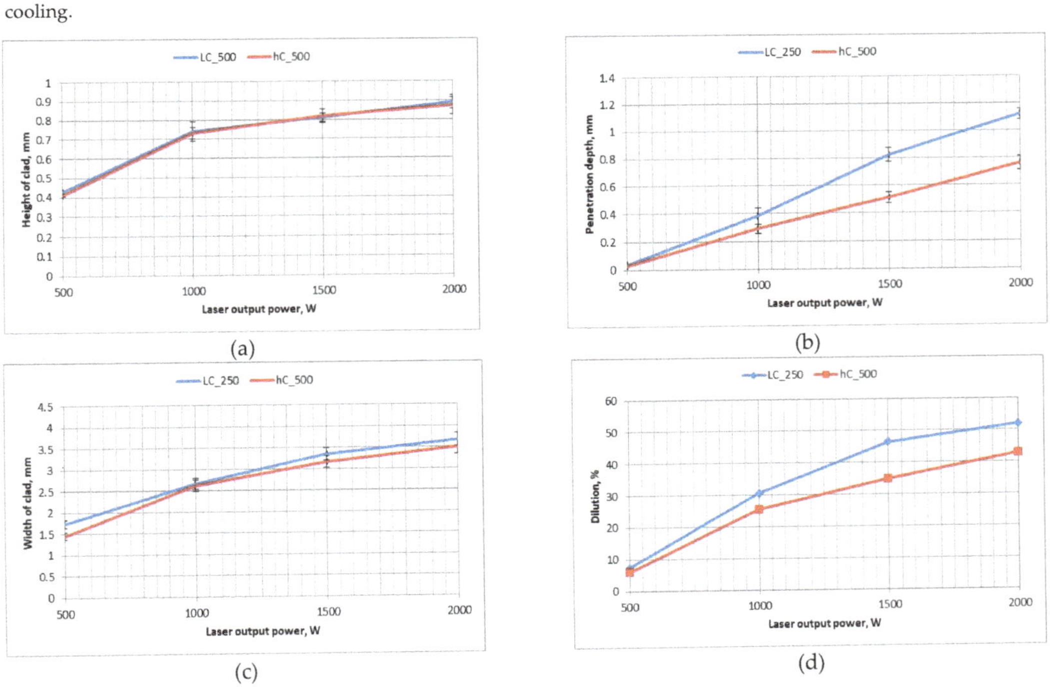

Figure 8. Influence of the process parameters and cooling conditions on the geometry and dimensions of the stringer beads produced by laser deposition of WC-Ni powder (Table 3): (**a**) height of the clad (reinforcement); (**b**) penetration depth into the substrate; (**c**) width of the single clad layer; and (**d**) calculated rate of dilution.

Based on the visual inspection and analysis of cross-sections morphology, no cracks neither tendency for cracking was observed both in the clad layers produced at free cooling and under cryogenic conditions. It is related with the relative high ductility of the nickel base matrix. However, single pores were found on cross-sections of the clads produced at free cooling in the range of heat input 120 ÷ 240 J/mm. The single pore with the highest diameter of 180 µm was found on the cross-section of the clad LC3. In turn, the tendency for porosity was found to be less in the case of clads produced under cryogenic conditions. Just a single pore was found just on the cross-section of the clad produced at the highest heat input of 240 J/mm but its diameter was lower, approx. 140 µm. Thorough analysis of the geometrical parameters on cross-sections of the clads showed that the localized forced cooling of the cladding region by a stream of nitrogen vapours has neglectable influence on the height of the clads. As can be seen in Figure 8a, the height of the clads produced under different conditions of cooling (free cooling and localized forced cryogenic cooling) is comparable. In turn, the average width of the clads produced under cryogenic conditions is approximately 5 ÷ 15% lower if compared to the clads produced at free cooling (conventional laser cladding), Figure 8c. A greater difference can be noticed in the case of penetration depth and dilution, Figure 8b,d. The average penetration depth of the clads produced at the same heat input but localized forced cooling (cryogenic conditions) is 24% to 38% lower than in the case of clads produced at free cooling conditions. While the average dilution of the clads produced at localized forced cooling is 20 ÷ 25% lower if compared to the clad produced by conventional laser cladding at free cooling.

The obtained results indicate that the localized cooling of the substrate in the region of cladding can effectively reduce the penetration depth by dissipating heat. However, the applied technique of localized cooling does not significantly reduce the height and width of the clads, thus also the area of the clads. Simultaneously, the applied technique of localized cooling has a beneficial effect on reducing the penetration depth and dilution, as can be seen in Figure 8b,d. This phenomenon is caused by the localized cooling just a narrow region along the axis of the stringer bead which leads to reducing the maximum temperature in the middle region of laser beam interaction and reduces the temperature gradient.

3.2. Microstructure

The comparative microstructures of the clad layers produced at maximum heat input of 240 J/mm (output power 2000 W, scanning speed 500 mm/min) and different cooling conditions (LC4 and HC4) are presented on optical micrographs in Figure 9, while the microstructures of the clad layers produced at minimum heat input of 60 J/mm and different cooling conditions (LC1 and HC1) are presented on optical micrographs in Figure 10. Comparing the Figure 9a,b, it can be seen, that the number and size of the carbides are different on the cross sections of the clads produced at the same heat input but different cooling conditions. Only single massive carbides can be found on the cross section of the clad layer produced at maximum heat input and free cooling conditions. In the case of the clad layer produced at the same heat input but at forced cooling by the stream of nitrogen vapours, the number of carbides is clearly higher. Closer view of the microstructure of the clad produced at maximum heat input and free cooling conditions (LC4) revealed that the regions between massive carbides are rich mainly in needle-like precipitations with a length up to approx. 40 µm, Figure 9c. For comparison, the microstructure in the regions between massive carbides of the clad produced at the maximum heat input but at localized forced cooling is different, Figure 9c. As can be seen, the number and size of precipitation is much smaller. The longest needle-like precipitations have a length up to approx. 15 µm, Figure 9d. The reason of this phenomenon is different cooling thus different thermal conditions during deposition and solidifying of the two comparative clad layers.

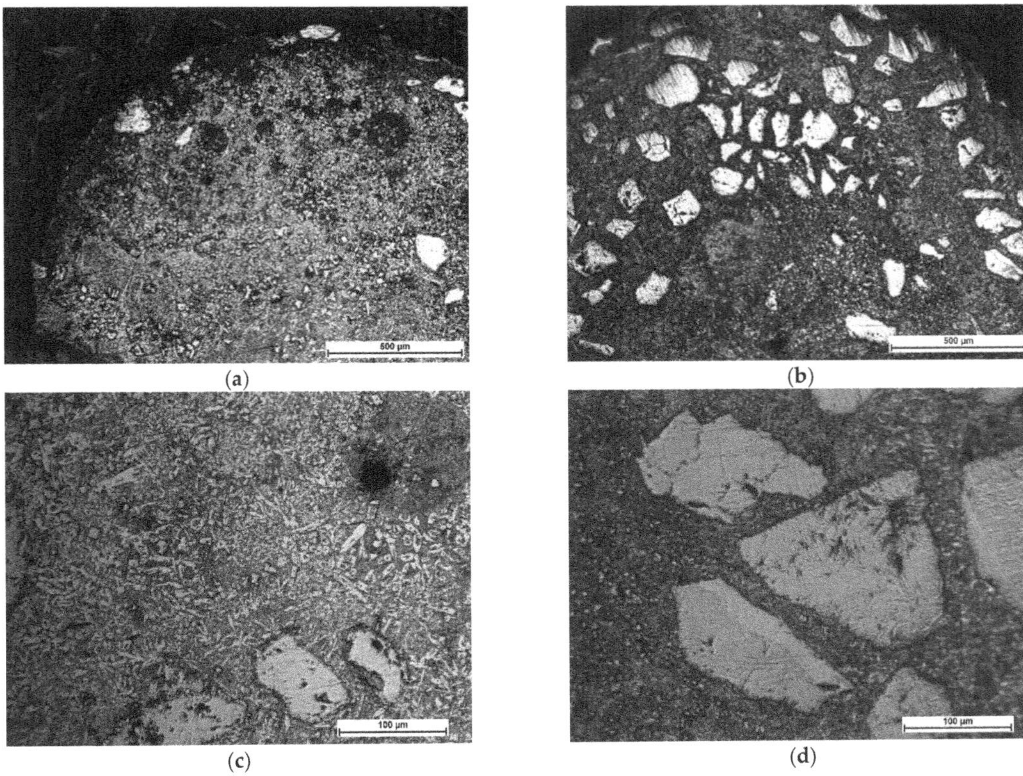

Figure 9. Microstructure of the comparative stringer beads LC4 and HC4 produced at the same maximum heat input of 240 J/mm (scanning speed 500 mm/min, laser output power 2000 W) but different cooling conditions (Table 3): (**a**,**c**) free cooling in ambient air—a view of the entire clad layer and the middle region of the clad respectively; and (**b**,**d**) forced localized cooling by nitrogen vapours stream under cryogenic conditions ("hybrid laser deposition")—a view of the entire clad layer and the middle region of the clad respectively.

The clad layer LC4 was produced by conventional laser cladding at free cooling in ambient and at relatively high heat input 240 J/mm, Table 3. Under such conditions, the massive carbides exposed to high temperature for a relatively long time, decompose and dissolve in the melt pool. It is worth noting that in addition to being exposed to the thermal influence of the melt pool, carbides are also exposed to the direct interaction of the laser beam radiation.

The carbide particles dissolved in the melt pool first enrich the liquid solution with carbon and tungsten and then recrystalize as secondary carbides in the form of dendrites, needle-like or block-like precipitations. So, the optical micrographs of the clad layer LC4 are typical for the clad produced at excessive energy and heat input, characterized by unfavourable microstructure with low share of massive tungsten carbides and high amount of secondary precipitations, Figure 9a,c. On the other hand, the clad layer HC4 produced at the same processing parameters and heat input 240 J/mm but additionally with the localized cooling showed clearly lower tendency for decomposition the massive carbides due to dissipation of the excessive heat by the nitrogen vapours stream under cryogenic conditions. Therefore, the amount and size of precipitations in the regions between massive carbides is lower. The massive carbides show "feather" structure, typical for eutectic W_2C/WC carbides, Figure 9d.

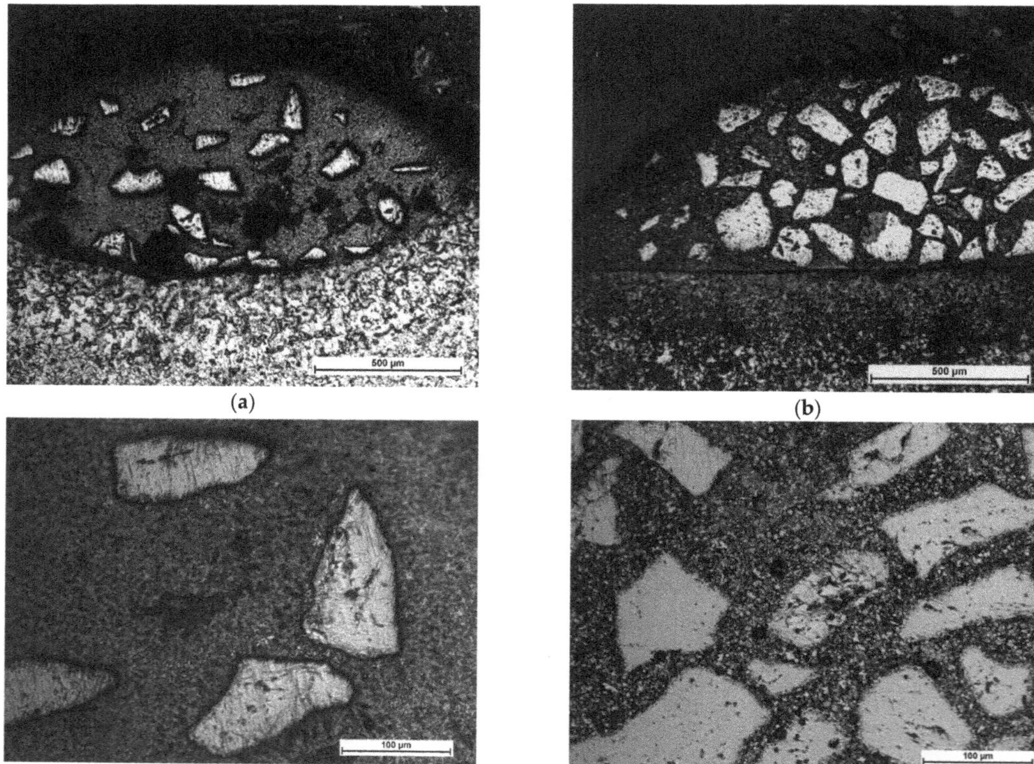

(**a**) (**b**) (**c**) (**d**)

Figure 10. Microstructure of the comparative stringer beads LC1 and HC1 produced at the same minimum heat input of 60 J/mm (scanning speed 500 mm/min, laser output power 500 W) but different cooling conditions (Table 3): (**a**,**c**) free cooling in ambient air—a view of the entire clad layer and the middle region of the clad respectively; and (**b**,**d**) forced localized cooling by nitrogen vapours stream under cryogenic conditions ("hybrid laser deposition")—a view of the entire clad layer and the middle region of the clad respectively.

Beside comparison of microstructure, the optical micrographs revealed a cluster of pores on the cross section of the clad LC4 produced by conventional laser cladding. The pores are placed in the upper part of the clad layer and the average diameter is approx. 20 ÷ 30 μm.

In turn, the optical micrographs of the comparative clad layers produced at minimum heat input of 60 J/mm and different cooling conditions (LC1 and HC1) also show differences in microstructure, Figure 10. First, if compared to the clad layers produced at the maximum heat input, these clad layers have more massive carbides on cross sections and their share in relation to the matrix is significantly greater, Figure 10. Additionally, the carbides are evenly distributed on the cross sections. However, the number and share of massive carbides is clearly greater in the case of the clad layer produced under forced cooling by the localized stream of nitrogen vapours, Figure 10a,b. Observations of the optical micrographs confirmed the tendency for porosity in the case of clad layers produced by conventional laser cladding at free cooling conditions, even at the minimum heat input of 60 J/mm, Figure 10a. The cluster of pores can be observed in the middle part of the clad layer LC1, while the average diameter of pores ranges between 20 to 80 μm. In turn, just two single pores with a diameter approx. 15 μm were observed on cross section of the clad produced under forced cooling conditions, Figure 10d.

In contrast to the clad layer produced at maximum heat input and free cooling, the microstructure of the clad LC1 in the regions between the massive carbides is characterized by a low number of individual precipitations not exceeding 10 μm in length, as can be clearly seen in Figure 10c. There are two explanations for this phenomenon. The first is related with lower amount of decomposed and diluted carbides due to four times lower laser output power (500 W) and thus the heat input. Therefore, the enrichment of the liquid by the W and C elements was also limited and insufficient for precipitation of carbides. The second reason is related with lower volume of the melt pool thus shorter time of material being at liquid state and shorter time for precipitation.

Close observation of the massive carbide's morphology on the optical micrographs in Figure 10d, showed that all these carbides are covered with a thin layer of block-like particles. Similar observations were reported by Zhang P. et al. [12] for spherical and irregularly shaped carbides in nickel-based matrix. Jones M. and Waag U. provided a comprehensive study on the effect of WC particle type on Ni-based composite coatings. They explained this phenomenon by partial melting of the WC particles during laser cladding. Next the C and W elements combine with Cr, Fe, and Ni to form the complex carbides $M_{23}C_6$, M_7C_3, and M_6C, which distribute in the matrix to form needle-like or block-like. However, a detailed analysis of this boundary and an attempt to explain this phenomenon is provided in the further part of the discussion.

To quantify the relationship between of laser output power of deposition process (proportional to heat input at constant scanning speed) and the cooling conditions and the morphology of the clad layer, the share of carbides relative to the matrix was determined on the cross sections, and the carbides size distribution for different clad layers was determined. The calculations were made by digitalising the micrographs images and using the NIS-Elements software (Nikon Corporation, Tokyo, Japan) for determining the area share of carbides, and the results are presented in Figure 11.

As can be seen, the calculated percentage share of carbides on cross section of the clad layer produced at maximum heat input of 240 J/mm and free cooling conditions is below 7%, while the share of carbides in the case of clad layer produced at the same heat input but with additional forced cooling is significantly higher, approx. over 22%, Figure 11a.

The relationship indicates generally the lower heat input the higher the share of carbides in the matrix. However, as can be seen, the effect of the forced localized cooling is advantageous, because it provides significantly higher share of carbides for every clad layer produced in the range of investigated parameters. In turn, the carbides size distribution determined for the comparative clad layers produced at minimum heat input of 60 J/mm show that the dominant size for the clad layer LC1 produce under free cooling conditions is approx. 150 ÷ 200 μm. While in the case of the clad layer HC1 produce under forced cooling conditions the carbides distribution is different and the share of carbides sized in the range 50 up to 150 μm is even and proportional, Figure 11b. The relationship clearly indicates that the cooling conditions at the same heat input affect significantly the thermal conditions and the tendency to dissolve carbides and thus reduce their size is lower when forced cooling is applied.

In turn the size distribution of carbides determined on the surface of the specimen indicates that the proportion of fine carbides is significant, higher than it would appear from the fraction share for the powder. The first reason for this phenomenon is the presence of secondary carbides precipitated in the liquid because of crystallization. However, these are carbides mainly in dendritic form. The share of such dendritic carbides is greater in the range of higher heat inputs (laser power), due to the conditions conducive to the dissolve of carbides in the liquid solution because of diffusion, decomposition due to direct heating with a laser beam, and then precipitation from the liquid solution. However, the presence of a large amount of very small carbide particles is also evident. The explanation for the presence of such many fine carbides is the primary form and morphology of WC-type carbides, which are in fact the eutectic of WC + W_2C manufactured as a result of melting and crushing. As can be seen from Figures 10d, 12–14 and 15a,b these massive eutectic carbides are coated by a layer of stoichiometric WC with high thermal stability.

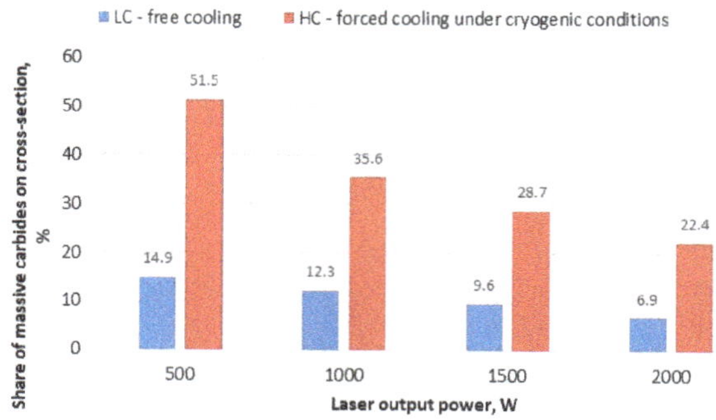

(**a**)

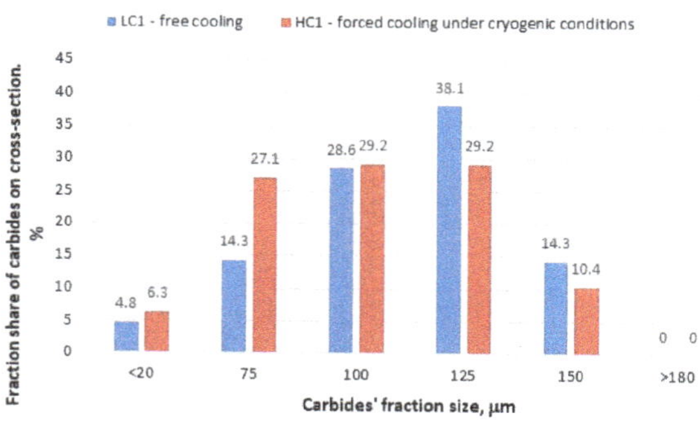

(**b**)

Figure 11. Influence of laser output power and cooling conditions on the share of massive carbides (**a**) and the fraction share of carbides on the cross-section of the clad layers produced at minimum laser out power 500 W at free and forced cooling under cryogenic conditions (**b**).

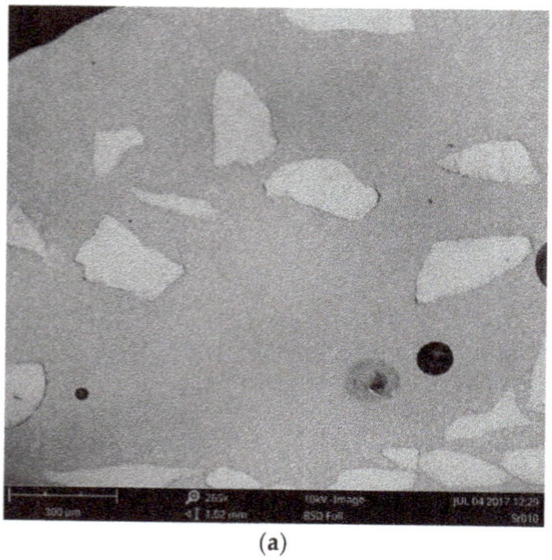

(**a**)

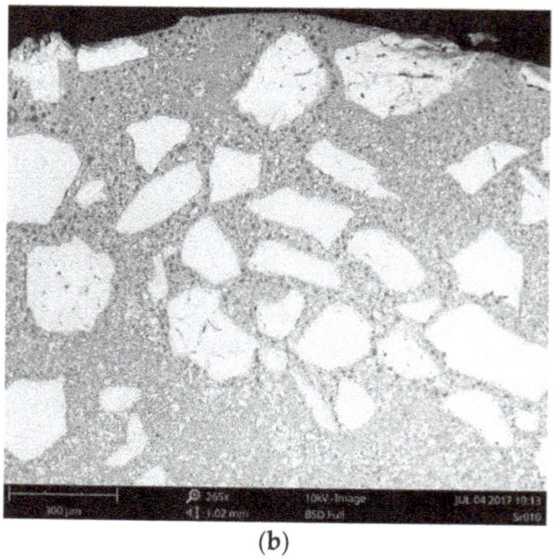

(**b**)

Figure 12. SEM micrographs of the comparative stringer beads LC1 and HC1 produced at the same minimum heat input of 60 J/mm (scanning speed 500 mm/min, laser output power 500 W) but different cooling conditions (Table 3): (**a**) free cooling in ambient air—a view of the entire clad layer and the middle region of the clad respectively; and (**b**) forced localized cooling by nitrogen vapours stream under cryogenic conditions ("hybrid laser deposition")—a view of the entire clad layer and the middle region of the clad respectively.

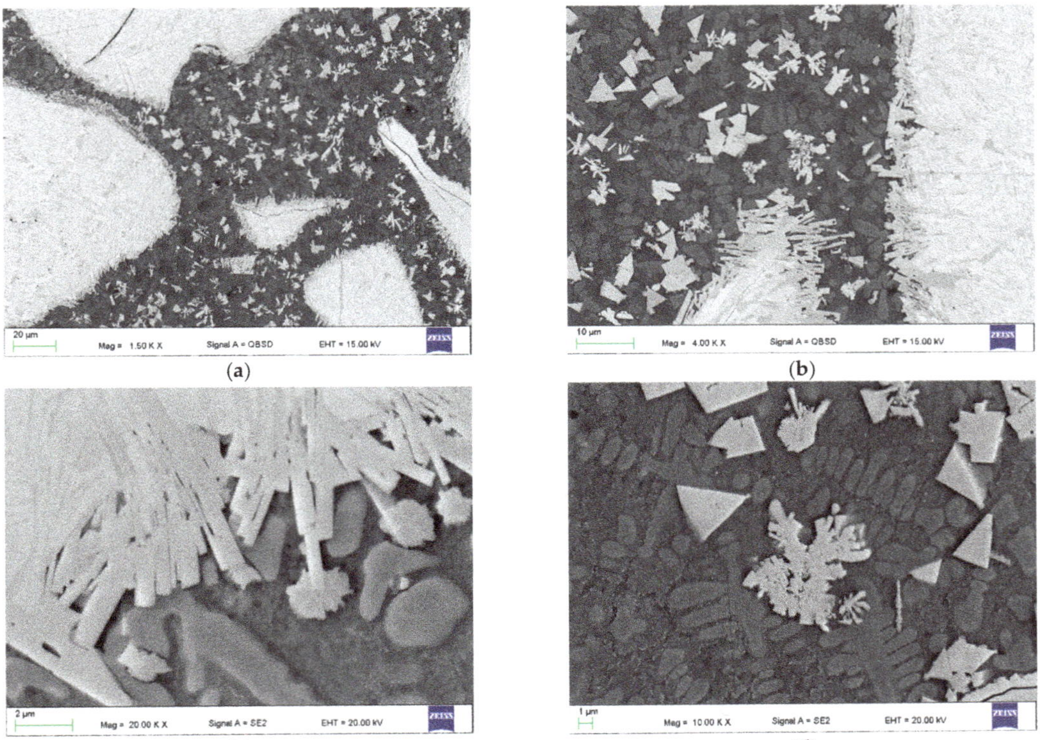

Figure 13. SEM microstructure of the stringer bead HC1 produced at the minimum heat input of 60 J/mm (scanning speed 500 mm/min, laser output power 500 W) and forced localized cooling by nitrogen vapours stream under cryogenic conditions ("hybrid laser deposition", Table 3): (**a**) general view of massive carbides and matrix; (**b**) a view of carbides boundaries; (**c**) partially recristalized needle-like carbides; (**d**) block-like particles in the dendritic matrix.

Table 4. Summary of EDS analysis on the cross section of the stringer bead HC1 produced at the minimum heat input of 60 J/mm and forced cryogenic cooling conditions (hybrid laser deposition process), Figure 14.

Point	W	Ni	Si	Fe	C	N	B	Possible Phase
				Atomic/Weight Concentration				
a_1	60.4/95.5	0.8/0.4	-	-	38.8/4.1	-	-	W_2C/WC
a_2	59.2/94.4	2.7/1.4	-	-	38.1/4.2	-	-	W_2C/WC
a_3	60.9/95.1	2.0/1.0	-	-	37.1/3.9	-	-	W_2C/WC
a_4	56.9/90.5	7.7/3.9	2.1/1.1	-	26.4/2.5	3.1/0.5	3.8/1.5	M_7C_3
a_5	61.4/92.9	8.0/4.0	-	-	30.6/3.1	-	-	M_2C
a_6	56.4/94.3	1.8/0.9	-	-	42.0/4.7	-	-	WC
b_1	58.1/94.6	2.0/1.0	-	-	39.9/4.4	-	-	WC
b_2	60.9/91.9	10.0/5.0	-	-	29.1/3.1	-	-	M_2C
b_3	55.7/90.2	11.5/6.2	-	-	32.8/3.6	-	-	M_2C
b_4	59.7/92.7	7.1/3.7	-	-	33.2/3.7	-	-	M_2C
b_5	21.1/54.0	49.8/40.7	-	-	21.9/3.7	7.2/1.6	-	Primary γ-Ni

Table 4. *Cont.*

Point	W	Ni	Si	Fe	C	N	B	Possible Phase
				Atomic/Weight Concentration				
b_6	21.0/54.5	48.3/39.9	-	-	23.2/3.9	7.5/1.7	-	Primary γ-Ni
b_7	6.0/25.7	37.8/52.3	4.5/2.9	0.9/1.1	17.4/5.0	27.7/10.6	3.0/2.36	Ni/Ni_3B eutectic, M_2C
b_8	9.9/33.3	54.4/58.3	-	-	26.8/5.9	-	8.8/2.6	Ni/Ni_3B eutectic, M_2C
b_9	64.4/93.6	7.7/3.7	-	-	27.9/2.7	-	-	M_7C_3
b_10	32.1/69.7	36.8/25.4	-	-	20.1/2.8	11.0/2.1	-	Line scan

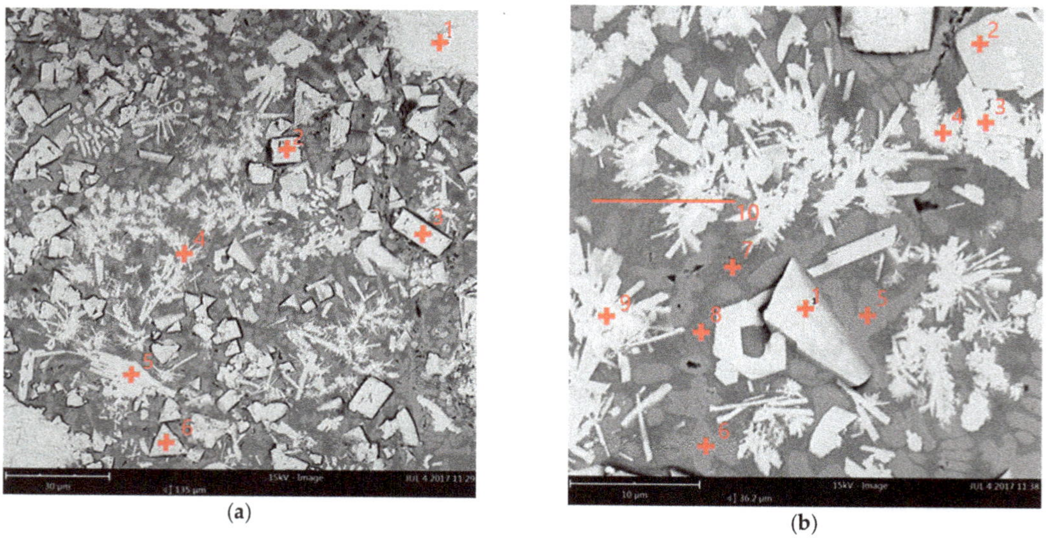

Figure 14. SEM microstructure of the selected representative regions between the massive tungsten carbides on cross section of the stringer bead HC1 produced at the minimum heat input of 60 J/mm (scanning speed 500 mm/min, laser output power 500 W) and forced localized cooling by nitrogen vapours stream under cryogenic conditions ("hybrid laser deposition", Table 3): (**a**,**b**) a view of the region between massive carbides; results of EDS for individual points and line scan provided in Table 4.

Table 5. Summary of EDS analysis on the cross section of the stringer bead LC1 produced at the minimum heat input of 60 J/mm and free cooling conditions (conventional laser cladding), Figure 15.

Point	W	Ni	Si	Fe	C	N	B	Possible Phase
				Atomic/Weight Concentration				
b_1	54.4/94.8	-	-	-	45.6/5.2	-	-	W_2C/WC
b_2	54.7/94.5	-	-	-	45.3/5.5	-	-	W_2C/WC
b_3	39.0/82.1	6.6/4.6	9.1/1.7	10.3/6.7	35.0/4.9	-	-	M_2C
b_4	8.2/31.2	20.5/24.8	5.9/1.8	27.6/31.9	28.9/7.2	-	8.8/3.1	Ni/Ni_3B eutectic, M_2C
b_5	22.6/60.7	16.3/14.0	-	21.7/17.7	26.7/4.7	-	12.5/2.9	M_3C in eutectic
d_1	35.0/80.0	7.6/5.5	11.5/2.3	9.7/6.8	36.2/5.4	-	-	M_2C
d_2	34.8/80.5	6.7/5.0	16.6/3.3	8.8/6.2	33.0/5.0	-	-	M_2C
d_3	19.4/57.3	17.6/16.6	-	18.5/16.6	30.0/5.8	-	14.6/3.7	M_3C in eutectic
d_4	15.8/51.4	17.6/18.3	-	19.0/18.9	28.5/6.0	-	19.1/5.4	M_3C in eutectic
d_5	18.0/53.5	18.9/17.9	9.6/2.4	21.2/19.2	22.5/4.3	-	10.1/2.7	line scan

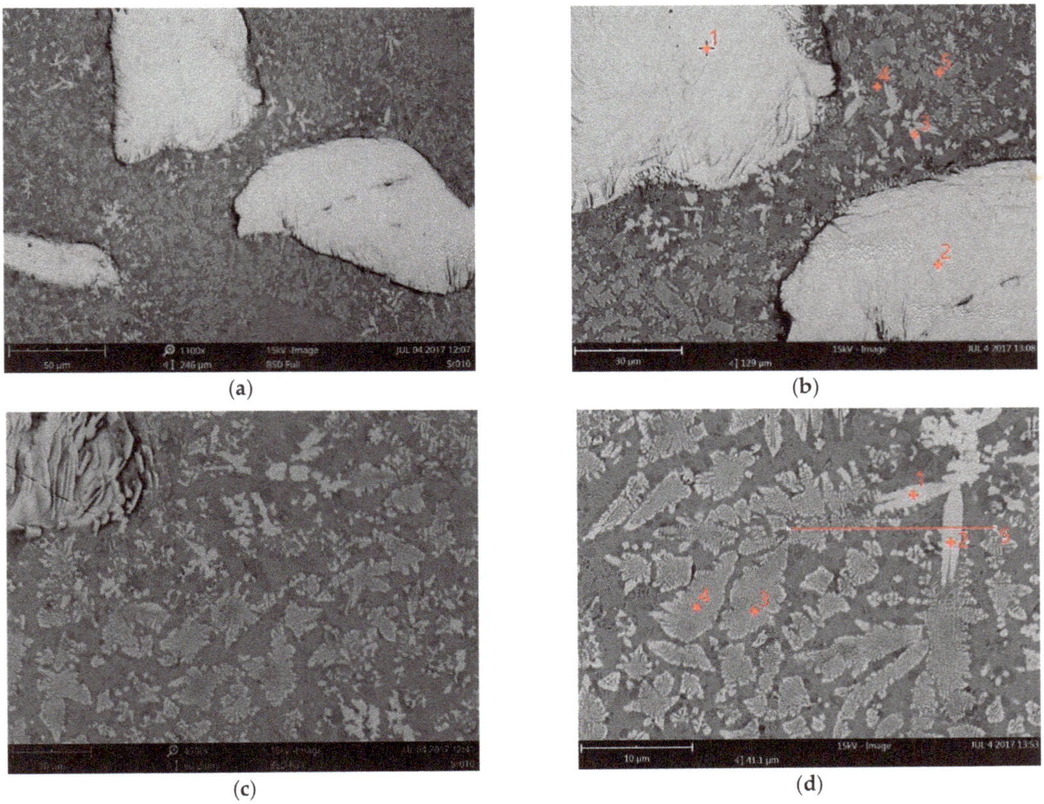

Figure 15. SEM microstructure of the stringer bead LC1 produced at the minimum heat input of 60 J/mm (scanning speed 500 mm/min, laser output power 500 W) and free cooling in ambient air (Table 3): (**a**,**b**) general view of massive carbides and matrix; (**c**) a view of carbides boundaries; (**d**) a view of precipitations in the matrix. Results of EDS for individual points and line scan provided in Table 5.

However, in the conditions of laser cladding the particles are exposed to direct action of the laser beam, and thus separation or even decomposition. If the carbide gets directly to the melt pool with a temperature below the carbides melting point, its decomposition will be slight, mainly due to the mutual diffusion of carbide and the liquid. When the carbide is exposed to direct laser beam radiation with high power density, it can directly lead to its partial or complete decomposition. In such a case, the carbide decomposes into fine carbides of needle-like or block-like form, as can be seen in Figures 9b and 14a,b.

The representative and comparative SEM micrographs were presented for the clad layers produced at the minimum heat input and different cooling conditions in Figure 12. The SEM micrograph of the clad layer produced at free cooling in Figure 12a show fine dendritic precipitations mainly near the massive carbides. In turn, on the SEM micrograph of the clad layer produced under forced cooling conditions in Figure 12b the regions between the massive carbides are densely filled mainly with fine block-like particles. The SEM micrographs with higher magnification of the clad layer HC1 produced at forced cooling revealed fine dendritic and needle-like particles, as well as trapezoidal blocks evenly distributed in dendritic matrix, Figure 13. While the massive eutectic W_2C/WC carbides with feather structure are composed of needle-like and trapezoidal blocks of carbides which are light grey and white, Figure 13a–c. The boundaries of the massive carbides are partially dissolved, as can be seen in Figure 13b,c. It is also characteristic that all needles and trapezoidal blocks on the boundary are light grey. Between the light grey particles at the massive eutectic carbide's boundary the grey dendrites and dark grey interdendritic regions of matrix can be clearly seen, Figure 13d. It can be also seen that some of the light grey particles, especially long needles (length up to 40 μm), go deep into the massive eutectic carbides, Figure 13b,c. The observed and described morphology of the light grey particles (needle-like and trapezoidal blocks) on the boundary of massive eutectic carbides indicates that the particles were not formed because of recrystallization or precipitation from the liquid phase of the melt pool, but the particles are the integrated part of the massive eutectic carbides, Figure 13.

Another evidence that the light grey particles on the boundaries of massive eutectic carbides were not recrystalized are the very fine crystals growing epitaxially on the tips of some of needles and tops of trapezoidal blocks, as shown in Figure 13c,d. Since the needles' tips have been exposed to longest exposure to the liquid and higher temperature, therefore, these regions were dissolved and recrystalized. In turn, smaller separate particles (both needles and blocks) almost completely recrystalized into dendrites, as shown in Figure 13b,d. The above observations indicate that just more thermodynamically stable phases (stoichiometric carbides) remained at the boundary of massive eutectic carbides. Since the massive eutectic carbides (eutectic composition lies at approx. 38 at.% C on W-C phase diagram) are originally composed of approx. 80% of W_2C (β phase) and 20% of stoichiometric WC (δ phase), it can be assumed that the remaining phases on boundary are WC carbides. According to a W-C phase diagram, the melting point of tungsten monocarbide WC is at least 50÷60 degree higher if compared to melting point of W_2C [32,33]. According to some literature data, the difference is even greater. The melting point for β-W_2C is often reported at 2730 °C, while for δ-WC at 2870 °C. Moreover, the δ-WC phase is metastable almost up to melting point. However, it should be noted that the conditions of phases transformation during laser deposition are far from the conditions of thermodynamic equilibrium, due to high rates of heating, cooling and high maximum temperature and also high temperature gradients.

In the case of chemical composition analysis with an EDS spectrometer, the quantitative content of elements such as nitrogen and carbon are usually not given due to the high uncertainty and measurement error. However, in the case of the presented results of the study on chemical composition, a satisfactory and sufficient convergence of the results of carbon and nitrogen content was obtained, allowing for the identification and confirmation of the presence of the expected constituents. For this reason, it was decided to include the results of the quantitative measurements of carbon and nitrogen content to ensure greater clarity of the results.

The EDS analysis conducted on the cross-section of the clad layer HC1 produced at the lowest heat input and forced cooling, confirmed that the composition of the massive carbides is typical for eutectic W_2C/WC, Figure 14, Table 4. In turn, the composition of the trapezoidal block indicates presence of stoichiometric monocrystalline tungsten carbides WC, Figure 14, Table 4. The determined composition of grey dendrites indicates for the matrix of primary γ-Ni, while the dark grey interdendritic regions show composition typical for Ni/Ni_3B eutectic, and possible share of M_2C type carbides, Figure 14, Table 4. The EDS analysis conducted for the needle-like constituents indicated composition typical for complex carbide M_7C_3 type, Figure 14, Table 4. It should be also noted that the content of iron Fe detected in the clad layer LC1 is relatively low. The Fe was confirmed just in the interdendritic regions, Figure 14, Table 4. It indicates that thanks low dilution of the clad by the substrate material, the composition of the clad is close to the applied powder.

Although the quantitative results of the measurement of nitrogen content should be considered with a great uncertainty, it is evident that the nitrogen was detected in the clad layer produced under cryogenic conditions, mainly in regions of matrix dendrites and interdendritic regions, Figure 14 and Table 4. It is most likely the result of nitrogen vapours stream application directly into the deposition region, and thus the presence of high amount of gaseous nitrogen in this region. In any case, the test results show that there is partial nitrogen absorption under the investigated conditions of laser deposition.

On the other hand, the morphology of the clad layer LC1 produced at minimum heat input but at free cooling conditions observed on the SEM micrographs is completely different that the morphology of the clad layer produced under forced cooling conditions, Figure 15. Beside significantly lower share of massive eutectic carbides, the morphology of precipitations in the regions between the massive eutectic carbides is also different. Closer observation of the massive carbides revealed another difference in the morphology. These carbides observed on the cross-section of the clad layer LC1 produced at free cooling have been dissolved much less in the boundary regions if compared with the carbides of clad layer produced at forced cooling, Figures 14 and 15.

Just little traces of local dissolution and subsequent resolidification can be found on the carbides' boundary. However, it should be noted that these are just few massive eutectic carbides that have not completely dissolved. The explanation is that the liquid solution has reached the solubility limit of carbon and tungsten, at least locally, which inhibited the further dissolution of remaining massive carbides in the melt pool. The resolidified regions of boundaries show fine light grey dendritic precipitation smaller than 10 μm, Figure 15b,d. In turn, the massive carbides are also composed of needle-like particles and trapezoidal blocks, however, highly integrated and embedded in the matrix-like phase, as can be seen in Figure 15b. It can be supposed that the matrix-like phase is the sub stoichiometric tungsten carbide W_2C (β phase) which is less stable than the stoichiometric monocarbide WC (δ phase). In Figure 15 there are no traces of fine needle-like particles and trapezoidal blocks typical for the clad layer HC1 produced at forced cooling under cryogenic conditions. This is another evidence that these particles do not nuclei nor solidify in the liquid melt pool under the investigated laser cladding conditions. The particles rather come from the decomposition and dissolution of massive eutectic W_2C/WC carbides. Since the particles are small and have low heat capacity, thus they heat up intensively and next completely dissolve in the liquid malt pool. However, it must be emphasised that the observations and findings are valid just for the specific conditions of the experiment, determined technological conditions such as the laser beam characteristic (energy density, intensity profile, wavelength, etc.), processing parameters, the way of powder delivery, thermal conditions.

The EDS analysis conducted on the cross-section of the clad layer LC1 produced at the lowest heat input and free cooling, beside the massive eutectic W_2C/WC, confirmed presence of M_2C type carbides as dendritic precipitations, Figure 15 and Table 5. In turn, the densely packed light grey precipitations between the massive carbides show composition typical for M_3C carbides in eutectic, Figure 15, Table 5.

The composition of dark grey matrix indicates presence of Ni/Ni_3B eutectic, possibly with some amount of M_2C type carbides, Figure 15, Table 5. In general, the dilution was higher from a dozen to several dozen percent in the case of clad layers produced at free cooling conditions, Figure 8. Therefore, the elements from the non-alloy steel substrate, mainly the Fe, passed into the clad. As a result of higher dilution, the composition of the clad is different than the original composition of the applied powder, Table 2.

Although the quantitative results of the measurement of nitrogen content should be considered with a great uncertainty, it is evident that in this case the nitrogen content is significantly lower, on the limit of the measurement error of EDS spectrometer, Table 5. This is due to application of shielding of argon during laser deposition tests, and thus protection against absorption of gases from the ambient. The X-ray diffraction pattern of the experimental powder is presented in Figure 3. It indicates the presence of tungsten carbides δ-WC and β-W_2C, as well as the γ-Ni. Additionally, clear picks for Ni_3B and complex carbides M_7C_3 type were identified, Figure 3.

In turn, the X-ray diffraction patterns for two comparative samples are presented in Figure 16. The comparative samples were prepared by multi-bead cladding of the disc's substrate intended for the tribological tests. The discs were produced by cladding the substrate with an overlap approx. 20 ÷ 25%, to produce the coating on the entire disc surface. The comparative coatings were produced at the minimum heat input (minimum laser output power) at free cooling and at forced cooling under cryogenic conditions (parameters for the single beads LC1 and HC1, Table 3). Surfaces of the discs were grinded with roughness Ra 0.8 ÷ 1.1 μm.

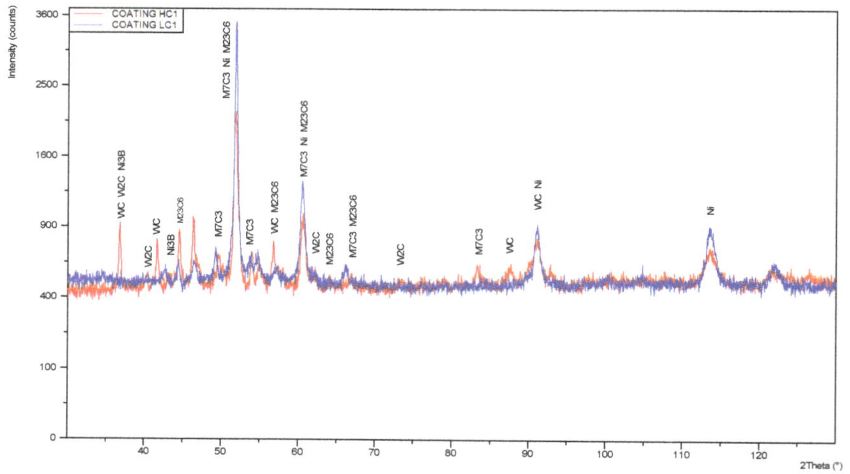

Figure 16. XRD patterns of the comparative coatings produced at the same minimum heat input of 60 J/mm (scanning speed 500 mm/min, laser output power 500 W) but different cooling conditions: LC1 at free cooling, and HC1 with forced localised cooling under cryogenic conditions, Table 3.

The initial and general comparison of the two X-ray diffraction patterns show a clear difference in the phase composition. The sample coated at forced cooling under cryogenic conditions (parameters for HC1) shows simply more peaks compared to the sample coated at free cooling (parameters for LC1), Figure 16. However, peaks for the LC1 are more intensive. It is evident that the peaks for γ-Ni are dominant in the case of the coating produced at free cooling (LC1), Figure 16. There are also clear peaks for complex carbides $M_{23}C_6$ and M_7C_3 type. Moreover, slight peaks for δ-WC and Ni_3B were identified. The obtained results are consistent with the previous observations of SEM micrographs and EDS analysis. The phase composition of the clad layer produced under free cooling LC1 is also affected by the higher dilution by the substrate material, mainly iron Fe, as shown the chemical composition analysis. In turn, the X-ray diffraction patterns indicate that the share of eutectic W_2C/WC is significantly higher in the coating produced under cryogenic conditions. The dominant γ-Ni phase in the coating produced at free cooling is a result of different thermal conditions of the deposition process, higher impact of the heat, and thus higher degree of carbides decomposition and dissolution, as well as higher dilution by the Fe from the substrate of non-alloy steel. Therefore, the carbides precipitated from the liquid solution are more complex, as shown in Figure 16.

It should be noted that the detection level of the applied XRD method is approx. 3% for the individual phase, therefore, it does not allow for precise identification of the fine precipitations at a low share. However, due to the conjunction with chemical composition analysis in the micro regions and individual constituents, precise detection of the phase composition was provided.

3.3. Hardness

The profiles of hardness were determined on cross-section of the clad layers produced at minimum and maximum heat input at free cooling and under cryogenic conditions. For the Vickers tests of the composite clad layers the applied load was set at 5N. The large load was chosen to avoid large scatter of results (low values for matrix and high for carbides) providing mean hardness value for the subsequent regions. However, the profiles of Vickers HV0.5 hardness presented in Figure 17 showed clear difference for the hardness distribution for individual clad layers. The difference is related mainly with the width of the zone with a high hardness. The clad layers produced at free colling show higher width of the zone with the high hardness ranged approx. from 500HV0.5 to 600 HV0.5. It is related directly with the shape of the individual clads, specifically with the height and penetration depth. The clad layers produced at free cooling conditions have higher penetration depth compared to the clads produced under cryogenic conditions. Therefore, the total depth (width on the graph) of the clad layers produced at free cooling conditions is higher.

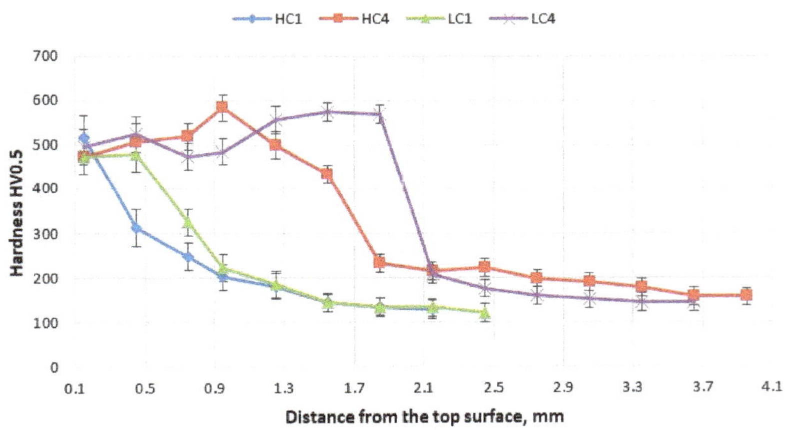

Figure 17. Hardness distribution on cross-sections of the representative clad layers produced by laser deposition of experimental WC-Ni composite powder (Table 3) at free cooling conditions (LC1, LC4—conventional laser cladding); and at forced localized cooling by nitrogen vapours stream under cryogenic conditions (HC1, HC4—hybrid laser deposition process).

On the other hand, despite the difference in the chemical and phase composition of the clad layers, the maximum values of hardness and the scattering of results are very similar. It should be noted that the presented results of hardness measurement for each point on the graphs are a mean value taken from three different sections for every tested clad layer. The values of standard deviation indicate small dispersion of results. This is due to relatively large size of Vickers's indenter imprint at the load of 5N. The clad layers HC1 produced al lowest heat input and under cryogenic conditions showed the highest value of hardness 520HV0.5 directly under the top surface (face) of the clad. In turn, the highest measured hardness approx. 600HV0.5 was detected for the clad layer produced at maximum het input and cryogenic conditions. For every tested clad layer, starting from a certain depth the hardness decreases gradually till the value of the base metal of non-alloy steel (S235JR) approx. 120 ÷ 15HV0.5. This finding contrasts with the phenomenon described in the previous manuscript on "Laser Deposition of Fe-based metallic powder under cryogenic conditions" [1]. However, different technique of cryogenic cooling was applied in the previous study, consisted of submerging the entire sample in the liquid nitrogen bath. Therefore, the undercooling of the substrate was higher in the previously tested "liquid nitrogen bath" technic.

3.4. Tribological Test

It must be noted that the task of the study is not a comprehensive analysis of the mechanisms of wear and tribological characteristics. Therefore, just basic results of tribological ball-on-disc tests were presented for the final comparison of the characteristics and wear resistance of the comparative coatings produced under different technological conditions and different thermal conditions. The values presented at the graphs are mean values taken from at least three individual measurements. The volume loss of the coatings was calculated by determining the wear track profile in four places around the circumference of the wear track. Comparison of the wear tracks for representative samples produced at the lowest heat input but different cooling conditions (coating produced at parameters for HC1 and LC1) is shown in Figure 18.

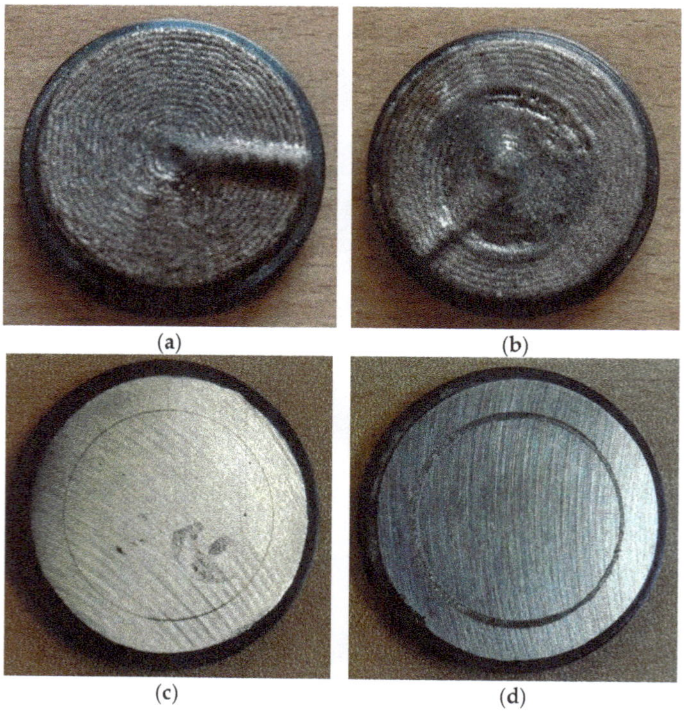

Figure 18. A view of the discs prepared for ball-on-disc wear tests with coating produced by laser deposition; (**a**) coating produced with parameters and technological conditions for the clad HC1 (cryogenic hybrid method), (**b**) coating produced with parameters and technological conditions for the clad LC1 (conventional laser cladding), and after the tribological test; (**c**) sample with coating HC1 type, and (**d**) sample with coating LC1 type.

The wear track of the sample HC1 produced under cryogenic conditions and the lowest heat input has the smallest width, as shown in Figure 18c. As expected, the determined volume loss was also the lowest for the coating HC1, Figure 19a. Since the lowest volume loss, the better wear resistance, it means that the composite coating produced at the lowest heat input and cryogenic conditions have the highest wear resistance under dry sliding conditions of the experiment. As can be seen in Figure 19a, even the coating produced at the maximum heat input but at localized forced cooling under cryogenic conditions showed lower volume loos and thus higher wear resistance than the coatings produced at free cooling and the lowest heat input. An explanation for this observed phenomenon can be found in Figure 11a, showing the share of massive carbides in the matrix. The share of massive carbides (eutectic W_2C/WC) in the matrix of every clad layer produced under cryogenic conditions is higher than for the clad layers produced at free cooling. Even in the case of the clad layer produced at the maximum heat input HC4 (laser output power 2000 W) but with simultaneous localized cooling the share of massive carbides is 22.4%, while in the case of the clad layer produced at the lowest heat input LC1 (500 W) but free cooling the share of carbides is less than 15%. Thus, the results show that under the test conditions, the wear resistance of the composite coatings is proportional to the massive carbides (eutectic W_2C/WC) content in the matrix. On the other hand, the volume loos of the counter face material of steel ball is directly proportional to the volume loos of the discs samples. This phenomenon may be explained by the values of the coefficient of friction determined during individual tests. The values of coefficient of friction presented at the graph in Figure 19b are the mean values, however, a clear relationship between the coefficient of friction and wear intensity can be observed. In general, the higher coefficient of friction the higher wear intensity, both the sample and the steel ball, Figure 19. The obtained results indicate that the coefficient of friction also depends on the phase composition of the coatings, mainly on the share of massive carbides in the matrix, as well as composition of the matrix.

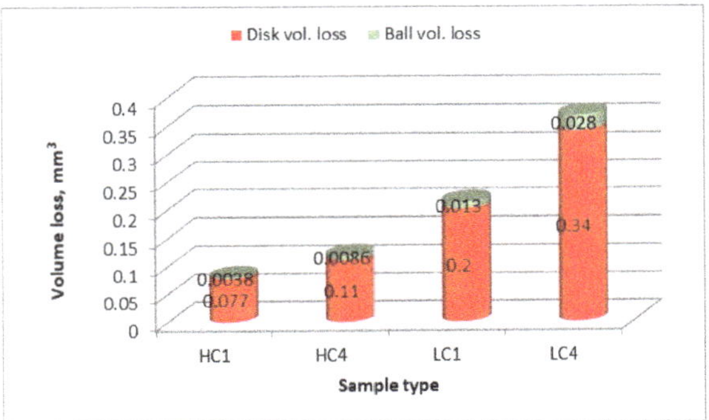

(**a**)

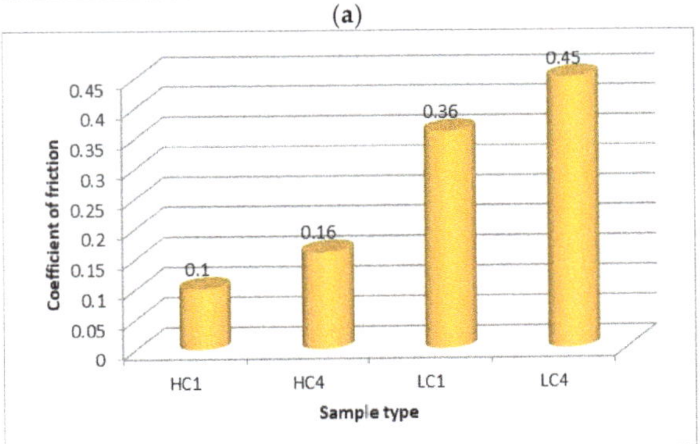

(**b**)

Figure 19. Volume loss of the counter body steel ball and the disc specimens of commercially pure titanium, titanium alloy and the nitrided surface layers and after sliding for a distance of 188.4 m at normal load of 30 N (**a**), and comparison of the coefficient of friction of the disc specimens sliding with a steel ball during ball-on-disc wear tests (**b**).

4. Conclusions

The novel technique of laser deposition of composite WC-Ni powder with forced localized cooling of the deposit by nitrogen vapours stream under cryogenic conditions was successfully demonstrated and its potential for shaping the geometry and the microstructure was presented. The results of the study confirmed the advantageous effect of the forced cooling in terms of chemical and phase composition, microstructure, and wear resistance under dry sliding conditions. The use of the forced cooling technique significantly reduces the tendency to decomposition and to dissolve carbides in the melt pool during laser deposition of composite WC/Ni powder. The share of carbides on cross section of the clad layer produced at minimum heat input (laser output power 500 W and scanning speed 500 mm/min) at conventional conditions of free cooling was less than 15%, while thanks to the application of forced localized cooling of the deposit the share of carbides was maintained over 51%. Moreover, the degree of dilution was significantly reduced thanks to the forced localized cooling. The chemical composition of clad layer produced at minimum heat input and with forced cooling under cryogenic conditions was very similar to that of experimental composite WC-Ni powder, except for the increased nitrogen content in the dendritic matrix and interdendritic regions. The increased content of nitrogen detected in the matrix of the clad layer is resulted by presence of high amount of gaseous nitrogen in deposition region if the nitrogen vapours stream is applied. Therefore, during the laser deposition of composite powder (laser cladding) the alloying process takes place additionally. The material of the clad layer (the deposit) is simultaneously enriched by nitrogen from the gaseous atmosphere, so the demonstrated process of laser deposition of composite powder with simultaneous localized cooling by the nitrogen vapor stream can be considered as hybrid process, combining the conventional laser cladding (LC) and also laser gas nitriding (LGN).

Author Contributions: Conceptualization, methodology, investigation, formal analysis, validation, supervision, writing—review and editing, A.L.; project administration, funding acquisition, D.Ś. All authors have read and agreed to the published version of the manuscript.

Funding: The research was supported by the National Centre for Research and Development, Poland, under the grant number POIR.01.01.01-00-0278/15-007-02, financed by EU funds.

Institutional Review Board Statement: Not applicable.

Informed Consent Statement: Not applicable.

Data Availability Statement: The data presented in this study are available on request from the corresponding author. The data are not publicly available due to know how protection.

Acknowledgments: The authors thank for technical support to all those who assisted in the preparation of samples and the conducting of measurements, especially to Marcin Żuk, Adrian Kukofka, and students who assisted in the laboratory tests.

Conflicts of Interest: Authors declare no conflict of interest.

References

1. Lisiecki, A.; Ślizak, D. Hybrid Laser Deposition of Fe-Based Metallic Powder under Cryogenic Conditions. *Metals* **2020**, *10*, 190. [CrossRef]
2. Lisiecki, A. Tribology and Surface Engineering. *Coatings* **2019**, *9*, 663. [CrossRef]
3. Czupryński, A. Comparison of Properties of Hardfaced Layers Made by a Metal-Core-Covered Tubular Electrode with a Special Chemical Composition. *Materials* **2020**, *13*, 5445. [CrossRef]
4. Górka, J. Assessment of the quality of Abrasion-Resistant Plates welded joint. *J. Min. Metall. Sect. B-Metall.* **2020**, *56*, 209–220. [CrossRef]
5. Kotarska, A.; Janicki, D.; Górka, J.; Poloczek, T. Solid Particle Erosion of Laser Surface Melted Ductile Cast Iron. *Arch. Foundry Eng.* **2020**, *3*, 105–111.
6. Czupryński, A.; Wyględacz, B. Comparative Analysis of Laser and Plasma Surfacing by Nickel-Based Superalloy of Heat Resistant Steel. *Materials* **2020**, *13*, 2367. [CrossRef] [PubMed]
7. Tomków, J.; Czupryński, A.; Fydrych, D. The abrasive wear resistance of coatings manufactured on high-strength low-alloy (HSLA) offshore steel in wet welding conditions. *Coatings* **2020**, *10*, 219. [CrossRef]
8. Górka, J.; Czupryński, A.; Żuk, M.; Adamiak, M.; Kopyść, A. Properties and Structure of Deposited Nanocrystalline Coatings in Relation to Selected Construction Materials Resistant to Abrasive Wear. *Materials* **2018**, *11*, 1184. [CrossRef] [PubMed]
9. Czupryński, A. Flame spraying of aluminum coatings reinforced with particles of carbonaceous materials as an alternative for laser cladding technologies. *Materials* **2019**, *12*, 3467. [CrossRef]

10. Górka, J.; Kik, T.; Czuprynski, A.; Foreiter, W. Technology of welding hard wearing plates. *Weld. Int.* **2014**, *28*, 749–755. [CrossRef]
11. Verwimp, J.; Rombouts, M.; Geerinckx, E.; Motmans, F. Applications of Laser Cladded WC-Based Wear Resistant Coatings. *Phys. Procedia* **2011**, *12*, 330–337. [CrossRef]
12. Zhang, P.; Pang, Y.; Yu, M. Effects of WC Particle Types on the Microstructures and Properties of WC-Reinforced Ni60 Composite Coatings Produced by Laser Cladding. *Metals* **2019**, *9*, 583. [CrossRef]
13. Wang, L.; Yao, J.H.; Hu, Y.; Zhang, Q.L.; Sun, Z.; Liu, R. Influence of electric-magnetic compound field on the WC particles distribution in laser melt injection. *Surf. Coat. Technol.* **2017**, *315*, 32–43. [CrossRef]
14. Liu, H.; Xu, Q.; Wang, C.; Zhang, X. Corrosion and wear behavior of Ni60CuMoW coatings fabricated by combination of laser cladding and mechanical vibration processing. *J. Alloys Compd.* **2015**, *621*, 357–363. [CrossRef]
15. Li, M.; Han, B.; Wang, Y.; Song, L.; Guo, L. Investigation on laser cladding high-hardness nano-ceramic coating assisted by ultrasonic vibration processing. *Optik* **2016**, *127*, 4596–4600. [CrossRef]
16. Huang, S.; Samandi, M.; Brandt, M. Abrasive wear performance and microstructure of laser clad WC/Ni layers. *Wear* **2004**, *256*, 1095–1105. [CrossRef]
17. Szczucka-Lasota, B.; Gajdzik, B.; Węgrzyn, T.; Wszołek, Ł. Steel Weld Metal Deposit Measured Properties after Immediate Micro-Jet Cooling. *Metals* **2017**, *7*, 339. [CrossRef]
18. Zhang, X.; Zhou, Y. Effect of Deep Cryogenic Treatment on Microstructure and Wear Resistance of LC3530 Fe-Based Laser Cladding Coating. *Materials* **2019**, *12*, 2400. [CrossRef]
19. Zieliński, A.; Jażdżewska, M.; Narożniak-Łuksza, A.; Serbiński, W. Surface structure and properties of Ti6Al4V alloy laser melted at cryogenic conditions. *J. Achiev. Mater. Manuf. Eng.* **2006**, *18*, 423–426.
20. Zieliński, A.; Serbiński, W.; Majkowska, B.; Jażdżewska, M.; Skalski, I. Influence of laser remelting at cryogenic conditions on corrosion resistance of non-ferrous alloys. *Adv. Mater. Sci.* **2009**, *9*, 21–28. [CrossRef]
21. Majkowska, B.; Serbiński, W. Cavitation wearing of the SUPERSTON alloy after laser treatment at cryogenic conditions. *Solid State Phenom.* **2010**, *165*, 306–309. [CrossRef]
22. Cui, Z.; Shi, H.; Wang, W.; Xu, B. Laser surface melting AZ31B magnesium alloy with liquid nitrogen-assisted cooling. *Trans. Nonferrous Met. Soc. China* **2015**, *25*, 1446–1453. [CrossRef]
23. Razavykia, A.; Delprete, C.; Baldissera, P. Correlation between Microstructural Alteration, Mechanical Properties and Manufacturability after Cryogenic Treatment: A Review. *Materials* **2019**, *12*, 3302. [CrossRef] [PubMed]
24. Lisiecki, A. Mechanisms of hardness increase for composite surface layers during laser gas nitriding of the Ti6A14V alloy. *Mater. Technol.* **2017**, *51*, 577–583.
25. Lisiecki, A.; Kurc-Lisiecka, A. Erosion wear resistance of Titanium-Matrix Composite Ti/TiN produced by diode-laser gas nitriding. *Mater. Technol.* **2017**, *51*, 29–34. [CrossRef]
26. Lisiecki, A.; Piwnik, J. Tribological characteristic of titanium alloy surface layers produced by diode laser gas nitriding. *Arch. Metall. Mater.* **2016**, *61*, 543–552. [CrossRef]
27. Lisiecki, A. Comparison of Titanium Metal Matrix Composite surface layers produced during laser gas nitriding of Ti6Al4V alloy by different types of lasers. *Arch. Metall. Mater.* **2016**, *61*, 1777–1783. [CrossRef]
28. Lisiecki, A. Titanium Matrix Composite Ti/TiN Produced by Diode Laser Gas Nitriding. *Metals* **2015**, *5*, 54–69. [CrossRef]
29. Kik, T.; Moravec, J.; Novakova, I. Numerical simulations of X22CrMoV12-1 steel multilayer welding. *Arch. Metall. Mater.* **2019**, *64*, 1441–1448. [CrossRef]
30. Kik, T.; Moravec, J.; Novakova, I. Application of Numerical Simulations on 10GN2MFA Steel Multilayer Welding. In *Springer Proceedings in Mathematics and Statistics, Proceedings of the 2018 International Conference on Dynamical Systems: Theory and Applications, DSTA 2017, Lodz, Poland, 11–14 December 2017*; Spinger: Cham, Switzerland, 2017; Volume 249, pp. 193–204. [CrossRef]
31. Kusinski, J.; Kac, S.; Kopia, A.; Radziszewska, A.; Rozmus-Górnikowska, M.; Major, B.; Major, L.; Marczak, J.; Lisiecki, A. Laser modification of the materials surface layer—A review paper. *Bull. Pol. Acad. Sci. Tech. Sci.* **2012**, *60*, 711–728. [CrossRef]
32. Koneshlou, M.; Asl, K.M.; Khomamizadeh, F. Effect of cryogenic treatment on microstructure, mechanical and wear behaviors of AISI H13 hot work tool steel. *Cryogenics* **2011**, *51*, 55–61. [CrossRef]
33. Garcia, J.; Cipres, V.C.; Blomqvist, A.; Kaplan, B. Cemented carbide microstructures: A review. *Int. J. Refract. Met. Hard Mater.* **2019**, *80*, 40–68. [CrossRef]

materials MDPI

Article

Study of Optical Properties of Surface Layers Produced by Laser Surface Melting and Laser Surface Nitriding of Titanium Alloy

Aleksander Lisiecki

Silesian University of Technology, Faculty of Mechanical Engineering, Department of Welding Engineering, Konarskiego 18A street, 44-100 Gliwice, Poland; aleksander.lisiecki@polsl.pl

Received: 9 August 2019; Accepted: 20 September 2019; Published: 24 September 2019

Abstract: This study measured optical properties, such as specular, diffuse, and total reflection for 808 nm wavelength, characteristic for high power diode lasers radiation, from the surface of titanium alloy Ti6Al4V at delivery conditions, polished, and oxidized. Moreover, the optical properties of surface layers produced by high power direct diode laser (HPDDL) melting and nitriding were determined. Additionally, a methodology for determining the value of absorption for 808 nm wavelength of the HPDDL radiation on the surface of a melt pool during laser surface melting and nitriding of titanium alloy was proposed. The results show that the distinct differences in absorption affect the heat transfer, thermal conditions of laser heating and thereby the penetration depth during laser melting and nitriding of the titanium alloy.

Keywords: titanium nitrides; reflectivity; absorption; high power diode laser

1. Introduction

Titanium alloys have excellent mechanical properties, but during the manufacture of machinery parts, components or tools it is often necessary to apply additional surface treatments to improve the tribological characteristic of working surfaces [1–11]. One of the methods of improving properties is laser gas nitriding (LGN) of titanium and its alloys. Laser gas nitriding involves heating or surface melting the titanium substrate in a nitrogen or nitrogen-rich atmosphere. Laser gas nitriding is a very efficient and flexible method for shaping the surface characteristics of commercially pure titanium and technical titanium alloys [1–11]. This is because of the large chemical affinity of titanium for nitrogen at elevated temperatures, particularly in a liquid state. Titanium has a strong tendency to absorb nitrogen at temperatures above 600 °C and precipitate nitrides ε-Ti_2N, even at a relatively low nitrogen content in the solid solution (approx. 4 at. % of nitrogen at temp. 500 °C), and also δ-TiN at a higher nitrogen content (23.0 at. % of nitrogen at 1050 °C). Due to this feature, a composite surface layer with in situ synthesized nitrides, embedded in the metallic matrix of titanium, can be formed during laser gas nitriding of titanium and titanium alloys [5–11]. Additionally, nitrogen has the effect of strengthening the structure, as an interstitial element, which distorts the lattice by occupying the available interstitial sites in the structure [7,11]. These types of titanium matrix composite (TMC) surface layers have significantly higher hardness and exhibit higher resistance to abrasive wear than the substrate of pure titanium or conventional technical titanium alloys [8,9,11].

So far, the process of titanium nitriding has been carried out mainly by solid state Nd:YAG (Yttrium-Aluminum Garnet doped with neodymium ions) and gaseous CO_2 lasers that emit radiation with a wavelength of 1.06 and 10.6 μm, respectively [7–21]. The next generation of lasers recently introduced to the industry are high power diode lasers. These lasers show some advantages in the field of laser surface treatment, compared to the above mentioned solid state and gaseous lasers [7,10,22,23].

Diode lasers provide a favorable rectangular shape of the laser beam spot, and uniform energy distribution across the spot, and also are characterized by a short wavelength in the very near infrared range from 0.808 to 0.960 μm, which is advantageous because of the relatively high absorptivity on the surface of most metals, particularly as compared to the gaseous CO_2 lasers with a wavelength of 10.6 μm [11]. However, to ensure stable and reproducible processing conditions, a precise control of heat input during laser processing is required. The "heat input" is often used as synonymous with the "energy input". However, the energy input is defined as the amount of energy per unit of a length:

$$Ev = P/v \text{ (J/mm)}, \tag{1}$$

where P is the power of the laser beam (W) and v is the scanning speed (mm/s).

In turn, the heat input defines the actual amount of heat transferred to the material, so it depends on the efficiency of heat transfer. In the case of heating the surface by a laser beam the efficiency of heat transfer is proportional to the surface absorption "A" of the laser radiation under such conditions. The surface absorption depends on many factors such as material type, surface roughness, surface temperature, wavelength of laser radiation, power density of laser beam, etc. Therefore, knowledge of the material surface absorption under different conditions is the necessary factor to ensure proper thermal conditions and control of the laser treatment process, and also it is necessary to develop a reliable numerical model of laser gas nitriding by the high power diode laser [24–26]. Knowledge of optical properties is also necessary for reliable non-contact temperature measurements by means of infrared cameras or pyrometers.

Previous studies, and the results published by other researches, also show that the penetration depth during laser melting of titanium and its alloys in a nitrogen atmosphere is significantly greater than in an inert gas atmosphere (e.g., in pure argon) [11,21,22]. This phenomenon has not yet been clearly elucidated. The key to explain this phenomenon is the analysis of thermal conditions during laser nitriding, efficiency of heat transfer and to determine the temperature gradients.

On the other hand, methods of non-contact temperature measuring during laser processing are commonly used for monitoring and controlling the process. In the case of non-contact temperature measuring of materials, the absolute value of temperature indicated by infrared camera or pyrometer depends strongly on the assumed value of the material emissivity at operating wavelength or in an operating spectral range [24–26]. Therefore, non-contact measurements of temperature require knowledge of the normal spectral emissivity ε_n of the material at a given wavelength or range of wavelengths. Moreover, the value of spectral emissivity depends on the material surface conditions such as presence of oxides, thickness of the oxides' layer, roughness, and temperature. Unfortunately, the literature data or materials databases usually provide just the information on hemispherical emissivity, in addition to surfaces of pure metallic or nonmetallic materials. Patsalas et al. developed an extensive review of the literature on the optical properties of TiN films grown by various techniques and on various substrates [24]. It provides wide data on optical reflectivity spectra of TiN at normal incidence but just at room temperature. In addition, the data reported by researchers differ significantly [24,25].

Therefore, the author attempts to determine optical properties such as emissivity, reflectivity, and absorption of surface layers produced on the substrate of titanium alloy Ti6Al4V by laser surface melting and nitriding using the high power direct diode laser emitting radiation with a wavelength of 808 nm.

In addition, the author attempts to estimate the absorption coefficients of the metal surface in a liquid state during laser melting of titanium alloy Ti6Al4V in a pure argon atmosphere and in a pure nitrogen atmosphere.

2. Materials and Methods

The aim of the first stage of the study was to determine and compare the optical properties of the titanium alloy Ti6Al4V (Grade 5 according to the standard ASTM B 265-99) with different surface conditions (delivery conditions, oxidized, and polished), as well as the surface layers produced by laser melting the titanium alloy in an argon atmosphere, and laser nitriding in a gaseous nitrogen atmosphere. The trials of laser surface melting and laser gas nitriding were conducted in a very wide range of processing parameters and technological conditions. The scanning speed varied from 200 to 800 mm/min as did the laser output power from 0.8 to 1.8 kW, thus the energy input was from 70 to 270 J/mm. Additionally, the trials of laser processing were at a constant energy input of 270 J/mm (400 mm/min and 1.8 kW) and different gaseous atmospheres (100% Ar, 80% Ar + 20% N_2, 70% Ar + 30% N_2, 60% Ar + 40% N_2, and 50% Ar + 50% N_2) were used (Figure 11). Therefore, representative samples with surface layers produced by laser melting in pure argon and pure nitrogen were selected for the study of optical properties. The selection criteria were an even surface with the lowest possible roughness.

Next, based on the results and findings of the first stage of the study, experiments were planned to determine the absorption for an 808 nm wavelength of high power diode laser radiation on the surface of a melt pool during laser surface melting and nitriding of titanium alloy Ti6Al4V.

These studies consisted of real-time measurements during laser melting or nitriding the titanium alloy substrate. Therefore, the processing parameters have been narrowed and adapted to the specificity of the temperature measurements in such a way as to ensure stability and repeatability of the results.

Real-time measurements were carried out at a constant scanning speed of 200 mm/min, different laser output power (determined precisely by setting the current supplying diode packages of the laser generator), and thus different energy input, starting with an energy input above 270 J/mm (Table 1).

The test surface layers for the study of optical properties, as well as for real-time measurements were produced as single stringer beads. All trials of laser melting and nitriding were carried out at the focal length of 82.0 mm and dimensions of the rectangular laser beam spot 1.8 × 6.8 mm. The focal plane was set at the top surface of the treated samples and the rectangular beam spot was positioned perpendicularly to the direction of scanning, in such a way that the spot width was 6.8 mm. To provide the proper and controlled gaseous atmosphere for laser processing, the specimens for study of optical properties were placed in a gas chamber filled up with argon or nitrogen or a mixture of argon and nitrogen of 99.999% purity. In turn, during the tests of real-time measurements the gas atmosphere was provided by a cylindrical nozzle with a diameter of 12.0 mm, and the gas flow rate was kept at 15.0 l/min.

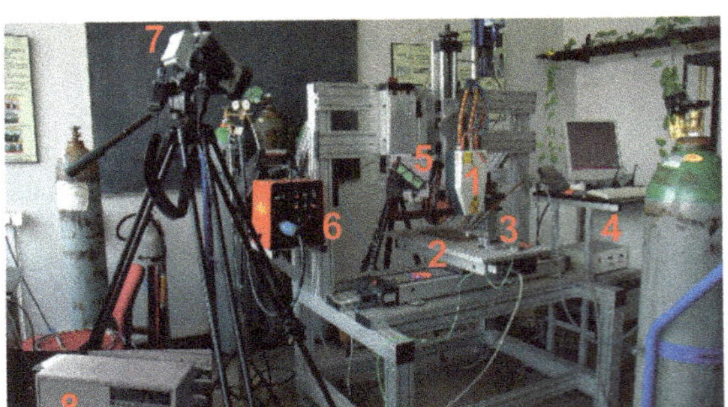

Figure 1. A view of experimental setup equipped with the high power direct diode laser (HPDDL) ROFIN SINAR DL020 (**1**), positioning system (**2**), mounting device (**3**), computer control system (**4**), pyrometer IMPAC IS 140 (**5**), electronic controller (**6**), infrared camera ThermaCAM P640 (**7**), and multichannel numerical recorder Agilent 34970A (**8**).

Table 1. Technical data of the continuous wave (CW) high power direct diode laser ROFIN DL 020 (Figure 1).

Parameter	Value
Wavelength of the laser radiation (nm)	808–940 (960) * (±5)
Range of laser power (kW)	0.1–2.2
Focal length (mm)	82/32
Laser beam spot size (mm)	1.8 × 6.8 or 1.8 × 3.8 **
Range of laser power intensity (kW/cm^2)	0.8–32.5

* The total spectral range is from 800–1000 nm but the dominant wavelengths (the strongest peaks in the spectral range) are 808, 940, and 960 nm. ** Size of the beam spot when an additional lens is applied with a focal length of 32 mm.

The high power direct diode laser (HPDDL) ROFIN DL 020 (ROFIN SINAR LASER, Hamburg, Germany) was applied in the study (Figure 1). This type of a laser emits the laser beam directly from the laser head, and it is called a "direct" laser. Additionally, the laser is characterized by a rectangular laser beam, uniform energy distribution across the beam spot, and a short wavelength. One of the characteristic features of this laser is its wavelength. Typically, it is given that the wavelength is 808 or 940 or even 960 nm. In fact, the emission spectral range of the specific HPDDL is relatively wide. The total emission spectral range is from 800 up to even 1000 nm but there are several dominant wavelengths such as 808, 940, and 960 nm. The most dominant is the 808 nm wavelength of the laser radiation, hence it is taken as the characteristic value for this type of an HPDDL.

The properties of the tested surfaces were characterized by roughness and the thickness of the surface layer (oxide or titanium nitrides layer). Measurements of surface roughness were carried out by means of a portable surface roughness tester SJ-210 Surftest (Mitutoyo Corporation, Kanagawa, Japan) according to the standard ISO 4288 standards (Figure 2). The thickness of titanium nitrides or surface oxide layers were determined on microstructure images, obtained by a metallographic light microscope (LM) Nikon Eclipse MA100 (NIKON CORPORATION, Tokyo, Japan) and then by scanning electron microscopy (SEM) (Carl Zeiss, Oberkochen, Germany) with an energy dispersive spectrometer (EDS) (Carl Zeiss, Oberkochen, Germany).

Figure 2. A view of the portable surface roughness tester SJ-210 Surftest Mitutoyo during measurements of roughness on the surface of stringer beads produced by laser gas nitriding of titanium alloy Ti6Al4V.

The measurements of reflectance of tested samples were performed by means of the Ocean Optics PC2000-ISA-PC Plug-in Spectrometer (Ocean Optics, Largo, FL, USA), ISP-REF Integrating Sphere, and WS-1 Diffuse Reflectance Standard. The main part of the measuring device is a so-called "Ulbricht sphere" ISP-REF Integrating Sphere, shown in Figure 3. The ISP-REF Integrating Sphere provides even surface illumination for reflectance measurements and also collects light to an optical fiber for emission experiments. The working wavelength range of the ISP-REF Integrating Sphere is 0.360–2.500 μm with the resolution ±0.5 nm. However, the scatter of results was significant in the measurement range wider than 0.450–1.130 μm, so it was decided to limit the measurement range to 0.500–1.100 μm.

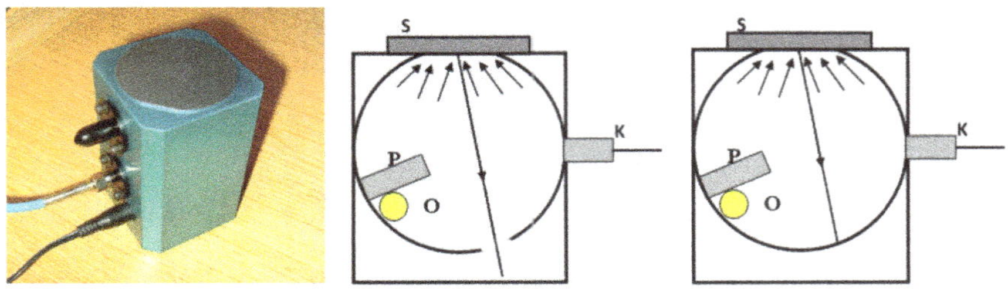

(a) (b) (c)

Figure 3. A view of the Ulbricht sphere applied for measurements of reflectance (**a**); and a scheme of the sphere design (O—source of white light, S—tested sample, P—aperture, K—fiber optics cable): (**b**) the aperture setting to measure the specular and diffuse reflections; (**c**) the aperture setting to measure the diffuse reflections only.

The measurements were carried out at room temperature (about 20 °C) and both the specular and diffuse reflectivity were determined for each sample. The test samples with dimensions of 40 × 70 mm were prepared from a titanium alloy sheet with a thickness of 3.0 mm. Surfaces of the samples were prepared and treated in different ways. One surface was untreated in as-received conditions with a thin natural layer of oxides, one surface was polished, and another surface was strongly oxidized at 900 °C in air atmosphere. In addition, surface layers after laser melting in argon and nitrogen characterized by different surface topography and roughness were applied for measurements of reflectance.

The temperature of the solid and liquid surface of the titanium alloy during laser processing was measured by the pyrometer IMPAC IS 140 (IMPAC Infrared GmbH, Frankfurt, Germany) with a measurement range from 550–3300 °C and spectral range 0.7–1.1 μm.

Additionally, high temperature, nonstandard thermocouple W-Mo with the measurement range up to 2400 °C, was used to calibrate the pyrometer. The characteristic of W-Mo thermocouple was determined based on the data provided by Caldwell, Michalski et al., and Blatt et al., as shown in Figure 4 [27–29].

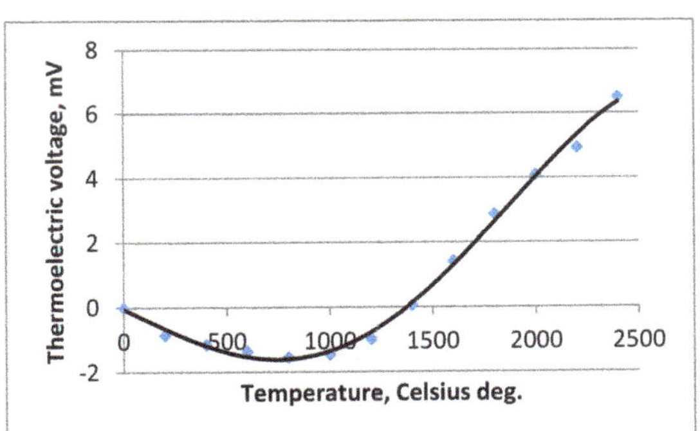

Figure 4. Characteristics of the non-standardized thermocouple W-Mo with a temperature range up to 2400 °C [27–29].

The multichannel numerical recorder Agilent 34970A (Agilent, Santa Clara, CA, USA) was applied for acquisition and recording of data. The procedure of calibration was performed on surfaces of as-received titanium alloy, on nitrided surface layers, and on a strongly oxidized surface. The samples were gradually heated in a protective gas atmosphere of argon inside an electronically controlled furnace (Figure 5).

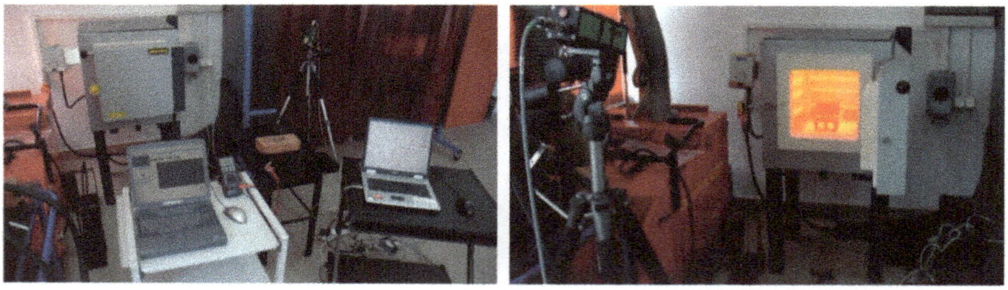

(**a**) (**b**)

Figure 5. A view of experimental setup: (**a**) with an electronically controlled furnace and high temperature pyrometer IMPAC IS 140; (**b**) a view of temperature measurements on the surface titanium alloy sample in a high temperature range.

During the gradual heating, temperature of the surface was measured simultaneously by the pyrometer and the thermocouple. Based on the temperature indications the values of emissivity in a spectral range of the IMPAC IS 140 were established for different surface conditions in a wide range of temperatures up to 1100 °C but below the melting point of titanium alloy. In the next step the temperature measurements of the liquid surface of the melt pool during laser melting were performed simultaneously by means of the pyrometer and the thermocouple (Figures 1 and 6). The thermocouples were mounted in cylindrical holes 0.5 mm in diameter and 1.5 mm deep from the bottom side of the sample. The thermocouples were arranged in rows, as shown in Figure 6.

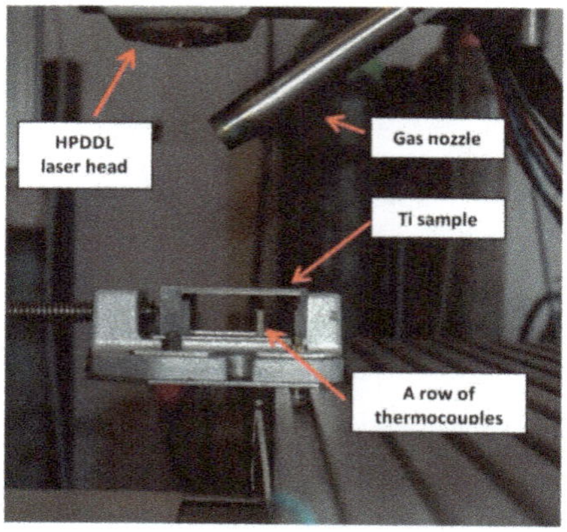

Figure 6. A view of the arrangement of non-standardized thermocouple W-Mo, mounted on the bottom surface of the sample used during the experimental study of heat conditions of HPDDL surface melting and nitriding.

The emissivity of the pyrometer was determined experimentally in such a way as to achieve compliance of indications between the pyrometer and thermocouple. Results of the tests, measurements, and analyses are given in Figures 7–13 and Table 2.

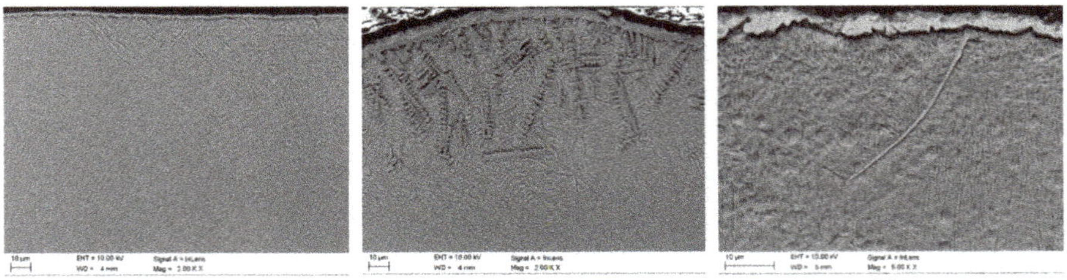

(a) (b) (c)

Figure 7. Scanning electron microscopy (SEM) micrographs at cross-sections in the near top surface region: (**a**) as-received titanium alloy Ti6Al4V sheet 3 0 mm thick (surface roughness Ra 0.4, thickness of the oxide surface layer approx. 1–2.5 µm); (**b**) nitrided surface layer produced at a laser power of 1.8 kW and scanning speed of 400 mm/min on a substrate of titanium alloy Ti6Al4V (surface roughness Ra 8.2, thickness of the titanium nitrides surface layer approx. 2–5 µm); (**c**) titanium alloy Ti6Al4V sheet 3.0 mm thick oxidized at 900 °C in pure air for 10 min (surface roughness Ra 25, thickness of the oxide surface layer approx. 2.5–8 µm).

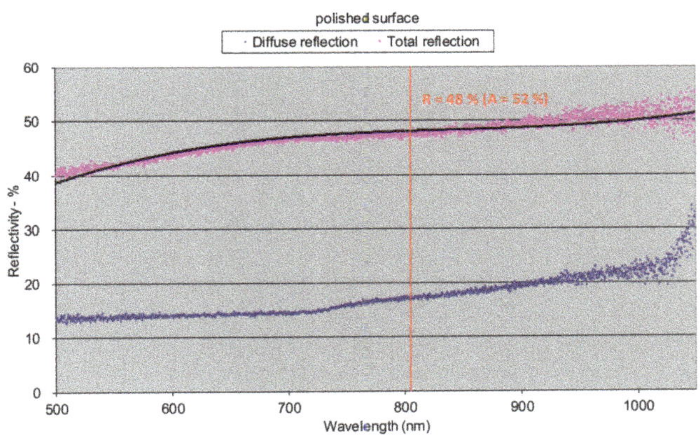

(**a**)

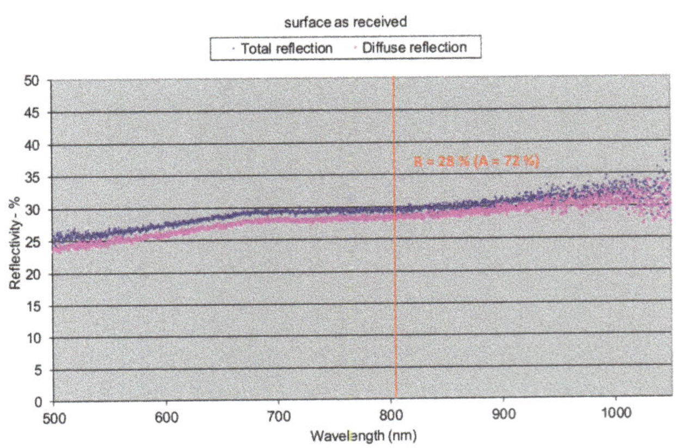

(**b**)

Figure 8. *Cont.*

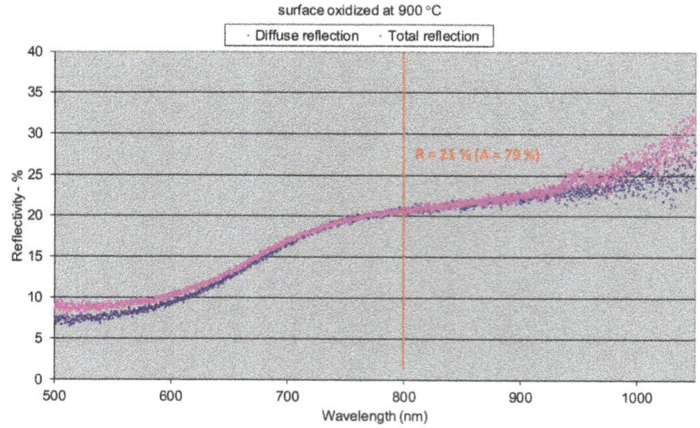

(c)

Figure 8. Reflectivity of the surfaces of titanium alloy Ti6Al4V determined at room temperature: (**a**) polished surface (surface roughness Ra 0.04); (**b**) as-received conditions (surface roughness Ra 0.4, thickness of the oxide surface layer approx. 1–2.5 μm); (**c**) oxidized at 900 °C (surface roughness Ra 25, thickness of the oxide surface layer approx. 2.5–8 μm).

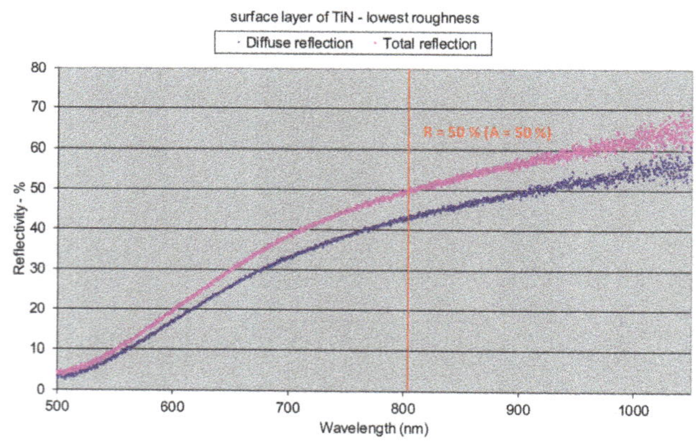

Figure 9. Reflectivity of the surface layer produced by nitriding of titanium Ti6Al4V by HPDDL in pure nitrogen, determined at room temperature (bead No. 5 in Figure 9, parameters of processing; laser power 1.8 kW, scanning speed 400 mm/min, energy input 270 J/mm, surface roughness Ra 8.2, thickness of the titanium nitrides surface layer approx. 2–5 μm).

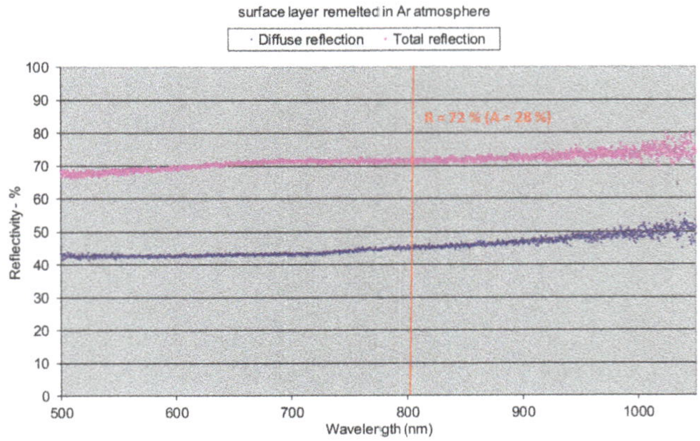

Figure 10. Reflectivity of the surface layer produced by laser melting of titanium alloy Ti6Al4V by HPDDL in pure argon, determined at room temperature (bead No. 8 in Figure 9, parameters of processing; laser power 1.8 kW, scanning speed 400 mm/min, energy input 270 J/mm, surface roughness Ra 1.1).

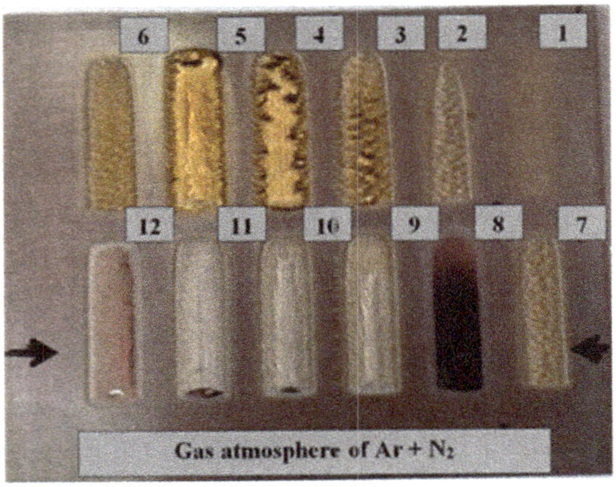

Figure 11. A view of the surface layers produced on titanium alloy Ti6Al4V by HPDDL melting in an argon and nitrogen atmosphere: bead No. 1: 800 mm/min, 0.9 kW, 70 J/mm, 100% N_2; bead No. 2: 600 mm/min, 0.9 kW, 90 J/mm, 100% N_2; bead No. 3: 400 mm/min, 0.9 kW, 136 J/mm, 100% N_2; bead No. 4: 200 mm/min, 0.9 kW, 270 J/mm, 100% N_2; bead No. 5: 400 mm/min, 1.8 kW, 270 J/mm, 100% N_2; bead No. 6: 400 mm/min, 1.2 kW, 180 J/mm, 100% N_2; bead No. 7: 400 mm/min, 0.8 kW, 120 J/mm, 100% N_2; bead No. 8: 400 mm/min, 1.8 kW, 270 J/mm, 100% Ar; bead No. 9: 400 mm/min, 1.8 kW, 270 J/mm, 50% Ar + 50% N_2; bead No. 10: 400 mm/min, 1.8 kW, 270 J/mm, 60% Ar + 40% N_2; bead No. 11: 400 mm/min, 1.8 kW, 270 J/mm, 70% Ar + 30% N_2; bead No. 12: 400 mm/min, 1.8 kW, 270 J/mm, 80% Ar + 20% N_2.

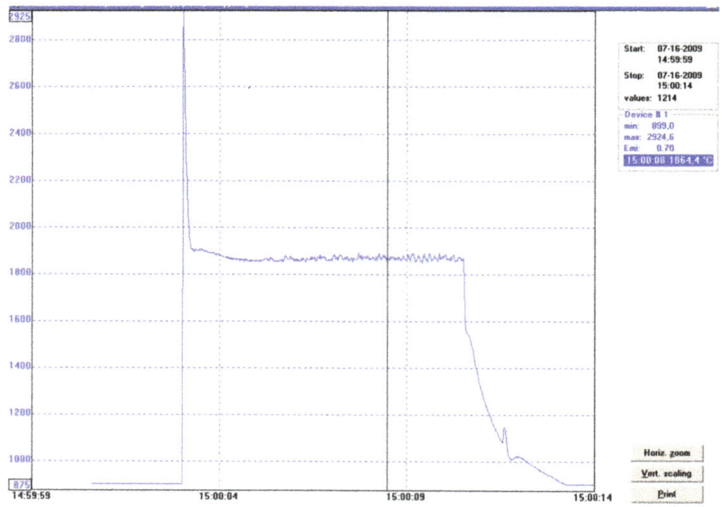

(**a**)

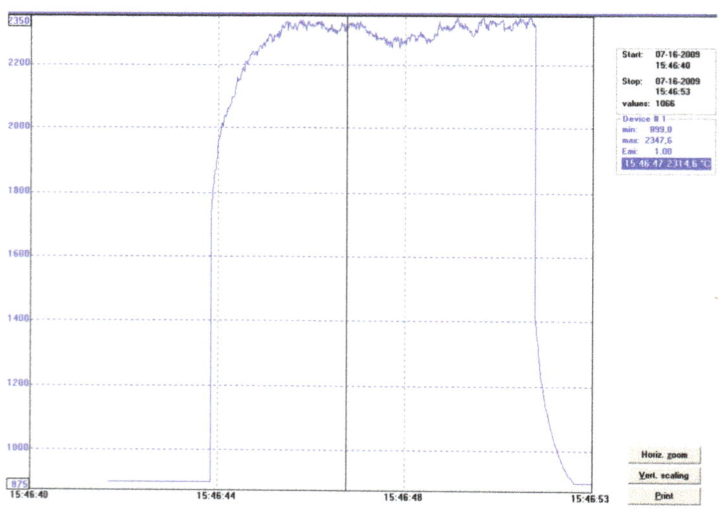

(**b**)

Figure 12. Results of temperature measurements on the surface of the melt pool during laser melting titanium alloy Ti6Al4V by HPDDL at output power of 1210 W and scanning speed 200 mm/min (200 mm/min): (**a**) in argon atmosphere, chosen value of emissivity ε = 0.7, mean temperature 1870 °C; (**b**) in nitrogen atmosphere, chosen value of emissivity ε = 1.0, mean temperature 2200 °C.

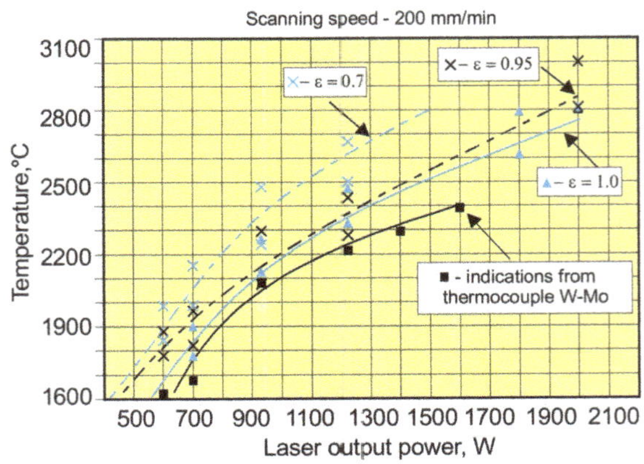

Figure 13. Results of temperature measurements on the surface of a melt pool obtained by the non-standardized thermocouple W-Mo and the pyrometer IMPAC IS 140 at different HPDDL output power of nitriding the Ti6Al4V substrate in pure nitrogen atmosphere. For comparison, the results of pyrometric measurements were determined for different value of the normal spectral emissivity.

Table 2. Influence of laser output power of diode laser ROFIN DL 020, emitting at 808–960 nm wavelength, on the temperature of the melt pool surface during laser melting of the titanium alloy Ti6Al4V sheet, 3.0 mm thick in argon atmosphere and during laser nitriding, determined by infrared pyrometer IMPAC IS140 with a spectral range 0.7–1.1 µm.

Gas Atmosphere	Diode Stack Current (A)	Laser Power (W)	Scanning Speed (mm/min)	Emissivity of the Melt Pool Surface *	Temperature of the Melt Pool Surface (°C)	
					Range	Mean Value
Ar	11	940	200	0.6–0.65 (approx. 0.7) **	1780–1800	1790
	13	1210			1850–1900	1875
	15	1490			1900–2000	1950
	17	1780			2045–2145	2095
N_2	11	940	200	0.9–0.98 (approx. 1.0) **	1925–2025	1975
	13	1210			2100–2300	2200

Remarks: diameter of the gas nozzle: 12 mm; argon or nitrogen feed rate: 15 l/min; dimensions of the rectangular laser beam spot of HPDDL: 1.8 × 6.8 mm; working distance: 82 mm; * emissivity for the spectral range 0.7–1.1 µm; ** simplified and constant values set for measurements and entered in the pyrometer IMPAC IS 140.

3. Results and Discussion

Since the roughness and the thickness of the surface layer of oxides have a significant influence on the optical properties, these features have been determined for all of the analyzed specimens (Figure 7). Results of the reflectivity measurements showed that the total reflection **R** (both specular and diffuse) of 808 nm wavelength radiation, characteristic for the HPDDL used in this study for surface treatment of titanium alloy, from the polished surface of titanium alloy Ti6Al4V reached about 48% at room temperature (Figure 8a). The significant difference between values of total reflection (48%) and diffuse reflection (18%) indicates a high proportion of the specular reflection, typical and characteristic for polished surfaces of metals.

Metals as electric conductors are non-transparent for laser radiation so the surface absorption **A** (Fresnel absorption) is the inverse reflection:

$$\mathbf{R} = 1 - \mathbf{A} \tag{2}$$

Thus, the value of absorption for the 808 nm wavelength radiation on the polished surface of titanium alloy Ti6Al4V was about 52%, as is apparent from Figure 8a. On the other hand, the total reflection from the surface of the as-received titanium alloy sheet was much lower and reached about 28% (Figure 8b). In this case the absorption was about 72%. Significantly lower reflection and simultaneously higher absorption were due to higher roughness and a thin layer of oxides that were present on the surface at delivery conditions. As can be seen there was no distinct difference between the diffuse and the total reflection in the case of as-delivered surface conditions (Figure 8b). It means that the radiation was strongly scattered after reflection from the surface of the titanium alloy.

The lowest reflection at about 21% showed the titanium alloy sample oxidized at 900 °C (Figure 8c). The phenomena can be explained by the presence of a relatively thick layer of complex oxides on the surface. Absorption in this case was very high and reached 79%. Similarly like in the as-delivered surface conditions, there was no distinct difference between the diffuse and the total reflection, indicating strong scattering of the radiation after reflection from the surface (Figure 8c). In the case of surface layers produced by laser gas nitriding of the substrate of titanium alloy, the total reflection of the 808 nm wavelength radiation varied in a range from 43% up to 50% (Figure 9). In turn, the absorption varied in a range from 50% up to 57% (Figure 9). The differences in reflection and also slight differences between the diffuse and the total reflection are attributed to the surface topography and roughness of the individual surface layers. The nitrided surface layers have a golden shine because the surfaces were covered by a homogeneous film of titanium nitrides TiN (Figure 11).

Therefore, the coefficient of reflection and the associated absorption in this case is associated with titanium nitrides, not the surface of the titanium alloy substrate. Results indicate that the absorption of radiation with a wavelength 808 nm is lower on the nitrided surface layer than on the surface of the as-received titanium alloy T6Al4V sample. The highest value of reflection for the 808 nm wavelength radiation was surprisingly determined in the case of the surface layer remelted in pure argon atmosphere and characterized by a silver color and metallic shine, as can be seen on bead No. 8 in Figure 11. In this case the total reflection was 72%, so the absorption was just 28% (Figure 10). The reflection is even higher than in the case of the polished surface of titanium alloy. Additionally, distinct differences between the diffuse reflection (45%) and the total reflection (72%) were found (Figure 10). All these results were obtained at room temperature of approx. 20–22 °C. It is obvious that the values of reflection and absorption will be different at elevated temperatures, especially over the melting point. However, these results may be a clue to explain the phenomenon of the significant increase in penetration depth during laser melting of titanium and titanium alloys in pure gaseous nitrogen or nitrogen rich gas mixtures, compared to penetration depth in pure argon gas atmosphere. This phenomenon was reported by many researches and also it was described in previous publications [5,7,21].

Thus, at room temperature (approx. 20–22 °C) the absorption of 808 nm wavelength radiation on the surface of a nitrided surface layer (50%–57%) was almost twice as high as the surface of titanium alloy remelted in pure argon (28%). It can therefore be assumed that these relationships are similar also at elevated temperatures.

The normal spectral emissivity of the tested surfaces of titanium alloy and surface layers for the spectral range 0.7–1.1 μm of the pyrometer IMPAC IS 140 was determined by adjusting the value of emissivity so as to obtain the compliance of the indication of the temperature by the pyrometer and the thermocouple at elevated temperatures and also during laser surface melting in argon and nitrogen atmospheres (Figures 1 and 13, Table 2). The results obtained during measuring the temperature of the melt pool were most interesting (Figures 12 and 13.)

It was found that the normal spectral emissivity of the melt pool surface during HPDDL melting of titanium alloy Ti6Al4V in pure argon atmosphere, for the spectral range 0.7–1.1 μm of the applied pyrometer, is in a range 0.6–0.65, depending on the processing parameters (Figure 12a, Table 2). However, during the laser gas nitriding tests the normal spectral emissivity was found to be significantly higher in a range 0.9–0.98 (Figures 12a and 13, Table 2).

It should be noted that the spectral range 0.7–1.1 μm of the applied pyrometer covers the wavelength of the applied HPDDL emitting at a dominant wavelength of 808 nm. However, the actual spectrum of radiation in the case of the applied HPDDL is relatively wide from 800 to about 960 or even 1000 nm (Table 1). On the other hand, Kirchhoff's law says that in the case of metals the absorption **A** is equal to the emissivity ε, for a given wavelength and temperature. Thus, the emissivity determined for the pyrometer should be proportional to the absorption of the laser beam radiation emitted by the HPDDL. This means that the absorption of the HPDDL beam during laser gas nitriding of titanium alloy was approx. 60% higher (**A** = 0.9–0.98) than the absorption on the melt pool surface of titanium alloy during laser melting in pure argon (**A** = 0.6–0.65).

These results confirm indirectly the thesis that the increase in the penetration depth during laser melting of titanium alloy in a nitrogen-rich atmosphere is inter alia due to the higher absorption of the laser beam on the melt pool surface.

4. Conclusions

Based on the results of reflectivity and absorption measurements at room temperature, results of temperature measurements and emissivity determining, the following findings can be summarized.

It was found that the absorption of the HPDDL radiation on the surface layer of titanium alloy remelted in argon atmosphere was higher than for the polished surface of titanium alloy, determined at room temperature. Absorption of the laser radiation with a wavelength of 808 nm (a dominant wavelength) for surface layer of titanium alloy Ti6Al4V remelted in pure argon reached only 28%, while the absorption for the polished surface was approx. 32%. Moreover, the absorption of 808 wavelength radiation on the nitrided surface layers (50%–57%), determined at room temperature, was almost twice as high as the absorption on the surface layer produced by laser melting of titanium alloy in pure argon (28%).

In turn, the estimated value of absorption on the melt pool surface during HPDDL nitriding of titanium alloy Ti6Al4V was about 60% higher than during HPDDL melting of the titanium alloy in a pure argon atmosphere.

Such significant differences in absorption influence the thermal conditions of the laser surface melting and nitriding process and therefore the penetration depth.

Funding: This research received no external funding.

Acknowledgments: The author is grateful to Andrzej Grabowski from the Institute of Physics and to Tadeusz Kruczek from for the Institute of Thermal Technology for experimental and substantive support.

Conflicts of Interest: The author declares no conflicts of interest.

References

1. Whittaker, M.T. Titanium Alloys. *Metals* **2015**, *5*, 1437–1439. [CrossRef]
2. Branco, A.; Moreira, V.; Reis, J.; Colaço, R.; Figueiredo-Pina, C.; Serro, A. Influence of contact configuration and lubricating conditions on the microtriboactivity of the zirconia-Ti6Al4V pair used in dental applications. *J. Mech. Behav. Biomed. Mater.* **2019**, *91*, 164–173. [CrossRef] [PubMed]
3. Ji, X.; Qing, Q.; Ji, C.; Cheng, J.; Zhang, Y. Slurry Erosion Wear Resistance and Impact-Induced Phase Transformation of Titanium Alloys. *Tribol. Lett.* **2018**, *66*, 64. [CrossRef]
4. Kao, W.-H.; Su, Y.-L.; Horng, J.-H.; Yang, S.-E. Tribological performance, electrochemical behavior and biocompatibility of high-temperature gas-nitrided Ti6Al4V alloy. *Ind. Lubr. Tribol.* **2018**, *70*, 1536–1544. [CrossRef]
5. Kamat, A.M.; Segall, A.E.; Copley, S.M.; Todd, J.A. Enhancement of CP-titanium wear resistance using a two-step CO2 laser-sustained plasma nitriding process. *Surf. Coat. Tech.* **2017**, *325*, 229–238. [CrossRef]
6. Fellah, M.; Aissani, L.; Samad, M.A.; Iost, A.; Zine, T.M.; Montagne, A.; Nouveau, C. Effect of Replacing Vanadium by Niobium and Iron on the Tribological Behavior of HIPed Titanium Alloys. *Acta Met. Sin. English Lett.* **2017**, *30*, 1089–1099. [CrossRef]
7. Lisiecki, A. Mechanisms of hardness increase for composite surface layers during laser gas nitriding of the Ti6A14V alloy. *Mater. Technol.* **2017**, *51*, 577–583.
8. Lisiecki, A.; Kurc-Lisiecka, A. Erosion wear resistance of titanium-matrix composite Ti/TiN produced by diode-laser gas nitriding. *Mater. Teh.* **2017**, *51*, 29–34. [CrossRef]

9. Lisiecki, A.; Piwnik, J. Tribological Characteristic of Titanium Alloy Surface Layers Produced by Diode Laser Gas Nitriding. *Arch. Met. Mater.* **2016**, *61*, 543–552. [CrossRef]
10. Lisiecki, A. Comparison of Titanium Metal Matrix Composite Surface Layers Produced During Laser Gas Nitriding of Ti6Al4V Alloy by Different Types of Lasers. *Arch. Met. Mater.* **2016**, *61*, 1777–1784. [CrossRef]
11. Lisiecki, A. Titanium Matrix Composite Ti/TiN Produced by Diode Laser Gas Nitriding. *Met.* **2015**, *5*, 54–69. [CrossRef]
12. Yang, C.; Liu, J. Intermittent vacuum gas nitriding of TB8 titanium alloy. *Vac.* **2019**, *163*, 52–58. [CrossRef]
13. Kgoete, F.; Popoola, A.; Fayomi, O.; Adebiyi, I. Spark plasma sintered Ti-6Al-4V-Si3N4-TiN ternary composites: Effect of combined microsized Si3N4 and TiN addition on microstructure and mechanical properties for aerospace application. *J. Alloy. Compd.* **2018**, *769*, 817–823. [CrossRef]
14. Kgoete, F.M.; Popoola, A.P.I.; Fayomi, O.S.I.; Adebiyi, I.D. Influence of Si3N4 on Ti-6A1-4V via spark plasma sintering: Microstructure, corrosion and thermal stability. *J. Alloy Compd.* **2018**, *763*, 322–328. [CrossRef]
15. Wolowiec-Korecka, E.; Michalski, J.; Kucharska, B. Kinetic aspects of low-pressure nitriding process. *Vac.* **2018**, *155*, 292–299. [CrossRef]
16. Liang, S.X.; Yin, L.X.; Liu, X.Y.; Wu, X.X.; Ma, M.Z.; Liu, R.P. Kinetics of thermodiffusion of TZ20 titanium alloy gas-nitride within temperature of 500 degrees C-650 degrees C. *J. Alloy Compd.* **2018**, *734*, 172–178. [CrossRef]
17. Kusinski, J.; Kąc, S.; Kopia, A.; Radziszewska, A.; Rozmus-Górnikowska, M.; Major, B.; Major, L.; Marczak, J.; Lisiecki, A. Laser modification of the materials surface layer – a review paper. *Bull. Pol. Acad. Sci. Tech. Sci.* **2012**, *60*, 711–728. [CrossRef]
18. Kulka, M.; Panfil, D.; Michalski, J.; Wach, P. The effects of laser surface modification on the microstructure and properties of gas-nitrided 42CrMo4 steel. *Opt. Laser Technol.* **2016**, *82*, 203–219. [CrossRef]
19. Haghighi, M.; Shaeri, M.H.; Sedghi, A.; Djavanroodi, F. Effect of Graphene Nanosheets Content on Microstructure and Mechanical Properties of Titanium Matrix Composite Produced by Cold Pressing and Sintering. *Nanomater.* **2018**, *8*, 1024. [CrossRef]
20. Zhao, X.; Zhang, P.; Wang, X.; Chen, Y.; Liu, H.; Chen, L.; Sheng, Y.; Li, W. In-situ formation of textured TiN coatings on biomedical titanium alloy by laser irradiation. *J. Mech. Behav. Biomed. Mater.* **2018**, *78*, 143–153. [CrossRef]
21. Dutta Majumdar, J. Laser Gas Alloying of Ti-6Al-4V. *Physcs. Proc.* **2011**, *12*, 472–477. [CrossRef]
22. Biswas, A.; Li, L.; Chatterjee, U.; Manna, I.; Majumdar, J.D. Diode Laser Assisted Surface Nitriding of Ti-6Al-4V: Properties of the Nitrided Surface. *Met. Mater. Trans. A* **2009**, *40*, 3031–3037. [CrossRef]
23. Kamat, A.M.; Copley, S.M.; Segall, A.E.; Todd, J.A. Laser-Sustained Plasma (LSP) Nitriding of Titanium: A Review. *Coatings* **2019**, *9*, 283. [CrossRef]
24. Patsalas, P.; Kalfagiannis, N.; Kassavetis, S. Optical Properties and Plasmonic Performance of Titanium Nitride. *Mater.* **2015**, *8*, 3128–3154. [CrossRef]
25. Shoji, E.; Takahashi, R.; Isogai, S.; Ito, N.; Kubo, M.; Tsukada, T.; Fukuyama, H. Compositional Dependence of Normal Spectral Emissivity of Molten Cu-Fe Alloy. *Met. Mater. Trans. A* **2019**, *50*, 2454–2458. [CrossRef]
26. Gopinath, M.; Thota, P.; Nath, A.K. Role of molten pool thermo cycle in laser surface alloying of AISI 1020 steel with in-situ synthesized TiN. *Surf. Coatings Technol.* **2019**, *362*, 150–166. [CrossRef]
27. Caldwell, F.R. *Thermocouple Materials*; United States. Government Printing Office: Washington, DC, USA, 1962.
28. Michalski, L.; Eckersdorf, K.; Kucharski, J.; McGhee, J. *Temperature Measurement*, 2nd ed.; Wiley and Sons: Hoboken, NJ, USA, 2001.
29. Blatt, J.; Schroeder, P.A.; Foiles, C.L.; Greig, D. *Thermoelectric Power of Materials*; Plenum Press: New York, NY, USA, 1976.

materials

MDPI

Article

Mechanical and Corrosion Properties of Laser Surface-Treated Ti13Nb13Zr Alloy with MWCNTs Coatings

Beata Majkowska-Marzec [1,*], Patryk Tęczar [1], Michał Bartmański [1], Bartosz Bartosewicz [2] and Bartłomiej J. Jankiewicz [2]

1 Department of Materials Engineering and Bonding, Faculty of Mechanical Engineering, Gdansk University of Technology, G. Narutowicza 11/22, 80-233 Gdansk, Poland; patryk.teczar@gmail.com (P.T.); michal.bartmanski@pg.edu.pl (M.B.)

2 Institute of Optoelectronics, Military University of Technology, gen. S. Kaliskiego 2, 00-908 Warsaw, Poland; bartosz.bartosewicz@wat.edu.pl (B.B.); bartlomiej.jankiewicz@wat.edu.pl (B.J.J.)

* Correspondence: beata.majkowska@pg.edu.pl

Received: 16 July 2020; Accepted: 4 September 2020; Published: 9 September 2020

Abstract: Titanium and its alloys is the main group of materials used in prosthetics and implantology. Despite their popularity and many advantages associated with their biocompatibility, these materials have a few significant disadvantages. These include low biologic activity—which reduces the growth of fibrous tissue and allows loosening of the prosthesis—the possibility of metallosis and related inflammation or other allergic reactions, as well as abrasion of the material during operation. Searching for the best combinations of material properties for implants in today's world is not only associated with research on new alloys, but primarily with the modification of their surface layers. The proposed laser modification of the Ti13Nb13Zr alloy with a carbon nanotube coating is aimed at eliminating most of the problems mentioned above. The carbon coating was carried out by electrophoretic deposition (EPD) onto ground and etched substrates. This form of carbon was used due to the confirmed biocompatibility with the human body and the ability to create titanium carbides after laser treatment. The EPD-deposited carbon nanotube coating was subjected to laser treatment. Due to high power densities applied to the material during laser treatment, non-equilibrium structures were observed while improving mechanical and anti-corrosive properties. An electrophoretically deposited coating of carbon nanotubes further improved the effects of laser processing through greater strengthening, hardness or Young's modulus similar to that required, as well as led to an increase in corrosion resistance. The advantage of the presented laser modification of the Ti13Nb13Zr alloy with a carbon coating is the lack of surface cracks, which are difficult to eliminate with traditional laser treatment of Ti alloys. All samples tested showed contact angles between 46° and 82° and thus, based on the literature reports, they have hydrophilic surfaces suitable for cell adhesion.

Keywords: Ti13Nb13Zr alloy; laser treatment nanoindentation; electrophoretic deposition; carbon nanotubes; potentiodynamic polarization

1. Introduction

In recent decades, a significant increase in the number of diseases of the osteoarticular system has been observed—especially the knee and hip joints. These diseases are an increasingly common cause of joint destruction, which often requires replacing the joints with artificial ones. This problem has increased the demand for implants and prostheses. Related research and modifications—as well as combinations of engineering materials used and their surfaces—constitute wide scientific interest [1].

Titanium and its alloys is one of the most widely used groups of materials for the production of orthopedic implants and prostheses. Due to the properties of titanium, titanium-based materials cover almost 40% of the current biomaterial market. This is particularly related to their mechanical and functional properties being similar to hard bone tissues (especially Young's modulus), good biocompatibility and better corrosion resistance than austenitic steels and many other bioalloys. However, the restrictions on their growth with fibrous tissue, the possibility of inflammation and allergic reactions, and poor wear resistance are still a drawback in their use [1–3].

The titanium alloys Ti6Al4V, Ti6Al7Nb and Ti13Nb13Zr are often used in medical applications. Alloys with the addition of zirconium and niobium eliminate the adverse effects of aluminum and vanadium on the nervous system, the possibility of metallosis and the initiation of diseases (including cancers or Alzheimer's disease). In addition, they have better corrosion resistance, and a Young's modulus value similar to longitudinal bone tissue. However, they still have insufficient abrasive resistance [3,4].

Due to the demand for more appropriate properties of titanium materials, many modification methods have been developed. Some of these are associated with the modification of surface topography, mechanical and physicochemical or biologic properties, resulting in increased adhesion. The simplest modification methods include chemical etching to increase roughness and osseointegration [2,5,6]. Among the many modification techniques used there are electrochemical oxidation which increases adhesion [7] or polishing, which has a significant impact on increasing the anti-corrosive properties [8]. Improvement of corrosion properties and reduction of bacterial adhesion have been achieved in the hydrophilic synthesis process [9]. These materials have also been subjected to anodizing processes [10] and processed in alkaline solutions, which resulted in high bioactivity of the Ti13Nb13Zr alloy [11]. In addition, titanium alloys have been oxidized in an acid solution in the presence of fluoride ions to obtain a nanoparticle oxide layer with good adhesion, high mechanical properties and increased bioactivity [12]. The improvement of titanium biomaterials properties has been obtained by subjecting them to ion implantation [13,14], plating [15,16] and nitriding [17,18]. A popular method of obtaining better properties of the titanium alloys tested is coating them using silicates [19], chitosan [20,21], or phosphates including hydroxyapatite (HAp) [22–25] and nanoHAp [7,26–28]—together with their composite combinations with other elements [26,28–33]—thus facilitating better adhesion and antibacterial properties. Carbon and diamond-like coatings [34–39] and their composite combinations with other elements [40–42], as well as carbon nanotubes (CNTs) [41,43–45] and their composite combinations with other elements [31,45,46] have been tested to obtain better mechanical (especially nanohardness) and anti-corrosive properties. The evolution of multi-walled carbon nanotubes (MWCNTs) for in situ TiC formation during spark-plasma sintering in titanium metal matrix composites allows significantly enhancing the mechanical and tribological properties of the composites [47]. The densification and microhardness of the sintered nanocomposites with MWCNTs additions have been shown improve tremendously with an increase in sintering temperatures [48]. Positive effects in the form of high value of material hardness and excellent wear resistance result from the laser-cladding of MWCNTs on Ti, leading to TiC-reinforced Ti-matrix composite layers [49]. Additionally, the layer of carbon nanotubes (MWCNTs) deposited on titanium via the electrophoretic (EPD) method provides better compatibility of the implant with the body tissues. Carbon coatings allow choosing the best time for the protein to form a stable connection with the surface of the titanium implant [50]. Modern laser technologies have been used in the modification of biocompatible materials such as titanium. These modification methods allow melting and remelting [51–53], alloying [17,44,54], direct laser deposition [55], texturing and marking [56,57] and laser ablation [58]. Interest in laser methods is increasing due to their numerous advantages including the possibility of local only processing. Laser modifications are characterized by high process efficiency, lower cost of industrial use and possibilities of process automation in industrial production, no deformation of the workpiece and great flexibility in modification of the processed material structure. The concentration of high-power densities of materials subjected to laser modification for just a few milliseconds allows observing structures different from equilibrium for this material. These results, among others, consist of improvement of mechanical properties, increase in their hardness and often improvement of their anti-corrosion properties. However, these positive changes are often accompanied by deterioration in surface quality, which is one of the disadvantages of laser surface modification techniques of most materials [17,23,33,44,52–54,56,57,59–62].

This study aims to assess the impact of laser modification using an neodymium-doped yttrium aluminum garnet laser (Nd:YAG) ((TruLaser Station 5004, TRUMPF, Ditzingen, Germany) with appropriately selected parameters on the microstructure, mechanical, physical and corrosion properties of the surface layer of the Ti13Nb13Zr titanium alloy with an electrophoretic deposited layer of multi-walled carbon nanotubes. The selected process parameters are the result of many melting and laser alloying processes carried out at the testing stage. The proposed laser modification of the Ti13Nb13Zr surface with a carbon nanotubes coating allows to obtain better mechanical properties and a hydrophilic surface.

2. Materials and Methods

2.1. Preparation of Materials

2.1.1. Preparation of Titanium Samples

A Ti13Nb13Zr alloy rod provided by a commercial supplier (Xi'an SAITE Metal Materials Development Co., Ltd., Xi'an, China). with chemical composition detailed in Table 1 was used. A rod (with a diameter of 40 mm) was cut using a precision cutter (Brillant 220, ATM GmbH, Mammelzen, Germany) into 4-mm-thick slices and divided into quarters. Each of the quarters was cut at the edge to fix the copper wire needed at a later stage of the work—the EPD process. Abrasive machining was carried out on a metallographic grinding machine (Saphir 330, ATM GmbH, Mammelzen, Germany) by the wet method to remove impurities and deformation as well as level the surface. Samples were sanded with SiC sandpaper with 220, 500 and 800 gradations. Along with the transition between individual sandpaper gradations, the samples were thoroughly washed and dried. The surface roughness of the substrates was adjusted to values within Ra 0.24 ± 0.08 µm.

The samples were washed with acetone (Chempur, Piekary Śląskie, Poland) and distilled water and etched for 20 s in 5% hydrofluoric acid (HF) (Chempur, Piekary Śląskie, Poland), then washed again with distilled water and dried. Etching was performed to thoroughly clean them from impurities and develop the surface of the prepared samples. The surface image of the Ti13Nb13Zr alloy samples after abrasive machining and etching is shown in Figure 1.

Table 1. Chemical composition of Ti13Nb13Zr titanium alloy in % by weight.

Chemical Element	C	Fe	N	O	Zr	Nb	Ti
Alloy	Chemical Composition (% wt.)						
Ti13Nb13Zr	0.04	0.05	0.019	0.11	13.0	13.0	72.78

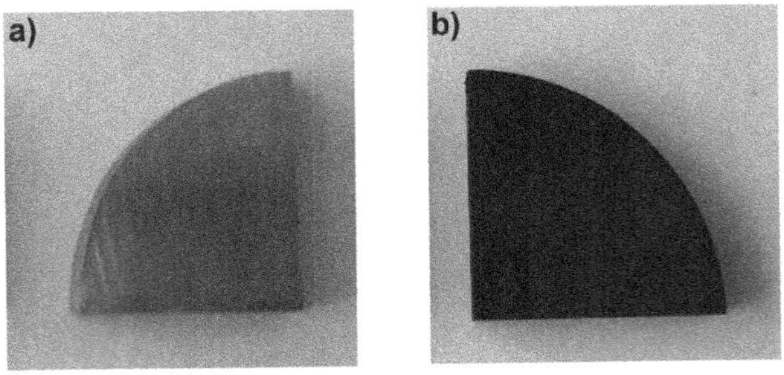

Figure 1. Alloy sample surfaces after (**a**) abrasive treatment and (**b**) etching in 5% hydrofluoric acid (HF).

2.1.2. Preparation of Suspension of Functionalized Carbon Nanotubes

The electrophoretic suspension was prepared from distilled water and multi-wall carbon nanotubes MWCNTs (3D Nano, PlasmaChem GmbH, Berlin, Germany) in loose form with an external diameter of 5–20 nm, number of walls of 3–15, internal diameter of 2–6 nm and length of 1–10 μm.

Before preparing the suspension, the carbon nanotubes were functionalized using a method described elsewhere [46]. The chemical modification allowed for the introduction of carboxyl groups on the surface of the nanotubes. These groups provide a negative charge for nanotubes and allow their electrophoretic seating on the anode. Functionalization of CNTs makes the process of their deposition by EPD faster and easier to occur. This process also affects the morphology of nanotubes, i.e., their length decreases [63], which results in better adhesion to the surface.

At first, 480 mg MWCNTs were annealed in a vacuum furnace (20VP-411/14 hV, Seco/Warwick SA, Świebodzin, Poland) at 400 °C for 8 h. Then, the nanotubes were dispersed by means of ultrasound in an ultrasonic homogenizer (Bandelin Sonopuls HD 2070, Berlin, Germany) in a small amount of distilled water and added to 200 mL of a 3:1 *v*/*v* mixture of H_2SO_4 and HNO_3 (Chempur, Piekary Śląskie, Poland). The resulting suspension was heated for 2 h at 70 °C on a heating plate. Then, after cooling to room temperature, the suspension was centrifuged (MPW-251, MPW MED. INSTRUMENTS, Warszawa, Poland) and washed several times with water until a neutral pH was reached. Finally, a suspension of MWCNTs in distilled water with a concentration of 0.19 wt% was obtained.

2.2. *Electrophoretic Deposition of Carbon Nanotube Coating*

The electrophoretic deposition (EPD) was carried out in a 0.19% suspension of MWCNTs in distilled water. Ti13Nb13Zr alloy samples constituted anodes in this process. The opposite electrode—the cathode—was platinum in this system. The electrodes were placed parallel to each other at a distance of about 10 mm and then connected to a DC power supply (MCP/SPN110-01C, MCP Corp., Shanghai, China). EPD was carried out at 20 V for 30 s at room temperature. After the deposition process, the samples were dried at 80 °C for 40 min in a laboratory dryer (SLN32, POL-EKO Aparatura, Wodzisław Śląski, Poland).

2.3. *Laser Modification*

The samples (T–Ti13Nb13Zr alloy after grinding and C–Ti13Nb13Zr alloy with MWCNTs coating) were subjected to laser treatment in order to develop the surface. This treatment was performed using a Nd:YAG pulse laser (TruLaser Station 5004, TRUMPF, Ditzingen, Germany) with the parameters described in Table 2. The term laser-melted corresponds in our manuscript to laser processing of titanium alloys, where the properties are changed without changing the chemical composition of the surface layer, while term laser-alloyed is used for laser processing of Ti13Nb13Zr alloy with a previously deposited nanotube coating leading to laser carburizing.

Table 2. Parameters of laser processing of the tested material.

Sample	Material	Power of the Laser Pulse (W)	Duration of the Laser Pulse (ms)	Feed Speed (mm/s)	Frequency (Hz)
T1, C1	T-Ti13Nb13Zr alloy, C-Ti13Nb13Zr alloy with MWCNTs coating	800	2.25	3.00	25
T2, C2		700	3.25	6.00	25
T3, C3		700	4.00	3.00	25
T4, C4		2000	2.25	3.00	25
T5, C5		1000	3.25	6.00	25
T6, C6		2000	2.25	3.00	10

This process was carried out under a protective gas with an argon (Ar) content of not less than 99.987% for the proper course of remelting the substrate.

2.4. Surface and Cross-Sections Evaluation of Samples

2.4.1. Preparation of Samples for Surface and Cross-Section Tests

The samples T(1 ÷ 6), C(1 ÷ 6) and native material—MR—were integrated in phenolic resin (Verte 602, Lamplan, Gaillard, France) in a laboratory press (Opal 410, ATM GmbH, Mammelzen, Germany) and subjected to grinding on a metallographic grinding machine (Saphir 330, ATM GmbH, Mammelzen, Germany) using sandpapers with a gradation of 180, 400, 600, 800, 1200, 2000 and 4000 and polishing using diamond paste (DP-Paste M 1 µm, Struers, Inc., Cleveland, OH, USA). The samples were then etched in Kroll's reagent (a mixture of HNO_3 and HF acids in distilled water in a 6/2/92 volume ratio).

2.4.2. Analysis of Surface Phase Composition

The samples MR, C, T(2,5), C(2,5) were subjected to phase structure tests using an X-ray diffractometer (X'Pert Pro, Philips, Amsterdam, The Netherlands) with a goniometer PW 3050/60 (θ–θ) with a Cu copper lamp ($K\alpha 1$ = 1, 5405980 Å, $K\alpha 2$ = 1.5444260 Å, $K\alpha 2/K\alpha 1$ ratio = 0.5). The test (with an 2θ angle range of 20–60° at 40 V, 30 mA and scanning speed 0.02/1.5 °/s) permits the determination of the crystal structure of the material.

2.4.3. Raman Spectroscopy

Raman measurements were carried out using Renishaw InVia Raman microscope (Renishaw plc., Wotton-under-Edge, UK) equipped with an EMCCD detector (Andor Technology Ltd, Oxford Instruments, Belfast, UK). The Raman signal was acquired using laser radiation with a wavelength of 532 nm (27 mW). The laser beam was directed to the sample through a 50× objective lens. The wavelength of the instrument was calibrated using an internal silicon wafer, and the spectrum was centered at 520.5 cm^{-1}. Raman spectra measurements were made for maps consisting of 100 points per sample. Based on this, an average spectrum was obtained and the standard deviation of the signal was determined—which is an indicator of the homogeneity of the layers.

2.4.4. Surface and Cross-Section Morphology and Topography and Chemical Analysis

The microstructure of the surface before and after corrosion tests and cross-sections of samples were observed using an optical microscope (UC50, Olympus Europa SE & Co. KG, Hamburg, Germany) and a scanning electron microscope SEM (JSM-7800F, JEOL, Tokyo, Japan). Microscopic images of specimens using the SEM were taken at an accelerating voltage of 1 kV. The thickness of the MWCNTs coating deposited on the Ti13Nb13Zr alloy was measured on the cross-section and after its local detachment using the adhesive force from the edge of the sample. Elemental composition analysis on the cross-section of the laser-modified samples was taken using an X-ray energy dispersion spectrometer (EDS) (Octane Elite 25, EDAX Ametek, Berwyn, PA, USA).

2.4.5. Surface Wettability

The water contact angle measurements—before and after laser processing of the Ti13Nb13Zr alloy with and without carbon coating—were made using an angular contact goniometer with a computer set (Contact angle goniometer, Zeiss, Ulm, Germany). Measurements were made by the falling drop method. The results comprise the arithmetic means of three measurements carried out at room temperature after 10 s.

2.4.6. Surface and Cross-Section Nanomechanical Properties

Nanoindentation tests were carried out on the surface and on cross-sections for MR, T(2,4) and C(2,4) using a nanoindenter equipped with a pyramidal trihedral diamond Berkovich indenter with an apical angle of 124.4° (NanoTest Vantage, Micro Materials, Wrexham, UK). The measurements were made with a maximum force of 50 mN. The rise time from zero value was 10 s, and the stop time with the maximum force was 5 s. The samples were unloaded within 4 s. A total of 115 measurements (5 × 13) were made every 30 μm for each sample on the cross-section and 115 measurements (A × B) every 30 μm on surfaces.

Hysteresis curves of load–strain were recorded during the study. The reduced Young's modulus (E_{IT}) and the nanohardness (H_{IT}) were determined based on the Olivier–Pharr method [64,65] using the NanoTest results analysis program. Calculations of the real Young's modulus (Es) were made using method developed by Olivier and Pharr [66].

2.5. Corrosion Tests

2.5.1. Preparation of Samples for Corrosion Tests

Copper wires were attached to uncoated and MWCNTs-coated Ti13Nb13Zr samples unmodified and modified using laser radiation. The back and side surfaces of the samples and their connection to the wire were isolated with a two-component epoxy ceramic coating (BELZONA 1321, Belzona, Inc., Harrogate, UK) to protect against corrosion. The coating was made of 4 parts base and 1 part of hardener. The partially isolated samples were allowed to dry for 48 h.

2.5.2. Corrosion Measurements Using the Potentiodynamic Polarization Method

Corrosion tests were carried out for samples MR, C, T2, T4, C2 and C4 in a three-electrode configuration: reference electrode (RE)—calomel electrode, auxiliary electrode (CE)—platinum electrode, working electrode (WE)—tested sample. Then, 250 mL of Ringer's solution, heated to 37 °C by a heating plate (MS-H-PRO +, Dragon Laboratory Instruments, Ltd., Beijing, China) was used as the electrolyte during the corrosion process.

The tests were carried out using a potentiostat (ATLAS 0531, ATLAS-SOLLICH, Rębiechowo Poland) equipped with AtlasCorr05 control and cooperating software (ATLAS 0531, ATLAS-SOLLICH, Rębiechowo, Poland). The corrosion tests began by determining the stationary potential as a function of time to determine the corrosion tendency of the material. After the potential values stabilized, potentiodynamic studies were started. The potential increased at a speed of 1 mV/s. A potential range 1–2 V was used. Individual polarization ranges were selected based on the stationary potential.

3. Results and Discussion

3.1. Measurement of MWCNTs Coating Thickness

The thickness of the electrophoretically deposited MWCNTs coatings was measured by using SEM. The measurement was made on the cross-section of the sample (Figure 2a) and its fracture (Figure 2b). The presence of uniformly deposited carbon nanotubes on the Ti13Nb13Zr substrate is shown in Figure 3b. The critical determinant of any coating is its adhesion to the substrate. Such measurements have been already performed for this type of coatings and published in [46], demonstrating this property suitable for coatings on titanium implants. The average thickness of MWCNTs coating was 7.88 μm ± standard deviation of 0.35 μm.

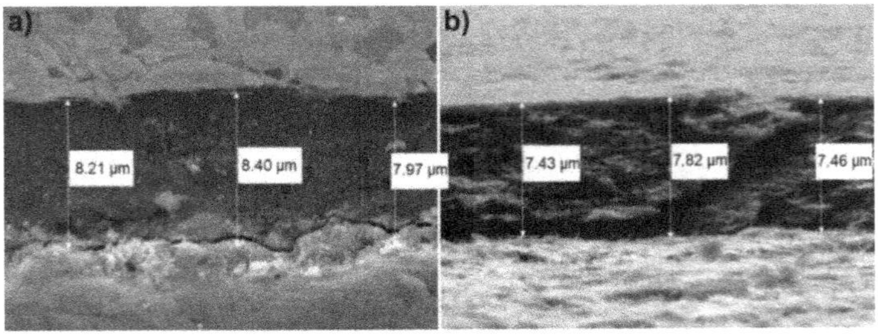

Figure 2. SEM images of Ti13Nb13Zr alloy MWCNTs coatings. (**a**) Cross-section; (**b**) fracture, with indicated measured thicknesses at three different points.

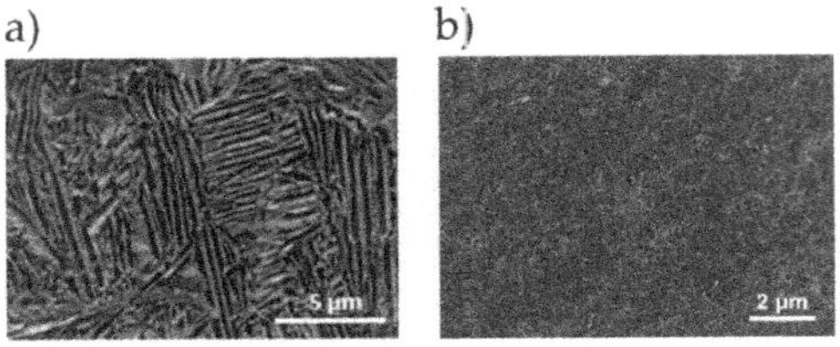

Figure 3. SEM images of Ti13Nb13Zr alloy before laser treatment. (**a**) MR; (**b**) MR with MWCNTs coatings.

3.2. Morphology and Topography Studies

The optical microscope and scanning electron microscope images of the surface and cross-sections of the tested samples of Ti13Nb13Zr alloy before and after laser treatment with and without a coating of carbon nanotubes are shown in Figures 3–5.

Based on the analysis of the surface morphology of the laser-processed samples (Figure 4), it can be seen that the T1 sample did not have the characteristic changes normally attributed to the laser processing processes, but only changes that can indicate a strong overheating of the material as in [52]. The small remains from remelting—which may constitute remelted material splinters—are visible on its surface. The remaining laser-melted layers had scratches and bumps characteristic for laser processes, which were the remaining effects of the laser beam pulse. This was especially visible for laser processed samples with higher laser pulse power (T4 ÷ T6). The laser beam penetrated deeper into the material shown on the image of cross-sections from the optical microscope in Figure 4, which caused greater mixing of the material and solidification in a more layered manner. No cracks were observed on the surface of the samples.

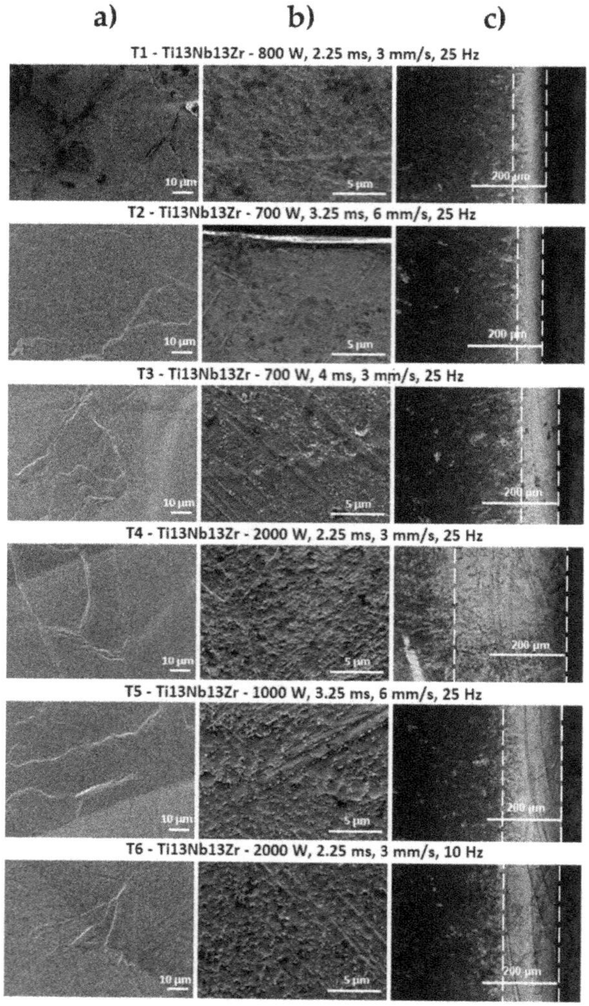

Figure 4. SEM images of Ti13Nb13Zr alloy after laser processing. (**a**) Surface topography; (**b**) cross-sections; (**c**) images from an optical microscope.

The SEM image of laser-processed surfaces with a previously deposited coating of multi-walled carbon nanotubes (Figure 5) reveals major changes occurring on the surface of the material. For samples treated with lower laser pulse powers—in which the laser beam penetrates the shallower layer of the substrate (shown in cross-sectional images from the optical microscope)—clusters remaining from the MWCNTs coating are visible. This indicates that the laser modification did not melt the entire area. In places where the laser-melted area is discernible, changes in structure were observed, which are represented by numerous folds, bulges and secretions. For samples where the laser beam with higher power penetrated the material deeper (T4 ÷ T6), these changes were less visible, which may have been associated with greater penetration of the carbon coating into the Ti13Nb13Zr alloy. Surface unevenness for these samples was greater than for laser-melted samples without coatings. No cracks were observed.

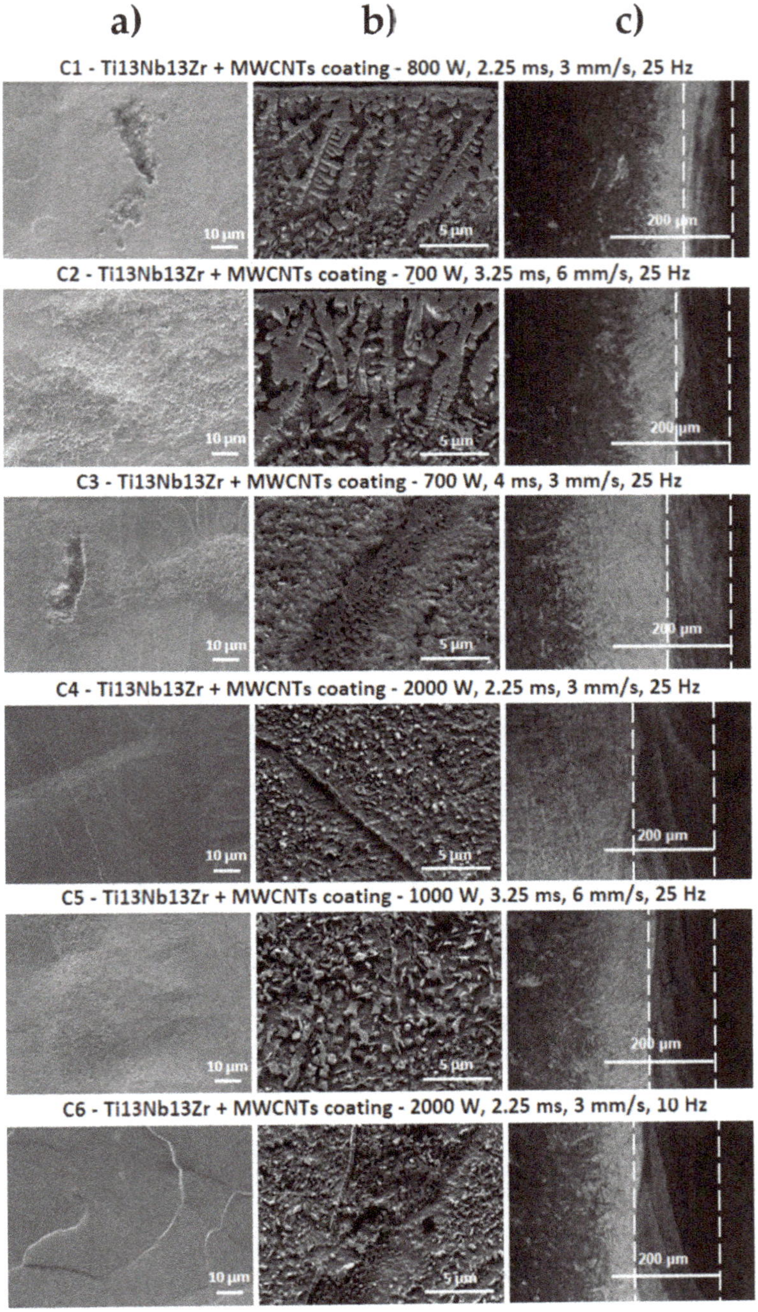

Figure 5. SEM images of Ti13Nb13Zr alloy with MWCNTs coating after laser processing. (**a**) Surface topography; (**b**) cross-sections; (**c**) images from an optical microscope.

The SEM image of the Ti13Nb13Zr alloy (Figure 3a) is characteristic for this material after chemical etching. The presence of one phase tiles in the matrix of the second phase can be observed. SEM images for laser-treated samples in the laser-melted layer directly at the sample front were taken (Figure 4b). For laser-modified samples with 700- and 800-W power, changes in the structure and grain refinement are seen. A typical martensitic transformation is noted. The microstructure of the T1 ÷ T3 samples was very similar, with the changes in the microstructure reaching the deeper material layers for the T3 sample. The microstructure of laser-melted layers made with greater power and laser pulse (T4 ÷ T6) indicates a much stronger mixing of components that are in a liquid state before crystallization in the laser beam pool. No visible MR phase plates were observed. This indicates a strong grain refinement. Despite the visible boundaries of the laser beam paths, no microstructure changes were observed in these places. Visible boundaries between the laser-fused layer, the layer below the fusion (which was affected by the heat of the laser beam) and the intact material were observed. We observed the deepest laser beam penetration in samples T4 and T6 visible as a bright band from the surface (Figure 4c).

As can be seen in the SEM images in Figures 4 and 5, laser-alloyed samples differed significantly in microstructure from laser-melted samples. This is the result of introducing a new element in the structure of the substrate (Ti13Nb13Zr alloy), which is carbon derived from the MWCNTs coating. Knowing about the results of XRD (Figure 6 and the Ti–C phase equilibrium plot, the newly formed phases are identified with TiC titanium carbide. They are formed due to the liquid state of the substrate and the coating applied to it under the influence of a high temperature laser beam of high power. Mixing of the two components and their crystallization results in the formation of grains characteristic for the surface zone after laser processing and structures resulting from the enrichment of the substrate with the coating element [67,68]. These structures are observed in the SEM images (Figures 4b and 5b) of each sample and they take different forms. Their presence is most densely observed in the shallowest laser-melted layers, where the coating material was melted into a smaller volume of the substrate material. In the C1 and C2 section (Figure 5b), TiC dendrites are observed at the entire penetration depth. TiC structures in the C3 microstructure turning into a structure consisting of spheres for which the pulse duration was almost twice as large. Based on [67], it is noted that a symmetrical dendritic structure in the form of titanium and carbon cubic subnets tends to grow into a spherical form to reduce interfacial energy. It is concluded that increasing the duration of the laser beam pulse and reducing its feed speed reduce the solidification rate of the material and smooth the convective flows in the alloy material. This promotes the formation of TiC in a spherical form which is also favored by the increase in laser pulse power. This is observed in the images for samples C4 ÷ C6 (Figure 5b), where the laser-melted layer is clearly deeper, and the substrate material was melted into the larger volume of the substrate material. The quantity of TiC phases is more numerous for areas of the laser-melted layer closer to the sample front. The smallest forms of TiC phases are observed in samples C4 and C6. In the SEM images of cross-sections one can see the boundaries of individual laser beam transitions (clear lines in samples C4 and C6 in Figure 5b). In these places, more spheres with smaller diameters were observed. This may have been the result of re-remelting the material during the next laser beam transition. For this method of machining, the Ti13Nb13Zr alloy combined with the enrichment of this material in carbon, the presence of cracks in the modified layer is not observed. This is an advantage of the presented modification, because traditional laser-processing of titanium alloys is most often associated with formation of surface cracks, which are difficult to eliminate.

The number of TiC structures in the form of dendrites was much smaller in the present study compared to [67,68]. This is probably due to the thickness of the carbon coating applied on the titanium alloy. In the cited studies [67,68], the coating was over 10-fold and 30-fold thicker than the EPD obtained in this process. The amount of carbonaceous material fused into a similar volume of substrate material was, therefore, clearly larger. The coating itself was made of a different carbon material and applied in a different manner. Carbon nanotubes were selected as a carburizing additive because of their mechanical and chemical stability and their tubular shape translating into increased mechanical properties of the laser-modified surface layers. In addition, the EPD process ensured obtaining repeatable coatings with a predefined thickness. It was also important that they are considered to be biocompatible materials with the human tissue.

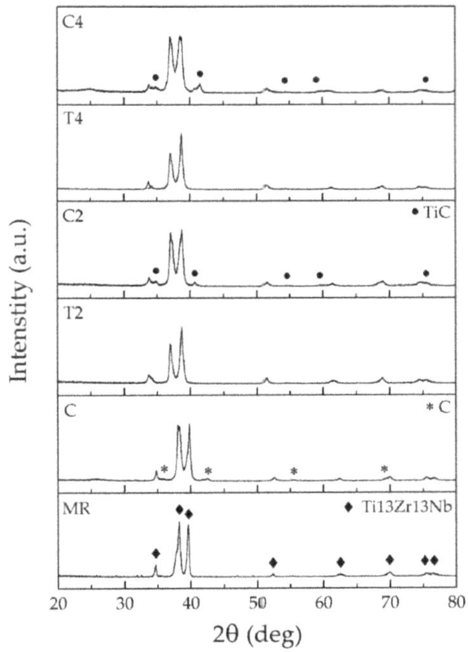

Figure 6. XRD spectra of Ti13Nb13Zr alloy for: MR, C (MR with MWCNTs coating), T2, T4—its alloys after laser treatment, C2, C4—its alloys with MWCNTs coating after laser treatment.

The analysis of the cross-sections observed on the optical microscope allowed the assessment of the penetration depth and layers of material that could be affected by laser processing. Dependence of material depth on the parameters used was observed. As the laser pulse power increased at a constant feed rate and frequency, more than a double deepening of the laser-melted layers was observed between T2 and T5. A fourfold increase in the depth of the modified layer was seen between T1 and T4, where a power increase was applied. No significant macroscopic changes were seen with a slight decrease in the pulse power between T1 and T2, a simultaneous increase of feed speed and extension of the pulse duration. By maintaining the same power between T2 and T3 and increasing the pulse duration and reducing the feed speed, a significant penetration deepening was seen. Between T4 and T6, a frequency reduction was used with other fixed parameters. A shallowing of the laser-melted layer was observed. Optical microscope images for laser carburized samples revealed laser-melted layers with a characteristic, darker color. This was the result of carburizing the material. Similar relationships were observed as for material modification without a carbon coating. Based on this analysis, it was decided that the next tests would be carried out on samples with the shallowest and deepest laser-melted layers: T2, T4, C2 and C4. Another aspect in their selection was the different form of TiC in the C2 and C4 samples.

3.3. Chemical and Phase Analysis, Raman Spectroscopy

The X-ray energy dispersion spectroscopy (EDS) spectra for the reference sample Ti13Zr13Nb and laser processed and carburized samples with the smallest and largest penetration depth obtained (T2, T4, C2, C4) are shown in Figure 5. The tests were carried out at a distance of 10 μm from the sample surface, for cross-sectional areas of 50 × 80 μm. Three main alloying elements were observed in the EDS spectrum of the reference sample: titanium (the most intense peak), zirconium and niobium. The spectra for laser melted samples (T2 and T4) did not differ much from the result for the reference sample. A quantitative analysis showed a smaller content of the alloying element Zr, which could indicate that this component was more chemically etched after laser processing. The smaller content of this alloying element can also result from temporary chemical composition fluctuation during laser melting and an appearance of lighter elements on the surface. For laser carburized samples (C2 and C4)—and especially for the sample with the shallowest penetration depth (C2)—a more pronounced peak for carbon in the EDS spectrum and an increase in the quantitative content of this element in the material were noted. The more intense carbon peak in EDS spectra, especially for sample C2 (Figure 7) and presence of dendritic and spherical TiC well visible in microstructure of surface zone (Figure 5b), confirmed additionally by XRD (Figure 6) and Raman (Figure 8) analysis, prove effect of laser carburization.

Figure 7. EDS spectra of Ti13Nb13Zr alloy: (**a**) MR, (**b**) MR with MWCNTs coating, (**c**,**d**) T2, T4—its alloys after laser treatment, (**e**,**f**) C2, C4—its alloys with MWCNTs coating after laser treatment with the shallowest penetration.

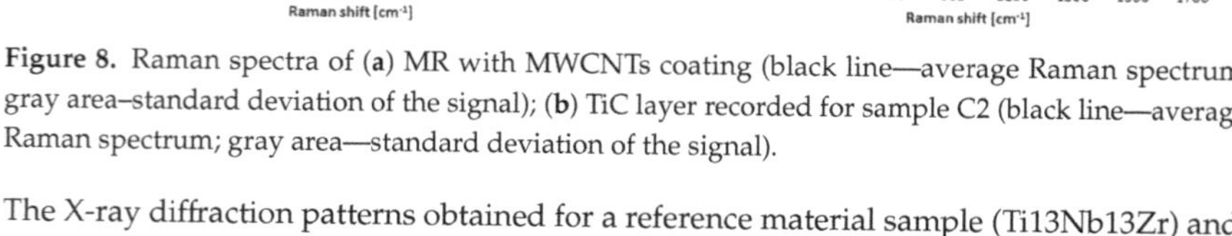

Figure 8. Raman spectra of (**a**) MR with MWCNTs coating (black line—average Raman spectrum; gray area–standard deviation of the signal); (**b**) TiC layer recorded for sample C2 (black line—average Raman spectrum; gray area—standard deviation of the signal).

The X-ray diffraction patterns obtained for a reference material sample (Ti13Nb13Zr) and with MWNCNTs coating as well as samples after melting and alloying with the shallowest and deepest penetration (T2, T4, C2, C4) are shown in Figure 6.

The highest peaks observed on the diffractograms are indexed as derived from the substrate material-Ti13Nb13Zr alloy. After laser processing, this material presented spectra similar to a laser-unprocessed material. For an alloy with a coating of carbon nanotubes, the presence of carbon-specific peaks (marked with an asterisk in Figure 6) is observed in addition to the peak characteristic of the substrate material. Samples after laser carbon alloying in their spectra revealed peaks suitable for the substrate alloy and characteristic for the new phase—titanium carbide (marked with a dot symbol in Figure 6). Such spectra were obtained previously in [67,68], where graphite was used as carburizing material.

The main mechanism of TiC formation in laser processing involves the diffusion of C atoms into liquid Ti, followed by nucleation and growth of TiC crystals. It is possible for titanium carbides to be formed in a range of about 10–19% by carbon weight in the temperature range 1648–3000 °C Naccording to the Ti–C phase equilibrium system [69,70].

In the Raman spectrum of the MWCNTs coating (Figure 8a) a typical spectral pattern was observed. The spectrum is characterized by the occurrence of the following bands: 1345 cm^{-1} (D band), 1580 cm^{-1} (G band) and 2690 cm^{-1} (2D band) [71]. In the Raman spectrum of sample after laser modification (Figure 8b) the characteristic spectral bands for titanium carbide were observed at 400 cm^{-1}, 610 cm^{-1}, 1340 cm^{-1} and 1560 cm^{-1} [72].

3.4. Wettability Analysis

The values of the average contact angle for uncoated and MWCNTs-coated Ti13Nb13Zr samples unmodified and modified using laser radiation with the obtained shallow and deepest laser-modified layer are listed in Table 3. All results of contact angle measurements prove the hydrophilicity of the tested materials, which is necessary for materials intended for use in implantology and prosthetics in order to improve their osseointegration with bone tissue [73]. Analyzing the wettability tests, it is noted that the laser melting of the Ti13Nb13Zr alloy causes an increase in the contact angle of the material as in [52] to a value of about 80°. It can be also mentioned that the parameters affecting the deepening of the laser-melted layer causes a smaller increase in the contact angle. The carbon enrichment of the surface layer of the Ti13Nb13Zr alloy reduced the dynamics of the contact angle increase compared to laser melting. A carburized sample with the deepest laser-melted layer showed only a slight increase in the contact angle relative to the MR reference. Studies [74–76] assess that contact angles between 35° and 80° are beneficial for materials cooperating with bone tissue. The best contact angle values according to other publications are 40–60° [77] depending on cells and for bone cells at 35 ÷ 85°, with the optimum value at 55° [78]. Therefore, all tested samples demonstrating the contact angles between 46° and 82° possess hydrophilic surfaces suitable for adhesion of cells. The difference between T2, T4, C2 and C4, in their wettability can be attributed to the positive and joined effect of the significant laser remelting at high laser power and the presence of MWCNTs. Therefore, positive effects of laser carburizing of the Ti13Nb13Zr alloy in the perspective of using this alloy as a biomaterial are seen.

Table 3. Average contact angle values for the tested samples.

Sample	Average Contact Angle (°)
MR	47.1 ± 1.1
T2	81.6 ± 0.9
T4	77.0 ± 0.8
C2	77.0 ± 1.1
C4	46.1 ± 3.6

3.5. Nanomechanical Studies

Table 4 presents the average values of the tested properties, i.e., nanohardness and Young's modulus (reduced and real) and the maximum penetration of the nanoindenter into the analyzed material for individual areas of the sample comprising the laser-melted layer, the layer affected by the heat of the laser beam and the material below that the laser did not affect. The values presented are accompanied by deviations consisting of the values of the largest and lowest recorded measurement.

Table 4. Mechanical and nanoindentation properties of the tested samples.

Area	Sample	Nanohardness (GPa)		Reduced Young's Modulus, Er (GPa)		Real Young's Modulus, E (GPa)		Maximum Depth of Indentation (nm)	
Laser-melted layer	T2	4.08	+0.72 −0.91	90.61	+10.58 −18.86	85.72	+10.88 −19.10	796.00	+31.12 −52.95
	T4	4.98	+1.03 −1.65	111.05	+13.52 −17.22	107.07	+14.57 −17.22	722.55	+154.92 −164.58
	C2	5.83	+2.25 −1.96	115.11	+13.23 −16.43	111.42	+14.37 −17.44	678.90	+138.87 −97.02
	C4	6.36	+2.37 −1.78	126.12	+25.09 −29.67	123.51	+28.09 −42.12	647.94	+128.02 −81.38
Layer affected by the heat of the laser beam	T2	3.92	+0.82 −0.57	104.96	+13.74 −9.14	100.62	+14.64 −9.61	812.05	+57.46 −78.07
	T4	3.79	+2.34 −1.46	116.21	+41.58 −26.49	112.75	+46.50 −28.03	816.70	+216.97 −237.58
	C2	4.51	+1.13 −1.53	109.33	+14.80 −20.16	105.28	+15.87 −21.13	765.10	+159.98 −87.53
	C4	5.80	+0.95 −1.20	127.76	+11.83 −13.66	125.19	+13.15 −14.91	667.52	+79.28 −48.62
Material that was not affected by the laser	T2	3.27	+1.26 −1.30	113.70	+29.82 −25.35	110.02	+32.77 −26.99	879.79	+238.07 −226.77
	T4	2.92	+1.57 −1.22	105.83	+26.35 −25.10	101.70	+28.35 −26.10	947.87	+259.87 −204.44
	C2	4.31	+2.02 −1.10	124.80	+29.48 −15.21	121.97	+33.20 −16.50	765.42	+111.24 −137.40
	C4	5.70	+1.86 −1.17	128.83	+23.80 −17.39	126.41	+28.84 −18.97	674.64	+79.26 −91.28

As the nanohardness increases, approaching the sample face, smaller penetration depths of the indenter are observed, confirming the hardening of the material. Approaching the forehead, a decrease in the Young's modulus value is also observed, which is beneficial because this value approaches the value of the Young's modulus of bone tissue. Analyzing the nanohardness, an increase is observed as the laser pulse power increases. For laser melting with the shallowest laser-melted layer, an increase in nanohardness of about 25% was obtained and a further 25% with a more than double increase in laser pulse power, where the deepest laser-melted layer was observed compared to MR. In the case of laser carburizing for a sample with the shallowest laser-melted layer, an increase in nanohardness of more than 40% is observed compared to a laser-melted sample with the same parameters. By more than doubling the laser pulse power, another 10% increase in nanohardness is achieved, and as a result, almost 30% relative to laser melting with the same parameters. Therefore, it is noted that laser carburizing increases the value of nanohardness more significantly than laser melting. In the case of Young's modulus, more desirable values were obtained with laser melting. With laser alloying, the decrease was small relative to the base material (material that was not affected by the laser in Table 4).

In [67,68], where the source of carbon was graphite, with higher laser pulse powers and more numerous titanium carbide structures observed, microhardness values of 4 GPa were obtained in zones with a direct penetration of up to 12 GPa in the layer at the penetration face. In [44], where a coating of carbon nanotubes from a solution with a higher percentage of MWCNTs was used, surface nanohardness values between 5 and almost 9 GPa were obtained using other laser parameters.

The 3D distributions of nanohardness and Young's modulus values recorded on the nanoindenter are shown in Figures 9 and 10. Observing the 3D distribution of nanohardness, several initial significantly different measurements should be omitted, which are the result of penetration of the depth gauge into the resin. By analyzing the proper material, it can be confirmed that for each tested sample there was an increase in nanohardness in the laser-modified layer. The shallowest cure occurred for the C2 carburized sample. In samples T2 and T4, despite the difference in the depth of the modified layer and its tested nanohardness, it can be seen that the depth of the hardened layer is similar. The deepest hardened area is observed in the C4 alloyed sample (Figure 9). By analyzing 3D distributions of Young's modulus values, disregarding the first 50 ÷ 100 μm which refers to the resin, it is confirmed that for each sample tested, the values of the elastic modulus decrease with the approach of the laser-melted front. Even in the T4 sample, for which the average values indicated a Young's modulus higher in the modified zone than the area below it, it is noted that this property reaches higher values as it moves away from the sample front. Smaller Young's modulus values occur in laser-modified layers without stopping with T2 and T4.

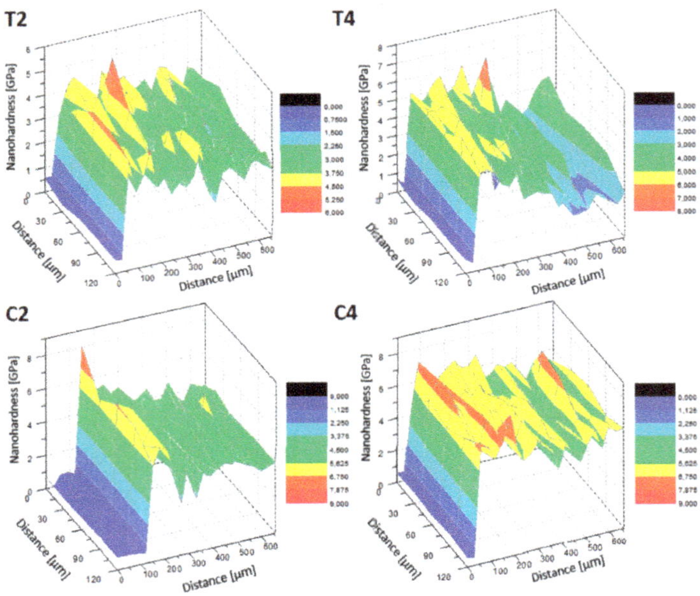

Figure 9. 3D distribution of nanohardness for tested samples T2, T4, C2 and C4.

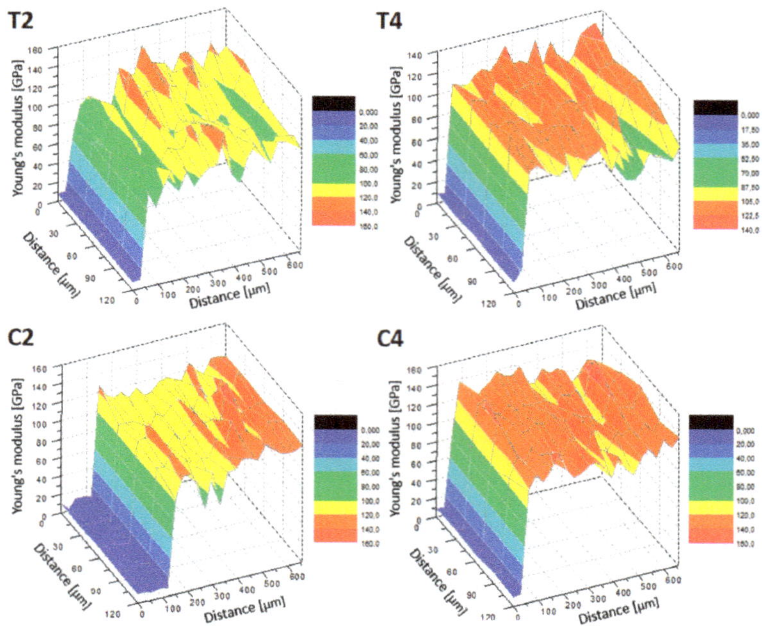

Figure 10. 3D distribution of reduced Young's modulus for tested samples T2, T4, C2 and C4.

3.6. Corrosion Tests Analysis

Corrosion tests were carried out for uncoated and MWCNTs-coated Ti13Nb13Zr samples unmodified and modified using laser radiation with the shallowest and deepest laser-melted layers (T2, T4, C2, C4). Potentiodynamic curves for the tests performed are shown in Figure 11. Corrosion current density (I_{corr}) and corrosion potential (E_{corr}) values are shown in Table 5.

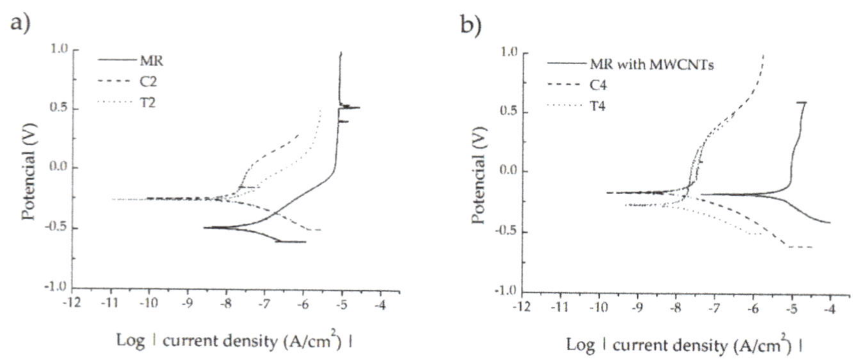

Figure 11. Potentiodynamic polarization curves in Ringer's solution at room temperature of: (**a**) reference sample Ti13Zr13Nb, C2, T2 laser-modified layers; (**b**) reference sample Ti13Zr13Nb with MWCNTs coating, C4, T4 laser-modified layers.

Table 5. Corrosion current density and corrosion potential of T2, T4, C2 and C4 laser-modified layers, non-laser-modified Ti13Zr13Nb substrate (MR) and MWCNTs coating.

Sample	Corrosion Parameters	
	E_{corr} (V)	I_{corr} (A/cm^2)
MR	-4.87×10^{-1}	5.19×10^{-8}
MR with MWCNTs coating	-1.80×10^{-1}	3.52×10^{-6}
T2	-2.55×10^{-1}	3.47×10^{-8}
T4	-2.69×10^{-1}	2.33×10^{-8}
C2	-2.45×10^{-1}	3.10×10^{-8}
C4	-1.63×10^{-1}	4.22×10^{-8}

The improvement of corrosion resistance is demonstrated by a decrease in the value of the corrosion current and an increase in the value of the corrosion potential. An increase in the value of corrosion potential, but also an increase in the corrosion current was obtained for the examined alloy with deposited carbon coating. Thus, covering the alloy with a carbon coating does not guarantee an improvement in corrosion resistance. The increase in the value of the corrosion potential and the decrease in the value of the corrosive current is observed for samples subjected to laser treatment with and without a carbon coating. This demonstrates the improvement of the corrosion resistance of Ti13Nb13Zr due to laser processing. In the case of the analysis of the change in the value of corrosion potential, a certain relationship is noticed regarding the laser treatment of the material with and without a carbon coating. Higher values are obtained for laser-alloyed samples relative to laser melting with the same parameters. The highest and most desirable value was obtained for a carburized C4 sample with the deepest laser-melted layer. When analyzing the change in the value of corrosive current density, no clear relationship between laser alloying and laser melting is seen. The smallest, and therefore the most desirable value of corrosive current is observed for sample T4.

4. Conclusions

Using a method of electrophoretic deposition, a uniform coating of functionalized multi-walled carbon nanotubes was successfully deposited on the Ti13Nb13Zr alloy. By using laser melting methods, changes in the microstructure of this alloy, primarily grain refinement, were successfully achieved. Carburizing successfully enriched the surface layer of this alloy with carbon, which resulted in the formation of a new component—titanium carbide—whose forms depend on the processing parameters.

The laser processing parameters have a significant impact on the depth and shape of the laser-melted layer. An increase in the laser pulse power and its duration affects the deepening of the modified layer, while an increase in feed speed and a decrease in frequency causes it to become shallower. No cracks were observed in the laser-modified areas for any of the investigations. Changes in the microstructure in the laser-melted layer caused an increase in nanohardness of about 25% over the base material. Laser alloying caused an additional increase in nanohardness of about 25% compared to laser melting with the same process parameters. A reduction of the Young's modulus value in laser-modified layers was achieved.

Corrosion tests showed that the applied laser treatment (laser remelting and laser alloying–carburizing) caused an increase in the value of corrosion potential and a decrease in the value of corrosive current density. This indicates better corrosion resistance, compared to the reference. Laser melting increased the contact angle to almost the hydrophilicity limit. For laser carburizing, a reduction in contact angle relative to laser melting was observed.

The results of the presented research lead to the conclusion that laser modification of the Ti13Nb13Zr alloy coated with carbon nanotubes may be an interesting alternative to the complicated and expensive techniques of producing coatings for applications in implants. Laser surface carburization promotes an increase in nanohardness, a decrease in Young's modulus and obtaining a hydrophilic surface—the properties desired for the indicated application. The success of further research will depend on the results of the biologic research that is critical to the foreseen application.

Author Contributions: Conceptualization, B.M.-M.; methodology, P.T., M.B., B.B. and B.M.-M.; software, P.T. and M.B.; formal analysis, B.M.-M. and P.T.; investigation, P.T., M.B., B.B. and B.M.-M.; writing—original draft preparation, P.T.; writing—review and editing, B.M.-M. and P.T.; supervision, B.M.-M. and B.J.J. All authors have read and agreed to the published version of the manuscript.

Funding: This research received no external funding.

Acknowledgments: The authors thank all those who contributed to this publication, i.e., the employees of the Biomaterials Group and the Department of Solid State Physics at the Gdansk University of Technology (in particular G. Gajowiec and A. Mielewczyk-Gryń) and the Institute of Optoelectronics at the Military University of Technology in Warsaw: for technical assistance in carrying out the tests. Mentoring activities and scientific support from A. Zieliński deserved special thanks.

Conflicts of Interest: The authors declare no conflict of interest.

References

1. Suchanek, K.; Bartkowiak, A.; Gdowik, A.; Perzanowski, M.; Kąc, S.; Szaraniec, B.; Suchanek, M.; Marszałek, M. Crystalline hydroxyapatite coatings synthesized under hydrothermal conditions on modified titanium substrates. *Mater. Sci. Eng. C* **2015**, *51*, 57–63. [CrossRef]
2. Souza, J.C.M.; Sordi, M.B.; Kanazawa, M.; Ravindran, S.; Henriques, B.; Silva, F.S.; Aparicio, C.; Cooper, L.F. Nano-scale modification of titanium implant surfaces to enhance osseointegration. *Acta Biomater.* **2019**, *94*, 112–131. [CrossRef] [PubMed]
3. Kaur, M.; Singh, K. Review on titanium and titanium based alloys as biomaterials for orthopaedic applications. *Mater. Sci. Eng. C* **2019**, *102*, 844–862. [CrossRef] [PubMed]
4. Oldani, C.; Dominguez, A. Titanium as a Biomaterial for Implants. In *Recent Advances in Arthroplasty*; Fokter, S., Ed.; InTech: Rijeka, Croatia, 2012; pp. 149–162.
5. Granato, R.; Bonfante, E.A.; Castellano, A.; Khan, R.; Jimbo, R.; Marin, C.; Morsi, S.; Witek, L.; Coelho, P.G. Osteointegrative and microgeometric comparison between micro-blasted and alumina blasting/acid etching on grade II and V titanium alloys (Ti-6Al-4V). *J. Mech. Behav. Biomed. Mater.* **2019**, *97*, 288–295. [CrossRef] [PubMed]
6. Khodaei, M.; Hossein Kelishadi, S. The effect of different oxidizing ions on hydrogen peroxide treatment of titanium dental implant. *Surf. Coat. Technol.* **2018**, *353*, 158–162. [CrossRef]
7. Bartmanski, M.; Zielinski, A.; Jazdzewska, M.; Głodowska, J.; Kalka, P. Effects of electrophoretic deposition times and nanotubular oxide surfaces on properties of the nanohydroxyapatite/nanocopper coating on the Ti13Zr13Nb alloy. *Ceram. Int.* **2019**, *45*, 20002–20010. [CrossRef]
8. Simka, W.; Mosiałek, M.; Nawrat, G.; Nowak, P.; Żak, J.; Szade, J.; Winiarski, A.; Maciej, A.; Szyk-Warszyńska, L. Electrochemical polishing of Ti–13Nb–13Zr alloy. *Surf. Coat. Technol.* **2012**, *213*, 239–246. [CrossRef]
9. Lorenzetti, M.; Biglino, D.; Novak, S.; Kobe, S. Photoinduced properties of nanocrystalline TiO2-anatase coating on Ti-based bone implants. *Mater. Sci. Eng. C* **2014**, *37*, 390–398. [CrossRef]
10. Lorenzetti, M.; Pellicer, E.; Sort, J.; Baró, M.; Kovač, J.; Novak, S.; Kobe, S. Improvement to the Corrosion Resistance of Ti-Based Implants Using Hydrothermally Synthesized Nanostructured Anatase Coatings. *Materials* **2014**, *7*, 180–194. [CrossRef]
11. Kazek-Kęsik, A.; Leśniak, K.; Zhidkov, I.; Korotin, D.; Kukharenko, A.; Cholakh, S.; Kalemba-Rec, I.; Suchanek, K.; Kurmaev, E.; Simka, W. Influence of Alkali Treatment on Anodized Titanium Alloys in Wollastonite Suspension. *Metals* **2017**, *7*, 322. [CrossRef]
12. Ossowska, A.; Sobieszczyk, S.; Supernak, M.; Zielinski, A. Morphology and properties of nanotubular oxide layer on the "Ti–13Zr–13Nb" alloy. *Surf. Coat. Technol.* **2014**, *258*, 1239–1248. [CrossRef]
13. Lin, Z.; Li, S.-J.; Sun, F.; Ba, D.-C.; Li, X.-C. Surface characteristics of a dental implant modified by low energy oxygen ion implantation. *Surf. Coat. Technol.* **2019**, *365*, 208–213. [CrossRef]

14. Acciari, H.A.; Palma, D.P.S.; Codaro, E.N.; Zhou, Q.; Wang, J.; Ling, Y.; Zhang, J.; Zhang, Z. Surface modifications by both anodic oxidation and ion beam implantation on electropolished titanium substrates. *Appl. Surf. Sci.* **2019**, *487*, 1111–1120. [CrossRef]
15. Gao, C.; Dai, L.; Meng, W.; He, Z.; Wang, L. Electrochemically promoted electroless nickel-phosphorous plating on titanium substrate. *Appl. Surf. Sci.* **2017**, *392*, 912–919. [CrossRef]
16. Wang, Y.; Gao, Y.; Takahashi, J.; Wan, Y.; Xiao, B.; Zhang, Y.; He, X.; Li, J. Titanium-modified graphite reinforced Cu-Ni composite by multi-arc ion plating technology. *Vacuum* **2019**, *168*, 108829. [CrossRef]
17. Zong, X.; Wang, H.; Li, Z.; Li, J.; Cheng, X.; Zhu, Y.; Tian, X.; Tang, H. Laser nitridation on Ti-6.5Al-3.5Mo-1.5Zr-0.3Si titanium alloy. *Surf. Coat. Technol.* **2020**, *386*, 125425. [CrossRef]
18. Zhu, W.; Shi, D.; Zhu, Z.; Sun, J. Effect of the surface oxidization and nitridation on the normal spectral emissivity of titanium alloys Ti–6Al–4V at 800–1100 K at a wavelength of 1.5 μm. *Infrared Phys. Technol.* **2016**, *76*, 200–205. [CrossRef]
19. Buga, C.; Hunyadi, M.; Gácsi, Z.; Hegedűs, C.; Hakl, J.; Schmidt, U.; Ding, S.-J.; Csík, A. Calcium silicate layer on titanium fabricated by electrospray deposition. *Mater. Sci. Eng. C* **2019**, *98*, 401–408. [CrossRef]
20. Avcu, E.; Baştan, F.E.; Abdullah, H.Z.; Rehman, M.A.U.; Avcu, Y.Y.; Boccaccini, A.R. Electrophoretic deposition of chitosan-based composite coatings for biomedical applications: A review. *Prog. Mater. Sci.* **2019**, *103*, 69–108. [CrossRef]
21. Gilabert-Chirivella, E.; Pérez-Feito, R.; Ribeiro, C.; Ribeiro, S.; Correia, D.M.; González-Martín, M.L.; Manero, J.M.; Lanceros-Méndez, S.; Ferrer, G.G.; Gómez-Ribelles, J.L. Chitosan patterning on titanium implants. *Prog. Org. Coat.* **2017**, *111*, 23–28. [CrossRef]
22. Suchanek, K.; Hajdyła, M.; Maximenko, A.; Zarzycki, A.; Marszałek, M.; Jany, B.R.; Krok, F. The influence of nanoporous anodic titanium oxide substrates on the growth of the crystalline hydroxyapatite coatings. *Mater. Chem. Phys.* **2017**, *186*, 167–178. [CrossRef]
23. Jażdżewska, M.; Majkowska-Marzec, B. Hydroxyapatite deposition on the laser modified Ti13Nb13Zr alloy. *Adv. Mater. Sci.* **2017**, *17*, 5–13. [CrossRef]
24. Chozhanathmisra, M.; Murugan, N.; Karthikeyan, P.; Sathishkumar, S.; Anbarasu, G.; Rajavel, R. Development of antibacterial activity and corrosion resistance properties of electrodeposition of mineralized hydroxyapatite coated on titanium alloy for biomedical applications. *Mater. Today Proc.* **2017**, *4*, 12393–12400. [CrossRef]
25. Saphronov, V.; Shishkovsky, I. Laser Annealing for Gas-Dynamical Spraying of HA Coating upon a Titanium Surface. *Crystals* **2015**, *5*, 447–457. [CrossRef]
26. Bartmanski, M.; Cieslik, B.; Glodowska, J.; Kalka, P.; Pawlowski, L.; Pieper, M.; Zielinski, A. Electrophoretic deposition (EPD) of nanohydroxyapatite—Nanosilver coatings on Ti13Zr13Nb alloy. *Ceram. Int.* **2017**, *43*, 11820–11829. [CrossRef]
27. Bartmanski, M.; Zielinski, A.; Majkowska-Marzec, B.; Strugala, G. Effects of solution composition and electrophoretic deposition voltage on various properties of nanohydroxyapatite coatings on the Ti13Zr13Nb alloy. *Ceram. Int.* **2018**, *44*, 19236–19246. [CrossRef]
28. Bartmański, M.; Pawłowski, Ł.; Strugała, G.; Mielewczyk-Gryń, A.; Zieliński, A. Properties of Nanohydroxyapatite Coatings Doped with Nanocopper, Obtained by Electrophoretic Deposition on Ti13Zr13Nb Alloy. *Materials* **2019**, *12*, 3741. [CrossRef]
29. Bose, S.; Vu, A.A.; Emshadi, K.; Bandyopadhyay, A. Effects of polycaprolactone on alendronate drug release from Mg-doped hydroxyapatite coating on titanium. *Mater. Sci. Eng. C* **2018**, *88*, 166–171. [CrossRef]
30. Ciobanu, G.; Harja, M. Cerium-doped hydroxyapatite/collagen coatings on titanium for bone implants. *Ceram. Int.* **2019**, *45*, 2852–2857. [CrossRef]
31. Długoń, E.; Niemiec, W.; Frączek-Szczypta, A.; Jeleń, P.; Sitarz, M.; Błażewicz, M. Spectroscopic studies of electrophoretically deposited hybrid HAp/CNT coatings on titanium. *Spectrochim. Acta. A. Mol. Biomol. Spectrosc.* **2014**, *133*, 872–875. [CrossRef]
32. Zhong, Z.; Qin, J.; Ma, J. Electrophoretic deposition of biomimetic zinc substituted hydroxyapatite coatings with chitosan and carbon nanotubes on titanium. *Ceram. Int.* **2015**, *41*, 8878–8884. [CrossRef]
33. Dziaduszewska, M.; Wekwejt, M.; Bartmański, M.; Pałubicka, A.; Gajowiec, G.; Seramak, T.; Osyczka, A.M.; Zieliński, A. The Effect of Surface Modification of Ti13Zr13Nb Alloy on Adhesion of Antibiotic and Nanosilver-Loaded Bone Cement Coatings Dedicated for Application as Spacers. *Materials* **2019**, *12*, 2964. [CrossRef] [PubMed]

34. Grabarczyk, J.; Batory, D.; Louda, P.; Couvrat, P.; Kotela, I.; Bakowicz-Mitura, K. Carbon coatings for medical implants. *J. Achiev. Mater. Manuf. Eng.* **2007**, *20*, 107–110.
35. Madej, M.; Piotrowska, K.; Ozimina, D. Properties of diamond-like carbon coatings on the titanium alloy Ti13Nb13Zr. *Bimon. Tribol.* **2019**, *6*, 39–46. [CrossRef]
36. Chen, X.; Chen, S.; Liang, L.; Hong, H.; Zhang, Z.; Shen, B. Electrochemical behaviour of EPD synthesized graphene coating on titanium alloys for orthopedic implant application. *Procedia CIRP* **2018**, *71*, 322–328. [CrossRef]
37. Chen, C.; Tang, W.; Li, X.; Wang, W.; Xu, C. Structure and cutting performance of Ti-DLC films prepared by reactive magnetron sputtering. *Diam. Relat. Mater.* **2020**, *104*, 107735. [CrossRef]
38. Strąkowska, P.; Beutner, R.; Gnyba, M.; Zielinski, A.; Scharnweber, D. Electrochemically assisted deposition of hydroxyapatite on Ti6Al4V substrates covered by CVD diamond films—Coating characterization and first cell biological results. *Mater. Sci. Eng. C* **2016**, *59*, 624–635. [CrossRef]
39. Ding, H.H.; Fridrici, V.; Guillonneau, G.; Sao-Joao, S.; Geringer, J.; Fontaine, J.; Kapsa, P. Investigation on mechanical properties of tribofilm formed on Ti–6Al–4V surface sliding against a DLC coating by nano-indentation and micro-pillar compression techniques. *Wear* **2019**, *432–433*, 202954. [CrossRef]
40. Jędrzejczak, A.; Batory, D. Powłoki węglowe domieszkowane krzemem wytwarzane metodą RF PACVD. *Inż. Mater.* **2013**, *34*, 459–462.
41. Wang, J.; Ma, J.; Huang, W.; Wang, L.; He, H.; Liu, C. The investigation of the structures and tribological properties of F-DLC coatings deposited on Ti-6Al-4V alloys. *Surf. Coat. Technol.* **2017**, *316*, 22–29. [CrossRef]
42. Chou, C.-C.; Lin, J.-S.; Wu, R. Microstructures and mechanical properties of an a-C:N film as the interlayer and the outmost layer of a DLC-deposited Ti bio-alloy. *Ceram. Int.* **2017**, *43*, S776–S783. [CrossRef]
43. Bhui, A.S.; Bains, P.S.; Sidhu, S.S.; Singh, G. Parametric optimization of ED machining of Ti-6Al-4V in CNTs mixed dielectric medium. *Mater. Today Proc.* **2019**, *18*, 1532–1539. [CrossRef]
44. Rogala-Wielgus, D.; Majkowska-Marzec, B. Wpływ stopowania laserowego z użyciem nanorurek węglowych stopu Ti13Nb13Zr do zastosowań biomedycznych na jego wybrane własności mechaniczne. *Przegląd Spaw. Weld. Technol. Rev.* **2018**, *90*, 18–23. [CrossRef]
45. Ba, J.; Zheng, X.H.; Ning, R.; Lin, J.H.; Qi, J.L.; Cao, J.; Cai, W.; Feng, J.C. Brazing of SiO 2-BN modified with in situ synthesized CNTs to Ti6Al4V alloy by TiZrNiCu brazing alloy. *Ceram. Int.* **2018**, *44*, 10210–10214. [CrossRef]
46. Majkowska-Marzec, B.; Rogala-Wielgus, D.; Bartmański, M.; Bartosewicz, B.; Zieliński, A. Comparison of Properties of the Hybrid and Bilayer MWCNTs—Hydroxyapatite Coatings on Ti Alloy. *Coatings* **2019**, *9*, 643. [CrossRef]
47. Munir, K.S.; Zheng, Y.; Zhang, D.; Lin, J.; Li, Y.; Wen, C. Microstructure and mechanical properties of carbon nanotubes reinforced titanium matrix composites fabricated via spark plasma sintering. *Mater. Sci. Eng. A* **2017**, *688*, 505–523. [CrossRef]
48. Okoro, A.M.; Machaka, R.; Lephuthing, S.S.; Oke, S.R.; Awotunde, M.A.; Olubambi, P.A. Evaluation of the sinterability, densification behaviour and microhardness of spark plasma sintered multiwall carbon nanotubes reinforced Ti6Al4V nanocomposites. *Ceram. Int.* **2019**, *45*, 19864–19878. [CrossRef]
49. Savalani, M.M.; Ng, C.C.; Li, Q.H.; Man, H.C. In situ formation of titanium carbide using titanium and carbon-nanotube powders by laser cladding. *Appl. Surf. Sci.* **2012**, *258*, 3173–3177. [CrossRef]
50. Wesełucha-Birczyńska, A.; Stodolak-Zych, E.; Piś, W.; Długoń, E.; Benko, A.; Błażewicz, M. A model of adsorption of albumin on the implant surface titanium and titanium modified carbon coatings (MWCNT-EPD). 2D correlation analysis. *J. Mol. Struct.* **2016**, *1124*, 61–70. [CrossRef]
51. Lisiecki, A. Study of Optical Properties of Surface Layers Produced by Laser Surface Melting and Laser Surface Nitriding of Titanium Alloy. *Materials* **2019**, *12*, 3112. [CrossRef]
52. Tęczar, P.; Majkowska-Marzec, B.; Bartmański, M. The Influence of Laser Alloying of Ti13Nb13Zr on Surface Topography and Properties. *Adv. Mater. Sci.* **2019**, *19*, 44–56. [CrossRef]
53. Xie, Z.; Dai, Y.; Ou, X.; Ni, S.; Song, M. Effects of selective laser melting build orientations on the microstructure and tensile performance of Ti–6Al–4V alloy. *Mater. Sci. Eng. A* **2020**, *776*, 139001. [CrossRef]
54. Yamaguchi, T.; Hagino, H. Formation of titanium carbide layer by laser alloying with a light-transmitting resin. *Opt. Lasers Eng.* **2017**, *88*, 13–19. [CrossRef]

55. Dolgun, E.; Zemlyakov, E.; Shalnova, S.; Gushchina, M.; Promahov, V. The influence of heat treatment on the microstructure of products manufactured by direct laser deposition using titanium alloy Ti-6Al-4V. *Mater. Today Proc.* **2020**. [CrossRef]
56. Kosec, T.; Legat, A.; Kovač, J.; Klobčar, D. Influence of Laser Colour Marking on the Corrosion Properties of Low Alloyed Ti. *Coatings* **2019**, *9*, 375. [CrossRef]
57. Pou, P.; Riveiro, A.; del Val, J.; Comesaña, R.; Penide, J.; Arias-González, F.; Soto, R.; Lusquiños, F.; Pou, J. Laser surface texturing of Titanium for bioengineering applications. *Procedia Manuf.* **2017**, *13*, 694–701. [CrossRef]
58. Feizi Mohazzab, B.; Jaleh, B.; Kakuee, O.; Fattah-alhosseini, A. Formation of titanium carbide on the titanium surface using laser ablation in n-heptane and investigating its corrosion resistance. *Appl. Surf. Sci.* **2019**, *478*, 623–635. [CrossRef]
59. Landowski, M. Influence of Parameters of Laser Beam Welding on Structure of 2205 Duplex Stainless Steel. *Adv. Mater. Sci.* **2019**, *19*, 21–31. [CrossRef]
60. Kusinski, J.; Kac, S.; Kopia, A.; Radziszewska, A.; Rozmus-Górnikowska, M.; Major, B.; Major, L.; Marczak, J.; Lisiecki, A. Laser modification of the materials surface layer—A review paper. *Bull. Pol. Acad. Sci. Tech. Sci.* **2012**, *60*, 711–728. [CrossRef]
61. Adesina, O.; Popoola, P.A.I.; Fatoba, O.S. Laser Surface Modification—A Focus on the Wear Degradation of Titanium Alloy. In *Fiber Laser*; Paul, M.C., Ed.; InTech: Rijeka, Croatia, 2016; pp. 367–381.
62. Jażdżewska, M. Effects of Co2 and Nd:YAG Laser Remelting of the Ti6Al4V Alloy on the Surface Quality and Residual Stresses. *Adv. Mater. Sci.* **2020**, *20*, 82–90. [CrossRef]
63. Fraczek-Szczypta, A.; Dlugon, E.; Weselucha-Birczynska, A.; Nocun, M.; Blazewicz, M. Multi walled carbon nanotubes deposited on metal substrate using EPD technique. A spectroscopic study. *J. Mol. Struct.* **2013**, *1040*, 238–245. [CrossRef]
64. Oliver, W.C.; Pharr, G.M. An improved technique for determining hardness and elastic modulus using load and displacement sensing indentation experiments. *J. Mater. Res.* **1992**, *7*, 1564–1583. [CrossRef]
65. Łatka, L.; Cattini, A.; Chicot, D.; Pawlowski, L.; Stefan, K.; Petit, F.; Denoirjean, A. Mechanical Properties of Yttria-and Ceria- Stabilized Zirconia Coatings Obtained by Suspension Plasma Spraying. *J. Therm. Spray Technol.* **2013**, *22*, 125–130. [CrossRef]
66. Oliver, W.C.; Pharr, G.M. Measurement of hardness and elastic modulus by instrumented indentation: Advances in understanding and refinements to methodology. *J. Mater. Res.* **2004**, *19*, 3–20. [CrossRef]
67. Hamedi, M.J.; Torkamany, M.J.; Sabbaghzadeh, J. Effect of pulsed laser parameters on in-situ TiC synthesis in laser surface treatment. *Opt. Lasers Eng.* **2011**, *49*, 557–563. [CrossRef]
68. Seo, D.M.; Hwang, T.W.; Moon, Y.H. Carbonitriding of Ti-6Al-4V alloy via laser irradiation of pure graphite powder in nitrogen environment. *Surf. Coat. Technol.* **2019**, *363*, 244–254. [CrossRef]
69. Courant, B.; Hantzpergue, J.-J.; Benayoun, S.; L'Huillier, J.-P. Melting and solidification processes in a moving graphite-covered titanium surface subjected to multi-pulse laser irradiation. *J. Phys. Appl. Phys.* **2001**, *34*, 1437–1446. [CrossRef]
70. ASM International. *ASM Handbook*, 10th ed.; ASM International: Materials Park, OH, USA, 1990; Volume 3.
71. Dobrzańska-Danikiewicz, A.D.; Łukowiec, D.; Cichocki, D.; Wolany, W. Carbon nanotubes manufacturing using the CVD equipment against the background of other methods. *Arch. Mater. Sci. Eng.* **2013**, *64*, 103–109.
72. Lohse, B.H.; Calka, A.; Wexler, D. Raman spectroscopy as a tool to study TiC formation during controlled ball milling. *J. Appl. Phys.* **2005**, *97*, 114912. [CrossRef]
73. Su, Y.; Luo, C.; Zhang, Z.; Hermawan, H.; Zhu, D.; Huang, J.; Liang, Y.; Li, G.; Ren, L. Bioinspired surface functionalization of metallic biomaterials. *J. Mech. Behav. Biomed. Mater.* **2018**, *77*, 90–105. [CrossRef]
74. Khandan, A.; Abdellahi, M.; Ozada, N.; Ghayour, H. Study of the bioactivity, wettability and hardness behaviour of the bovine hydroxyapatite-diopside bio-nanocomposite coating. *J. Taiwan Inst. Chem. Eng.* **2016**, *60*, 538–546. [CrossRef]
75. Menzies, K.L.; Jones, L. The Impact of Contact Angle on the Biocompatibility of Biomaterials. *Optom. Vis. Sci.* **2010**, *87*, 387–399. [CrossRef] [PubMed]
76. Lee, J.H.; Khang, G.; Lee, J.W.; Lee, H.B. Interaction of Different Types of Cells on Polymer Surfaces with Wettability Gradient. *J. Colloid Interface Sci.* **1998**, *205*, 323–330. [CrossRef] [PubMed]

77. Heise, S.; Forster, C.; Heer, S.; Qi, H.; Zhou, J.; Virtanen, S.; Lu, T.; Boccaccini, A.R. Electrophoretic deposition of gelatine nanoparticle/chitosan coatings. *Electrochim. Acta* **2019**, *307*, 318–325. [CrossRef]
78. Cordero-Arias, L.; Cabanas-Polo, S.; Gao, H.; Gilabert, J.; Sanchez, E.; Roether, J.A.; Schubert, D.W.; Virtanen, S.; Boccaccini, A.R. Electrophoretic deposition of nanostructured-TiO2/chitosan composite coatings on stainless steel. *RSC Adv.* **2013**, *3*, 11247. [CrossRef]

materials

MDPI

Article

Laser Surface Modification of Aluminium Alloy AlMg9 with B_4C Powder

Marek Sroka *, **Ewa Jonda** and **Wojciech Pakieła**

Department of Engineering Materials and Biomaterials, Silesian University of Technology, Konarskiego St. 18a, 44-100 Gliwice, Poland; ewa.jonda@polsl.pl (E.J.); wojciech.pakiela@polsl.pl (W.P.)
* Correspondence: marek.sroka@polsl.pl; Tel.: +48-32-237-18-47

Received: 5 December 2019; Accepted: 10 January 2020; Published: 15 January 2020

Abstract: This paper presents the effects of laser treatment (fiber laser YLS-4000) on the microstructure and selected mechanical properties of the surface layer of AlMg (AlMg9) foundry alloy obtained by alloying with boron carbide (B_4C). The correlation between laser alloying process parameters and selected properties of the formed layer was discussed. The studies were supported by microstructural analysis of the remelted zone (RZ), heat affected zone (HAZ), undissolved carbide particles, substrate material, and precipitates formed during rapid solidification. Metallographic investigations of the laser-treated layer were performed using optical microscopy and scanning electron microscopy (SEM). The elemental composition and a detailed analysis of chemical composition in micro-areas were carried out using energy dispersive X-ray spectroscopy (EDS). The remelting thickness, heat-affected zone (HAZ), and amount of base material in surface layers were determined. Microhardness tests were performed on transverse cross-sections of the remelted zone to obtain the hardness profiles in the base material (BM), remelted zone (RZ), and heat affected zone (HAZ). The hardness, roughness, and wear resistance measurements showed that the highest tribological properties of the obtained surface layer were achieved using 0.5 Bar protective gas (Ar) during alloying with B_4C powder.

Keywords: aluminium alloy; laser surface alloying; microstructure; wear resistance; boron carbide

1. Introduction

The good ductility and lightweight nature of aluminium and its alloys have permitted their broad use in aerospace, automotive, and transportation industries. Aluminium also has high thermal and electrical conductivities, good machinability, and is easily recycled; however, this group of materials has relatively low mechanical and wear properties. Because of this, there is a need to improve the functional properties of these materials [1–4]. One of the primary methods to obtain materials with better mechanical properties is surface treatment technology. Laser surface alloying (LSA) is used for lightweight metals to improve their properties because the surface layer formed on the metal has different properties than the substrate material, for example, higher hardness, fatigue, and corrosion resistance; however, the surface is usually rougher than the original alloyed material [2]. LSA consists of enriching the surface layer with alloying elements, accompanied by structural changes. Usually, the alloying elements used in the laser treatment are metal alloys, superalloys, stellits, carbides, borides, and nitrides. LSA involves simultaneously melting and mixing the alloying material containing the alloyed additions with the treated material (base material) [5–10]. The laser beam fuses the base material, and a pool of remelted materials is created. Owing to convection and gravitation movements and the pressure of the laser beam, the materials are intensively mixed, and the properties of the formed layer depend on the microstructure, porosity, and chemical composition of the base material [11–16]. Boron carbide (B_4C) has a high hardness (the third hardest material behind diamond and boron nitride), wear resistance, thermal conductivity, and melting temperature; however, its low strength (about

200–400 MPa) and low fracture toughness (2–3 MPa/m0.5), as well as poor sinterability, can significantly restrict its industrial applications [17,18].

Tian et al. [19] reported the effect of laser treatment parameters on the microstructure, microhardness, and wear resistance of pure titanium alloyed with B_4C and Ti. The authors found that the depth of the remelted zone increased at lower scanning speeds. Additionally, the microhardness of the surface layer measured in cross-sections as a function of distance from the sample front decreased as the remelting zone depth decreased in a gradient. It has been reported that, compared with non-laser-treated surface materials, alloyed layers have excellent wear resistance, as well as a lower friction coefficient. Yilbas at al. [20] studied laser controlled melting of pre-prepared H12 hot work tool steel surface with B_4C particles. The authors found that laser treatment reduced the friction coefficient of the surface layer, and the microhardness of the alloyed layer increased owing to the formation of nitrides, fine grains, and microstresses near B_4C particles. In another case, Yilbas at al. [21] investigated the effect of laser surface modification treatment of aluminium bronze with B_4C. The authors showed that the laser-treated surface was free of cracks, voids, and cavities, and the microhardness of the treated surface was significantly higher. Hlawka at al. [22] investigated chromium–molybdenum steel AISI 4135H surface hardened by laser melting with injected hard particles or by laser alloying using boron carbide or boron. They showed that laser melting of boron coatings produced very fine, uniformly-distributed microstructures in a remelting zone (RZ), and the surface had a good homogeneity without pores or cracks. The hardness was also higher than the substrate before laser treatment.

Tests were also conducted to cover the product with corrosion-resistant and harder phases (e.g., Al_2O_3) [23]. The effect of simultaneous melting and feeding of biphasic tungsten carbide WC/W_2C particles into the molten pool on the structure and mechanical properties of ENAC-AlMg9 aluminium alloy was investigated [24]. For laser alloying, Cu, Mg, and Mn powders added to 98.6% aluminium using a CO_2 laser were also used [25]. In contrast, Irek [26] presented the results of research on aluminium alloy AlSi7Cu4MgMn subjected to laser alloying using silicon carbide.

In spite of plenty of research, there is still not enough information about the microstructure and properties of the modification of surface layers ENAC-AlMg9 by laser alloying with the use of boron carbide (B_4C). This manuscript is going to be an attempt to fill this gap as a current topic, from both a scientific and an application point of view.

2. Experimental Procedure

Investigations were carried out on test pieces from the casting aluminium alloy with magnesium ENAC-AlMg9 (Institute of Non-Ferrous Metals in Gliwice, Skawina, Poland). The chemical composition of the alloy is shown in Table 1, and the microstructure of the aluminium alloy used in the laser surface treatment is shown in Figure 1. The microstructure of the AlMg9 alloy in the casting state consists of the primary aluminium phase α-Al, which is the matrix of the alloy eutectic phase (Al + Mg_2Si) and β-Al_8Mg_5. The development of the Al_8Mg_5 phase was observed at the boundaries of eutectic cells—between primary aluminium dendrites. The stoichiometric composition of the Mg_2Si phase is 66.6 at% Mg and 33.4 at% Si [27]. To improve the properties of the surface layer, boron carbide (B_4C) (Kamb Import-Export, Warsaw, Poland) powder was applied, which had the properties listed in Table 2. The average particle size of the powder was in the range of 63–106 μm. The carbide shapes determined with scanning electron microscopy (SEM) + energy dispersive X-ray spectroscopy (EDS) analysis, are shown in Figure 2a,b.

Table 1. Chemical composition of aluminium alloy EN AC–AlMg9 (in wt. %).

Elements	Si	Mn	Zn	Mg	Al
AlMg9	1.32	0.50	0.20	9.24	REST

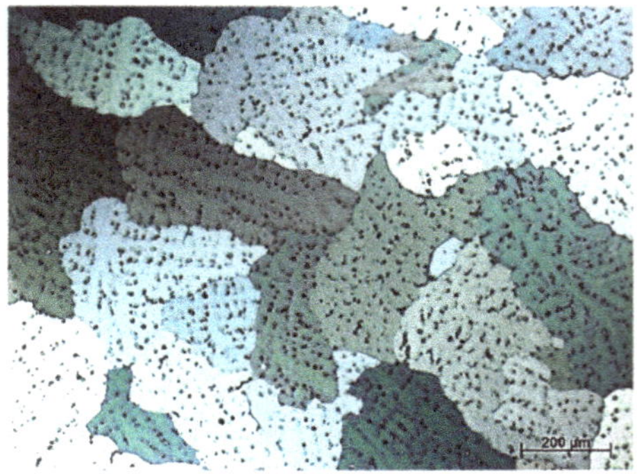

Figure 1. Microstructure of the AlMg9 aluminium alloy.

Table 2. Typical properties of B_4C boron carbide powder [28].

Property	Value
Density, g/cm^3	2.52
Melting point, °C	2445
Knoop hardness (100g), kg/mm^2	2900–3580
Young's modulus, GPa	450–470
Electrical conductivity (at 25 °C)	1.40

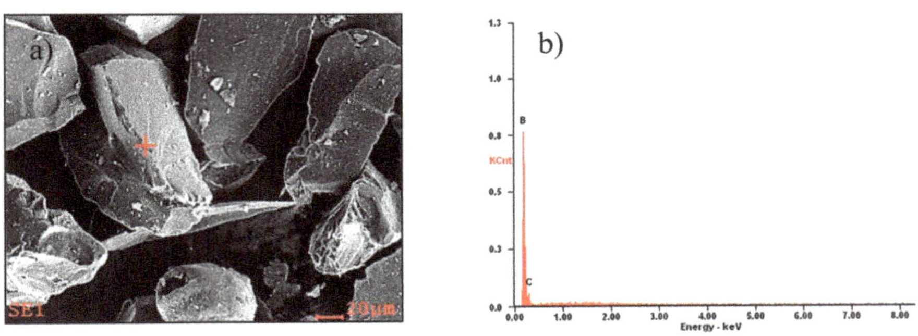

Figure 2. (**a**) Microstructure of the B_4C powder (scanning electron microscopy (SEM)); (**b**) energy dispersive X-ray spectroscopy (EDS) analysis of the chemical composition of the B_4C particle.

A fiber laser (FL) Ytterbium Laser System YLS-4000 (IPG Photonics Corporation, Oxford, MA, USA) was used for surface alloying, with a wavelength $\lambda = 1070$ nm, and a maximum laser beam power of 4000 W mounted on a six-axis robot REIS RV30-26 (Reis Robotics, Obernburg, Bavaria, Germany). The laser surface treatment was carried out under a shielding Ar gas to protect the molten weld pool. On the basis of preliminary experimental research regarding the impact of the shielding gas used on the depth and depth of the melted zone and the heat-affected zone for further studies, the best parameters were selected. The laser surface treatment was carried out using a constant alloying scanning rate of laser 0.2 m/min and laser beams power of 1.5 kW. The laser alloying parameters are shown in Table 3.

Table 3. Laser alloying parameters.

Parameter	Value
Laser beam power, kW	1.5
Protective gas, Bar	0.4
	0.5
The share of supplied powder, g/min	15
Laser beam scanning speed, m/min	0.2
Circle spot, mm	5
Wavelength λ, nm	1070

The topography of the alloyed surface was observed using a Zeiss stereomicroscope SteREO Discovery (Zeiss, Oberkochen, Germany) with magnification in the range of 10–100X. Specimens for metallographic observations were prepared by standard polishing techniques. Grinding at 25 N load successively on papers with grain gradation 120, 600, 1200, and 4000 and polishing with a colloidal suspension based on silicon oxide on a disc made of nephron rubber (MD-Chem). Electric etching was done in HBF_4 acid (5% solution) for 20 s. Metallographic investigations done made using light microscopy with an Axio Observer and a Zeiss Supra 35 SEM (Zeiss, Oberkochen, Germany) using secondary electron and backscattered detectors. The chemical composition was analyzed by EDS. Hardness changes across the laser runs versus distance from the surface were investigated using the Vickers microhardness test method with a force of 500 gf. Hardness tests were performed along lines perpendicular to specimen surfaces, along the run face axis.

The resistance of the surface layers without laser treatment and after alloying with boron carbide was analyzed and compared using the "ball-on-plate" tribological test. As a counter-specimen, a 6 mm diameter ball of aluminium oxide Al_2O_3 was used. During the test, the friction coefficient between the investigated surface and ceramic counter was recorded. The test was performed at room temperature using the testing conditions in Table 4. The wear track dimensions after tests were measured by a Sutronic 25-Taylor Hobson profilometer (Taylor Hobson Ltd., Leicester, England), and the topography was analyzed using SEM to locate rifts and deformations on the surface layer owing to laser alloying with B_4C carbide. The roughness of the investigated surface layer was also measured by a Sutronic 25-Taylor-Hobson profilometer.

Table 4. Testing conditions of the "ball-on-plate" method.

Parameter	Value
Load, N	15
Linear speed, cm/s	5
Distance, m	50
Measuring distance, mm	6
Counter specimen	ball Al_2O_3

3. Results and Discussion

On the basis of the analysis performed here, the surface layer obtained owing to alloying an aluminium alloy with B_4C powder was composed of three zones: a laser remelting zone (RZ), an enriched in boron carbide zone, a melted and rapidly solidified zone, and a heat-affected zone (HAZ). On the basis of preliminary experimental research regarding the impact of the shielding gas used on the depth and depth of the melted zone and the heat-affected zone for further studies, the best parameters were selected. The depth of the remelting layer obtained using 0.4 Bar of protective gas was about 1450 ± 18 μm, and the width was about 3833 ± 78 μm. When 0.5 Bar protective gas was used, the depth of the remelted layer was about 1642 ± 44 and the width was about 4132 ± 61 μm. The total surface layer thickness and width of both of the remelted zone and heat-affected zone grew when the pressure of the applied protective gas increased. Preliminary investigations of the alloyed aluminium ENAC–AlMg9 showed a clear effect of the laser treatment on the shape of the remelted material, the obtained run face showed characteristic flashes at the borders; however, no pores, cavities, or cracks were observed. The topography of the layers obtained by laser alloying with B_4C powder are presented in Figure 3a,b.

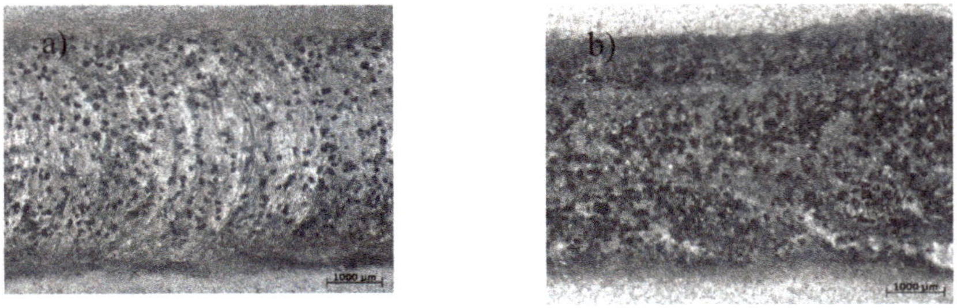

Figure 3. Topography of the layers obtained during laser alloying with B_4C powder: (**a**) 0.4 Bar of protective gas; (**b**) 0.5 Bar of protective gas.

The roughness measurements show that during alloying ENAC–AlMg9 with B_4C, the obtained run faces at both protective gas pressures had a higher roughness than the base material (BM) (average roughness of base material—*Ra* = 0.27 μm). The average roughness after alloying with 0.4 Bar of protective gas was 2.25 μm, and 4.82 μm with 0.5 Bar (Figure 4). The increase in roughness is closely related to the carbide amount introduced into the treated surface of the substrate material and the effect of shielding gas on the liquid metal, thus causing an increase in waviness. It should be emphasized that the roughness of the surfacing layer can be reduced to the desired value using grinding procedures and not as it is the case with physical vapour deposition (PVD) or chemical vapour depositinon (CVD) layers [29].

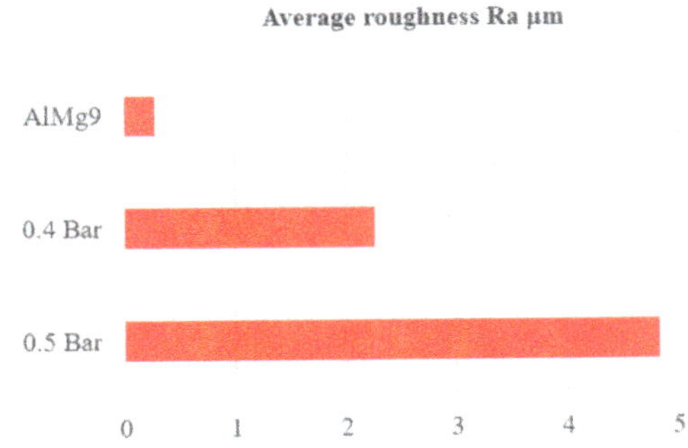

Figure 4. Effect of the laser alloying on surface layer roughness of the AlMg9 alloyed with B_4C powder.

The microstructure of the solidified material after laser alloying contained areas with diverse morphologies owing to crystallization of the alloyed material (Figure 5a–d). When using 0.5 Bar protective gas, more presence of carbides as compared with 0.4 Bar Ar was observed. SEM observations showed that the applied B_4C powder was evenly distributed in the remelted area (Figure 6), and the precipitates contained 90.83 wt% boron (Figure 7). In addition, around the disclosed carbides, zones of new separate phases were observed. EDS analysis showed that it consists of 46.70% at. coal and 53.30% at. silicon, which corresponds to the SiC phase (Figure 6 "C" and "Si"). Phases derived from substrate material rich in Mg and Al (in a ratio of 41.8/58.2% at) and Mg, Al, and Si (in a ratio of 39.2/37.9/22.9% at) were also disclosed, which correspond to the phases Al_8Mg_5 and Mg_2Si.

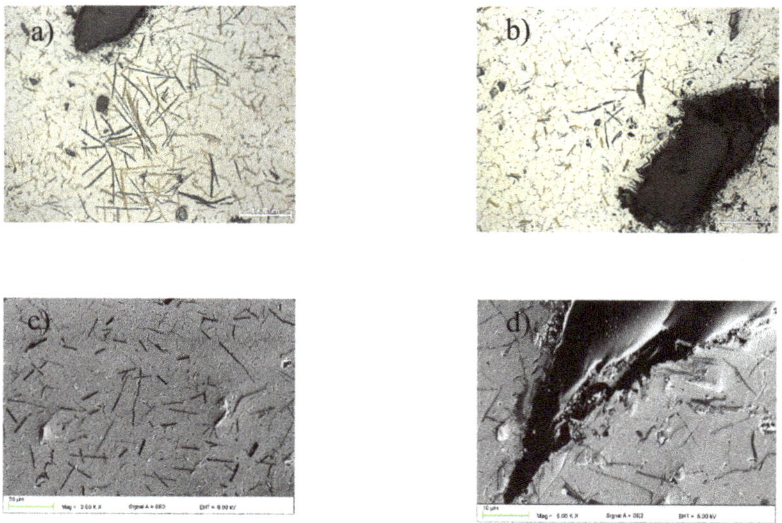

Figure 5. Microstructure of the layer obtained during the laser alloying with B_4C powder (**a**,**c**) 0.4 Bar; (**b**,**d**) 0.5 Bar.

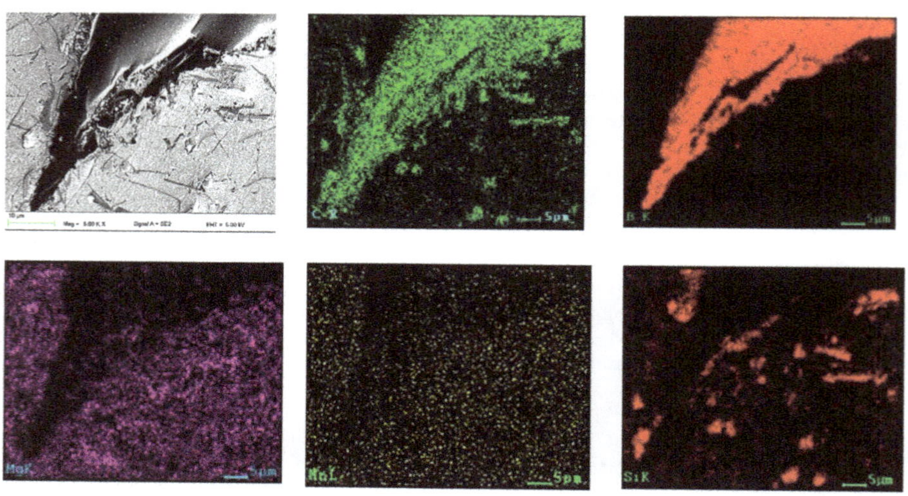

Figure 6. Elemental distribution maps of alloying elements in the analyzed area of the layer obtained during the laser alloying with B_4C powder.

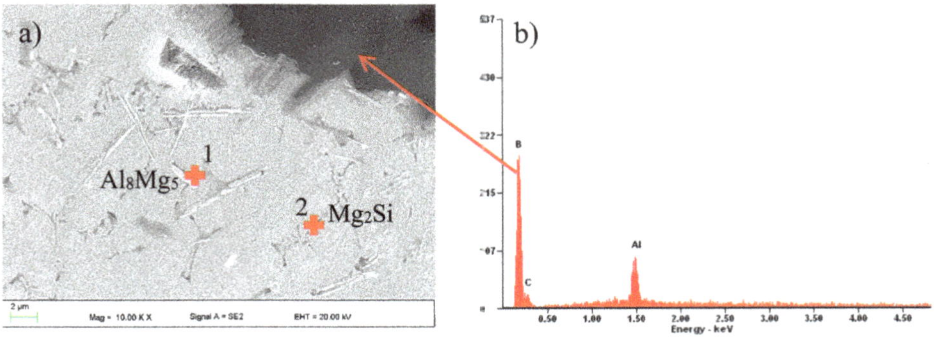

Figure 7. (**a**) Microstructure of the alloyed with B_4C layer; (**b**) EDS analysis of the chemical composition of the analyzed point.

AlMg9 without laser treatment showed a minimal friction factor, μ, of 0.72. The results showed that the coefficient of friction significantly decreased in the layers obtained during laser alloying with B_4C. In the specimen alloyed with boron carbide, the average μ value was approximately 0.52 (for 0.4 Bar protective gas), whereas for 0.5 Bar protective gas, the average μ was approximately 0.42. On the basis of the investigation results, the friction coefficient was lower and fluctuated less in samples alloyed with B_4C powder embedded in the surface layer compared with the native material. Initially, owing to the presence of partially-embedded boron carbide on the alloyed surface layer for all samples treated by a laser beam, the friction coefficient increased slightly. A decrease in the coefficient of friction was also observed for composites Al–B_4C by Mazaheri et al. [30].

The SEM topography observations of the substrate material abrasion showed various tribological wear mechanisms (Figures 8 and 9). The most intense and dominant mechanism in all cases was abrasion wear (Figures 8a and 9a,b). In addition, very intense delamination and plastic deformation were observed in the substrate (Figure 8a). The above mechanisms were also observed on the wear debris surface (Figure 8b). The analysis of the wear trace of the layer and wear debris (powder wear from the layer/wear product) showed no abrasion of large boron carbide particles debonded from the substrate, indicating good adhesion to aluminium. Wear debris observations showed significant differences in shape and size. Wear debris of the substrate occurred in the form of large flakes with sizes larger than 500 μm (Figure 8b), whereas the layers showed a mixture of fine dust and flakes smaller than 100 μm (Figure 9d), indicating much more even and stable wear. In addition, EDS microanalysis confirmed the occurrence of oxidation both on the surface of the wear trace of the layer and substrate material (Figure 10). Additionally, numerous agglomerations of fine oxidized wear debris smaller than 0.5 μm were observed on the surface of the wear trace of the composite layer. The wear trace dimensions after the "ball-on-plate" wear tests are shown in Table 5. Baradeswaran et al. [31] also showed a significant decrease in wear along with the increase in the participation of B_4C carbide in alloy AA7075. Tribological wear of the composite with 10% carbide accounted for about 11% of the base material wear, which confirms the trend occurring in the case of laser alloying. In addition, the disappearance of plastic deformation was observed with the increase of B_4C.

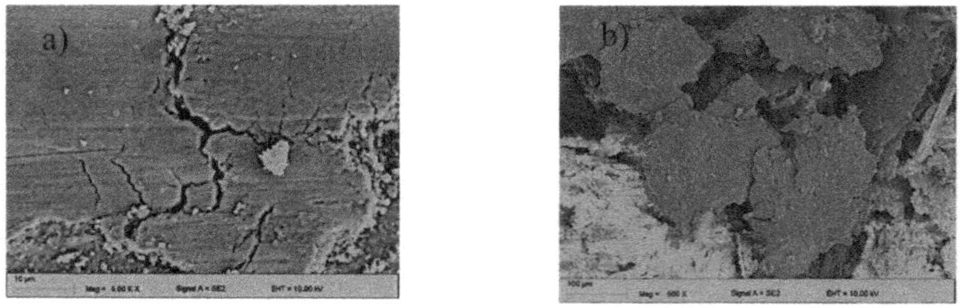

Figure 8. Wear trace (**a**,**b**) and the wear product (**b**) of the AlMg9 after the "ball-on-plate" wear test.

Table 5. The dimensions of the wear track of analyzed materials after the "ball-on-plate" wear test.

Substrate			**Dimensions of the Wear Track**		
			Volume, μm^2	**Width, mm**	**Depth, μm**
AlMg9			193,603	1.99	55.5
AlMg9 +B_4C	Protective gas, Bar	0.4	38,619	1.15	48.7
		0.5	26,313	1.2	43.2

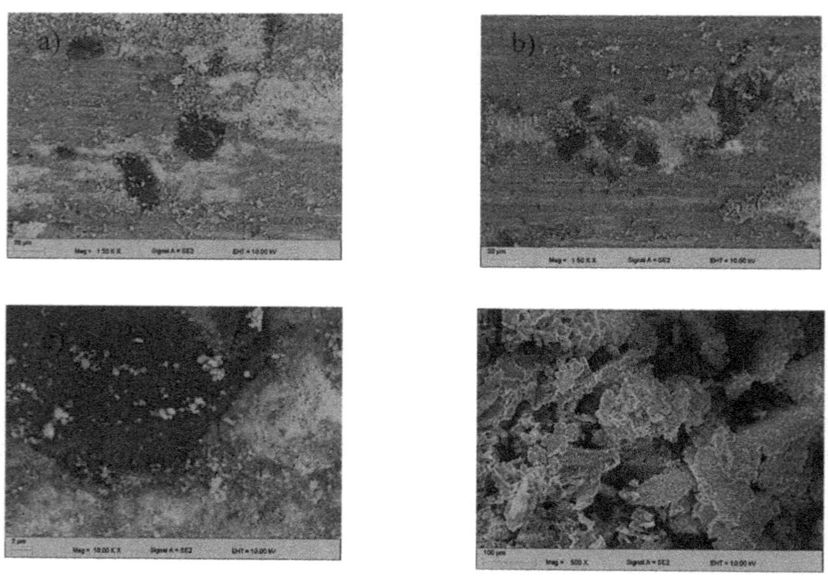

Figure 9. Wear track (**a**) 0.4 Bar, (**b**) 0.5 and the wear product, (**c**) 0.4 Bar, and (**d**) 0.5 of the sample alloyed with B_4C powder after the "ball-on-plate" wear test.

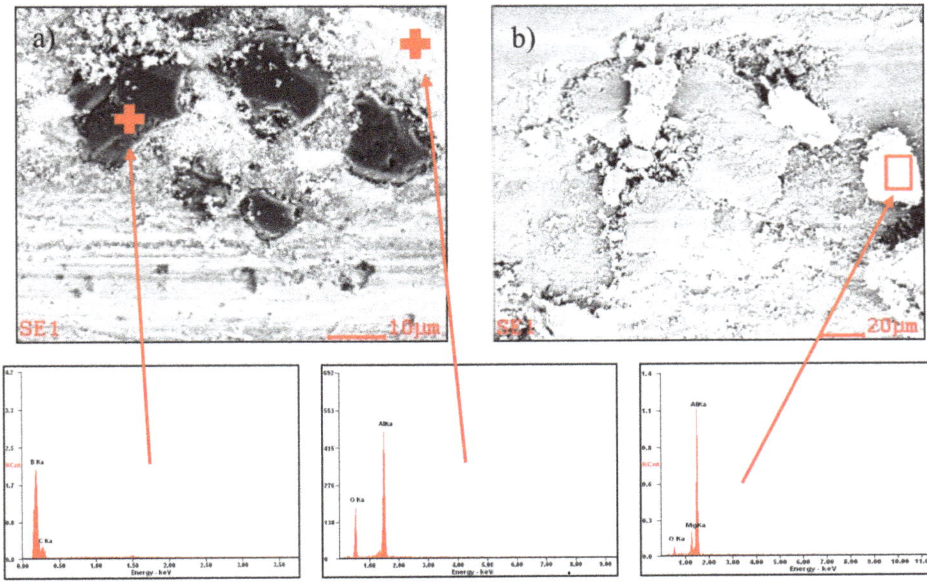

Figure 10. EDS microanalysis of the composite layer wear trace (**a**) and parental material (**b**).

The hardness tests revealed that, when laser alloying with boron carbide, the resulting hardness was higher than the material before laser treatment. The measured microhardness along the depth of the cross-section of the solidified pool indicated a hardness increase to 128 $HV_{0.5}$ (for 0.4 Bar protective gas) and 131 $HV_{0.5}$ (for 0.5 Bar protective gas) only in carbide-containing areas at a depth of about 0.15 mm. The hardness drastically decreased in the entire heat-affected zone (HAZ) and along the border of the base material (BM). The hardness at a depth of 1.75 mm across the top surface layer ranged from 99–96 (for 0.4 Bar protective gas) and 93–99.9 (for 0.5 Bar protective gas) (Figure 11). The increase in hardness in the remelting area is caused by the presence of evenly distributed carbides and the fragmentation of precipitations coming from the substrate material. Baradeswaran et al. [31], introducing boron carbide particles into the 7057 alloy, obtained finally about 225 HB for the composite with 20% carbide. Boron carbide hardness is in the range of 2900–3900 kg/mm2, while the base material is slightly above 90 HV [32].

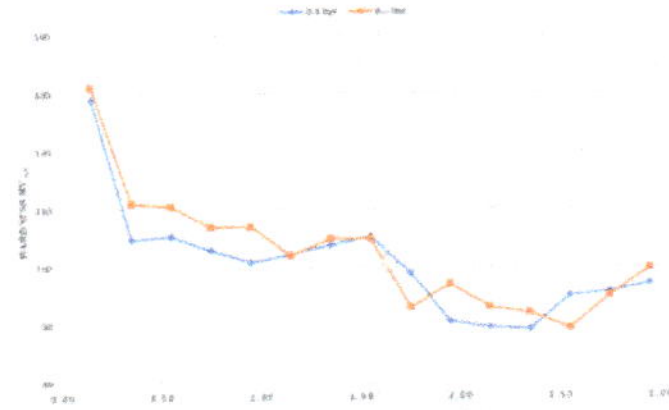

Figure 11. Profile of microhardness changes of the AlMg9 surface layer after laser alloying with B_4C.

4. Conclusions

On the basis of the tests carried out on the AlMg9 alloy subjected to laser surface modification by rapid remelting and rapid solidification, the following conclusions can be made:

- The surface layer rich in alloying elements had a higher hardness than the substrate (128 $HV_{0.5}$ for 0.4 Bar protective gas) and 131 $HV_{0.5}$ (for 0.5 Bar protective gas).

- The abrasion resistance increased owing to an increase in the hardness of the surface layer. The obtained layers had friction coefficients of 0.52 (for 0.4 Bar protective gas) and 0.42 (for 0.5 Bar protective gas).
- An increase in tribological properties and a rougher surface (Ra = 4.82 μm) were found compared with the substrate before alloying (Ra = 0.27 μm).
- The hardness, roughness, and wear resistance measurements showed that the highest tribological properties of the obtained surface layer were achieved using 0.5 Bar protective gas (Ar) during alloying with B_4C powder.

Author Contributions: E.J. conceived and designed the experiments; W.P. performed the experiments; M.S. analyzed the data; E.J. and M.S. wrote the paper—review and editing. All authors have read and agreed to the published version of the manuscript.

Funding: The publication was partially financed by rector's grant in the area of scientific research and development works, Silesian University of Technology, 10/010/RGJ18/0201.

Conflicts of Interest: The authors declare no conflicts of interest.

References

1. Labisz, K.; Tański, T.; Brytan, Z.; Pakieła, W.; Wiśniowski, M. Aluminium surface treatment with ceramic phases using diode laser. *Appl. Phys. A Mater.* **2016**, *122*, 1–10. [CrossRef]
2. Tański, T.; Labisz, K.; Brytan, Z.; Jonda, E.; Sroka, M. Thermal fatigue influence of laser treated tool steel surface. *Procedia Eng.* **2014**, *74*, 429–442. [CrossRef]
3. Tański, T.; Snopiński, P.; Prusik, K.; Sroka, M. The effects of room temperature ECAP and subsequent aging on the structure and properties of the Al-3%Mg aluminium alloy. *Mater. Charact.* **2017**, *133*, 185–195. [CrossRef]
4. Pakieła, W.; Tański, T.; Pawlyta, M.; Pakieła, K.; Brytan, Z.; Sroka, M. The structure and mechanical properties of AlMg5Si2Mn alloy after surface alloying by the use of fiber laser. *Appl. Phys. A Mater.* **2018**, *124*, 263. [CrossRef]
5. Jonda, E.; Brytan, Z.; Labisz, K.; Drygała, A. The influence of laser surface alloying on the thermal fatigue resistance of the hot work tool steels. *Arch. Metall. Mater.* **2016**, *61*, 963–968. [CrossRef]
6. Kulka, M.; Mikolajczak, D.; Makuch, N.; Dziarski, P.; Miklaszewski, A. Wear resistance improvement of austenitic 316L steel by laser alloying with boron. *Surf. Coat. Technol.* **2016**, *291*, 292–313. [CrossRef]
7. Kwok, C.T.; Wong, P.K. Laser surface alloying of various engineering alloys for sliding wear and corrosion resistance. *J. Laser Micro/Nanoeng.* **2010**, *5*, 90–96. [CrossRef]
8. Lisiecki, A. Titanium Matrix Composite Ti/TiN Produced by Diode Laser Gas Nitriding. *Metals* **2015**, *5*, 54–69. [CrossRef]
9. Kulka, M.; Makuch, N.; Pertek, A.; Piasecki, A. Microstructure and properties of borocarburized and laser-modified 17CrNi6-6 steel. *Opt. Laser Technol.* **2012**, *44*, 872–881. [CrossRef]
10. Rozmus-Gornikowska, M.; Kusinski, J.; Blicharski, M. The influence of the laser treatment on microstructure of the surface layer of an (X5CrNi18-10) austenitic stainless steel. *Arch. Metall. Mater.* **2011**, *56*, 717–721. [CrossRef]
11. Lisiecki, A. Study of Optical Properties of Surface Layers Produced by Laser Surface Melting and Laser Surface Nitriding of Titanium Alloy. *Materials* **2019**, *12*, 3112. [CrossRef] [PubMed]
12. Janicki, D. Improvement of wear resistance of stainless steel AISI 304L by diode laser surface alloying with chromium carbide. *Appl. Mech. Mater.* **2015**, *809–810*, 363–368. [CrossRef]
13. Tanski, T.; Snopinski, P.; Borek, W. Strength and structure of AlMg3 alloy after ECAP and post-ECAP processing. *Mater. Manuf. Process.* **2017**, *32*, 1368–1374. [CrossRef]
14. Król, M.; Staszuk, M.; Mikuszewski, T.; Kuc, D. Refinement effect of RE in light weight Mg-Li-Al alloys. *J. Therm. Anal. Calorim.* **2018**, *134*, 333–341. [CrossRef]
15. Shinn, M. Basics of Lasers and Laser Optics. In *Laser Processing of Materials Fundamentals Applications and Developments*, 1st ed.; Schaaf, P., Ed.; Springer: Berlin/Heidelberg, Germany, 2010; Volume 139, pp. 5–20.
16. Tański, T. Determining of laser surface treatment parameters used for light metal alloying with ceramic powders. *Mater. Werkst.* **2014**, *45*, 333–343. [CrossRef]

17. Rao, S.; Padmanabhan, G. Fabrication and mechanical properties of aluminium–boron carbide composites. *Int. J. Mater. Biomater. Appl.* **2012**, *2*, 15–18.
18. Ogunlana, M.O.; Akinlabi, E.T. Surface Effect of Laser Power on Microstructural Evolution and Hardness Behaviour of Titanium Matrix Composites. In Proceedings of the World Congress on Engineering, London, UK, 29 June–1 July 2016; Volume 2.
19. Tian, Y.S.; Chen, C.Z.; Chen, L.B.; Liu, J.H. Wear properties of alloyed layers produced by laser surface alloying of pure titanium with B4C and Ti mixed powders. *J. Mater. Sci.* **2005**, *10*, 4387–4390. [CrossRef]
20. Yilbas, B.S.; Patel, F.; Karatas, C. Laser controlled melting of H12 hot-work tool steel with B_4C particles at the surface. *Opt. Laser Technol.* **2015**, *74*, 36–42. [CrossRef]
21. Yilbas, B.S.; Matthews, A.; Leyland, A.; Karatas, C.: Akhtar, S.S.; Aleem, B.J. Laser surface modification treatment of aluminium bronze with B4C. *Appl. Surf Sci.* **2012**, *263*, 804–809. [CrossRef]
22. Hlawka, F.; Cornet, A. Laser surface alloying with a CO2 laser: Surface hardening of AISI 4135 H chromium–molybdenum steel. *Adv. Eng. Mater.* **2003**, *5*, 641–647. [CrossRef]
23. Borowski, J. Obróbka laserowa warstw wierzchnich stopów aluminium. *Projektowanie i Konstrukcje Inżynierskie* **2013**, *3*, 1–4. (In Polish)
24. Pakieła, W.; Dobrzański, L.A.; Labisz, K.; Tański, T.; Basa, K.; Roszak, M. The effect of laser Surface treatment on structure and mechanical properties aluminium alloy ENAC-AlMg9. *Arch. Metall. Mater.* **2016**, *61*, 1343–1350. [CrossRef]
25. Jiru, W.G.; Sankar, M.R.; Dixit, U.S. Laser Surface Alloying of Copper, Manganese, and Magnesium with Pure Aluminium Substrate. *J. Mater. Eng. Perform.* **2016**, *25*, 1172–1181. [CrossRef]
26. Irek, P. Laser surface alloying of AlSi7Cu4MgMn aluminium alloy with silicon carbide. *Inst. Weld. Bull.* **2014**, *58*, 42–48.
27. Snopiński, P.; Król, M.; Tański, T.; Krupińska, B. Effect of cooling rate on microstructural development in alloy ALMG9. *J. Therm. Anal. Calorim.* **2018**, *133*, 379–390. [CrossRef]
28. Boron Carbide (B4C)—Properties and Information about Boron Carbide. Available online: https://www.azom.com/article.aspx?ArticleID=75 (accessed on 28 December 2019).
29. Harlin, P.; Carlsson, P.; Bexell, U.; Olsson, M. Influence of surface roughness of PVD coatings on tribological performance in sliding contacts. *Surf. Coat. Technol.* **2006**, *201*, 4253–4259. [CrossRef]
30. Mazaheri, Y.; Meratian, M.; Emadi, R.; Najarian, A.R. Comparison of microstructural and mechanical properties of Al–TiC, Al–B_4C and Al–TiC–B_4C composites prepared by casting techniques. *Mater. Sci. Eng. A Struct.* **2013**, *560*, 278–287. [CrossRef]
31. Baradeswaran, A.; Elaya Peruma, A. Influence of B4C on the tribological and mechanical properties of Al 7075–B4C composites Composites. *Engineering* **2013**, *54*, 146–152. [CrossRef]
32. Zhu, X.; Dong, H.; Lu, K. Coating different thickness nickel–boron nanolayers onto boron carbide particles. *Surf. Coat. Technol.* **2008**, *202*, 2927–2934. [CrossRef]

materials

MDPI

Article

Investigation on the Microstructure and Wear Behavior of Laser-Cladded High Aluminum and Chromium Fe-B-C Coating

Jingjing Li [1,2], Jiang Ju [1,2], Weiwei Chang [3], Chao Yang [1,2] and Jun Wang [1,2,*]

1 School of Materials Science and Engineering, Shanghai Jiao Tong University, Shanghai 200240, China; lijingjing-41422134@sjtu.edu.cn (J.L.); jujiang1990@sjtu.edu.cn (J.J.); yangchao1987@sjtu.edu.cn (C.Y.)

2 Shanghai Key Laboratory of Advanced High-temperature Materials and Precision Forming, Shanghai Jiao Tong University, Shanghai 200240, China

3 Institute for Advanced Materials and Technology, University of Science and Technology Beijing, Beijing 100083, China; s20181310@xs.ustb.edu.cn

* Correspondence: junwang@sjtu.edu.cn

Received: 2 May 2020; Accepted: 25 May 2020; Published: 27 May 2020

Abstract: In this study, a high aluminum and chromium Fe-B-C coating was prepared using laser cladding on 2Cr13 steel substrate. The microstructure, microhardness, and wear resistance of the high aluminum and chromium Fe-B-C coating were investigated. The results show that this dense coating possesses good metallurgical bond with the substrate. The microstructure is mainly composed of α-(Fe, Cr, Al) lath martensite, orthorhombic M_2B boride, orthogonal M_3C_2, and orthorhombic M_7C_3 carbides. The microhardness of the coating can reach 620 HV which is 3.3-times higher than that (190 HV) of the substrate. The coating shows a lower friction coefficient of 0.75 than that of the substrate (1.08). The wear rates of the substrate and the coating are 0.295 mg/min and 0.103 mg/min, respectively, indicating the coating exhibits excellent wear resistance. The wear mechanism transforms severe adhesive wear and abrasive wear of the substrate to slight abrasive wear of the coating. The results can provide technical support to improve the properties of the Fe-based laser cladded coating.

Keywords: Fe-based coating; laser cladding; transmission electron microscopy (TEM); microstructure; wear resistance

1. Introduction

With the rapid development of the world economy, global energy demand is steadily on the increase [1,2]. The non-renewable fossil fuels accounts for more than 80% of the total energy, which puts forward higher request to the mining equipment [3,4]. The wear of the metal parts is one of the main problems affecting the service life of mining equipment. Thus, it is urgent and crucial to improve the wear resistance of mining equipment [5].

Laser cladding technology is a material surface modification method for improving the wear resistance of alloys [6]. It can fabricate a coating with special physical, chemical, or mechanical properties using a high energy laser beam [7]. Compared with traditional spraying processes, the laser cladding technology possesses four advantages: (1) Good metallurgical bonding with low dilution; (2) High yield and small margin for subsequent processing; (3) precise control, online control and remote manipulation for the repair of local and critical parts; and (4) low cost [8–10].

The coatings for laser cladding mainly consist of Fe-based, Co-based, Ni-based, ceramics, etc. [11]. Co-based and Ni-based coatings are characterized by high hardness, excellent wear resistance, heat resistance, and oxidation resistance. However, the cost of Co-based and Ni-based alloys is very high [12,13]. Ceramic coatings have much better properties in comparison with metal coatings, but its brittleness has always been an obstacle to its wide application [14]. Hence, Fe-based coatings attract more and more attention due to their low cost and high performance [15]. Many scholars have studied the improvement of wear resistance of materials by Fe-based coatings via laser cladding [16]. W.J. Wang et al. fabricated Fe-Cr-Ni-B composite coating with high hardness and good wear resistance on wheel and rail materials using laser cladding [17]. Zhang et al. studied the microstructure, wear resistance and corrosion resistance of Fe-Ti-V-Cr-C-CeO_2 coating, and found the coating possessed good wear and corrosion resistance [18].

H.G. Fu et al. found the addition of Al and Cr can significantly enhance the microhardness, wear resistance and oxidation resistance Fe-B coatings [19–21]. In this study, the high aluminum and chromium Fe-B-C coating was fabricated using laser cladding technology on the 2Cr23 steel substrate. The microstructure, microhardness and wear resistance of the high aluminum and chromium Fe-B-C coating was investigated.

2. Materials Fabrication and Experimental Methods

2.1. Materials Fabrication

The 2Cr13 steel (corresponding to 420 steel, US ASTM Standards) with a dimension of 150 mm × 150 mm × 10 mm was used to the substrate, its chemical compositions are 0.22 wt.% C, 12.9 wt.% Cr, 0.4 wt.% Si, 0,61 wt.% Mn, 0.016 wt.% S, 0.018 P wt.%, and the balance Fe. The microstructure of the 2Cr13 substrate is mainly composed of the α-Fe matrix and the $M_{23}C_6$ carbides distributed along the grain boundaries, as shown in Figure 1. The 2Cr13 substrate was grinded using the 1000# SiC sandpaper, and then cleaned by alcohol and acetone. High aluminum and chromium Fe-B-C powder was proportioned with pure alloy powder with a size of 60–80 μm. The prepared high aluminum and chromium Fe-B-C powder was dried in a drying oven to remove moisture, and then put into a ball mill pot in a glove box and ground at 120 r/min for 4 h. The chemical compositions of high aluminum and chromium Fe-B-C powder after mixing were measured by inductively coupled plasma mass spectrometry (ICP-MS), as shown in Table 1.

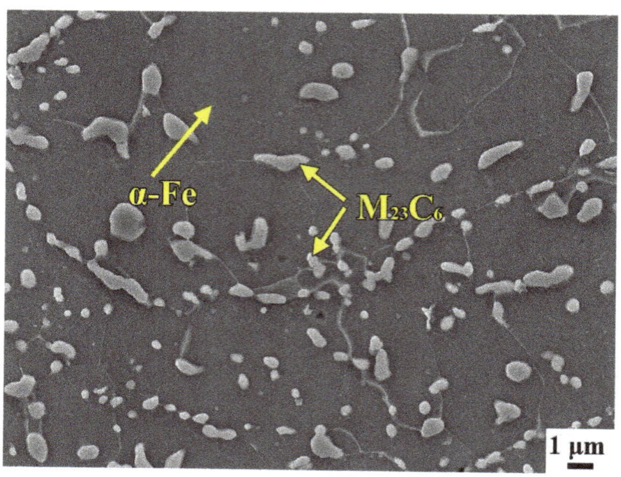

Figure 1. The microstructure image of the 2Cr13 steel substrate.

Table 1. The chemical compositions of the high aluminum and chromium Fe-B-C cladding powder (wt.%).

Elements	Al	Cr	B	Si	Mn	C	Fe
Content	6.0	10.0	1.5	0.6	0.7	0.3	Bal.

In this study, IPG-6000 optical fiber laser cladding system was used for laser cladding. First, a layer of high aluminum and chromium Fe-B-C powder with a thickness of 1 mm was prefabricated on 2Cr13 steel plate. The binder was cellulose acetate ($(C_6H_8O_3(COOCH_3)_2)_n$. Before laser cladding, the 2Cr13 substrate was preheated to 80 °C to reduce cracking. The main parameters are 2.5 kW laser power, 4 mm/s scanning speed, 15 L/min-Ar gas flow speed, and 5 mm × 5 mm spot size. The metallographic sample with a size of 10 mm × 10 mm × 5 mm was prepared by wire cutting. After grinding and polishing, it was cleaned using acetone and alcohol for 60 s.

2.2. Phase Diagram Calculation

Thermo-Calc® software was used to calculated the phase diagram. The mass percent of each element was C-0.3 wt.%, Cr-10.0 wt.%, Al-6.0 wt.%, Si-0.6 wt.%, Mn-0.7 wt.%, and the balance Fe. The content of B varied from 0.0 to 3.0 wt.%. The POLY-3 module provided initial conditions for the phase diagram calculations, such as composition, temperature, activity etc. The thermodynamic data for each substance was provided by the TCFE8 database. Furthermore, multivariate and multiphase equilibrium calculations were performed. The temperature range was 0–1750 °C, and the pressure was 105 kPa.

2.3. Microhardness and Wear Resistance Tests

The microhardness was measured using the MICRO MET-5103 Digital micrometer. The normal load was 4.9 N, the load time was 10 s. The UMT-3 reciprocating wear tester was conducted to test the wear resistance. The Al_2O_3 ceramic ball with Ø 4 mm was the grinding material, its microhardness is 1300 HV [21]. The load was 20 N, the reciprocating stroke was set as 5 mm, the frequency was 1 Hz. The wear time and the sliding speed were 60 min and 0.1 m/s, respectively. The wear resistance was evaluated using wear rate (Wr, $g^3 \cdot N^{-1} \cdot min^{-1}$) which can be calculated according to the following Equation (1) [22]:

$$W_r = \frac{w_{loss}}{F_n \times t} \quad (1)$$

where w_{loss} represents the total wear loss (mg), F_n is the normal load (N) and t is the wear time (s). The microhardness and wear tests were repeated three times for each sample to ensure more accurate experimental data.

2.4. Characterization

The samples were ground on SiC papers with granulations of 240, 400, 600, 800, 1200, 1500, 2000, and 3000 and polished using diamond suspensions. The samples were etched using 5% $FeCl_3$ solution. The microstructure and compositional analyses of various phases were performed using a Mira 3 scanning eletron microscope (SEM), transmission electron microscopy (TEM) and a JXA-8230 electron probe micro-analyzer (EPMA). The EPMA parameters were set to the 15 kV acceleration voltage, 10 nA beam current and 1 μm beam spot diameter. The standard sample was pure material (such as pure Al, Fe etc.). X-ray diffraction (XRD) was performed on a SHIMADZU Japan XRD-7000 diffractometer with copper Kα radiation coupling continuous scanning at 40 kV and 200 mA as the X-ray source. The specimen was scanned in the angular 2θ ranging from 20° to 80° with a step size of 0.2° and a collection time of 10 s.

3. Results and Discussions

3.1. Phase Diagram and Phase Structure

The vertical sections of high aluminum and chromium Fe-B-C coatings were calculated using Themo-Calc®software, as shown in Figure 2. From Figure 2, under the condition of the equilibrium solidification, Cr_2B phase precipitates directly from the liquid phase when the concentration of B is 1.5 wt.%. The eutectic reaction of L→δ-Fe + Cr_2B occurs at about 1400 °C, precipitating δ-Fe phase.

As the temperature continues to reduce, the remaining liquid phase transform into the γ-Fe, forming the δ-Fe + γ-Fe + Cr_2B three-phase zone. When the temperature drops to about 900 °C, the M_7C_3 phase gradually precipitates. It can be seen that the equilibrium microstructure consists mainly of α-Fe + Cr_2B + M_3C_2 at room temperature. In addition, the M_7C_3 and Fe_3Al phase may be obtained when the B content is less and greater than 1.5 wt.%, which also may be obtained under the rapid cooling.

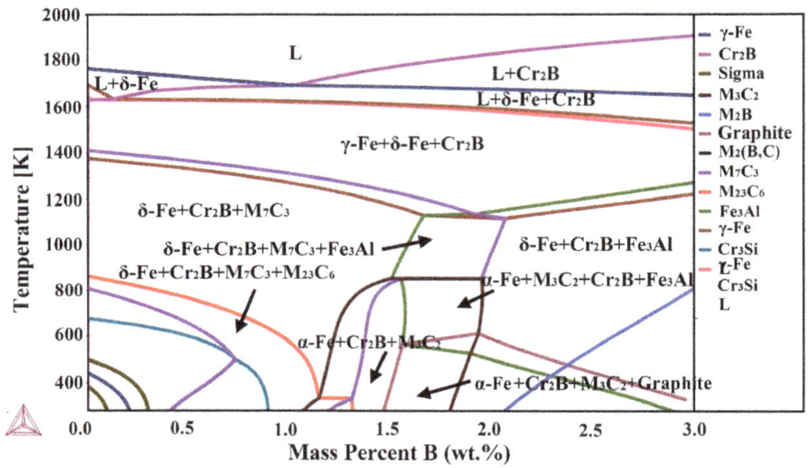

Figure 2. The calculated vertical sections of Fe-Cr-B-Al-C high aluminum and chromium Fe-B-C coatings.

3.2. Microstructure

Figure 3 shows the microstructure morphologies of high aluminum and chromium Fe-B-C coating. From Figure 3a, the coating is uniform and dense, and no defects such as pores and cracks are found. Figure 3b shows the coating has a good metallurgical bonding with the substrate. The interface bonding zone is a thick planar crystal, which gradually develops into columnar crystal and dendritic crystal of the cladding layer. The microstructure of coating mainly depends on the ratio value of temperature gradient G to the solidification rate R [23]. There is a large temperature gradient at the interface between the substrate and the molten pool, the solidification rate R tends to zero, this makes the G/R value tends to infinity. Therefore, the microstructure grows epitaxy as a planar crystalline from the substrate. As the solidification proceeds, the temperature gradient decreases due to the decreasing molten pool temperature, leading to the dropping of the G/R value. The microstructure changes from planar crystal to columnar crystal. The dendritic crystals are formed in the middle and upper part of the coating with a decrease of the G/R value, as shown in Figure 3c.

Figure 4 presents the X-ray pattern results of the coating. The matrix occupies the peak location (2θ = 44.49°, 64.31°, and 81.97°) of the α-Fe, which is slightly less than the peak values (2θ = 44.69°, 64.96°, and 82.36°) of α-Fe in 2Cr13 steel [24]. The compositions of various phases were detected by EPMA, as shown in Table 2. It can be seen that the elements Cr, Al, etc. are dissolved into the α-Fe. Moreover, the atomic radii of Fe and Cr are 118 and 117 pm respectively, which is very helpful for Cr to be solidly dissolved in α-Fe, resulting in lattice mismatches, leading to a left shift of 2θ [25]. The Al and Cr elements dissolve into the matrix to form α-(Fe, Cr, Al) solid solution. The eutectic microstructure is mainly composed of M_2B borides and M_3C_2 and M_7C_3 carbides, M mainly represents the elements of Fe, Cr, etc. [26]. The contents of Cr, B, and C in the eutectic phase zone 2 and zone 3 are much higher than that in dendrite (zone 1), while the contents of Fe, Al, and Si are higher in dendrite. This indicates that Cr element is easy to form boroncarbides among dendrites, while Al element are easier to be solidly dissolved into matrix to form matrix α-(Fe, Cr, Al) solid solution, which is conducive to improving the hardness and wear resistance of the coating.

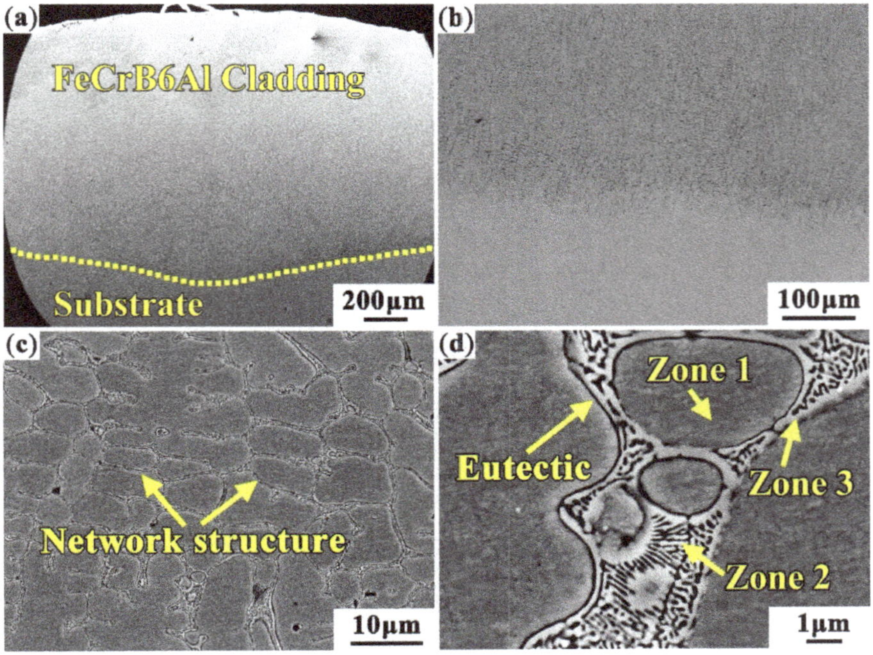

Figure 3. The microstructure morphologies of high aluminum and chromium Fe-B-C coating: (**a**) Macromorphology; (**b**) Bottom of cladding layer; (**c**,**d**) Middle part of cladding layer.

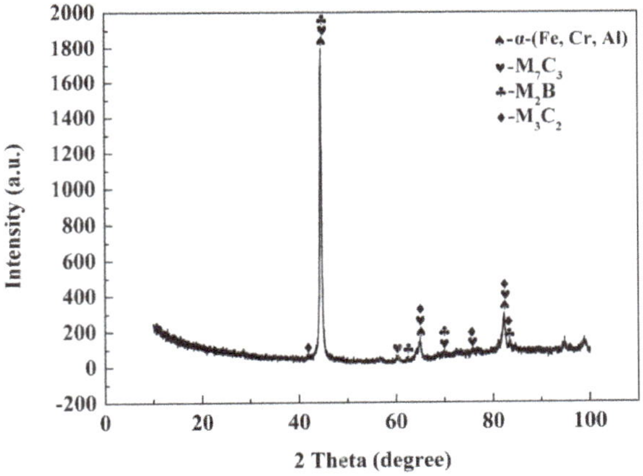

Figure 4. The X-ray pattern of high aluminum and chromium Fe-B-C coating.

Table 2. Chemical compositions of the phases with different morphologies in high aluminum and chromium Fe-B-C coating (wt.%).

Content	Fe	Cr	B	Al	Si	Mn	C
Zone1	84.175	12.271	0.44	2.271	0.494	0.628	0.554
Zone2	80.647	13.756	0.926	1.808	0.465	0.664	1.259
Zone3	70.813	21.483	3.623	0.692	0.193	0.837	3.105

The TEM tests are conducted to further determine the type of matrix and interdendritic boroncarbides in coating, as shown in Figure 5. From Figure 5a,d, the matrix of the coating is identified as the lath martensite with face-centered-cubic (fcc) structure. The large-sized block phase formed between dendrites are orthorhombic M_2B, and the small rod-shaped phase between the block M_2B are M_3C_2 with orthogonal structure and M_7C_3 with orthorhombic structure carbides. This network structure of eutectic structure is very similar to the microstructure of the cast Fe-Cr-B-Al alloy [27]. The size and distribution of the M_2B borides, M_3C_2 and M_7C_3 carbides in coating mostly depend on the microstructure evolution mechanism from the initial reactant powders. It is very interesting to study the microstructure evolution mechanism during the laser cladding. The powders dissolve into atoms in the molten pool under the laser beam, the Cr_2B phase firstly precipitates from the molten pool during the solidification according to the results in Figure 2. From Figure 3d, it can be speculated that as solidification process progresses, the remaining atoms begin to nucleate and grow on the Cr_2B and the M_7C_3 and M_3C_2 precipitates.

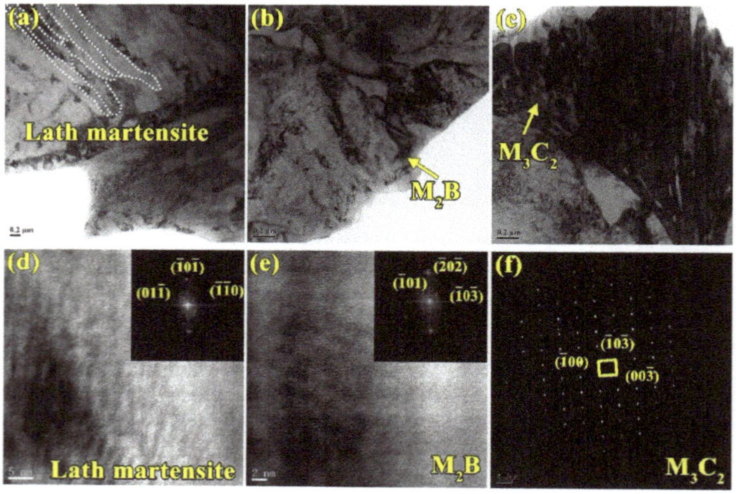

Figure 5. Transmission Electron Microscopy (TEM) of high aluminum and chromium Fe-B-C coating: (**a**–**c**) bright field images; (**d**) high resolution image of lath martensite and corresponding selective area diffraction (SAD) pattern; (**e**) high resolution image of M_2B and corresponding SAD pattern; and (**f**) the SAD pattern of M_3C_2.

3.3. Microhardness

Figure 6 shows the microhardness profile along the cross-section of the coating. It can be seen that the microhardness exhibits a gradient variation from the substrate to the coating, gradually increases from the interface bonding zone, heat-affected zone to the coating, which is mainly connected with microstructure evolution of the coating. According the results of microstructure analysis, the microstructure gradually transforms columnar crystals to equiaxed grains from the substrate to the coating. H. Xie et al. reported the hardness of the equiaxed grains is much higher than that of the columnar grains [28]. J. Li et al. thought the heat-affected zone (HAZ), as a transition zone, can effectively decrease the stress concentration and enhance the interface bonding [29]. It also can be seen that the microhardness of the coating has no major fluctuation and reaches 620 HV, which is 3.3 times that (190 HV) of the substrate. This indicates the microstructure is very evenly distributed and no obvious cracks and holes defects were produced. According to the microstructure analysis, the matrix of the coating is the α-(Fe, Cr, Al) solid solution martensite. The Al and Cr elements plays a solid solution strengthening role. The microhardness of martensite (480–560 HV) is higher than that of the ferrite (160–220 HV) [30]. Furthermore, the M_2B borides, M_3C_2 and M_7C_3 carbides with high hardness precipitates in the coating. Z.F. Huang found the microhardness of Fe_2B is about 1500 HV [31], furthermore, J. Lentz found the hardness of the M_2B is lightly higher than that of the Fe_2B [32,33]. The hardness of M_7C_3 is reported to be about 1800 HV by Tassin [34]. H.G. Fu et al. found the hardness of the M_3C_2 is slightly lower than that of the Fe_2B about 1300 HV [35]. The martensitically hardened Fe matrix and the high hardness of the hard phases lead to an improvement of microhardness of the coating.

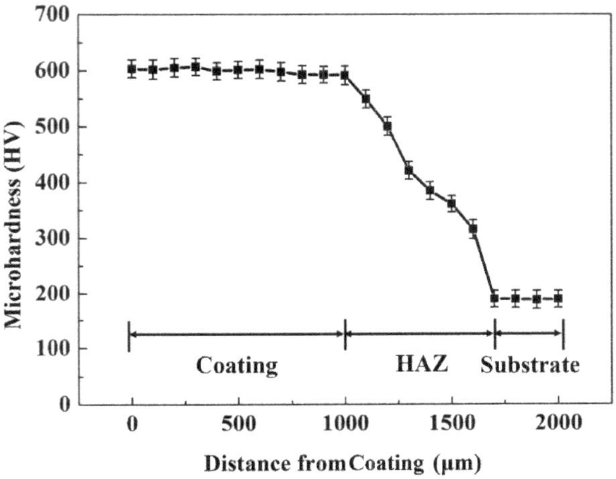

Figure 6. The microhardness profile along the cross-section of the coating.

3.4. Wear Resistance

Figure 7 shows the wear loss and wear rate of the substrate and coating. As can be seen from Figure 7a, the wear loss of substrate reaches 17.7 mg, while that of the coating is only 6.2 mg. According to Equation (1), the calculated wear rates of the coating and substrate are 0.295 mg/min and 0.103 mg/min, respectively, as shown in Figure 7b. The wear resistance of laser cladded coating is ~2-times higher than that of substrate. H. Berns studied the effect of B/C on wear resistance of the as-cast Fe-B-C alloy, and found the as-cast Fe-B-C alloy exhibited excellent wear resistance due to the main $Fe_3(C,B)$ phase with high hardness (1300 ± 50 HV) when the B/C was 1/3 [36]. According to the microstructure analysis, the Al and Cr elements dissolve into the matrix to form α-(Fe, Cr, Al) solid solution, furthermore, the M_2B borides and M_7C_3 and M_3C_2 carbides of the coating have high hardness, which is higher than the $Fe_3(B,C)$ phase, exhibiting better wear resistance. The high hardness borides and carbides can inhibit the wear of the grinding ball on the matrix, while the matrix with high hardness can also prevent the boron-carbons from falling off. The boroncarbide network structure in the coating makes it more stable, while the $M_{23}C_6$ carbides in the substrate are mainly distributed along grain boundaries, and the microhardness of α-Fe matrix is lower than that of the martensite. After the application of the load, the $M_{23}C_6$ carbides with high hardness may be pressed into the matrix, resulting in severe cutting to the matrix. The reciprocating wear can further aggravate the wear of the substrate.

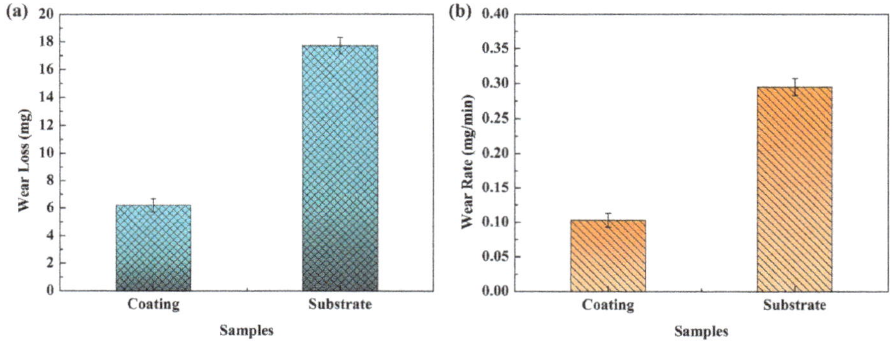

Figure 7. The wear loss (**a**) and wear rate (**b**) of the tested samples.

Figure 8 presents the variation of friction coefficient with time and average friction coefficient of the tested samples. It can be seen that the friction coefficient of the substrate increases quickly and remain stable in the first 12 min, and exhibits large fluctuation in the subsequent wear process which is caused by the incomplete occlusal of samples and grinding balls. Comparing with the friction coefficient of the substrate, the fluctuation of friction coefficient of the coating decreases significantly, as shown in Figure 8a. From Figure 8b, it is clear that the average friction coefficient of the coating is 0.75, which is much lower than that of the substrate (1.08).

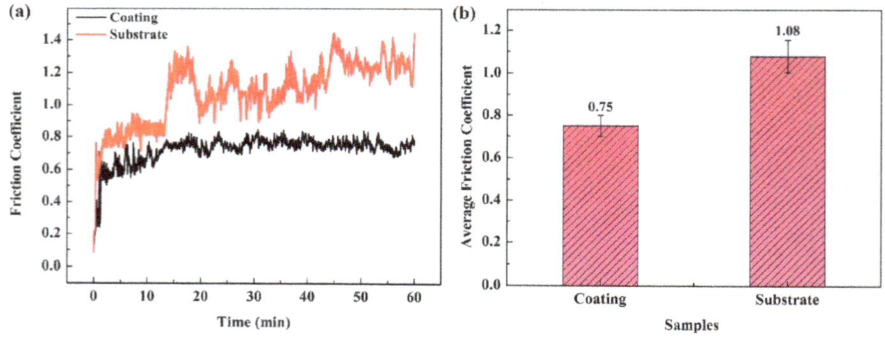

Figure 8. The variation of friction coefficient with time (**a**) and average friction coefficient (**b**) of the tested samples.

3.5. Wear Mechanisms

Figure 9 shows the morphologies of the worn surface of the tested samples. It can be seen that large pits and abrasive particles can be observed on the worn surface of the substrate, as shown in Figure 9a, which indicates the wear mechanism of the substrate is composed of the adhesive wear and abrasive wear. However, the worn surface of the coating shows a completely different morphology. A lot of shallow furrows appear on the worn surface, and contains some abrasive particles with small size. The formation of the abrasive particles causes the three body abrasion, which may be the main reason for forming the furrows. This suggests the wear mechanism of the coating is mainly comprised of slight abrasive wear.

The wear resistance of materials is mainly related to morphology and microstructure [37]. The substrate is composed of soft α-Fe matrix and little hard $M_{23}C_6$ carbides distributed along the grain boundary [38]. The soft matrix does not provide adequate protection for hard $M_{23}C_6$ carbides, resulting in the peeling off of hard carbides. This also is the main reason to bring about the fluctuation of friction coefficient of substrate. The martensite and network structure comprising of M_2B borides and M_7C_3 and M_3C_2 carbides are observed in the coating, as shown in Figure 5c,d. The martensite is stiffer than α-Fe ferrite, which suppresses the falling off of the hard phase. Moreover, the hard phases such as M_2B borides and M_7C_3 and M_3C_2 carbides exhibit network structure. Compared with granular and block $M_{23}C_6$ carbides in the substrate, these continuous network structures are more resistant to abrasion, which can protect the martensitic matrix, thus the wear resistance of the coating is improved.

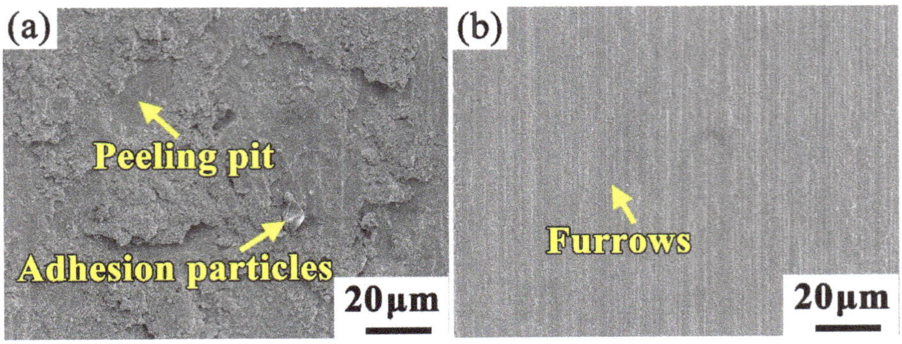

Figure 9. The morphologies of the worn surface of (**a**) substrate and (**b**) high aluminum and chromium Fe-B-C coating.

4. Conclusions

A novel high aluminum and chromium Fe-B-C coating which compositions are designed using the Thermo-Calc® software was fabricated using laser cladding technology. The microstructure, microhardness and wear resistance were systematically investigated. The main conclusions of this research can be drawn as follows:

1. The coating is dense and has good metallurgical bond with the substrate. The microstructure is mainly composed of α-(Fe, Cr, Al) lath martensite, orthorhombic M_2B borides, orthogonal M_3C_2, and orthorhombic M_7C_3 carbides.
2. The microhardness of the coating can reach to 620 HV which is 3.3-times higher than that of the substrate (190 HV). This is mainly due to the formation α-(Fe, Cr, Al) lath martensite, M_2B borides, and M_3C_2 and M_7C_3 carbides with high hardness.
3. The coating shows lower friction coefficient of 0.75 than that of the substrate (1.08). The wear rate of the substrate and the coating are 0.295 and 0.103 mg/min, respectively, indicating the coating exhibits excellent wear resistance. The wear mechanism transforms severe adhesive wear and abrasive wear of the substrate to slight abrasive wear of the coating.

Author Contributions: Investigation, formal analysis, writing—original draft preparation, J.L.; software, writing—review and editing, J.J.; writing—review and editing, W.C. and C.Y.; supervision, writing—review and editing, project administration, funding acquisition, J.W. All authors have read and agreed to the published version of the manuscript.

Funding: The authors appreciate the financial support for this work from The National Key Research and Development Program of China (2016YFB0701405), Aeronautical Science Foundation of China (2016ZE57010), New Young Teachers Initiation Program Fund (18X100040027) and the China Postdoctoral Science Foundation (19Z102060057).

Conflicts of Interest: The authors declare that they have no known competing financial interests or personal relationships that could have appeared to influence the work reported in this paper.

References

1. Kenneth, H.; Paivi, K.R.; Pirita, H.; Kati, V.; Ali, E. Global energy consumption due to friction and wear in the mining industry. *Tribol. Int.* **2017**, *115*, 116–139.
2. Ge, S.R.; Wang, Q.L.; Wang, J.X. The impact wear-resistance enhancement mechanism of medium manganese steel and its applications in mining machines. *Wear* **2017**, *376–377*, 1097–1104. [CrossRef]
3. Fleury, R.M.N.; Salvati, E.; Nowell, D.; Korsunsky, A.M.; Silva, F.; Tai, Y.H. The effect of surface damage and residual stresses on the fatigue life of nickel superalloys at high temperature. *Int. J. Fatigue* **2019**, *119*, 34–42. [CrossRef]
4. Li, M.Y.; Han, B.; Song, L.M.; He, Q.K. Enhanced surface layers by laser cladding and ion sulfurization processing towards improved wear-resistance and self-lubrication performances. *Appl. Surf. Sci.* **2020**, *503*, 144226. [CrossRef]
5. Makwana, D.; Bhingole, P.P. Electrochemical and plasma surface modification of magnesium and its alloy. *Mater. Today Proc.* **2019**, *5*, 18260–18267. [CrossRef]
6. Li, D.; Gou, G.Q.; Liu, Y.; Zhu, T.H.; Ji, H.; Chen, H. Effect of Al on microstructure and impact property of surfacing repair layer of rail. *Surf. Eng.* **2012**, *28*, 548–553. [CrossRef]
7. Zhan, X.; Qi, C.; Zhou, J.; Liu, L.; Gu, D. Effect of heat input on the subgrains of laser melting deposited Invar alloy. *Opt. Laser Technol.* **2019**, *109*, 577–583. [CrossRef]
8. Luo, X.; Li, J.; Li, G.J. Effect of NiCrBSi content on microstructural evolution, cracking susceptibility and wear behaviors of laser cladding WC/Ni–NiCrBSi composite coatings. *J. Alloys Compd.* **2015**, *626*, 102–111. [CrossRef]
9. Lv, Y.H.; Li, J.; Tao, Y.F.; Hu, L.F. High-temperature wear and oxidation behaviors of TiNi/Ti_2Ni matrix composite coatings with TaC addition prepared on Ti_6Al_4V by laser cladding. *Appl. Surf. Sci.* **2017**, *402*, 478–494. [CrossRef]
10. He, X.; Song, R.G.; Kong, D.J. Microstructures and properties of Ni/TiC/La_2O_3 reinforced Al based composite coatings by laser cladding. *Opt. Laser Technol.* **2019**, *117*, 18–27. [CrossRef]
11. Wang, K.M.; Du, D.; Liu, G.; Chang, B.H.; Ju, J.; Sun, S.T.; Fu, H.G. Microstructure and property of laser clad Fe-based composite layer containing Nb and B_4C powders. *J. Alloys Compd.* **2019**, *802*, 373–384. [CrossRef]
12. Ma, Q.; Li, Y.; Wang, J.; Liu, K. Investigation on cored-eutectic structure in Ni60/ WC composite coatings fabricated by wide-band laser cladding. *J. Alloys Compd.* **2015**, *645*, 151–157. [CrossRef]
13. Sun, S.; Fu, H.; Chen, S.; Ping, X.; Wang, K.; Guo, X.; Lin, J.; Lei, Y. A numerical experimental investigation of heat distribution, stress field and crack susceptibility in Ni60A coatings. *Opt. Laser Technol.* **2019**, *117*, 175–185. [CrossRef]
14. Zhang, J.B.; Zhang, J.J.; Xiao, G.C.; Chen, Z.Q.; Yi, M.D.; Zhang, Y.B.; Xu, C.H. Orentational effect of graphene on the friction and wear behavior of Si3N4/TiC based composite ceramic tool materials. *Ceram. Int.* **2020**, *46*, 3550–3557. [CrossRef]
15. Zhao, Y.; Yu, T.B.; Guan, C.; Sun, J.Y.; Tan, X.F. Microstructure and friction coefficient of ceramic (TiC, TiN and B4C) reinforced Ni-based coating by laser cladding. *Ceram. Int.* **2019**, *45*, 20824–20836. [CrossRef]
16. Wang, C.; Zhang, S.; Zhang, C.H.; Wu, C.L.; Zhang, J.B.; Abdullah, A.O. Phase evolution and wear resistance of in situ synthesized V8C7 particles reinforced Fe-based coating by laser cladding. *Opt. Laser Technol.* **2018**, *105*, 58–65. [CrossRef]
17. Fu, Z.K.; Ding, H.H.; Wang, W.J.; Liu, Q.Y.; Guo, J.; Zhu, M.M. Investigation on microstructure and wear characteristic of laser cladding Fe-based alloy on wheel/rail materials. *Wear* **2015**, *330–331*, 592–599. [CrossRef]
18. Zhang, H.; Zou, Y.; Zou, Z.D.; Wu, D.T. Microstructure and properties of Fe-based composite coating by laser cladding Fe-Ti-V-Cr-C-CeO_2 powder. *Opt. Laser Technol.* **2015**, *65*, 119–125. [CrossRef]
19. Ju, J.; Fu, H.G.; Lei, Y.P. Effect of Al addition on microstructure and properties of an Fe-B-Al alloy. *Mater. Test.* **2016**, *58*, 753–762. [CrossRef]
20. Ju, J.; Kang, M.D.; Wang, K.M.; Zhou, Y.; Jiang, M.; Zhao, Y.Y.; Zhang, Z.; Wang, J.; Fu, H.G. Studies on as-cast microstructure and oxidation behavior of the Fe-Cr-B-Al alloys at 1073 K. *Vacuum* **2019**, *164*, 436–448. [CrossRef]

21. Ju, J.; Yang, C.; Ma, S.Q.; Kang, M.D.; Wang, K.M.; Li, J.J.; Fu, H.G.; Wang, J. Effect of temperature on oxidation resistance and isothermal oxidation mechanism of novel wear-resistant Fe-Cr-B-Al-C-Mn-Si alloy. *Corros. Sci.* **2020**, *170*, 108620. [CrossRef]
22. An, Q.; Huang, L.J.; J, S.; Wang, S.; Zhang, R.; Sun, F.B.; Moridi, A.; Geng, L. Wear rate of titanium matrix composite coating at high temperature further increased by non-stoichiometric Ti_xC oxidation. *Ceram. Int.* **2020**, *46*, 8068–8074. [CrossRef]
23. Ju, J.; Zhou, Y.; Kang, M.D.; Wang, J. Optimization of process parameters, microstructure and properties of laser cladding Fe-Based Alloy on 42CrMo steel roller. *Materials* **2018**, *11*, 2061. [CrossRef] [PubMed]
24. Monson, P.J.E.; Steen, W.M. Comparison of Laser Hardfacing with Conventional Processes. *Surf. Eng.* **1990**, *6*, 185–193. [CrossRef]
25. Lyu, Y.Z.; Sun, Y.F.; Jing, F.Y. On the microstructure and wear resistance of Fe-based composite coatings processed by plasma cladding with B_4C injection. *Ceram. Int.* **2015**, *41*, 10934–10939. [CrossRef]
26. Wang, M.M.; Gao, R.; Gao, H.Y.; Zhou, Y.; Fan, Y.Y.; Zhao, Y.Y.; Ju, J.; Liu, Y.H.; Kang, M.D.; Wang, J. Improved corrosion resistance of Ni-modified Fe-Cr-B steel in molten zinc via phase transformation and microstructure control. *Surf. Coat. Technol.* **2013**, *235*, 811–818. [CrossRef]
27. Lv, Z.; Fu, H.G.; Xing, J.D.; Ma, S.Q.; Hu, Y. Microstructure and crystallography of borides and mechanical properties of Fe-B-C-Cr-Al alloys. *J. Alloys Compd.* **2016**, *662*, 54–62. [CrossRef]
28. Luo, X.X.; Cao, J.; Meng, G.H.; Chuan, Y.Y.; Yao, Z.J.; Xie, H. Systematical investigation on the microstructures and tribological properties of Fe-Al laser cladding coatings. *Appl. Surf. Sci.* **2020**, *516*, 146121. [CrossRef]
29. Jiang, J.S.C.; Li, J.B.; Lee, S.L.; Chang, Y.S.; Jian, S.R.; Huang, J.C.; Nieh, T.G. Prominent plasticity of Mg-based bulk metallic glass composites by ex-situ spherical Ti particles. *Intermetallics* **2012**, *30*, 25–29. [CrossRef]
30. Das Bakshi, S.; Shipway, P.H.; Bhadeshia, H.K.D.H. Three-body abrasive wear of fine pearlite, nanostructured bainite and martensite. *Wear* **2013**, *308*, 46–53. [CrossRef]
31. Huang, Z.F.; Xing, J.D.; Lv, L.L. Effect of tungsten addition on the toughness and hardness of Fe_2B in wear-resistant Fe–B–C cast alloy. *Mater. Charact.* **2013**, *75*, 63–68. [CrossRef]
32. Jonathan, L.; Arne, R.; Felix, G.; Werner, T. Enhancement of hardness, modulus and fracture toughness of the tetragonal $(Fe,Cr)_2B$ and orthorhombic $(Cr,Fe)_2B$ phases with addition of Cr. *Mater. Des.* **2018**, *156*, 113–124.
33. Jonathan, L.; Arne, R.; Werner, T. Hardness and modulus of Fe_2B, $Fe_3(C,B)$ and $Fe_{23}(C,B)_6$ borides and carboborides in the Fe-C-B system. *Mater. Charact.* **2018**, *135*, 192–202.
34. Tassin, C.; Lroudie, F.; Pons, M.; Lelait, L. Carbide-refinforced coatings on AISI 326L stainless steel by laser surface alloying. *Surf. Coat. Technol.* **1995**, *76–77*, 450–455. [CrossRef]
35. Yang, Y.W.; Fu, H.G.; Lei, Y.P.; Wang, K.M.; Zhu, L.L.; Jiang, L. Phase diagram calculation and analyze on cast high-boron high-speed steel. *J. Mater. Eng. Perform.* **2016**, *25*, 409–420. [CrossRef]
36. Berns, H.; Anastasia, S.; Arne, R.; Dietrich, H. Wear Protection by Fe-B-C Hard Phases. *Steel Res. Int.* **2011**, *82*, 786–794. [CrossRef]
37. Janicki, D. Microstructure and sliding wear behavior of in-situ TiC-reinforced composite surface layers fabricated on ductile cast iron by laser alloying. *Materials* **2018**, *11*, 75. [CrossRef]
38. Liu, X.; Yang, J.C.; Yang, L.; Gao, X.Z. Effect of Ce on inclusions and impact property of 2Cr13 stainless steel. *J. Iron Steel Res. Int.* **2010**, *17*, 59–64. [CrossRef]

materials

MDPI

Article

Experimental Analysis of Tribological Processes in Friction Pairs with Laser Borided Elements Lubricated with Engine Oils

Janusz Lubas [1,*], Wojciech Szczypiński-Sala [2], Paweł Woś [1], Edyta Zielińska [1] and Krzysztof Miernik [3]

1 Faculty of Mechanical Engineering and Aeronautic, Rzeszow University of Technology, Powstańców Warszawy 8, 35-959 Rzeszów, Poland; pwos@prz.edu.pl (P.W.); ezielins@prz.edu.pl (E.Z.)

2 Faculty of Mechanical Engineering, Cracow University of Technology, al. Jana Pawła II 37, 31-864 Cracow, Poland; ws@mech.pk.edu.pl

3 Faculty of Materials Engineering and Physics, Cracow University of Technology, al. Jana Pawła II 37, 31-864 Cracow, Poland; kmiernik@pk.edu.pl

* Correspondence: lubasj@prz.edu.pl

Received: 17 October 2020; Accepted: 16 December 2020; Published: 19 December 2020

Abstract: The present study discusses the influence of engine oils on the tribological parameters of sliding couples with laser borided surface layer. The borided layer was formed on specimens made from AISI 5045 steel by laser remelting of a surface layer coated with amorphous boron. The sliding friction and wear process was carried out on the pairs with AISI 5045 steel and SAE-48 bearing alloys which were lubricated with 5W-40 and 15W-40 engine oils. The investigation showed significant differences in the friction coefficient and temperature in the tested pairs with the laser borided surface layer. In the couples lubricated with 5W-40 engine oil, the tested parameter of friction was higher than in the couples lubricated with 15W-40 engine oil. The couples lubricated with 5W-40 engine oil showed more intensive wear of SAE-48 bearing alloy in contact with the laser borided surface layer than the pairs lubricated with 15W-40 engine oil. The laser borided surface layer used in friction pairs leads to the destruction of the lubricating properties of engine oils and reduces its resistance to scuffing.

Keywords: laser boriding; engine oil; wear; friction; surface layer

1. Introduction

The structural elements of machines are subjected to various unfavorable operating factors that shorten their service life, which, in turn, affects the reliability of the entire structure. The friction pairs used in internal combustion engines are exposed to an intensive destruction processes caused by the abrasive wear, adhesion, oxidation and cavitation processes. The intensity of these processes is influenced by the structure of the friction pair, materials of the friction elements, surface treatment, loading and lubrication conditions.

In the currently used constructions of friction pairs, friction and wear are the result of appropriate shaping of the contacting surface layers. In the case of most components operating under frictional conditions, a surface layer of high hardness and wear resistance is required. Sometimes, they also require increased fatigue resistance, corrosion resistance, heat resistance and creep resistance. One of the surface treatment methods which enables the formation of the required structure of the surface layer of constructional elements operating under friction conditions is laser treatment [1–6]. The surface layers produced in this way show favorable tribological properties in the form of the required hardness, fracture toughness, abrasion resistance and corrosion resistance [7–15]. Improving the microstructure of surface layers and coatings is also mentioned as one of the advantages of using laser techniques in surface treatment [14,16,17]. The study of wear of the laser-treated steel shows lower values of friction coefficients and lower wear of samples [18,19].

One of the elements that can be used in the laser remelting process is boron. This element forms stable and hard phases with iron, and laser treatment allows to eliminate the textured, coniferous structure of the surface layer [2,14,20,21]. Other authors indicate that this process also provides a lower coefficient of friction, good oxidation and resistance to erosion [14,20,22]. Boron content in the laser treatment steel layer significantly reduces the wear under dry friction conditions [15,22–24]. Tests in dry sand/rubber wheel conditions showed that the abrasive wear of $Fe_{50}Mn_{30}Co_{10}Cr_{10}$ was reduced by up to 30% compared to the material without laser treatment [22]. The results of EN25 steel tests showed that laser boriding increases corrosion resistance and causes uniform corrosion of the surface of the test sample, which results from the formation of iron borides during laser processing [20,22]. However, some authors indicate that replacing diffusion drilling in the case of elements requiring high fatigue strength by laser drilling is not recommended [24].

The surface cracks occurring after laser remelting are then the cause of the fatigue crack. In the case of laser borided carbon steels, the carbon content influences the heating and cooling of the material. It has been shown that higher concentration of carbon in the laser borided steel influences the increase in peak temperature and cooling rate, and also increases the depths of re-melted zone and heat-affected zone [25]. The process of boriding nickel alloys is difficult when using classical methods or requires the use of proprietary boriding agents [26]. The use of the laser beam allows the melting of the boron layer and the creation of a boron zone consisting of iron, nickel and chromium borides. The hardness of this layer is comparable to that obtained in the case of diffusion drilling, with its much greater thickness (346 or 467 μm depending on the power of the laser used) [27–29]. Laser boriding of nickel alloy increases its wear resistance, which may be ten times higher compared to the untreated material [26,27,29].

Although real components usually work in a mixed lubrication mode [30], most publications limit the scope of their research to abrasive wear [15,19,21,22]. Garcia et al. [31] showed that partially laser remelted surfaces of plasma spray coatings reduce wear only under certain pressures and sliding velocities. Moreover, they observed that higher percentage of the remelted surface may cause inadequate lubrication conditions and increase the wear of elements [31,32]. Most friction pairs require lubrication, it is important to determine the influence of laser remelting for tribological processes under limited lubrication conditions.

The aim of the present study is to investigate the influence of the selected classes of engine oils on friction and wear processes in friction pairs containing elements with laser borided surface layer. The use of boriding causes the formation of hard borides, which are resistant to wear, and the use of laser boriding assures that surface layer does not show a tendency to crack. Such surface layers can be used in friction pairs in internal combustion engines lubricated with 5W-40 and 15W-40 engine oils. This requires an answer to the question of what is the influence of engine oils on tribological processes under limited lubrication conditions. This paper also describes the anti-seizure parameters of the engine oils used for lubrication of the sliding pairs with elements containing boron.

2. Materials and Methods

AISI 5045 steel is widely used in mechanical engineering, including the construction of heavily loaded elements in internal combustion engines (Table 1). AISI 5045 steel ring samples with dimensions of ϕ 35 × 9 mm were prepared and were heat-treated (40 ± 2) and polished (Figure 1). The boron layer was produced by a CO_2 laser with a power of 2 kW. Steel samples were covered with mixed amorphous boron with water glass and fused with a laser beam to protect argon. The laser treatment parameters used in the surface layer remelting process were determined by the spot size of the laser beam of 4 mm, the processing speed of 16 mm/s and the path coverage of 0.5 mm. After the laser treatment, the surface layer of steel samples was polished. The counterpart was cut from the the

SAE-48 alloy journal bearing (Tables 1 and 2) with dimensions of 15.75 × 6.35 mm (Figure 1). As a result of the laser treatment, a boron layer was produced with a content of 1–1.5% boron, a maximum thickness of 25 μm and a maximum hardness of 1800 HV (Figure 2).

Tribological comparative studies were carried out in a conformal contact with the lubrication of the friction area with the engine oil, which is widely used for lubrication of petrol and diesel engines of passenger cars and delivery vans, working with and without turbochargers that require this level of quality (Table 3).

Table 1. Chemical composition of specimens (wt.%).

Material	C	Cr	Mn	Si	Fe	Pb	Cu
AISI 5045	0.46	0.5	0.65	<0.4	Balance	-	-
SAE-48	-	-	-		-	26–33	Balance

Table 2. Mechanical properties of SAE-48 bearing alloy (CuPb30).

Operating temperature Tmax (°C)	170
Load Pmax (N/mm^2)	140
Tensile Strength (N/mm^2)	200
Sliding Speed V-oil Lubricated (m/s)	8
Alloy Hardness (HB)	30–45

Table 3. Characteristics of engine oils.

Parameter	5W-0 Synthetic Oil	15W-40 Mineral Oil
Kinematic Viscosity at 100 °C	13.7 mm^2/s	14.5 mm^2/s
Viscosity Index	178	133
HTHS Dynamic Viscosity at 150 °C	3.6 mPa·s	3.8 mPa·s
Specification	API-SL/SJ/CF/CD ACEA-A3-98/B3-98/B4-98 ACEA-A3/B4-04	API SJ/CF ACEA A3-02/B3-98

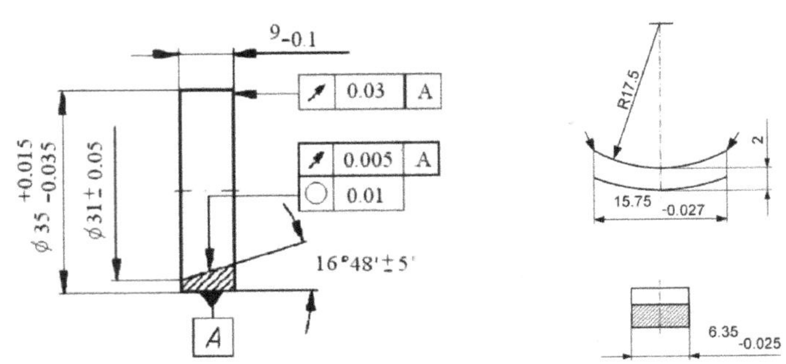

Figure 1. Dimensions of elements of the tested pair (mm).

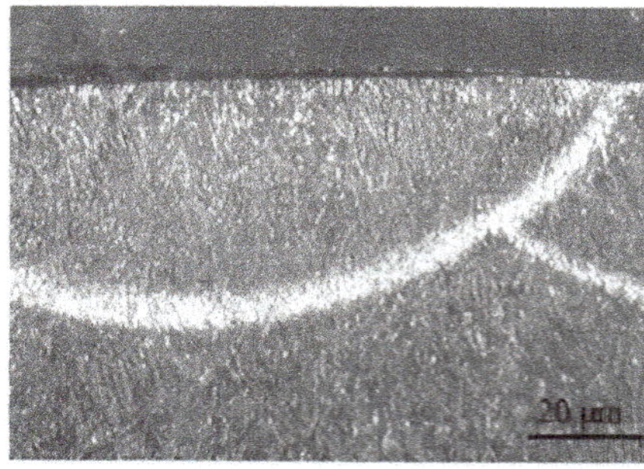

Figure 2. Microstructure of the steel specimen with laser borided surface layer.

The measurements of the friction pair co-operating under lubrication conditions were taken at the ring specimen rotational speed of 100 rpm and at a changeable unit pressure of 5, 10, 15 and 20 MPa. The measured parameters, such as friction coefficient, temperature in the friction area and linear wear as a function of changeable load were registered at real-time during the tests. The tribological tests were conducted on the T-05 block on ring tester (Figure 3) and the steel specimen was immersed in lubricants (engine oils) (Figure 3).

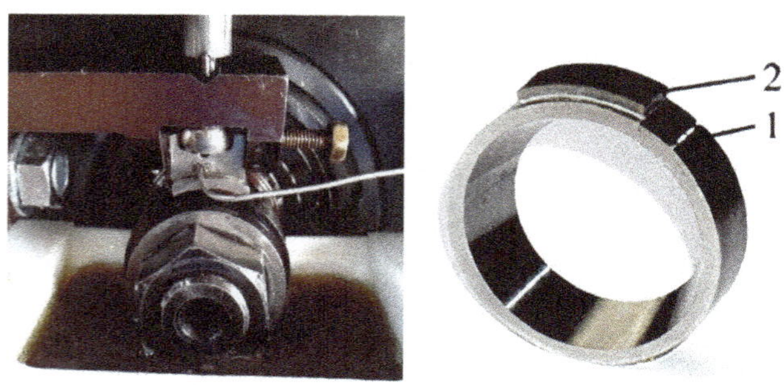

Figure 3. The friction pair in block on ring tester (**a**) and composition of friction pair: 1—ring specimen, 2—counterpart (**b**).

Scuffing test of engine oils in the sliding movement was performed utilizing the four-ball testing machine. In the tests, the balls with a diameter of 12.7 mm (0.5 in.) were used. The surface roughness, expressed as Ra parameter equaled 0.032 μm, and hardness amounted to 62 HRC ± 2. All tests during the current research were repeated three times.

3. Results

The measurements of surface roughness of the borided ring samples and counterparts with the bearing alloy after tests revealed significant changes, as compared to the roughness prior to the tests. The measurements of the Ra, Rz and Sm parameters of surface roughness indicated an increase of surface roughness. The measurements of Ra parameter of ring specimens showed the change of 3% in the pairs lubricated with synthetic oil and 7% in the pairs lubricated with mineral oil (Table 4).

Changes of a similar character, but on a much larger scale occurred in the case of the measurement of the Sm parameter, which increased by 62% in the friction pairs lubricated with the 15W-40 mineral oil, and by 49% in those lubricated with the 5W-40 synthetic oil. When measuring the Rz parameter, lubrication with the 15W-40 engine oil caused smaller changes in the geometric structure of the surface (19%) than in the case of using the 5W-40 engine oil (26%).

Table 4. Roughness of steel ring specimens after the test under pressure of 20 MPa (DIN 4768, ISO 4287).

Engine Oil	5W-40 Synthetic Oil		15W-40 Mineral Oil	
Parameter	Value [μm]	Change [%]	Value [μm]	Change [%]
Ra	0.31	3	0.32	7
Rz DIN	3.4	26	3.2	19
Sm	82	49	89	62

The surface roughness of the counterparts from SAE-48 bearing alloy showed more extensive changes in the measured parameters than of ring specimens with laser borided layer (Table 5). The changes of Ra, Rz and Sm parameters of the counterparts exceeded even several dozen percent. The 15W-40 mineral oil lubrication showed smaller percentage changes in surface roughness, as opposed to lubrication with the 5W-40 synthetic oil. Particularly significant changes relate to the Sm parameters, which in the friction pairs lubricated with mineral oil increased by 156% and with synthetic oil by 192%. Minor changes were observed for the Rz parameter and were 83% and 137%, respectively. The Ra parameter was increased by 44% in the pairs lubricated with mineral oil and by 78% in those lubricated with synthetic oil.

Table 5. Roughness of counterparts from SAE-48 after the test under pressure 20 MPa (DIN 4768, ISO 4287).

Engine Oil	5W-40 Synthetic Oil		15W-40 Mineral Oil	
Parameter	Value [μm]	Change [%]	Value [μm]	Change [%]
Ra	0.89	78	0.72	44
Rz DIN	7.1	137	5.5	83
Sm	140	192	123	156

During the start-up of the friction pair, the registration of the friction coefficient allows for the determination of the energy demand necessary to start the friction pair, specification of frictional resistance in variable sliding conditions and their stabilization areas depending on the load conditions. An important parameter characterizing kinematic pairs is the maximum moment of friction during start-up of the tested friction pairs. In the tested pairs with a boron surface layer, the start-up moment was lower when lubricating the friction area with mineral oil than when lubricating it with synthetic oil (Figure 4). An important change is observed in the entire tested load range from 5 to 20 MPa. A significant difference in the start-up moment of the friction pair is observed at a load of 10 MPa, which is then 22%. Smaller difference is observed at the lowest load of 5 MPa and then the difference does not exceed 15%. At loads of 15–20 MPa, the start-up moment value is similar for both tested engine oils and does not exceed 3%.

The changes of friction resistance in the friction pairs with the laser borided surface layer in the start-up period showed a lower friction coefficient in pairs lubricated with the 5W-40 synthetic oil than in pairs lubricated with the 15W-40 mineral oil (Figure 5). For loads of 10–20 MPa, after the initial increase in the friction coefficient at the moment of start-up of the friction pair, a further increase in the friction resistance is observed along with the duration of the test. The value of the friction coefficient stabilizes only at a load of 5 MPa. In the pairs lubricated with mineral engine oil, at the load of 15–20 MPa, the course of the friction coefficient is similar and reaches similar values for all three tested loads, so that in the final stage of the test, the coefficient value is about 0.17. At the load of 15 MPa, the friction coefficient stabilizes at the level of 0.15. The variable course of the friction coefficient at the load of 5 MPa, in the initial phase of cooperation of the friction pair, leads to its stabilization after 350 s, at the level of 0.12. Under lubrication conditions with synthetic oil, at the loads of 10–20 MPa, there is a systematic increase in the value of the friction coefficient to its final value of 0.13 for the load of 10 MPa,

0,14 for 15 MPa, and over 0.15 for 20 MPa. At the lowest load of 5 MPa, the friction coefficient stabilizes below 0.09, after less than 100 s.

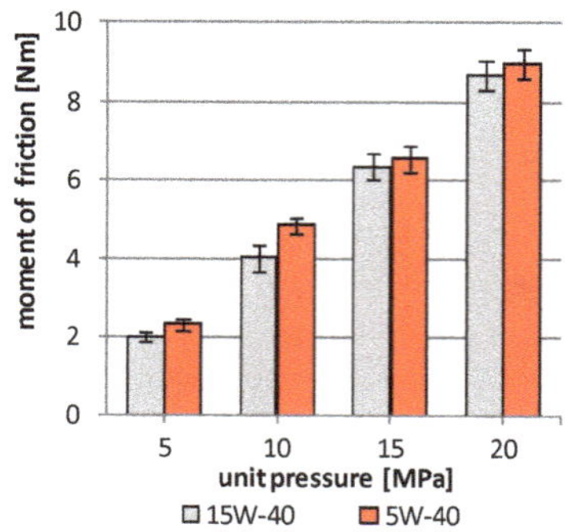

Figure 4. Moment of friction during start-up of the pair vs unit pressure when lubricating with 15W-40 mineral engine oil and 5W-40 synthetic engine oil.

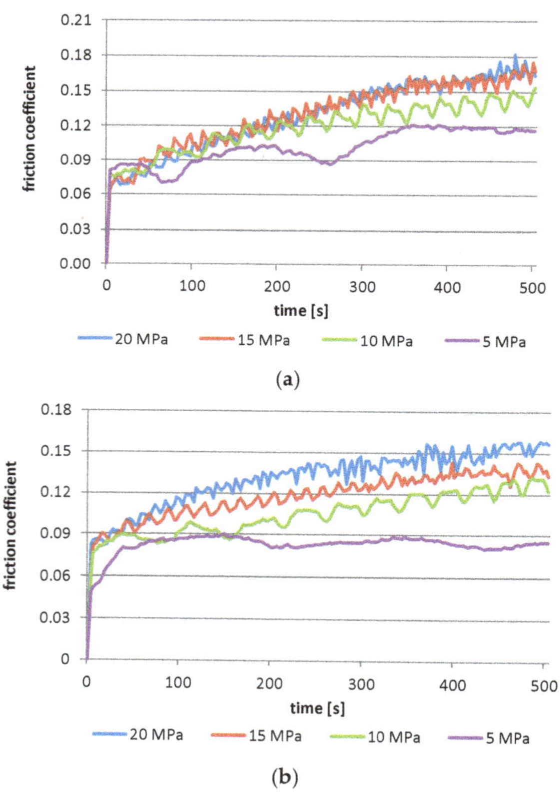

Figure 5. Friction coefficient in a sliding couple with laser borided surface layer vs rotation speed, lubricated with (**a**) 15W-40 mineral engine oil, (**b**) 5W-40 synthetic engine oil.

The friction force and temperature in the friction area allows to determine the working conditions of the friction pair, and if they show lower values, such a system ensures greater stability of operation and greater safety area in the event of an overload of the friction pair, which may lead to its seizure and destruction.

The friction force and temperature in the friction area for the pairs lubricated with the 15W-40 mineral oil are higher than in the case of the pairs lubricated with the 5W-40 synthetic oil (Figure 4). The greatest difference in the friction force occurs at the load of 5 MPa of the friction pair and it is 19% lower for the pair lubricated with the 5W-40 oil, as compared to the pair lubricated with 15W-40 oil. In the case of the remaining loads of 10–20 MPa, the difference amounts to a few percent, and in the case of the 15 MPa load, the difference is 2%, which is within the limits of the measurement method. The measured temperature in the friction area also shows a very similar course, while the differences between the tested friction pairs, depending on the engine oil used, are much smaller than those calculated in the friction force (max 4%). In the case of the15 MPa load, the temperature value for both used engine oils is identical and amounts to 98 °C (Figure 6).

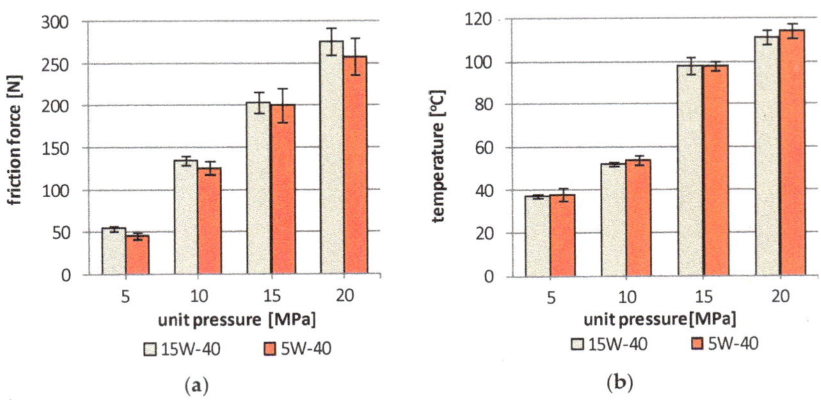

Figure 6. Friction forces (**a**) and temperature in friction area (**b**) vs unit pressure (at 100 rpm and after 500 s).

The analysis of wear of the friction pair elements makes it possible to determine the durability of the friction pair, which allows to determine the period of correct operation of the entire device, as well as periods of periodic inspections and replacement of consumables.

The ring samples with the laser borided surface layer did not show the measurable linear wear, but it was possible to observe the intense wear process of the SAE-48 bearing alloy. The wear of the SAE-48 bearing alloy in the friction pair lubricated with 5W-40 synthetic oil was higher than the in the friction pair lubricated with 15W-40 mineral oil, at the load 10–20 MPa (Figure 7). The wear of the bearing alloys increases with the pressure increase in the contact area between the surface layers of both elements of the friction pairs. The difference in wear of the alloy ranges from a few percent (at the load of 20 MPa) to several percent (at the load of 10–15 MPa). At the pressure of 5 MPa, the wear of SAE-48 alloy was lower by 22%. In the friction pair lubricated with 5W-40 synthetic oil.

The behavior of a lubricant under scuffing conditions determines the areas in which the oil retains its properties, enabling the separation of the lubricated surface layers of the cooperating elements of the friction pair. During the tests on the T-02U tester, two loads can be registered: one marking the beginning of the scuffing area and the other, the so-called limiting pressure of seizure. Limiting pressure of seizure depends on the load at which the balls seize and the average value of the wear-scar diameters measured on the stationary balls.

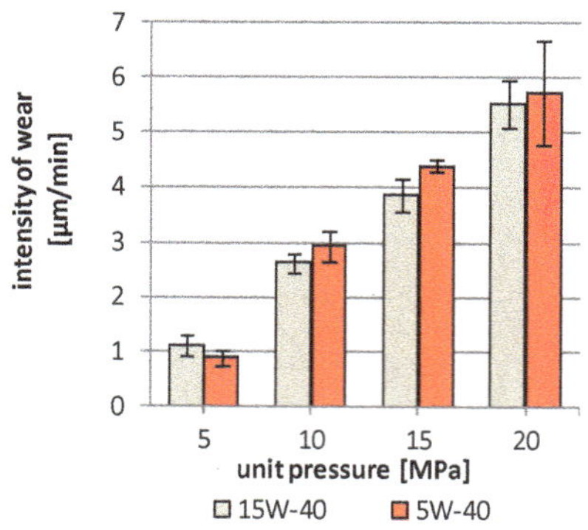

Figure 7. Intensity of wear of SAE-48 bearing alloy during lubrication by engine oils (at 100 rpm).

Measurements of the seizing load of the used and unused 15W-40 and 5W-40 engine oils show a reduction in the seizure resistance of the oil used during the wear tests of the friction pair, as compared to the unused oil (Figure 8). In the tests carried out at the temperature of 40 °C, it can be observed that unused 5W-40 oil shows the highest resistance to scuffing, while at the temperature of 100 °C, both unused oils show similar values. After the cooperation, a significant reduction in resistance to scuffing of the 5W-40 oil is observed at 40 °C (13%), and for the 15W-40 oil, it decreases to 4%. At the temperature of 100 °C, the seizing load of 5W-40 oil is higher than that of 15W-40 oil, while the value of the 5W-40 oil seizing load is similar to the scuffing load of the unused oil.

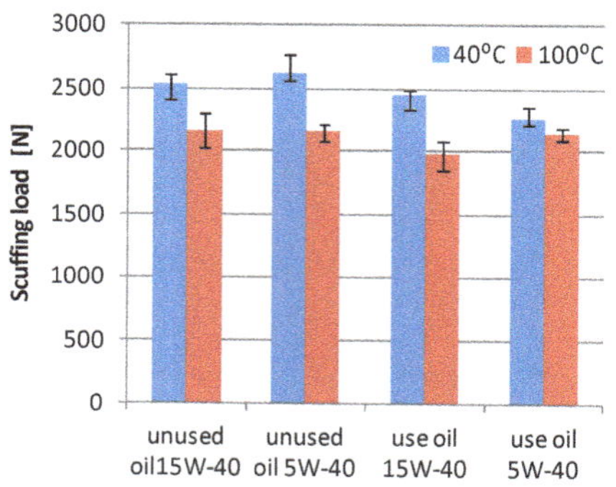

Figure 8. Scuffing load of the unused and used oil; 15W-40 and 5W-40.

The seizure load measurements for the tested oils show significant differences and it can be concluded that the 15W-40 oil shows higher load values than the 5W-40 oil (Figure 9). In the case of 15W-40 oil, the load values for used and unused oils are similar at 100 °C. On the other hand, at 40 °C the load is lower than that for the used oil. The examination of the 5W-40 oil shows a much greater decrease in the seizure load at 100 °C, which is 11%, and at 40 °C, the seizure load value is similar.

The pressure of seizure shows a significant deterioration in the performance properties of the tested engine oils, which may have been caused by contamination of the oil with wear products and a reduction in the amount of anti-seizure additives in the oil. In the conditions of friction with limited lubrication in the friction area, the anti-seizure additives contained in the engine oil are responsible for the formation of boundary layers preventing seizure, which leads to their wear. On the other hand, stabilization or slight increases in the scuffing load can be explained by the migration of boron, exhibiting lubricating properties, into the oil. Another factor is the decrease in the amount of other additives refined from engine oils, which in the new oils cause the degeneration of the lubricating properties (viscosity modifiers, corrosion or rust inhibiting, detergent additives).

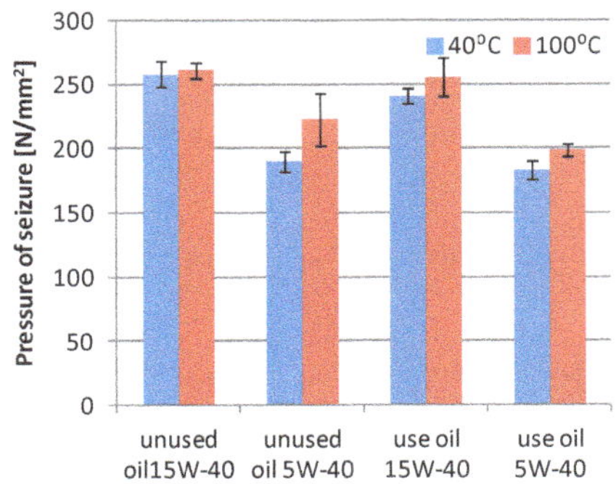

Figure 9. Pressure of seizure of the unused and used oil; 15W-40 and 5W-40.

An observation of the sliding surface of SAE-48 bearing alloy samples after wear tests shows non-uniformity of the friction surface. On the surface of the bearing alloy, there are areas of the surface layer with significant deformation, which results from the variable hardness of the laser treatment layer and the lines formed on it, resulting from the arrangement of overlapping traces of the laser beam (Figure 10). In these areas, significant damage to the sliding layer of the bearing alloy can also be observed—cracks and flaking. The maps of the elements in the designated friction areas show significant differences in the distribution of the selected elements Cu, Pb, S, Mo and B (Figure 11).

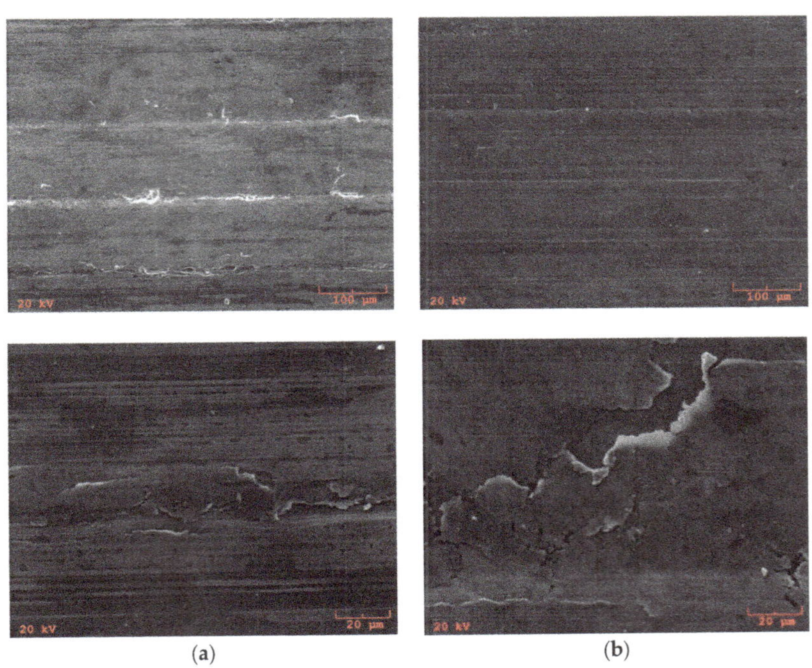

Figure 10. Micrographs of SAE-48 bearing after tests; (**a**) 15W-40 engine oil; (**b**) 5W-40 engine oil.

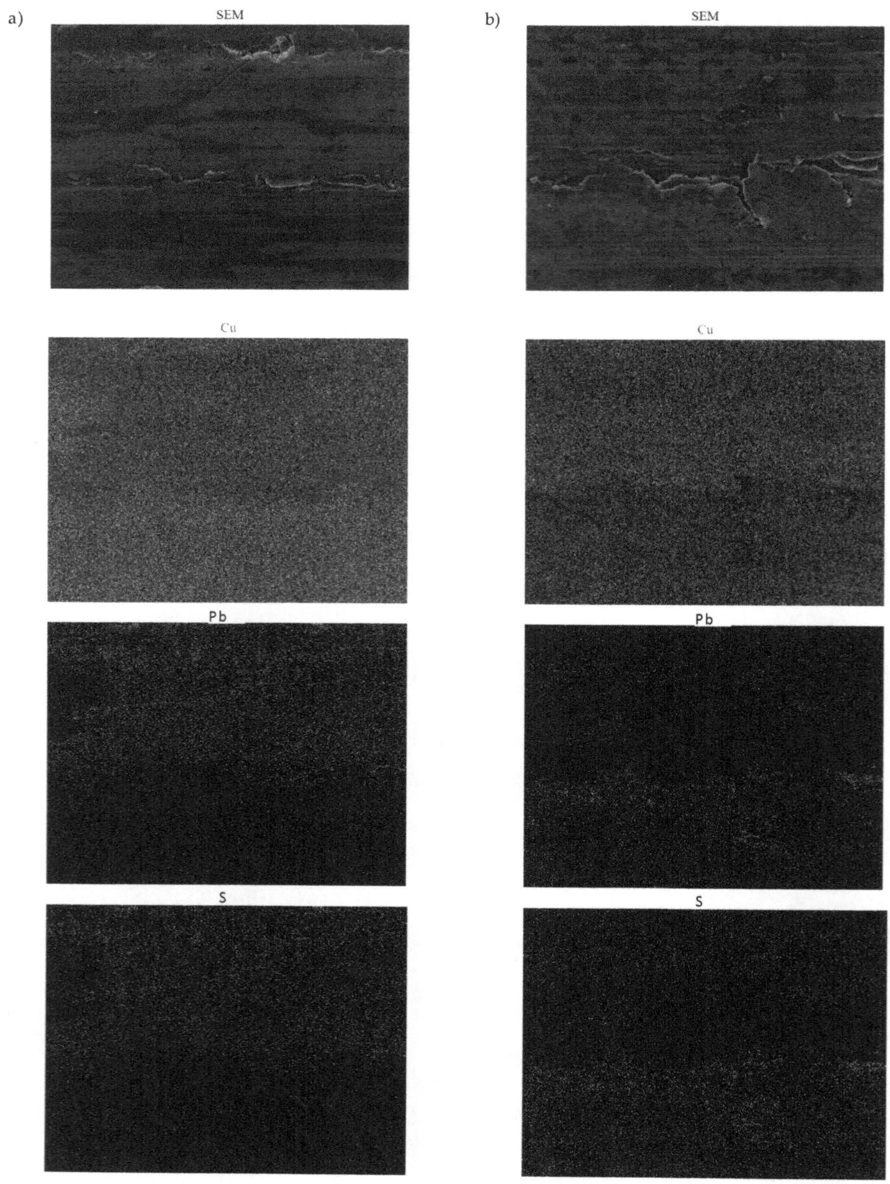

Figure 11. *Cont.*

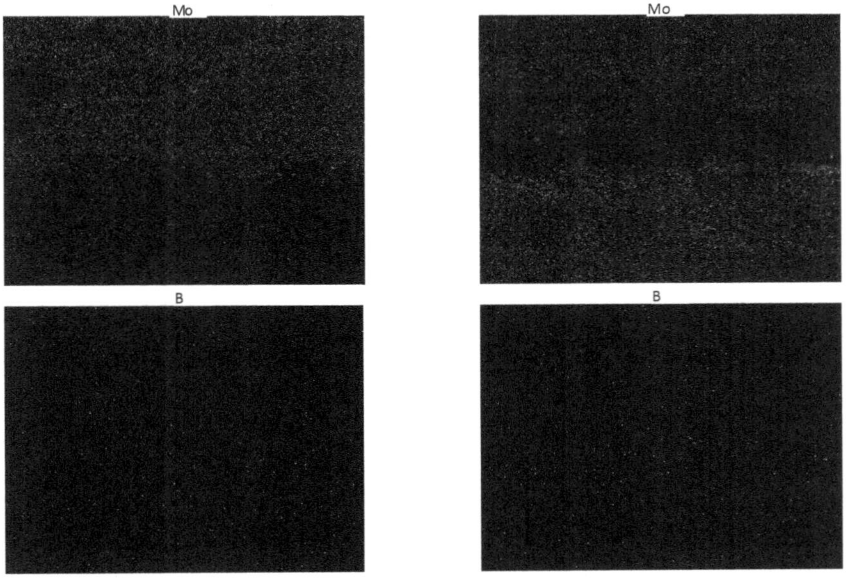

Figure 11. Maps of the distribution of elements in the surface layer SAE-48 bearing alloy; (**a**) 15W-40 engine oil, (**b**) 5W-40 engine oil.

An observation of the base elements of the SAE-48 bearing alloy shows a uniform distribution of Cu, while in the case of Pb, significant differences can be observed, which may result from the migration of lead from the heavily loaded areas to the boundary layers formed in the cooperation area. The analysis of the distribution of S and Mo elements, which are components of the engine oils, also shows their uneven distribution. Intensive wear processes occur in the areas with low presence of the S and Mo elements, which impede the formation of permanent boundary layers containing these elements. The maps of boron distribution shows single atoms in the friction area, which may indicate that in this type of friction pair composition boron transfer is negligible or this element is washed away by the lubricant.

4. Discussion

Lubrication of friction pairs with laser remelting structural elements significantly shapes the processes of friction and wear by formation of boundary layers, as a result of the reaction of the surface layers of these elements and the lubricant. The observation of the geometric structure of the surface layers shows a significant reorganization, the effect of which is an increase in the surface roughness of both layers, especially the surface of counter-samples of the SAE-48 bearing alloy [33].

Changes in the Ra parameter show a comprehensive picture of the surface, and the changes in the surface roughness result mainly from the adaptation of both surface layers of the cooperating elements to the specific load conditions of the friction pair. Changes in this parameter significantly affect the load-bearing capacity of the operating surface layers formed in the process of friction. The+ Rz and Sm parameters are parameters that allow the assessment of the local surface unevenness field, and significant changes in their size allow for the identification of changes that arise as a result of drawing, grooves, micro-cutting to protruding surface irregularities or hard wear products occurring in the area of friction or grafting processes occurring as a result of the disappearance of the lubricating layer. These changes result from the processes of rapid destruction of the existing lubricating layers or difficulties in shaping them as a result of friction processes and direct contact of the cooperating layers of the friction pair elements [34]. The unfavorable effect of the laser borided layer may be the result of the variable hardness of these surface layers, which is the result of a specific method of its shaping, by overlapping successive laser beam (Figure 10). The observation of the course of the friction coefficient shows that low unit pressures influence the shaping of favorable lubrication conditions, and the system changes the existing geometric structures of both elements into a composition ensuring the most favorable cooperation conditions. The resulting structure provides the given pair with optimal functionality, caused by the creation of stable operational surface layers on both elements of the friction pair [35,36].

The friction and wear processes taking place in the contact area of the sliding pairs depend on unit pressure between cooperating surface layers, microstructural changes of surface layer, chemical reactions between materials, material transformation and material properties variation in surface layers [33,37,38]. The wear of bearing alloy is mainly caused by the hard areas in the surface layer of the second material and hard wear particles, which leads to the interaction between the two surface layers and a more intense abrasion of the softer material [37,38]. Also, an increase in roughness of the harder material results in an increase in the wear particles detached from the softer material [33]. The hard wear products created in the friction process induce chipping, slicing and grinding, which intensify the wear process [36,39].

During operation, engine oil changes its properties and accumulates impurities, which worsens the lubrication conditions and increases the intensity of wear processes. The lubrication process can remove wear products from the friction area. However, as a result of tribochemical reactions, corrosion processes, cavitation wear, or fatigue wear may occur, the latter caused by dynamic changes in the pressure in the friction area. The assessment of the phenomena is complex and only on the basis of the surface observation of the elements can it be concluded that the oil affects the processes occurring in the contact area.

5. Conclusions

Based on the experimental test, the following conclusions can be drawn:

1. The surface roughness of the SAE-48 bearing alloy and laser borided ring specimen shows tendencies to increase and a greater change was observed in the pairs lubricated with 5W-40 engine oil.
2. Friction force in the pair lubricated with the 15W-40 engine oil is considerably higher than in the pair lubricated with the 5W-40 oil, and temperature in the friction area for both tested oils is of a comparable value level.
3. The application of the 5W-40 oil in the sliding pair with the laser borided surface layer causes a significant increase of wear of the SAE-48 bearing alloy, together with the increase of unit load in the friction area.
4. The processes of wear lead to the destruction of engine oil and the deterioration of their resistance to scuffing, especially during the operation under low temperature.
5. Lubrication of the friction pair with laser borided surface layer with the 15W-40 oil is more advantageous because it causes smaller changes in the geometrical structure of the cooperating surface layers, lower friction pair starting resistance and lower wear at a comparable temperature in the friction area.

Author Contributions: Conceptualization, J.L. and W.S.-S.; methodology, J.L., W.S.-S., P.W., E.Z. and K.M.; formal analysis, J.L., W.S.-S., P.W., E.Z. and K.M.; investigation, J.L., W.S.-S. and K.M.; resources, P.W. and E.Z.; data curation, P.W. and E.Z.; writing—original draft preparation, J.L., W.S.-S., P.W., E.Z. and K.M.; writing and editing, J.L.; supervision, J.L. All authors have read and agreed to the published version of the manuscript.

Funding: This research received no external funding.

Conflicts of Interest: The authors declare no conflict of interest.

References

1. Yu, J.; Song, B. Effects of heating time on the microstructure and properties of an induction cladding coating. *Results Phys.* **2018**, *11*, 212–218. [CrossRef]
2. Makuch, N.; Dziarski, P.; Kulka, M.; Piasecki, A.; Tuliński, M.; Majchrowski, R. Influence of niobium and molybdenum addition on microstructure and wear behavior of laser borided layers produced on Nimonic 80A alloy. *Trans. Nonferrous Met. Soc. China* **2019**, *29*, 322–337. [CrossRef]
3. Sui, Q.; Zhou, H.; Bao, H.; Zhang, P.; Yuan, Y.; Meng, C. Wear behavior of quenched iron with various shapes and unit processed through two step laser alloying of C powder. *Opt. Laser Technol.* **2018**, *104*, 103–111. [CrossRef]
4. Liu, Q.; Chen, C.; Zhang, M. Effect of heat treatments on the microstructural evolution and tribological behaviors of a NiCrSiB/Y2O3 alloy fabricated by laser additive manufacturing. *Mater. Character.* **2020**, *165*, 110401. [CrossRef]
5. Zhang, D.; Cui, X.; Jin, G.; Feng, X.; Lu, B.; Song, Q.; Yuan, C. Effect of in-situ synthesis of multilayer graphene on the microstructure and tribological performance of laser cladded Ni-based coatings. *Appl. Surf. Sci.* **2019**, *495*, 143581. [CrossRef]
6. Xu, X.; Lu, H.F.; Luo, K.Y.; Yao, J.H.; Xu, L.Z.; Lu, J.Z ; Lu, Y.F. Mechanical properties and electrochemical corrosion resistance of multilayer laser cladded Fe-based composite coatings on 4Cr5MoSiV1 steel. *J. Mater. Proc. Tech.* **2020**, *284*, 116736. [CrossRef]
7. Yan, S.X.; Dong, S.Y.; Xu, B.S.; Wang, Y.J.; Ren, W.B.; Fang, J.X. Effect of preheating temperature on microstructure and property of laser clad ni-based alloy coating on gray cast iron substrate. *J. Mater. Eng.* **2015**, *43*, 30–36.
8. Zhou, S.F.; Dai, X.Q.; Zheng, H.Z. Microstructure and wear resistance of Fe-based WC coating by multi-track overlapping laser induction hybrid rapid cladding. *Opt. Laser. Technol.* **2012**, *44*, 190–197. [CrossRef]
9. Mohan, D.G.; Gopi, S. Induction assisted friction stir welding: A review. *Aust. J. Mech. Eng.* **2020**, *1*, 119–123. [CrossRef]
10. Zhang, S.; Wu, C.L.; Zhang, C.H.; Guan, M.; Tan, J.Z. Laser surface alloying of FeCoCrAlNi high entropy alloy on 304 stainless steel to enhance corrosion and cavitation erosion resistance. *Opt. Laser Technol.* **2016**, *84*, 23–31.
11. Adebiyi, D.I.; Popoola, A.P.I. Mitigation of abrasive wear damage of Ti–6Al–4V by laser surface alloying. *Mater. Des.* **2015**, *74*, 67–75. [CrossRef]
12. Yan, H.; Wang, A.; Xiong, Z.; Xu, K.; Huang, Z. Microstructure and wear resistance of composite layers on a ductile iron with multicarbide by laser surface alloying. *Appl. Surf. Sci.* **2010**, *256*, 7001–7009. [CrossRef]
13. Li, J.; Liu, H.; Li, G. Surface performance and physical properties of nanoscale ceramics reinforced laser alloying composite coatings. *Compos. Part B* **2014**, *57*, 86–90. [CrossRef]
14. Prince, M.; Arjun, S.L.; Surya, R.; Gopalakrishnan, P. Experimental Investigations on the Effects of Multicomponent Laser Boriding on steels. *Mater. Today Proc.* **2018**, *5*, 25276–25284. [CrossRef]
15. Klopotov, A.A.; Ivanov, Y.F.; Potekaev, A.I.; Abzaev, Y.A.; Kalashnikov, M.P.; Chumaevskii, A.V.; Volokitin, O.G.; Teresov, A.D.; Vlasov, V.A.; Klopotov, V.D. The use of low-temperature plasma in a combined technology for the formation of wear-resistant boron-containing coatings. *Surf. Coat. Technol.* **2020**, *389*, 125576.
16. Misra, D.; Dhakar, B.; Anusha, E.; Shariff, S.M.; Mukhopadhyay, S.; Chatterje, S. Evaluation of nanomechanical and tribological properties of laser surface alloyed boride-nitride-carbide ceramic matrix composite coatings. *Cer. Int.* **2018**, *44*, 17050–17061. [CrossRef]
17. Hu, G.; Meng, H.M.; Liu, J.Y. Microstructure and corrosion resistance of induction melted Fe-based alloy coating. *Surf. Coat. Technol.* **2014**, *251*, 300–306. [CrossRef]
18. Gök, M.S.; Küçük, Y.; Erdoğan, A.; Öge, M.; Kanca, E.; Günen, A. Dry sliding wear behavior of borided hot work tool steel at elevated temperatures. *Surf. Coat. Technol.* **2017**, *328*, 54–62. [CrossRef]
19. Telasang, G.; Majumdar, J.D.; Padmanabham, G.; Manna, I. Wear and corrosion behaviour of laser surface engineered AISI H13 hot working tool steel. *Surf. Coat. Technol.* **2015**, *261*, 69–78. [CrossRef]
20. Sashank, S.; Dinesh Babu, P.; Marimuthu, P. Experimental studies of laser borided low alloy steel and optimization of parameters using response surface methodology. *Surf. Coat. Technol.* **2019**, *363*, 255–264.

21. Wei, X.; Chen, Z.; Zhong, J.; Wang, L.; Hou, Z.; Zhang, Y.; Tan, F. Facile preparation of nanocrystalline Fe_2B coating by direct electrospark deposition of coarse-grained Fe_2B electrode material. *J Alloys Comp.* **2017**, *717*, 31–40. [CrossRef]
22. Aguilar-Hurtado, J.Y.; Vargas-Uscategui, A.; Paredes-Gil, K.; Palma-Hillerns, R.; Tobar, M.J.; Amado, J.M. Boron addition in a non-equiatomic $Fe^{50}Mn_{30}Co_{10}Cr_{10}$ alloy manufactured by laser cladding: Microstructure and wear abrasive resistance. *Appl. Surf. Sci.* **2020**, *515*, 146084. [CrossRef]
23. Kulka, M.; Mikołajczak, D.; Makuch, N.; Dziarski, P.; Miklaszewski, A. Wear resistance improvement of austenitic 316L steel by laser alloying with boron. *Surf. Coat. Technol.* **2016**, *291*, 292–313. [CrossRef]
24. Kulka, M.; Makuch, N.; Dziarski, P.; Mikołajczak, D.; Przestacki, D. Gradient boride layers formed by diffusion carburizing and laser boriding. *Opt. Lasers Eng.* **2015**, *67*, 163–175.
25. Bendoumi, A.; Makuch, N.; Chegroune, R.; Kulka, M.; Keddam, M.; Dziarski, P.; Przestacki, D. The effect of temperature distribution and cooling rate on microstructure and microhardness of laser re-melted and laser-borided carbon steels with various carbon concentrations. *Surf. Coat. Technol.* **2020**, *387*, 125541. [CrossRef]
26. Günen, A.; Kanca, E.; Çakir, H.; Karakaş, M.S.; Göke, M.S.; Küçük, Y.; Demir, M. Effect of borotitanizing on microstructure and wear behavior of Inconel 625. *Surf. Coat. Technol.* **2017**, *311*, 374–382. [CrossRef]
27. Kulka, M.; Dziarski, P.; Makuch, N.; Piasecki, A.; Miklaszewski, A. Microstructure and properties of laser-borided Inconel 600-alloy. *Appl. Surf. Sci.* **2013**, *284*, 757–771. [CrossRef]
28. Dziarski, P.; Kulka, M.; Mikołajczak, D.; Makuch, N. Corrosion resistance of laser-borided Inconel 600 alloy. *Inżyn. Mater.* **2017**, *3*, 149–156.
29. Makuch, N.; Piasecki, A.; Dziarski, P.; Kulka, M. Influence of laser alloying with boron and niobium on microstructure and propertiesof Nimonic 80A-alloy. *Opt. Laser. Technol.* **2015**, *75*, 229–239. [CrossRef]
30. Eder, S.J.; Ielchici, C.; Krenn, S.; Brandtner, D. An experimental framework for determining wear in porous journal bearings operated in the mixed lubrication regime. *Tribol. Intern.* **2018**, *123*, 1–9.
31. García, A.; Cadenas, M.; Fernandez, M.R.; Noriega, A. Tribological effects of the geometrical properties of plasma spray coatings partially melted by laser. *Wear* **2013**, *305*, 1–7. [CrossRef]
32. Habib, K.A.; Cano, D.L.; Heredia, J.A.; Mira, J.S. Effect of post coating technique on microstructure; microhardness and the mixed lubrication regime parameters of thermally sprayed NiCrBSi coatings. *Surf. Coat. Technol.* **2019**, *358*, 824–832. [CrossRef]
33. Akchurin, A.; Bosman, R.; Lugt, P.M.; Drogen, M. Analysis of wear particles formed in boundary-lubricated sliding contacts. *Tribol. Lett.* **2016**, *63*, 16. [CrossRef]
34. Nehme, G.; Mourhatch, R.; Aswath, P.B. Effect of contact load and lubricant volume on the properties of tribofilms formed under boundary lubrication in a fully formulated oil under extreme load conditions. *Wear* **2010**, *268*, 1129–1147. [CrossRef]
35. Schouwenaars, R.; Jacobo, V.H.; Ortiz, A. Microstructure aspect of wear in soft tribological alloys. *Wear* **2007**, *263*, 727–735. [CrossRef]
36. Dudek, K.; Szczypinski-Sala, W.; Lubas, J. Influence of laser boriding on the friction and wear of sliding couples with bearing alloys. *Surf. Rev. Lett.* **2020**, *27*, 1950084.
37. Piasecki, A.; Kotkowiak, M.; Makuch, N.; Kulka, M. Wear behavior of self-lubricating boride layers produced on Inconel 600-alloy by laser alloying. *Wear* **2019**, *426*, 919–933. [CrossRef]
38. Dziedzic, A.; Lubas, J.; Bochnowski, W.; Adamiak, S. The mechanisms of tribological wear in lubricated sliding bearings composed of soft bearing alloys with borided steel. *Proc. Inst. Mechan. Eng. Part J J. Eng. Tribol.* **2016**, *230*, 350–363. [CrossRef]
39. Lubas, J. Practical application of boron-modified sliding pairs in I.C. engine. *Tribol. Int.* **2010**, *43*, 2046–2050. [CrossRef]

materials

MDPI

Article

Feasibility Evaluation of Local Laser Treatment for Strengthening of Thin-Walled Structures from Low-Carbon Steel Subjected to Bending

Oleksandr Kapustynskyi , Nikolaj Višniakov *, Darius Zabulionis and Artur Piščalov

Faculty of Mechanics, Vilnius Gediminas Technical University, Saulėtekio al. 11, 10221 Vilnius, Lithuania; oleksandr.kapustynskyi@vgtu.lt (O.K.); darius.zabulionis@vgtu.lt (D.Z.); artur.piscalov@vgtu.lt (A.P.)

* Correspondence: nikolaj.visniakov@vgtu.lt; Tel.: +370-5-2745053

Received: 10 June 2020; Accepted: 3 July 2020; Published: 10 July 2020

Abstract: This paper is devoted to investigating numerically, by finite element analysis (FEA), and analytically the influences and effects of laser processing of the surface of thin-plate, low-carbon structural steel. The plate mechanical properties—axial and flexural stiffnesses, force-deflection behavior and cross-section force-strain behavior—are investigated after different laser treatments. An analytical methodology of the estimation of the cross-section area of the laser-processed metal is also proposed in the present article, that can be applied to choosing the reasonable distance between the centers of the laser-processed tracks. The methodology takes into account the width of the laser-processed tracks and the distances between these tracks. The experimental, finite element numerical and analytical analyses showed that the laser treatments of the surface of the steel plate increase the yield point of the laser-processed metal and the axial and flexural stiffnesses of the plate.

Keywords: reinforcing; laser treatment; strengthening; bending; low-carbon steel; thin steel plate

1. Introduction

High-quality carbon and low-alloyed steels are widely used to manufacture thin-walled metal products like tubes, vessels, beams, shells, panels, folds, membrane structures and other forms used in metalworking. However, application of these cheap steels has been limited due to inadequate mechanical strength, stiffness and low resistance to corrosion [1]. Reducing the weight of steel structures and the amount of metal required by improving the strength, hardness and stiffness of steel is a major task in the metalworking sector.

Thin-walled shell structures are lightweight constructions which are mostly designed to carry only tension loads, without taking into account compression or bending. However, such three-dimensional tensile structures can be subjected to unplanned bending loads and bending stresses due to concentrated transversal forces. These forces can be accidental impact, concentrated external or support/reaction forces, which may cause significant damage or failure of the whole construction [2].

Different design strategies are useful to avoid failure or the risk of collapse of the metal structure, and to increase the resistance to collapse or sensitivity of the structure to unplanned heavy loads. Heavily loaded thin-walled structural elements, such as the shell structures of buildings, rockets, chemical reactors, thin-walled pipes, vessels and other metal structures, are designed using special calculation methods that take plastic deformation into account. Then, calculations are carried out beyond the elasticity of the metals. The ultimate load method or limit-state design allows metal consumption to be reduced and gives access to additional reserves of structural strength in thin-walled steel structures. It also increases available workloads or ultimate bending moments in comparison with the results achieved by traditional working stress methods in the elastic stage [3,4].

Different heat-treatment or thermochemical treatment processes are very popular finishing techniques in metalworking, and can be useful to increase the yield, tensile strength and hardness of steel structures [5].

In practice, only steels with carbon content above 0.3% are reliably heat-treatable to improve their mechanical properties by formation of martensitic structures. Theoretically, steels with carbon content below 0.3% can be strengthened by refining grain size and using a sufficient cooling rate during traditional quenching. The critical cooling rate for low-carbon steels with 0.19 wt.% must be around 1925 °C/s [6], but the rate at which they must be cooled to produce martensite is so high that it cannot be attained by traditional quenching in water or induction hardening.

Furthermore, heat-treatment and thermomechanical treatment techniques are very expensive and complicated. The strengthening of the metal sheets by means of strengthening elements and reinforcing ribs, special steel profiles and sections of complex geometry or local thickenings of ribs, is a well-known engineering solution [2,3]. However, the application of geometric improvements increases the manufacturing costs and the weight of metal parts, and requires complex manufacturing methods, expensive metalworking equipment and tooling, special designs and very experienced engineering staff.

An alternative way to manufacture metal structures with better strength and stiffness can involve methods for fabricating rib-strengthened or grid-strengthened composite structures. All dual-phase steels have heterogeneous microstructures, similar to natural composite materials, the properties of which are determined by the properties of the individual phases, following the rule of mixtures [7]. The structure of steel in the necessary place can be modified by special local treatment, where creation of strengthening or a reinforcing phase due to local phase transformation and changes of microstructure become available. The creation of structural strengthening ribs can significantly affect the overall strength and elasticity of thin-walled steel parts which have insufficient or very low bending strength and stiffness. The bending stress increases linearly away from the neutral axis until the maximum values at the top and bottom of the bent plate. Therefore, the reinforcement of metal surface layers can be an effective method of strengthening thin-walled metal structures affected by bending loads.

Laser-assisted methods of surface treatment are one of the most popular and well-established methods of surface modification of metals. Laser treatment technologies allow local heating, high temperature gradients and cooling speeds of 105–106 °C/s due to the thermal conductivity of metal. Therefore, the hardening of low-carbon steels with less than 0.3 % carbon is also possible, whereas these steels cannot be hardened effectively by other heat treatment methods [8].

Transformation hardening, nitriding and carburizing by laser allows changes to the microstructure and properties such as strength, hardness, roughness, coefficient of friction, wear resistance, chemical resistance and corrosion resistance of the surface of various metals [9,10]. The typical thickness of the hardened layer after laser transformation hardening by CO_2 laser, without melting of the surface, usually does not exceed 0.3 mm, applying a laser pulse of 0.15 mm [11]. The efficiency of laser transformation hardening and the final strength of the laser-processed structure depends on the total laser-processed area and the depth of hardening.

Therefore, application of laser transformation hardening without melting for strengthening of low-carbon steel structures with less than 0.3% carbon is limited by the low thickness of the hardened layer, because a very large surface area must be processed. In contrast, laser processing with melting gives a thicker laser-processed layer.

The main objective of this research was to test and model the bending of laser-processed steel plates, evaluating the efficiency of the application of laser treatment with surface melting to strengthening thin sheet components from unalloyed and structural carbon steels containing less than 0.2% carbon.

The results of a study of the influence of laser treatment on the microstructure and mechanical properties of bent plates of low-carbon steel are presented. The results of modelling of elastoplastic deformations of differently laser-processed samples are compared with results obtained by bending tests on real samples. The research results show that local laser treatment with surface melting can be used to increase the bending strength of thin-walled structural elements from steel (1.0402) and to decrease its deflection under identical workloads.

2. Object of the Study

One of the most popular grades of structural high-quality-carbon steel (1.0402) containing less than 0.3% carbon was used (Tables 1 and 2). The microstructure and mechanical properties of this hypoeutectoid steel, according to standard EN 10250-2, can vary depending on the treatment applied [12]. Thin metal plates of size 20 mm × 150 mm × 2 mm were used for the bending test and structure analysis.

Table 1. Chemical composition of samples from steel (1.0402).

Chemical Elements [wt.%]					
C	Si	Mn	P	S	Cr + Mo + Ni
0.18	0.18	0.45	0.02	0.02	0.15

Table 2. Mechanical properties of samples from steel (1.0402).

Elastic Modulus *E*, [GPa]	Yield Strength $\sigma_{0.2}$, [MPa]	Tensile Strength σ_B [MPa]	Relative Extension *A*, [%]	Hardness, [HV]
200	256	410	30	125

First, before laser treatment, all the samples were subjected to tempering for internal stress relief. The microstructure of the steel (1.0402) samples after tempering demonstrated a typical hypoeutectoid ferrite-pearlite microstructure, with hardness 135 HV. The surface of the samples was subjected to additional blast cleaning to ensure uniform laser energy absorption. A "Power Plus Tools" (Shanghai, China) sandblaster and quartz sand, "Sakret" (Berlin, Germany) with a grain size of 0.1–0.5 mm were used for surface preparation until roughness Ra measured less than 5 μm.

3. Research Methodology

3.1. Laser Surface Treatment

The Nd:YAG 4-axis laser-welding machine BMM400 (Boaolaser, Beijing, China) was used for local laser treatment. The metal samples were tightly fixed to the work table of the laser machine. A shielding gas mixture Ar-CO_2 (20% CO_2) with a shielding gas flow rate 20 L/min was used for Nd:YAG laser treatment. The depth of penetration should be the minimum necessary to provide sufficient hardening and avoid excessive embrittlement, cracking and deformation of the laser-processed, thin-walled steel plate [13,14]. Melting by laser or laser transformation hardening, without melting of most metal, proceeds at low energy density in the range of 10^3–10^5 W/cm^2 [15].

One of the criteria for the selection of the optimum processing mode was the required depth of laser penetration. A depth of penetration of about 0.35 mm, amounting to 20% of the total thickness of the metal plate, was used in the present experiments. The width of the track of the laser-processed metal was approximately 0.7 mm. The most appropriate laser processing parameters were calculated, using the methodology presented in reference literature [15], considering the physical properties of the steel (Table 3) and the technical characteristics of the 400 W laser equipment. The laser processing parameters, other than the number of processed sides, number of laser tracks and distance between them, remained constant throughout all experiments (Table 4).

The dimension of the laser-processed surface area in all cases was 40 mm × 20 mm. Four different laser-processing cases were adopted that differed in the number of laser tracks. In case I, no laser tracks were made at all, i.e., the metal plate was left unprocessed. In case II, 38 laser tracks were made, with the tracks overlapping in approximately 30% of their width. In case III, 17 laser tracks were made, with a distance between the tracks approximately equal to 50% of their widths. In case IV, 14 laser tracks were made, with a distance between the tracks equal to the track width.

Table 3. Physical properties of low-carbon steel (1.0402) [16].

Melting Temperature, T_m, [K]	Ambient Temperature, T_0, [K]	Absorptivity, A	Reflection Coefficient, R	Density, ρ, [kg/m^3]	Specific Heat, c, [J/kg·K]
1733	295	0.09	0.91	7800	462

Table 4. Parameters of local laser treatment.

Depth h, [mm]	Speed v, [mm/min]	Frequency f, [Hz]	Single Pulse Energy E_p, [J]	Spot Size d, [mm]	Peak Power P_p, [kW]	Overlap Coefficient P_{er}, [%]	Critical Power P_{d1}, [W/cm^2]
0.346	240	10	14.4	3	2.8	98.6	$4.11 \cdot 10^5$

All four cases were applied to both one-side-processed and two-side-processed plates. In case II, case III and case IV, the laser-processed tracks were oriented along the plate, in the direction in which the greatest stresses appear, by analogy with the principle of the strengthening of composite materials. A general view of the plate specimen and the laser-treated area is shown in Figure 1a. The plate cross-sections and the laser-processed layers are shown in Figure 1b,c. The laser processing cases, track width, depth and positions are depicted in Figure 1d.

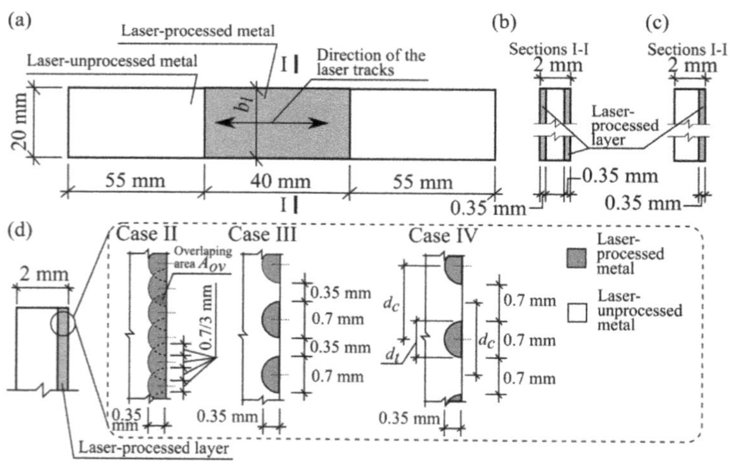

Figure 1. The general view of the plate specimen (**a**); sketches of their cross-sections: of the one-side laser-processed plate (**b**), and of the two-side laser-processed plate (**c**); and cases of the laser tracks (**d**).

3.2. Experimental Methods of Investigation of the Materials Properties and the Bending Stiffness

The mechanical properties of the base, i.e., laser-unprocessed metal, i.e., stress-strain diagram, modulus of elasticity E_b, offset yield stress (proof stress) $\sigma_{0.2,B}$, ultimate and fracture stresses as well as the fracture strains $\epsilon_{ul,b}$, were determined by the standard tensile test according to LST EN ISO 6892-1 [17]. The corresponding mechanical properties of the laser-processed metal, i.e., E_l, $\sigma_{0.2,l}$ and $\sigma_{0.2,B}$ were determined indirectly by using the empirical relationships between the corresponding mechanical properties and hardness which was determined according to EN ISO 6507-1 [18]. The series of three samples were used for the mechanical tests—the estimated means $m_x = \sum_{i=1}^{N} x_i$ are presented hereafter and were used for the calculations presented below. Unmachined samples of type B, with initial gauge length $L_0 = 80$ mm, length $L_c = 120$ mm and initial cross-section $S_0 = 40$ mm^2 were used for the tensile test.

A universal tensile-testing machine TIRAtest 2300 (TIRA, Schalkau, Germany) with a bending test tool and Catman-Express software (version 5.1, HBM, Germany) was used for the tensile and bend tests. A tension dynamometer up to 50 kN was used for the tensile test, while a compression dynamometer up to 1 kN was used for the bending test, described in the last paragraph of this subchapter. Steel hardness was determined using a Zwick/Roell ZHU (Ulm, Germany) universal hardness tester using the Vickers method. Steel hardness was determined on the surface with a load of 10 N. The hardness of the laser-processed layer was determined using a Zwick/Roell ZHμ (Ulm, Germany) hardness tester with a diamond, square-based tetrahedral pyramid tip with a load of 2.942 N. Hardness was measured on the surface of the laser-processed layers and on its cross-section.

The chemical composition of the steel was determined using the PMI Master PRO Oxford Instruments (High Wycombe, UK) optical emission spectrometer.

The metallographic examination of the steel and the laser-processed layer and analysis of the geometry and dimensions of the laser-processed layer was carried out using a Nikon Eclipse MA200 optical microscope (Tokyo, Japan) with a Lumenera Infinity 2-2 video camera and a JEOL JSM-7600 (Tokyo, Japan) scanning microscope with an energy dispersive spectrometer (EDS) Oxford INCA Energy X-Max20 (Oxford, UK) at different magnifications (up to ×1500).

The qualitative X-ray diffraction (XRD) analysis of the phase composition of materials was carried out by applying diffractometer DRON-7 (Burevestnik, Saint-Petersburg, Russia). Graphite-monochromated Cu Kα radiation (λ = 0.154178 nm) was used. The parameters of the tests were as follows: voltage—30 kV; current—12 mA; the range of the diffraction angle—from 4° to 120°, with detector movement steps of 0.04°; the duration of the intensity measuring in each step—2.0 s.

The elastoplastic bending of the thin plate was conducted by using a 3-point bending device (see Figure 2), analogous to the device used in the metal bending tests, according to LST EN ISO 7438 [19]. The calculation scheme of the bending test is shown in Figure 2b. The distance between supports was 76 mm, the rounding radius and the Poisson's width were 10 mm. The test was conducted by using the strain-control mode with a velocity of displacement of 1.5 mm/s. The load F_{exp} was imposed at the middle of the span, at point B; see Figure 2b. The load F_{exp} ranged from 0 N up to the maximum load, corresponding to the greatest deflections of the experiment, which equal 2 mm. In general, $F_{exp} < 500$ N. The deflections were measured at the middle of the span of the plate, i.e., at point B; see Figure 2b. It should be noted that the distance between the supports, 76 mm, was greater than required according to LST EN ISO 7438 [19] to decrease the influence of shear strains on the deflections.

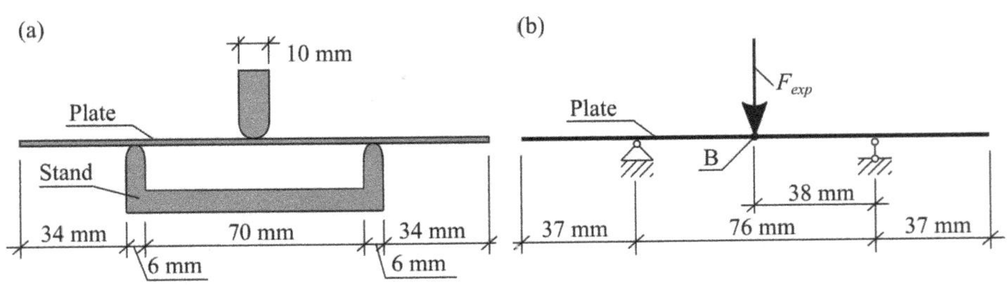

Figure 2. The principal view of the bending test device (**a**) and corresponding calculation scheme (**b**).

In the case of bending the one-side laser-processed plate, the laser-processed layer may be under tension or under compression. To evaluate the influence of the stress-strain state of the laser-processed layer on the deflections or the bending stiffness of the plate, 3 different loading variants were considered. In Variant A, the bending load F_{exp} was imposed on the laser-processed side of the one-side laser-processed plate, so that the laser-processed layer was under compression.

In Variant B, F_{exp} was imposed on the laser-unprocessed side of the one-side laser-processed plate, so that the laser-processed layer was under tension. In Variant C, F_{exp} was imposed on the two-side laser-processed plate. In this case, the top laser-processed layer was under compression, while the bottom laser-processed layer was under tension. Also, 4 already mentioned laser-processing cases were considered with each loading variant: case I, laser-unprocessed plate; case II, 38 laser tracks; case III, 17 laser tracks; and finally, case IV, 14 laser tracks. In total, 10 different force–deflection experiment cases were conducted. These cases are summarized in Table 5. In general, three measurements were performed to determine each quantity. Three plates were tested for case I, and 9 plates for the rest of the laser-processed plate cases: case II–case IV. Thus, in general, 39 plates were tested for the bending experiment. The estimated mean values $m_x = \sum_{i=1}^{N} x_i$ are presented and analyses hereafter are in the present article.

Table 5. Force–deflection experiment cases.

Experiment Case Notation	Laser Processing Case and Number of One-Side Laser Tracks	Loading Variant	Position and Stress-Strain State of the Laser-Processed Layer
Case I	Case I, 0 tracks		there are no laser-processed layers
Case II A		A	at the top under compression
Case II B	Case II, 38 tracks	B	at the bottom under tension
Case II C		C	at the top under compression, and at the bottom under tension
Case III A		A	at the top under compression
Case III B	Case III, 17 tracks	B	at the bottom under tension
Case III C		C	at the top under compression, and at the bottom under tension
Case IV A		A	at the top under compression
Case IV B	Case IV, 14 tracks	B	at the bottom under tension
Case IV C		C	at the top under compression, and at the bottom under tension

3.3. Finite Element Analysis (FEA) Simulation of Bending of Thin Metal Plate

The bending of the thin metal plate was modeled by finite element analysis (FEA), using the Ansys workbench software package, version 16.0. A numerical model of the bending test was created identical to the experimental bend test device; see sub-chapter 3.2. The general view of the numerical model is shown in Figure 3a. The geometry and dimensions of the FEA models of the plate were identical to the plates used in the experimental investigation; see Figure 1. The shape and depth of the laser-processed metal (0.35 mm) were identical for all FEA models of the plates (see Figure 3 and Table 6). The distance between supports (76 mm), punch width (10 mm) and fixation conditions in the FEA simulation were the same as those used during the real experiments. The plates in FEA modeling were supported simply, as it is shown in Figure 2b. For all three variants, the load was imposed in the middle of the span of the stand; see Figure 3a. The laser processing cases and loading variants used in the present FEA are summarized in Table 6. As in the case of experimental investigation in the numerical modeling in general, 10 different simulations were conducted; see Table 6.

The three-dimensional solid brick and tetrahedral elements were used for the discretization of the complex geometry of the modelled plate with laser-processed layers [20]. Large-scale finite elements with a maximum size of 0.7 mm were used to mesh the laser-unprocessed parts of the plate, while a finer mesh, with finite element sizes up to 0.12 mm, was adopted for the discretization of the laser-processed layer.

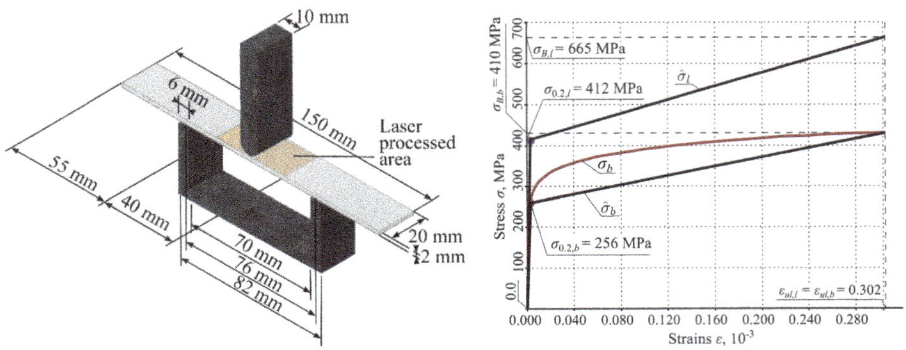

(a) (b)

Figure 3. Part (**a**) shows: geometry, dimensions, stand and the loading scheme of the laser-processed plate of the numerical model of the bending simulation; and in (**b**): σ_b denotes the base metal $\sigma - \epsilon$ curve; $\hat{\sigma}_b$ denotes the bilinear approximations of σ_b; $\hat{\sigma}_l$ denotes the bilinear approximation of the $\sigma - \epsilon$ curve of unknown laser-processed metal; yield strengths, ultimate strength, and fracture strains of the base metal and the laser-processed metal are denoted by $\sigma_{0.2,b}$, $\sigma_{0.2,l}$, $\sigma_{B,b}$, $\sigma_{B,l}$, $\epsilon_{ul,b}$ and $\epsilon_{ul,l}$ respectively.

Table 6. The laser processing and variants of geometry of the laser-processed area and bending loading.

Laser Processing Case	Loading Variants	Number of the Laser-Processing Tracks [pc]	Volume of the Laser-Processed Metal [mm^3]	Ratios of Volumes of the Laser-Processed Metal to the Total Bended Volume [%]
Case I		0	0	0
Case II	A	38	277.3	9.9
	B	38	277.3	9.9
	C	76	554.6	19.8
Case III	A	17	261.5	9.3
	B	17	261.5	9.3
	C	34	523	18.6
Case IV	A	14	215.4	7.7
	B	14	215.4	7.7
	C	28	430.8	15.4

The different mechanical properties of the base metal and laser-processed layer were used in the FEA models (treatment cases IIA—IIC, IIIA—IIIC, IVA—IVC). The yield strength, ultimate strength and modulus of elasticity of the steel were determined by mechanical tensile and bending testing of the specimens. The approximate yield strength and ultimate strength of the metal of the laser-processed layers were obtained by applying the relations between hardness and strength that are explained in [21–23]. The modulus of elasticity, Poisson's ratio and other required properties of materials are taken from reference literature [5]. The bilinear isotropic hardening plasticity model was used for the bending case numerical investigation [8].

The so-called bilinear constitutive laws of the laser-unprocessed and laser-processed metals of the plates were adopted for the finite element analysis of the bending of the plates; see $\hat{\sigma}_b$ and $\hat{\sigma}_l$ curves depicted in Figure 3b. These constitutive laws are approximations of the real physical stress-strain ($\sigma - \epsilon$) curves of the laser-processed and unprocessed metals. The direct investigation of the $\sigma - \epsilon$ curve of the laser-processed metal is very complicated; therefore, the simplified bilinear $\sigma - \epsilon$ diagrams already mentioned were adopted for the finite element analysis. Below, explanations are given of the obtained and adopted mechanical parameters of the laser-unprocessed and processed metals of the plates.

The $\sigma - \epsilon$ curve of the base, i.e., laser-unprocessed, metal was obtained experimentally, as described above in Sub-Chapter 3.2. From this $\sigma - \epsilon$ curve, the yield strength $\sigma_{0.2,b} = 256$ MPa, the ultimate strength $\sigma_{B,b} = 410$ MPa and the ultimate strains $\epsilon_{ul,b} = 0.302$ of the base metal were obtained directly. The $\sigma - \epsilon$ curve, denoted as σ_b, is depicted as a smooth curve in Figure 3b.

As can be seen from Figure 3b, the $\sigma - \epsilon$ curve does not exhibit a clear yield plateau. After reaching the yield strength, so-called work hardening is specific to the base metal. Due to the difficulty of experimental investigation of the mechanical properties of the laser-processed metal, the $\sigma - \epsilon$ curve of this metal is unknown. Only indirect estimations of the yield strength, $\sigma_{0.2,l} = 412$ MPa, and the ultimate strength, $\sigma_{B,l} = 665$ MPa, were obtained, using the J. R. Cahoon equations, [21,22], which establish the relations between the hardness and strength of metals. However, the ultimate strains, $\epsilon_{ul,l}$ of the laser-processed metal are still unknown. Therefore, it is assumed that the ultimate strains of the laser-processed metal are the same as for the laser-unprocessed metal, i.e., it is assumed that $\epsilon_{ul,l} = \epsilon_{ul,b} = 0.302$.

The moduli of elasticity, $E_b = 200$ GPa and $E_l = 210$ GPa, Poisson's ratios $v_b = v_l = 0.28$ and the shear moduli, $G_b = 78.1$ GPa and $G_l = 82$ GPa of the base metal and the laser-processed metal respectively, were taken from the reference literature [23]. Having the yield strength, $\sigma_{0.2,i}$, and the moduli of elasticity, E_i, $i \in \{b, l\}$, the strains $\epsilon_{1,b} = 1.28 \cdot 10^{-3}$ and $\epsilon_{1,l} = 1.962 \cdot 10^{-3}$ that correspond to the yield strength $\sigma_{0.2,i}$ were calculated by the formula $\epsilon_{1,i} = \sigma_{0.2,i}/E_i$.

The strength coefficients of the laser-processed and unprocessed metals, $E_{1,b} = 512$ MPa and $E_{1,l} = 843$ MPa, respectively, were calculated by the formula $E_{1,i} = (\sigma_{B,i} - \sigma_{0.2,i})/\left(\epsilon_{ul,i} - \epsilon_{1,i}\right)$, $i \in \{b, l\}$, where $\sigma_{0.2,b} = 256$ MPa and $\sigma_{0.2,l} = 412$ MPa are the yield strengths of the base metal and the laser-processed metal, respectively, while $\sigma_{B,b} = 410$ MPa and $\sigma_{B,l} = 665$ MPa are the ultimate strengths of the same metals.

On the basis of the obtained values of the mechanical properties of the metals, the bilinear approximations of $\sigma - \epsilon$ curves of the laser-unprocessed and processed metals $\hat{\sigma}_b$. and $\hat{\sigma}_l$, respectively, are depicted in Figure 3b. All the values of the mechanical properties of the two metals discussed above are summarized in Table 7.

Table 7. Parameters of base metal and laser-processed layer used to simulate elastoplastic deformation of samples [24,25].

Material	Modulus of Elasticity, E [GPa]	Shear Modulus, G [GPa]	Yield Strength, $\sigma_{0.2}$ [MPa]	Ultimate Strength, σ_B [MPa]	Poisson's Ratio ν	Strength Coefficient E_1 [GPa]	Yield Stress Strains ϵ_1 (10^{-3})	Ultimate Strains ϵ_{ul} (10^{-3})
Base metal	200	78.1	256	410	0.28	0.512	1.280	302.0
Laser-processed layer	210	82	412	665	0.28	0.843	1.962	302.0

3.4. Analytical Analysis of the Stiffness Properties of the Laser-Unprocessed and Processed Plates

The geometry and dimensions of the modeled plate are given in Figure 1. The material properties are given in Table 7. The main assumptions are as follows: (1) the laser-processed metal has a perfect bond with the base metal; that is, no slip occurs between the laser-processed layer and the layer of the base metal; (2) the hypotheses of the plane sections are valid; (3) the influence of Poisson's ratio is ignored; (4) only the normal stresses σ acting along the laser tracks (see Figure 1a) are taken into account; (5) the stress-strain diagram for the laser-processed layer and the base metal is bilinear, as shown in Figure 3b.

Then, under the accepted assumptions, we can write that the total axial force acting on the cross-section of the plate, $N(\epsilon) = N_b(\epsilon) + N_l(\epsilon)$; where $N_b(\epsilon) = A_b\hat{\sigma}_b(\epsilon)$ and $N_l(\epsilon) = A_l\hat{\sigma}_l(\epsilon)$ are the axial forces acting in the base metal and laser-processed metal of the plate, respectively; A_b and A_l are the cross-sectional areas of the base metal and the laser-processed metal, respectively; and $\hat{\sigma}_l$ are the stress functions of the base and the laser-processed metals, respectively, depending on the strains, ϵ.

When the bilinear stress–strain diagrams are applied (see Figure 3b), then these functions $\hat{\sigma}_b$ and $\hat{\sigma}_l$ can be expressed as follows:

$$\hat{\sigma}_i(\epsilon) = \begin{cases} E_i\epsilon, \text{ as } \epsilon \leq \epsilon_{1,i}, \\ E_i\epsilon_{1,i} + E_{1,i}(\epsilon - \epsilon_{1,i}), \text{ as } \epsilon_{1,i} < \epsilon \leq \epsilon_{ul,i}, \; i \in \{b, l\}, \\ 0, \text{ as } \epsilon > \epsilon_{ul,i}. \end{cases} \tag{1}$$

where E_b, $E_{1,b}$, E_l, $E_{1,l}$, $\epsilon_{ul,b}$ and $\epsilon_{ul,l}$ are the moduli of elasticity, the stiffness moduli and the ultimate strains of the base and the laser-processed metals, respectively; see Table 7 and Figure 3b.

Since $\epsilon_{1,b} \leq \epsilon_{1,l}$ and $\epsilon_{ul,b} \leq \epsilon_{ul,l}$, then by putting functions $\hat{\sigma}_b$ and $\hat{\sigma}_l$ (see Equation (1)), in $N(\epsilon) = \sum_{i\in\{b,l\}} A_i\hat{\sigma}_i(\epsilon)$, we obtain an explicit function for the axial force, N:

$$N(\epsilon) = \begin{cases} \epsilon \sum_{i\in\{l,b\}} A_iE_i, \text{ as } \epsilon \leq \epsilon_{1,b}, \\ A_b\big(E_b\epsilon_{1,b} + E_{1,b}(\epsilon - \epsilon_{1,b})\big) + A_lE_l\epsilon, \text{ as } \epsilon_{1,b} < \epsilon \leq \epsilon_{1,l}, \\ \sum_{i\in\{l,b\}} A_i(E_i\epsilon_{1,i} + E_{1,i}(\epsilon - \epsilon_{1,i})), \text{ as } \epsilon_{1,l} < \epsilon \leq \epsilon_{ul,b}, \\ A_l\big(E_l\epsilon_{1,l} + E_{1,l}(\epsilon - \epsilon_{1,l})\big), \text{ as } \epsilon_{ul,b} < \epsilon \leq \epsilon_{ul,l}, \\ 0, \text{ as } \epsilon > \epsilon_{ul,l}. \end{cases} \tag{2}$$

3.5. Evaluation of the Cross-Sectional Area of Laser-Processed Metal Track

The geometry of the laser-processed layer is not rectangular, but of more difficult geometry, as shown in Figure 1. Therefore, it is worth discussing further the evaluation of the area of the cross-section of the laser-processed metal, A_l. As can be seen from Figures 1d and 4, the cross-sections of the laser-processed layer and a single laser-processed track are not rectangular. If the laser tracks do not overlap, i.e., $d_c \geq d_t$, where d_t is the laser-processed track width and d_c is the distance between the tracks of the laser-processed metal (see Figure 1d, case IV), and the number of laser tracks, n_{tr}, is known, then the cross-section of the laser-processed metal can be calculated easily.

$$A_l = n_s n_{tr} A_{tr,1}, \text{ if } d_c - d_t \geq 0, \text{ and } n_{tr} \text{ is known} \tag{3}$$

where $A_{tr,1} = \pi d_t^2/8$ is the cross-sectional area of one laser-processed track, assumed to be a semicircle; $n_s \in \{1, 2\}$ is the number of laser-processed sides of the plate. It should be noted that, in Equation (3) and hereafter in the article, the number of laser tracks, n_{tr}, may not be an integer; i.e., in general, $n_{tr} \in \mathbb{R}$.

However, if n_{tr} is not known or the laser-processed tracks overlap, i.e., $d_c < d_t$, then the exact evaluation of A_l can be complicated. Some suggestions are made below on the following assumptions: the cross-section of a laser track is a semicircle whose area $A_{tr,1} = \pi d_t^2/8$, where d_t is the width of the laser track (see Figure 1d); the width of the tracks, d_t, is constant for all laser tracks and for the entire length of all laser tracks; the distance between centers of the laser tracks, d_c (see Figure 1d) is also constant for all pairs of the vicinal laser tracks.

The calculation of A_l depends on the distances between the tracks, d_c. If the difference $d_c - d_t > 0$ then the laser tracks do not overlap each other and, in general, the exact evaluation of A_l is impossible without knowledge of the exact position of the laser tracks. However, the lower and upper bounds $A_{l,inf}$ and $A_{l,sup}$ can be suggested to estimate A_l, $A_{l,inf} \leq A_l \leq A_{l,sup}$:

$$A_{l,inf} = n_s A_{tr,1} floor(b_l/d_c), \text{ if } d_c - d_t > 0, \tag{4}$$

$$A_{l,sup} = n_s A_{tr,1} ceil(b_l/d_c), \text{ if } d_c - d_t > 0. \tag{5}$$

where $floor(x) = max\{y \in \mathbb{Z},\ y \leq x\}$ and $ceil(x) = min\{y \in \mathbb{Z},\ y \geq x\}$ are ceiling and floor functions; $\mathbb{Z}$ is the set of the integer numbers; b_l is the width of the laser-processed area. When b_l/d_c increases, then the relative differences of the estimations $\left(A_{l,sup} - A_{l,inf}\right)/A_{l,sup}$ and $\left(A_{l,inf} - A_{l,sup}\right)/A_{l,sup}$ decreases. For the present case, when $d_t = 0.7$ mm, and $d_c \in \{1.05,\ 1.4\}$ mm (see Figure 1d), Equations (3) and (4) give:

$A_{l,inf} = 14A_{tr,1}$ and $A_{l,sup} = 15A_{tr,1}$ when $d_c = 1.4$ mm, and $A_{l,inf} = 19A_{tr,1}$ and $A_{l,sup} = 20A_{tr,1}$ when $d_c = 1.05$ mm.

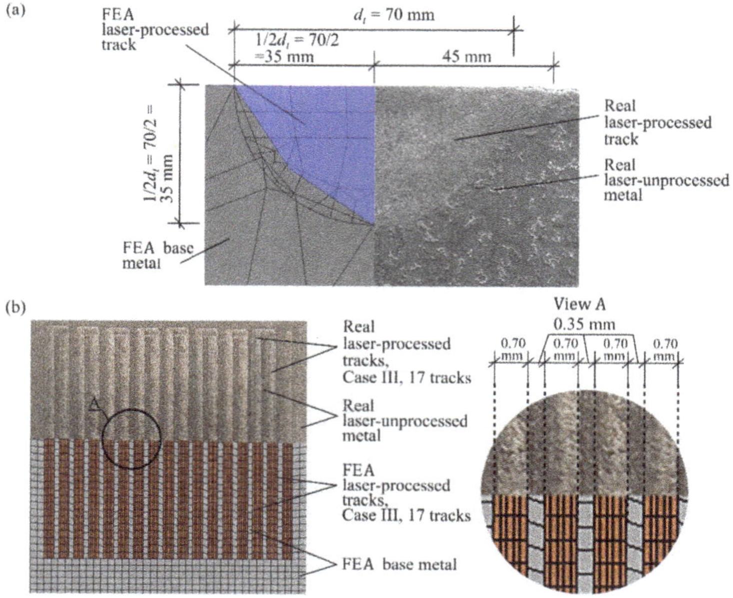

Figure 4. Cross-sections of a laser-processed track and the base metal of a FEA and a real specimen respectively (**a**) and the general view of the laser-processed tracks of case III of the real specimen and its FEA model respectively (**b**).

If n_{tr} is unknown, then it can be estimated as follows:

$$\hat{n}_{tr} = b_l/d_c \tag{6}$$

It is clear that Equation (6) can be used to evaluate the distance between the laser-processed tracks $d_c = b_l/\hat{n}_{tr}$.

When $d_c - d_t = 0$ and n_{tr} is known, then A_l can be calculated by Equation (3). Otherwise, when the number of laser-processed tracks, n_{tr} is not known, then A_l can be calculated by assuming that there are no laser-unprocessed bands between the laser-processed tracks by the following formulae:

$$A_l = n_s A_{tr,1} b_l/d_c, \text{ if } d_c - d_t = 0. \tag{7}$$

When $d_c - d_t < 0$, then the laser tracks overlap each other and we have to take into account the overlapped areas of the tracks, see Figure 1d, case II. Therefore, Equations (3)–(5) and (7) are not valid. On the basis of the above assumptions, the following bounds infimum $A_{l,inf}$ and supremum $A_{l,sup}$, $A_{l,inf} \leq A_l \leq A_{l,sup}$, of the cross-section area A_l of the laser-processed metal are derived when n_{tr} is not known

$$A_{l,inf} = n_s \hat{n}_{tr}(A_{tr,1} - A_{ov}), \text{ if } d_c - d_t < 0, \tag{8}$$

$$A_{l,sup} = \min\left\{n_s(\hat{n}_{tr}A_{tr,1} - (\hat{n}_{tr} - 1)A_{ov}), b_l 1/2d_t\right\}, \text{ if } d_c - d_t < 0, \tag{9}$$

where $\hat{n}_{tr}$ is an estimation of the number of laser-processed tracks:

$$\hat{n}_{tr} = floor\left(\frac{b_l}{d_c}\right) - 1 + \left(1 + frac\left(\frac{b_l}{d_c}\right)\right)\frac{d_c}{d_t} \tag{10}$$

where $frac(x) = x - floor(x)$ is the fractional part of a number x.

When the number of the tracks n_{tr} is known:

$$A_l = n_s(n_{tr}A_{tr,1} - (n_{tr} - 1)A_{ov}),\ \text{if } d_c - d_t < 0 \text{ and } n_{tr} \text{ is known} \tag{11}$$

In Equations (8), (9) and (11), A_{ov} is the overlapping area of two adjacent laser tracks; see Figure 1d, case II. Under the accepted assumptions, A_{ov} can be calculated as the area of the circle segment $A_{ov} = d_t^2(\alpha - \sin(\alpha))/8$, where $\alpha = 2\tan^{-1}\left(\sqrt{d_t^2 - d_c^2}/d_c\right)$ is the sector angle, $\tan^{-1}$ is the inverse tangent function. It should be noted that $A_{l,inf} \leq A_{l,sup} \leq A_{l,upp} = 1/2\ b_l d_t$, where $A_{l,upp}$ is the upper bound of the cross-section of the laser-processed layer.

4. Results

The results of the structural analysis and hardness of the laser-processed surface layer are given in Section 4.1. The comparative analyses of the experimental and numerical FEA results of the bending of the laser-unprocessed (case I) and laser-processed plates are given in the Sections 4.2 and 4.3. In Section 4.2 the numerically obtained von Mises stresses and forces F_{exp} at the different vertical displacements $w \in \{0.5, 1.0, 1.5, 2.0\}$ mm imposed at midpoint B are presented, see Figure 2b. The relation between the experimentally determined vertical forces F_{exp}, imposed at the midpoint B, and the deflection w is analyzed in Section 4.3. Furthermore, a comparison of the experimental and calculated-by-FEA vertical forces F_{exp} and F_{calc} is given in Section 4.3. In Section 4.4.1, the results of the analytical analysis of the cross-section areas A_l of the laser-processed metal depending on the track width of the laser-processed metal d_t and the distances between these tracks centers d_c are given. In Section 4.4.2, the results of the analytical analysis of the influence of the laser processing of the plate metal on the axial stiffness of the cross-section and the force-strain behavior of the plates under investigation are given. The obtained results definitely showed that the axial and flexural stiffnesses of the laser-processed cross-sections are bigger than the stiffnesses of the laser-unprocessed cross-sections.

4.1. Results of the Structural Analysis of the Laser-Processed Surface Layer

The thickness of the laser-processed layer, which is established by metallographic examination of cross-sections of samples of metal processed by laser, was about 0.35 mm (see Figure 4). There are no unacceptable inclusions, porosity or internal defects in the remelted area and transition area. The microstructure of the base metal consists of 70% ferrite and 30% pearlite (see Figure 5a), according to GOST 8233 [26]. Granularity in the processed zone decreases from G8 (average grain diameter about 18 μm) to G10 (average grain diameter about 10 μm), according to ISO 643 [27]. According to the XRD data analysis (see Figure 6), the X-ray diffraction pattern of the laser-processed layer is typical for the ferrite microstructure family (including sorbite and troostite) with BCC crystal structure). The same crystal structure has martensite and bainite. The Fe_3C peak and other carbide peaks were not observed. This is typical, so carbon content in the steel is low or peak sensitivity is below the limit of detection sensitivity. Therefore, XRD analysis confirms only that there is no unstable, retained austenite in the laser-processed area, because the austenite has the FCC crystal structure and other XRD diffraction peaks. In low-carbon steel, high-retained austenite contents are usually found together with the martensite or bainite phase in the quenched steel after cooling. The measurement of microhardness shows that the hardness of the laser-processed layer increases up to 200 HV (by 60%) compared to the laser-unprocessed base metal hardness (see Figure 7). According to the hardness measurement results, there are no hard and brittle bainite and martensite microstructures in the laser-processed areas or their proportion is very low, because such quenching microstructures have the highest hardness: bainite—about 400 HV and martensite—above 450 HV [28,29]. Consequently, to achieve a hardness of about 200 HV, it is necessary to have a minimum of about 25% of bainite or martensite structure in the present low-carbon steel after rapid cooling of the microstructure of the melted area, or the steel's microstructure must be strengthened by refining the grain size and creating a finely dispersed perlite (sorbite, troostite) structure.

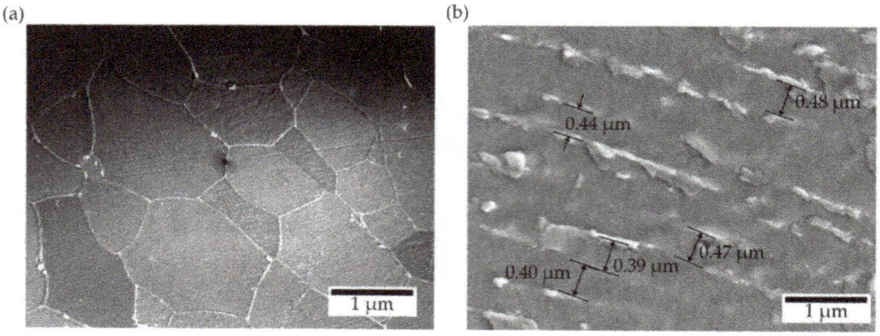

Figure 5. SEM images of microstructure: (**a**) ferrite-pearlite structure of base metal (magnification ×2000); (**b**) distance between lamella in the sorbite structure of laser-processed layer (magnification ×10,000).

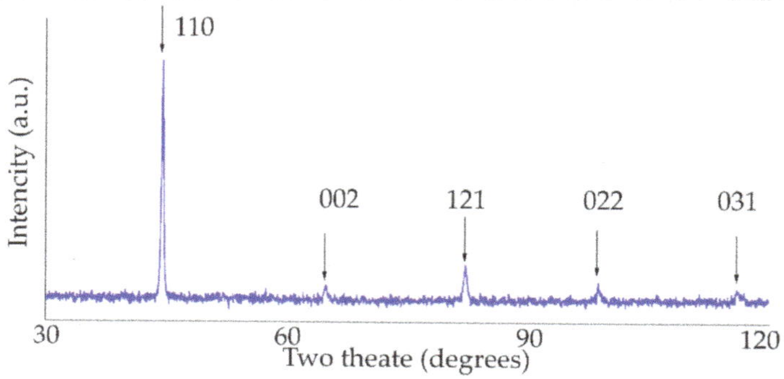

Figure 6. XRD pattern of the laser-processed layer.

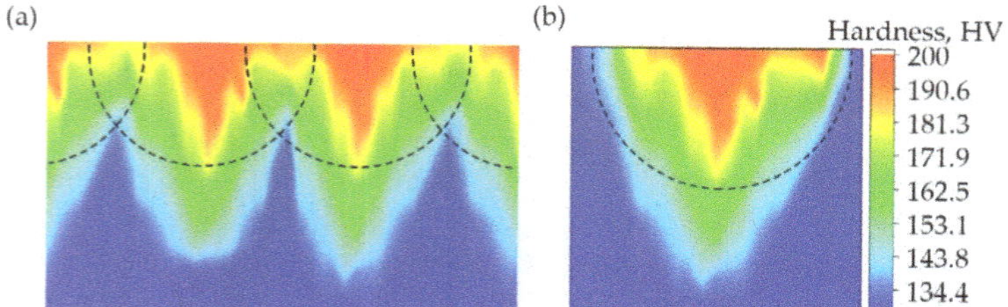

Figure 7. Distribution of hardness through the cross-section of the laser-processed surface of sample IIA: (**a**) and (**b**) show the distributions of three- and one-laser-processed tracks respectively.

The microstructure located at the laser-processed area of the low-carbon steel sample demonstrated a typical sorbite structure (Figure 5b). The distance between lamella, which is measured by SEM, is less than 0.3–0.4 µm. This distance is typical for sorbite, because perlite has a distance between lamella of about 0.6–0.7 µm, troostite—about 0.1 µm, martensite and bainite—about 0.2 µm and thickness of retained austenite lamellae—0.05–0.2 µm [30]. The hardness of the laser-processed layer also corresponds to the typical hardness of sorbite, which must be in the range 200–300 HV [31].

The sorbite structure of the laser-processed layer was formed due to the applied high overlap coefficient of laser spots, with inevitable additional heating and partial remelting of the crystallized pool during the next laser pulse. This effect allows the cooling rate of the melted pool to be reduced and prevents the formation of more brittle, quenched structures in the laser-processed layer.

The sorbite structure has many advantages in this case, because sorbite has a finer texture, with higher dispersity and stiffness than pearlite, which increases the strength and wear resistance of the laser-processed metal parts, without loss in plasticity, that is typical for hard and brittle quenching (martensite or bainite) microstructures [32]. SEM-EDS element mapping, line scan and point analysis of the distribution and concentration of chemical elements (C, Mn, Si) in the remelted area show that there is no significant chemical heterogeneity in the laser-processed layer (see Figures 8 and 9, Table 8). This uniform distribution of the chemical elements and the homogeneous structure of the remelted layer can positively influence the mechanical properties of the laser-processed layers and the entire plate.

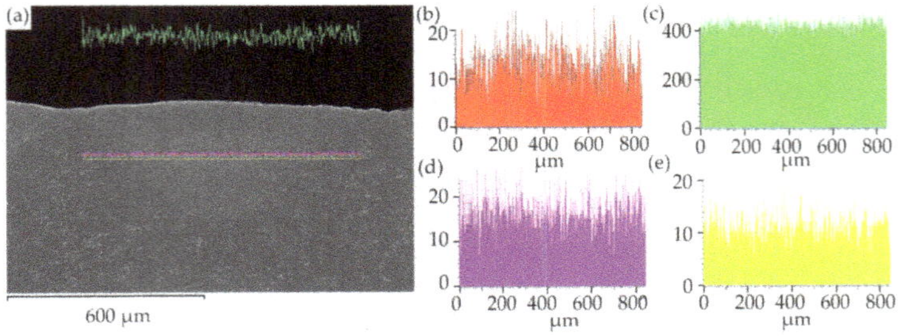

Figure 8. Line scan element analysis along the surface of laser-processed layer: (**a**) place of line scan in cross-section of layer; (**b**) C distribution; (**c**) Fe distribution; (**d**) Si distribution; (**e**) Mn distribution.

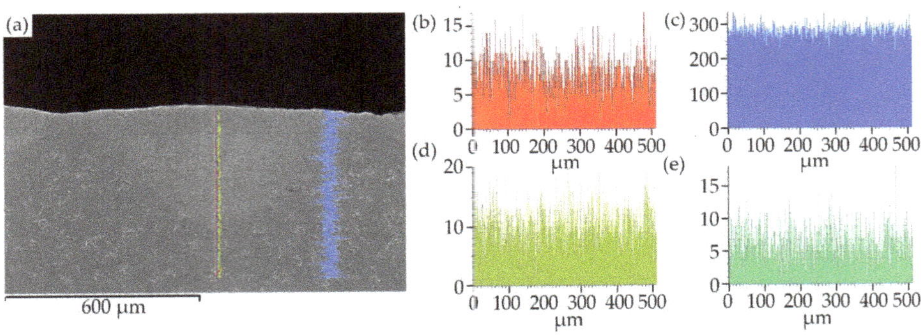

Figure 9. Line scan element analysis normal to the laser-processed surface: (**a**) place of line scan in cross-section of layer; (**b**) C distribution; (**c**) Fe distribution; (**d**) Si distribution; (**e**) Mn distribution.

Table 8. Chemical compositions (by energy-dispersive X-ray spectroscopy analysis) of surface regions.

Region	Chemical Elements [wt.%]				
	C	Fe	Si	Mn	Other
Laser-processed area	1.49	97.28	0.20	0.42	0.61
Basic metal	1.13	97.67	0.21	0.43	0.56

4.2. Results of the FEA Simulations of the Bending of the Laser-processed Plates

The increase in the bending force varied, depending on the plate treatment case and deflection level: for 0.5 mm deflection, the force was 11–27%; for 1.0 mm deflection about 43–49%; for 1.5 mm deflection—32–55%; for 2.0 mm deflection—30–52%. The increase in the maximum equivalent stresses varied according to the plate treatment case and deflection level: for 0.5 mm deflection—13–30%; for 1.0 mm deflection—13–32%; for 1.5 mm deflection—8–26%; for 2.0 mm deflection—8–25%.

The maximum equivalent stress at the bent plates depends on the volume and position of the laser-processed layer. The greatest increase in the required bending load and available stresses, according to the modeling results, was in the double-sided laser-processed samples, while the volume of treatment was greatest from 15.4% to 19.8% of the volume of the bent plate part. The increased bending load required for similar deflection depends also on the position of the laser-processed layer. Better results were obtained in the bent samples where the zone of the laser-processed layer was subjected to tension, rather than compression.

It was determined that double-side laser-processed metal plates with 30% overlay of laser tracks had the highest resistance to bending, compared to the other treatment modes; see Table 9. However, modeling results (see Table 9 and Figure 10) show that it is possible to use a non-continuous laser with either one-side or double-side processing with a certain distance between tracks, because the difference in the efficiency of such surface treatments is small. The difference of the required bending loads for one deflection level of the plate, applied to different positions of the laser-processed layer (cases A, B, C) did not reach 20%.

Table 9. The maximum Von Mises equivalent stresses (in MPa), and corresponding forces F_{calc} (in N) of the finite element analysis (FEA) simulation of the bending of the differently laser-processed plates at different imposed vertical displacements $w \in \{0.5, 1.0, 1.5, 2.0\}$ mm of the middle point B.

Imposed Vertical Displacement of the Point B, see Figure 2b, [mm]	**0.5**	**1.0**	**1.5**	**2.0**
Laser-Unprocessed Plate (Case I)				
Maximum von Mises stress, [MPa]	176	293	395	409
Bending force F_{calc}, [N]	179	231	269	290
Laser-Processed Plate (Case IIA)				
Maximum von Mises stress, [MPa]	217	369	460	468
Bending force F_{calc}, [N]	215	318	390	409
Laser-Processed Plate (Case IIB)				
Maximum von Mises stress, [MPa]	219	372	463	470
Bending force F_{calc}, [N]	217	320	392	410
Laser-Processed Plate (Case IIC)				
Maximum von Mises stress, [MPa]	228	385	497	510
Bending force F_{calc}, [N]	228	345	417	442
Laser-Processed Plate (Case IIIA)				
Maximum von Mises stress, [MPa]	209	350	431	460
Bending force, [N]	205	311	375	394
Laser-Processed Plate (Case IIIB)				
Maximum von Mises stress, [MPa]	210	352	432	462
Bending force F_{calc}, [N]	206	312	378	398
Laser-Processed Plate (Case IIIC)				
Maximum von Mises stress, [MPa]	223	382	485	505
Bending force F_{calc}, [N]	222	334	411	428
Laser-Processed Plate (Case IVA)				
Maximum von Mises stress, MPa	199	332	425	448
Bending force F_{calc}, N	184	298	354	378
Laser-Processed Plate (Case IVB)				
Maximum von Mises stress, MPa	200	338	426	452
Bending force F_{calc}, N	187	300	356	381
Laser-Processed Plate (Case IVC)				
Maximum von Mises stress, MPa	220	380	480	502
Bending force F_{calc}, N	219	328	403	419

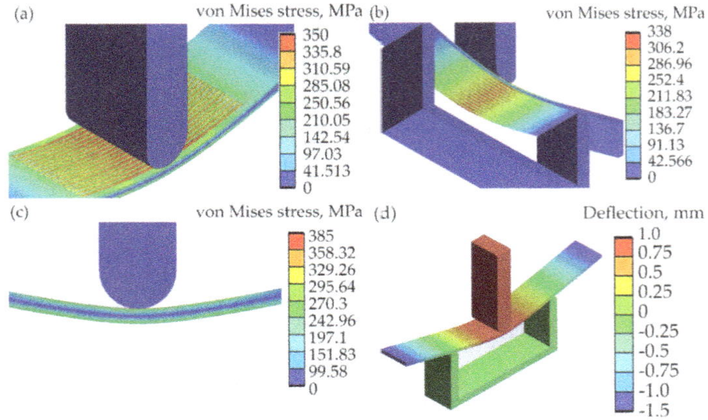

Figure 10. FEA results of the bending modeling of the laser-processed plates when the imposed vertical displacement of the plate middle point B, see Figure 2b, equals 1 mm; von Mises stresses are shown in (**a**–**c**): (**a**) for case IIIA, (**b**) for case IVB and (**c**) for case IIC; while (**d**) shows the deflections or vertical displacements of the plate of case IIIA.

4.3. Commparison of the Experimental and Modeling Results of the Bending of the Laser-Processed and Unprocessed Plates

The mechanical bend tests showed that local laser processing increases the load required to reach the same level of deflection of laser-processed samples compared to unprocessed ones; see Figure 11 and Table 10. The experimental bending load, F_{exp}, increases depending on the laser-processed case: for 0.5 mm deflection, the F_{exp} increases from 3% up to 27%; for 1.0 mm deflection, F_{exp} increases from 29% up to 48%; for 1.5 mm deflection, F_{exp} increases from 32% up to 55%; for 2.0 mm deflection, F_{exp} increases from 30% up to 52%.

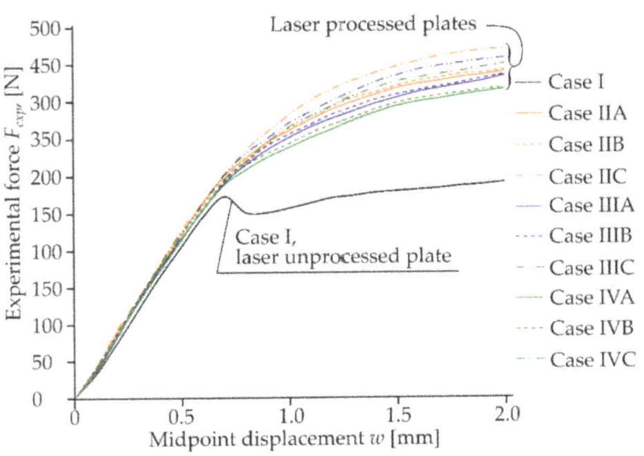

Figure 11. Experimental bending force F_{exp} vs. deflections w of the middle point of the differently laser-processed plates.

Experimental data confirmed that the effect of strengthening and total resistance of samples to bending was influenced by the position of the laser-processed area, the distance between laser tracks, the volume of the hardened phase and its ratio to the volume of laser-untreated material in the area of maximum stress from the applied load. The difference between modeling and bending test results was less than 15% (Table 10).

Table 10. Experimental and modeled bending forces F_{exp} and F_{calc} respectively, and their relative differences $d_{F,r}$ at different deflections w.

Bending Forces F_{exp}, F_{calc} and Relative Differences $d_{F,r}=(F_{exp}-F_{calc})/F_{calc}$	**Case I**	**Case IIA**	**Case IIB**	**Case IIC**	**Case IIIA**	**Case IIIB**	**Case IIIC**	**Case IVA**	**Case IVB**	**Case IVC**
Deflection, 0.5 mm										
F_{exp}, [N]	200	226	228	242	220	221	236	209	211	230
F_{calc}, [N]	179	215	217	228	205	206	222	184	187	219
$d_{F,r}$, [%]	12	5	5	6	7	7	6	13	13	5
Deflection, 1 mm										
F_{exp}, [N]	261	365	368	387	352	356	376	333	336	370
F_{calc}, [N]	231	318	320	345	311	312	334	298	300	328
$d_{F,r}$, [%]	13	15	15	13	13	14	13	12	12	13
Deflection, 1.5 mm										
F_{exp}, [N]	282	425	426	455	412	413	437	402	403	431
F_{calc}, [N]	269	390	392	417	375	378	411	354	356	403
$d_{F,r}$, [%]	5	9	9	9	10	9	6	13	13	7
Deflection, 2 mm										
F_{exp}, [N]	296	445	447	475	439	440	463	421	423	455
F_{calc}, [N]	290	409	410	442	394	398	428	378	381	419
$d_{F,r}$, [%]	2	9	9	7	11	10	8	11	11	9

4.4. Results of the Analysis of the Axial Stiffness of the Laser-Processed Plate

The analysis of the results is presented mainly in terms of the relative quantities to make the analysis, discussion and results valid, not only for the particular cases considered in the present article, but also to extend the results and conclusions to other cases.

4.4.1. Area A_l of the Laser-Processed Metal

Figure 12 shows the dependencies of the bounds of the cross-sectional area of the laser-processed metal, $A_{l,inf}$ and $A_{l,sup}$, as well as their ratios $A_l/(1/2b_lb_t)$, where $A_l \in \{A_{l,inf}, A_{l,sup}\}$ on the estimated number of the laser-processed tracks, $\hat{n}_{tr}$, calculated by Equations (6) or (10), and on the ratio d_c/d_t. The areas $A_{l,inf}$ and $A_{l,sup}$ were calculated by Equations (4), (5), (7)–(10) of the plate, shown in Figure 1, when the width of the laser-processed track $d_t = 0.7$ mm and the width of the laser-processed area of the plate under consideration, $b_l = 0.02$ m. The vertical dotted lines correspond to the minimum and maximum of $\hat{n}_{tr}$ or d_c/d_t and laser-processing cases: case II when $d_c/d_t = 2/3$; case III when $d_c/d_t = 1.5$; and case IV when $d_c/d_t = 2$; see Table 5.

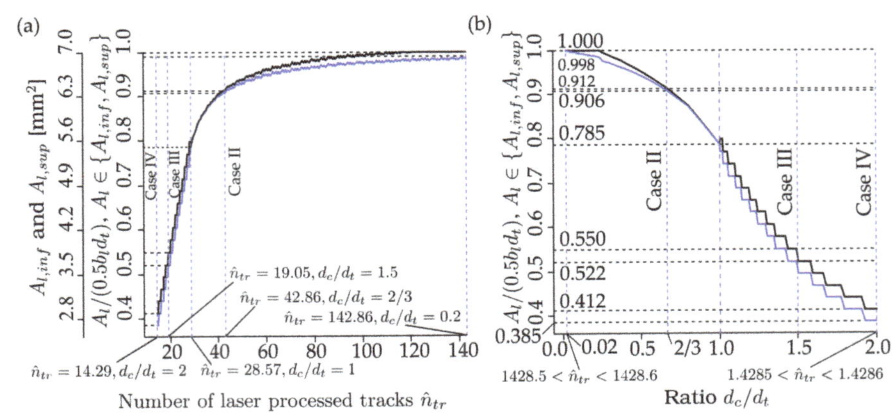

Figure 12. The dependencies of the bounds of the cross-section of the laser-processed metal $A_{l,inf}$ and $A_{l,sup}$ and the ratios $A_{l,inf}/(1/2b_ld_t)$ and $A_{l,sup}/(0.5b_ld_t)$ on the number of tracks $\hat{n}_{tr}$ and ratio d_c/d_t of the considered plate; see Figure 1: (**a**) when $\hat{n}_{tr} \in [14.29, 142.86]$, corresponding to $d_c \in \left[1.4\times10^{-4}, 1.4\times10^{-3}\right]$ or $d_c/d_t \in [0.2, 2]$; and (**b**) when $d_c/d_t \in [0.02, 2]$, corresponding to $d_c \in \left[1.4\times10^{-5}, 1.4\times10^{-3}\right]$ or $\hat{n}_{tr} \in [14.286, 1428.571]$.

Please note: in the present calculations, the numbers of tracks, n_{tr}, for cases II–IV were evaluated by Equations (6) or (10), i.e., n_{tr} was not taken from Table 5.

As can be seen from Figure 12, $A_{l,inf}$ approaches $A_{l,sup}$ with increasing numbers of laser-processed tracks, $\hat{n}_{tr}$, or with a decrease in the ratio, d_c/d_t or d_c. Also from Figure 12, it can be seen that the rate of increase of the areas, $A_l \in \{A_{l,inf}, A_{l,sup}\}$, is practically constant with respect to increasing n_{tr} within the interval $n_{tr} \in [\min\{n_{tr}\}, 28, 57]$, corresponding to the interval, $d_c/d_t \in [1, max\{d_c \backslash d_t\}]$, or when $d_c/d_t \geq 1$.

More remarkably, the rate of increase of $A_l \in \{A_{l,inf}, A_{l,sup}\}$ even increases with respect to decreasing d_c/d_t within the interval $d_c/d_t \in [max\{d_c \backslash d_t\}, 1]$. However, as shown in Figure 12, the increase in the areas $A_l \in \{A_{l,inf}, A_{l,sup}\}$ becomes slower with increasing n_{tr} or decreasing d_c/d_t or d_c at $d_c/d_t < 1$.

Finally, the increase of $A_l \in \{A_{l,inf}, A_{l,sup}\}$ practically does not change at a very large $\hat{n}_{tr}$. It should be noted that the biggest difference $max\{A_{l,sup} - A_{l,inf}\} = 0.1924$ mm^2 and the $(max\{(A_{l,sup} - A_{l,inf})/A_{l,sup}\} \cdot 100 = 6.66\%)$ occurs at the largest value of ratio $d_c/d_t = 2$, or at the smallest $\hat{n}_{tr} = 14.29$.

Figure 13 shows the dependency of the ratio of the area of the remelted metal to the supremum bound of the area of the laser-processed metal $A_{l,rem}/A_{l,sup}$ (Figure 13a), and the dependency of the ratio of the overlapping area on the cross-sectional area of the single laser-processed track $A_{ov}/A_{tr,1}$ on the ratio d_c/d_t of the plate under consideration; see Figure 1, where the total area of the remelted metal is calculated as follows: $A_{l,rem} = (n_{tr} - 1)A_{ov} = (b_l/d_c)A_{ov}$.

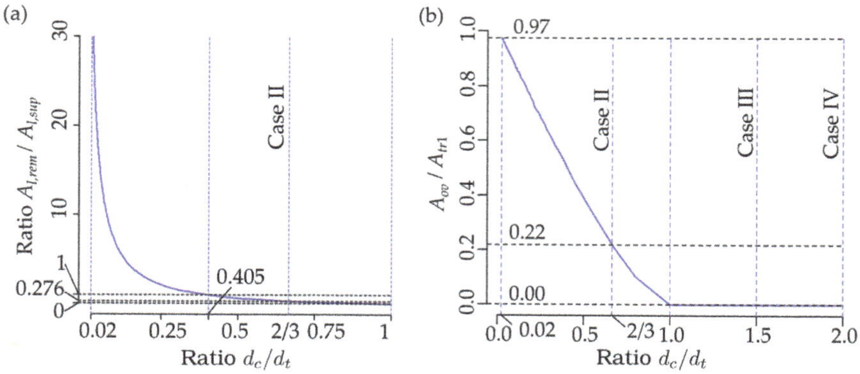

Figure 13. The dependencies of the ratios $A_{l,rem}/A_{l,sup}$ and $A_{ov}/A_{tr,1}$ on the ratio d_c/d_t of the considered plate, see Figure 1: (**a**) $A_{l,rem}/A_{l,sup}$ versus $d_c/d_t \in [0.02, 1]$; and (**b**) $A_{ov}/A_{tr,1}$ versus $d_c/d_t \in [0.02, 2]$.

It is clear that $A_{l,rem} = A_{ov} = 0$ as $d_c > d_t$ or $d_c/d_t > 1$. As we can see from Figure 13a,b, the rate of the increase of the ratio $A_{ov}/A_{tr,1}$ on d_c/d_t jumps sharply at $d_c/d_t = 1$, and this rate is almost constant within the interval $d_c/d_t \in [0.02, 1]$, since $A_{ov}/A_{tr,1}$ on d_c/d_t depends very similarly to the line. However, the dependence of the ratio of the total area of the remelted metal to the supremum bound of the laser-processed metal $A_{l,rem}/A_{l,sup}$ increases very slowly at $d_c/d_t = 1$, but becomes very steep with a higher ratio of d_c/d_t.

The behavior of $A_{l,rem}/A_{l,sup}$ with respect to d_c/d_t differs from $A_{ov}/A_{tr,1}$ due to the influence of the number of laser-processed tracks, n_{tr}. At relatively large d_c/d_t, the number n_{tr} is small. Therefore, the contribution of A_{ov} to $A_{l,rem}$ is also small. With decreasing d_c/d_t the number n_{tr} increases and, hence the contribution of increased A_{ov} to $A_{l,rem}$ increases also.

As we can see from Figure 13a, the ratio $A_{l,rem}/A_{l,sup} = 1$ as $d_c/d_t = 0.405$. Therefore $A_{l,rem} > A_{l,sup}$ as $d_c/d_t < 0.405$, and the difference $A_{l,rem} - A_{l,sup}$ or the ratio $A_{l,rem}/A_{l,sup}$ increases very quickly with increasing $d_c/d_t \geq 1$. It should be noted that the cross-sectional area of the remelted metal, $A_{l,rem}$ can be many times, ten- or even twenty-times, bigger than the supremum bound of the cross-sectional area of the laser-processed metal $A_{l,sup}$ at small values of d_c/d_t. Throughout this subchapter, a decrease in d_c/d_t corresponds to increasing the number of laser-processed tracks, n_{tr} and vice versa, while increasing d_c/d_t corresponds to a decrease in n_{tr}.

The analysis above allows us to conclude that making a laser-processed layer is very inefficient at small values of d_c/d_t or large values of n_{tr}, since a very large volume of the metal is melted repeatedly. For practical applications, it is reasonable to take the ratio, $d_c/d_t = 1$. In this case, as we can see from Figure 12b, $A_l/(1/2b_lb_t) = 0.785$. This means that the increase in the number of laser-processed tracks n_{tr} or decrease of the distance between the tracks, d_c or the ratio d_c/d_t can increase the cross-sectional area of the laser-processed metal only up to $(1 - 0.785) \cdot 100 = 21.5\%$. However, as shown above, the cross-section of the remelted metal, or the ratio $A_{l,rem}/A_{l,sup}$, increases very quickly with decreasing d_c or increasing n_{tr}. Hence, the efficiency of laser processing decreases very quickly.

For practical applications, the optimal limit of ratio d_c/d_t can be 0.405, since at this point $A_{l,rem}/A_{l,sup} = 1$ and ratios $A_{l.inf}/(1/2b_lb_t) \approx 0.96$ and $A_{l.sup}/(1/2b_lb_t) \approx 0.97$. Therefore, a further increase in n_{tr} or decrease in d_c/d_t can increase the ratios $A_l/(1/2b_lb_t)$ only up to 4%. To increase the cross-sectional area of the laser-processed metal, A_l, it is better to increase the width, and hence the depth, of the laser-processed track d_t than to increase n_{tr} or decrease d_c/d_t.

The considerations given above concerning the ratios $A_l/(1/2b_ld_t)$, $A_{l,inf}/(1/2b_ld_t)$, $A_{l,sup}/(1/2b_ld_t)$, $A_{l,rem}/A_{l,sup}$, $A_{ov}/A_{tr,1}$ and the corresponding cross-section areas $A_{l,inf}$, $A_{l,sup}$, $(1/2b_ld_t)$, A_{ov}, $A_{tr,1}$, $A_{l,upp} = 1/2b_ld_t$ are also valid for the corresponding ratios of the volumes $V_l/V_{l,upp}$, $V_{l,inf}/V_{l,upp}$, $V_{l,sup}/V_{l,upp}$, $V_{l,rem}/V_{l,sup}$, $V_{ov}/V_{tr,1}$ and the corresponding volumes $V_{l,inf}$, $V_{l,sup}$, V_{ov}, $V_{tr,1}$, $V_{l,upp}$ of the laser-processed metal, where V stands for the volume of the laser-processed metal and its indexes "l, inf", "l, sup", "l, upp", "$V_{l,rem}$", "ov", and "$tr, 1$" means the same as the corresponding indexes of the cross-sectional areas denoted by A.

Since the above analysis, results, conclusions and recommendations are expressed in relative terms, then these considerations are also valid for other cases of plates, laser-processed areas, their widths b_l, the laser-processed track widths d_t, the distances between the track centers d_c, the number of laser tracks n_{tr} and so on.

4.4.2. Axial Stiffness and Force—Strain Behavior of the Laser-Processed Plate

The dependencies of the ratios of the axial forces of the laser-processed plates (see Figure 1, cases II, III and IV) to the unprocessed plate, case I: $N_i/N_{\text{Case I}}$, $i \in$ {CaseII, Case III, Case IV} on strain $\epsilon \in [0, \epsilon_{ul}] = \left[0,\ 302 \cdot 10^{-3}\right]$ are shown in Figure 14. The axial forces N_i, $i \in$ {Case I, ..., Case IV} were calculated according to Equations (1) and (2), using the properties of the base and laser-processed metals given in Table 7. The cross-sectional areas of the laser-processed metal of plate A_l were calculated by the proposed Equations, (3), (7) and (11) when the number of the laser-processed sides $n_s = 2$ and the number of the laser-processed tracks of one side of the plate, corresponding to cases I, II, III and IV $n_{tr} \in \{0,\ 14,\ 17,\ 38\}$ is given in Table 5. The following cross-sectional areas were obtained: $A_l = 0$ for case I, $A_l = 1.1504 \times 10^{-5}$ m^2 for case II, $A_l = 6.5423 \times 10^{-6}$ m^2 for case III, $A_l = 5.3878 \cdot 10^{-6}$ m^2 for case IV. The cross-sectional areas of the base metal of the plate corresponding to cases I–IV, $A_b = A_{pl} - A_l$, where $A_{pl} = b_{pl} = 4 \times 10^{-5}$ m^2 is the cross-sectional area of the plate; see Figure 1b.

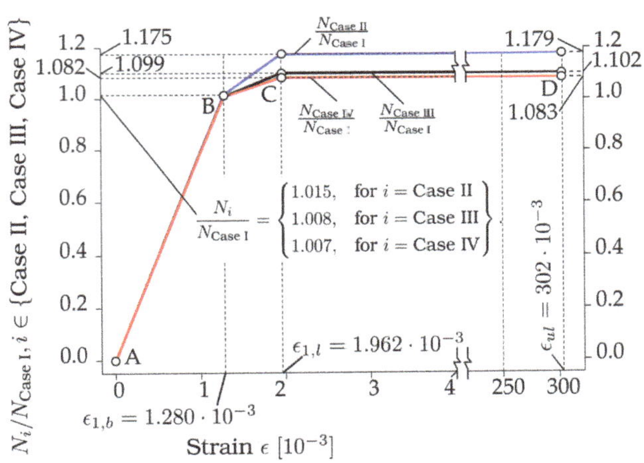

Figure 14. The dependencies of the ratios of the axial forces of the laser-processed plates, cases II, III and IV, to the unprocessed plate, case I: $N_i/N_{Case\ I}$, $i \in$ {CaseI, Case III, Case IV} on strain $\epsilon \in [0, \epsilon_{ul}] = \left[0,\ 302 \cdot 10^{-3}\right]$ on strain ϵ.

As we can see from Figure 14, the influence of laser processing is almost infinitesimal when $\epsilon \in \left[0,\ \epsilon_{1,b}\right] = \left[0,\ 1.28 \times 10^{-3}\right]$; see section AB in Figure 14. The maximum ratio, $\max\left\{N_i/N_{Case\ I},\ i \in \{\text{CaseII, Case III, Case IV}\},\ \epsilon \in \left[0,\ \epsilon_{1,b}\right]\right\} = 1.0015$. If the axial stiffness of the cross-section of the plate is $B = N/\epsilon$, then we can say that laser processing does not have any influence on the axial stiffness of the plate when $\epsilon \in \left[0,\ \epsilon_{1,b}\right]$. However, the influence of laser processing on the ratio of the axial forces $N_i/N_{Case\ I}$ increases with increasing strain within the interval $\epsilon \in \left[\epsilon_{1,b},\ \epsilon_{1,l}\right] = \left[1.28\times10^{-3}, 1.962\times10^{-3}\right]$ and remains constant when $\epsilon \in \left[\epsilon_{1,l},\ \epsilon_{ul}\right] = \left[1.962\times10^{-3}, 302\times10^{-3}\right]$, where hereafter $\epsilon_{ul} = \epsilon_{ul,b} = \epsilon_{ul,l} = 302\times10^{-3}$; see sections BC and CD in Figure 13.

As shown in Figure 14, laser processing increases the axial forces N_i and hence the axial stiffness B_i, $i \in$ {Case II, Case III, Case IV}, about 17.5% for case II, 9.9% for case III, and 8.2% for case IV. For practical structural calculations, it is reasonable to assume that the plate's bearing capacity corresponds to the axial force N_i at $\epsilon = \epsilon_{1,l}$, since the increase of the ratio $N_i/N_{Case\ I}$ within the interval $\left[\epsilon_{1,l},\ \epsilon_{ul}\right]$ is very low: (1.179–1.175) $=$ 2.5% for case II; $1.102 - 1.009 = 0.9\%$ for case III; and $1.083 - 1.082 = 0.1\%$ for case IV.

The dependencies of the ratios of the axial forces $\eta_{\epsilon,A_l/A_b} = N(A_l/A_b,\epsilon)/N(0,\epsilon)$, at $\epsilon \in \left\{\epsilon_{1,b}, \epsilon_{1,l}, \epsilon_{ul}\right\}$ with respect to the ratio of the cross-sectional areas of the laser-processed and base metals $A_l/A_b \in [0,\ 0.538]$ of the plate shown in Figure 1 are shown in Figure 15. In the situation when $A_l = 0$, since $A_l/A_b = 0$ also corresponds to the laser-unprocessed plate, i.e., case I, then $N(0,\epsilon) = N_{\text{Case I}}$. In these calculations, it was assumed that the cross-sectional area of the laser-processed metal attains values from the interval: $A_l \in \left[0,\ A_{l,up}\right] = [0,\ 2\times1/2b_ld_t] = \left[0,\ 1.4\times10^{-5}\right]$. Since $A_b = A_{pl} - A_l$, where $A_{pl} = b_{pl}\times2\times10^{-3} = 4\times10^{-5}$, then $max\{A_l/A_b\} = 1.4/(4-1.4) = 0.5385$. Thus, $max\{A_l/A_b\} = 0.5385$ corresponds to the upper bound of the cross-sectional area of the laser processed metal when $A_{l,up} = n_s1/2b_ld_t = 20\times10^{-3}\times0.7\times10^{-3} = 1.4\times10^{-3}$ of the two-side laser-processed plate; see Figure 1b. Other parameters used to draw Figure 15 were the same as in Figure 14; see the explanations given at the beginning of the present subchapter.

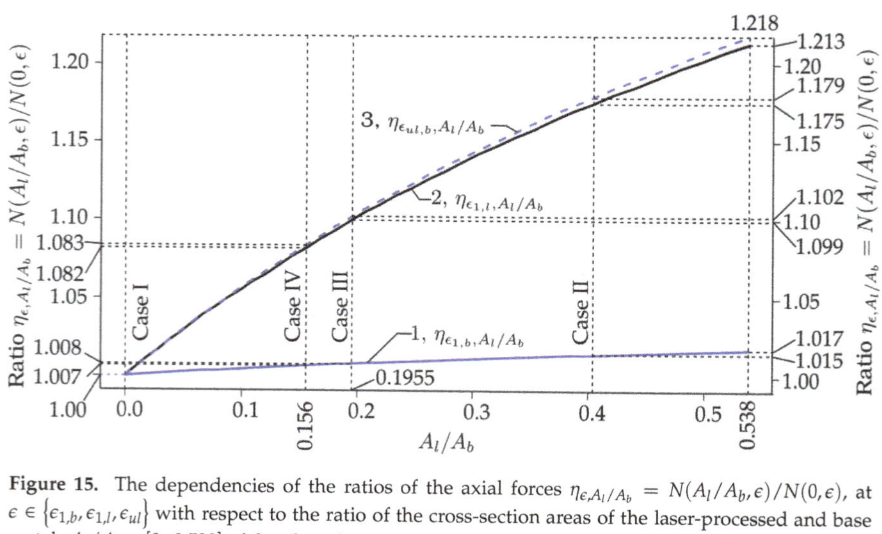

Figure 15. The dependencies of the ratios of the axial forces $\eta_{\epsilon,A_l/A_b} = N(A_l/A_b,\epsilon)/N(0,\epsilon)$, at $\epsilon \in \left\{\epsilon_{1,b}, \epsilon_{1,l}, \epsilon_{ul}\right\}$ with respect to the ratio of the cross-section areas of the laser-processed and base metals $A_l/A_b \in [0,\ 0.538]$ of the plate shown in Figure 1.

As we can see from Figure 15, the ratios $\eta_{\epsilon,A_l/A_b}$ on A_l/A_b do not change according to the linear law, as in the case of the dependence of the N on A_l when A_b is constant, according to Equation (2). This happens since the cross-sectional area of the base metal $A_b = A_{pl} - A_l$, i.e., A_b also changes when A_l changes. Also, it is clearly visible that the ratio A_l/A_b affects the ratios $\eta_{\epsilon,A_l/A_b}$, especially when $\epsilon \in \left[\epsilon_{1,b},\ \epsilon_{1,l}\right]$. Thus, when $\epsilon \in \left\{0, \epsilon_{1,b}\right\}$, then the relative difference $\Delta_{max}\eta_{\epsilon,A_l/A_b} = max\left\{\eta_{\epsilon,A_l/A_b} - \eta_{\epsilon,0}, A_l/A_b \in \{0,\ 0.538\}\right\} = \eta_{\epsilon,0.538} - \eta_{\epsilon,0} = 1.7\%$, while when $\epsilon \in \left\{\epsilon_{1,b}, \epsilon_{1,l}\right\}$, then $\Delta_{max}\eta_{\epsilon,A_l/A_b} = 21.3\%$; and when $\epsilon \in \left\{\epsilon_{1,b}, \epsilon_{ul}\right\}$, then $\Delta_{max}\eta_{\epsilon,A_l/A_b} = 21.8\%$. When the strain increases from $\epsilon_{1,l}$ to ϵ_{ul} and $A_l/A_b = 0.538$, then the relative difference $\Delta_{max}\eta_{\epsilon,A_l/A_b} = \eta_{\epsilon_{ul},0.538} - \eta_{\epsilon_l,0.538} = 0.5\%$. From Figure 15, we can see that the biggest possible increase of the axial force is when $A_l/A_b = A_{l,sup}/\left(A_{pl} - A_{l,sup}\right) = 0.538$ is $\eta_{\epsilon_{ul},0.538} - \eta_{\epsilon_l,0.538} = 1.213 - 1.017 = 19.6\%$.

5. Conclusions

1. According to the results of these experiments and computer simulations of elastoplastic deformation, it is established that the local laser processing can be used to increase the flexural stiffness of structural elements made from thin-sheet steel (1.0402) up to 5% and to decrease its deformation or deflection under similar workload up to 13–24%, instead of applying complex geometric shapes, additional stiffening elements or heat treatment.
2. The steel hardness and strength in the processed zones increased by up to 50% compared to untreated areas, which led to an increase in the flexural rigidity of the laser-processed metal plate and the maximum level of endured load without plastic deformation.
3. It has been established that the stiffness of thin-sheet steel depends on and can be altered by the parameters of laser processing, the area treated, the number of laser tracks, and the distance between tracks.
4. The conducted analytical modeling showed that, when the ratio of the distance between the centers of the laser-processed tracks and their width equals 1, i.e., no gaps exist between the laser-processed tracks, then the volume ratio of the laser-processed metal to its greatest possible volume is 0.785.
5. For practical applications, the smallest reasonable ratio of the distance between the centers of the laser-processed tracks to their width is 0.405. In this case, the volume of the repeatedly melted metal equals the volume of laser-processed metal; and the ratio of the volume of laser-processed metal to its greatest possible volume is 0.97.
6. To produce a greater volume of laser-processed metal, we suggest increasing the width of the laser-processed metal track by increasing the laser power, instead of decreasing the distance between the laser-processed tracks centers below 0.405 of their width.
7. The analytical modelling of the axial force-strain behavior of the laser-processed plates showed that the axial stiffness of the plate cross-section increases up to 21.3% at the strain corresponding to the yield point of the laser-processed metal when the distance between the laser-processed tracks equals 2/3 of their width.

Author Contributions: Methodology, N.V.; investigation, O.K.; data curation, N.V.; writing—Original draft preparation, O.K.; writing—Review and editing, N.V.; visualization O.K.; supervision, N.V.; analytical modeling, calculation and writing, D.Z.; calculation and writing, A.P. All authors have read and agreed to the published version of the manuscript.

Funding: This research received no external funding.

Conflicts of Interest: The authors declare no conflict of interest.

Nomenclature

A_l and A_b	The total areas of the cross-sections of the laser-processed and unprocessed metals respectively
$A_{l,sup}$ and $A_{l,inf}$	The supremum and infimum bounds of the cross-section area of the laser-processed metal
A_{pl}	The cross-sections area of the plate
$A_{l,rem}$	The cross-sections area of the remelted laser-processed metal due to overlapping the adjacent laser-processed metal tracks

A_{ov}	The cross-sections area of the two adjacent overlapping laser-processed metal tracks
$A_{tr,1}$	The cross-sections area of the single track of the laser-processed metal
B	Cross-section axial stiffness of a plate
b_{pl}	Width of a plate
b_l	Width of the laser-processed area
d_t and d_c	The width of the laser-processed tracks and the distance between centers of two adjacent laser-processed tracks respectively
E_i, $E_{1,i}$, G_i, and ν_i, $i \in \{b, l\}$	Modulus of elasticity, strength coefficient, shear modulus and Poisson's ratio of the base metal, when $i = b$, and the laser-processed metal, when $i = l$ respectively
FEA	Finite element analysis
F_{calc}	The vertical force imposed in the middle of the plate span calculated by FEA
F_{exp}	Experimentally determined vertical force imposed in the middle of the plate span
N, N_b and N_l	The axial forces acting in the total cross-section of the plate, in the cross-section of the base metal and in the cross-section of the laser-processed metal respectively
$n_s \in \{1, 2\}$	The number of laser-processed sides
n_{tr} and $\hat{n}_{tr}$	Number and estimated number of the laser-processed tracks
w	The deflection or the vertical displacement of the plate midpoint
ϵ	Linear strain
$\epsilon_{1,i}$, $i \in \{b, l\}$	Yield stress strains of the base, i.e., laser-unprocessed, and laser-processed metals respectively
σ	The normal stress
ϵ_{ul}, $\epsilon_{ul,i}$, $i \in \{b, l\}$	Ultimate strains of the base, i.e., laser-unprocessed, and laser-processed metals respectively
σ_b and $\hat{\sigma}_b$	The stress-strain function and its approximation of the base, i.e., laser-unprocessed, metal respectively
$\hat{\sigma}_l$	The approximation of the stress-strain function of the laser-processed metal
$\sigma_{0.2,i}$, $\sigma_{B,i}$, $i \in \{b, l\}$	Yield strengths and ultimate strength of the base metal and the laser-processed metal respectively

References

1. Tylek, I.; Kuchta, K. Mechanical properties of structural stainless steels. *Tech. Trans. Civ. Eng.* **2014**, *4-B*, 59–80.
2. Guo, L.; Yang, S.; Jiao, H. Behavior of thin-walled circular hollow section tubes subjected to bending. *Thin Walled Struct.* **2013**, *73*, 281–289. [CrossRef]
3. Kurkin, S.A.; Nikolaev, G.A. *Welded Structures. Manufacturing Technology, Mechanization, Automation and Quality Control in Welding Production*; High School: Moscow, Russia, 1991; p. 390. (In Russian)
4. Kapustynskyi, O.; Višniakov, N.; Černašėjus, O.; Golovko, L.; Chayeuski, V. Optimization of the Parameters of Local Laser Treatment for the Creation of Reinforcing Ribs in Thin Metal Sheets. In Proceedings of the 24th International Conference "Mechanika 2019", Kaunas, Lithuania, 17 May 2019; pp. 71–75.
5. Zhukov, A.A.; Luzhnikov, L.P.; Dynkina, S.Y. *Engineering Materials, Reference Book*; Mashinostroenie: Moscow, Russia, 1967; p. 191. (In Russian)
6. Trichevskyi, I.S.; Klepanda, V.V. *Metallic Lightweight Constructions. Reference Book*; Budivelnik: Kiev, Ukraine, 1978; p. 112. (In Russian)
7. Rana, R.; Singh, S.B. *Automotive Steels. Design, Metallurgy, Processing and Applications*, 1st ed.; Woodhead Publishing, Elsevier: Cambridge, UK, 2017; p. 478.
8. Jones, R.M. *Deformation Theory of Plasticity*; Bull Ridge Publishing: Blacksburg, VA, USA, 2009; p. 615.
9. Ion, J.C. Laser Transformation Hardening. *Surf. Eng.* **2002**, *18*, 14–31. [CrossRef]
10. Höche, D.; Kaspar, J.; Chaaf, P. Laser nitriding and carburization of materials. In *Laser Surface Engineering: Processse and Applications*; Woodhead Publishing, Elsevier: Cambridge, UK, 2015; pp. 33–58.
11. Nath, A.K.; Sarkar, S. Laser Transformation Hardening of Steel. In *Advances in Laser Materials Processing*, 2nd'ed.; Woodhead Publishing, Elsevier: Cambridge, UK, 2018; pp. 257–298.
12. European Committee for Standardization. *Open Die Steel Forgings for General Engineering Purposes, Part 2: Non Alloy Quality and Special Steels*; EN 10250-2:1999; CEN: Brussels, Belgium, 1999.

13. Montealegre, M.A.; Castro, G.; Rey, P.; Arias, J.L.; Vazquez, P.; González, M. Surface treatments by laser technology. *Contemp. Mater.* **2010**, *1*, 19–30. [CrossRef]
14. Abboud, J.H.; Khaled, Y.B.; Julifkar, H.; Hashmi, M.S.J Material Response with High Power Laser in Surface Treatment of Ferrous Alloys. In *Materials Science and Materials Engineering*; Elsevier: Oxford, UK, 2017; p. 12.
15. Višniakov, N.; Mikalauskas, G.; Černašėjus, O.; Škamat, J. Laser welding of copper-niobium microcomposite wires for pulsed power applications. *Mater. Werkst.* **2019**, *50*, 646–662. [CrossRef]
16. Neimark, B.A. *Physical Properties of Steels and Alloys for Energetics*; Energija: Saint Petersburg, Russia, 1967; p. 235. (In Russian)
17. International Organization for Standardization. *Metallic Materials—Tensile Testing—Part 1: Method of Test at Room Temperature*; ISO 6892-1:2019; ISO: Geneva, Switzerland, 2019.
18. International Organization for Standardization. *Metallic Materials—Vickers Hardness Test—Part 1: Test Method*; ISO 6507-1:2018; ISO: Geneva, Switzerland, 2018.
19. International Organization for Standardization. *Metallic Materials—Bend Test*; ISO 7438:2016; ISO: Geneva, Switzerland, 2016.
20. FEA-CAE Engineering Analysis & Design. Available online: http://fea-cae-engineering.com/fea-cae-engineering/element_types.htm (accessed on 16 January 2020).
21. Cahoon, J.R.; Broughton, W.H.; Kutzak, A.R. The Determination of Yield Strength from Hardness Measurements. *Metall. Trans.* **1971**, *2*, 1979–1983.
22. Cahoon, J.R. An Improved Equation Relating Hardness to Ultimate Strength. *Metall. Mater. Trans.* **1972**, *3*, 3040. [CrossRef]
23. Pavlina, E.J.; Van Tyne, C.J. Correlation of Yield Strength and Tensile Strength with Hardness for Steels. *J. Mater. Eng. Perform.* **2008**, *17*, 888–893. [CrossRef]
24. Dvorkin, L.I. *Building Materials Science. Reference Book*; Infra-Inzheneriya: Moscow, Russia; Vologda, Russia, 2017; pp. 249–294. (In Russian)
25. Tsiklis, D.S. *Technique of Physical and Technical Research at High and Ultrahigh Pressures*; Himiya: Moscow, Russia, 1965; pp. 15–17. (In Russian)
26. *Steel Microstructure Standards*; GOST 8233-56; IPK Izdatelstvo Standartov: Moscow, Russia, 1956. (In Russian)
27. International Organization for Standardization. *Steel—Micrographic Determination of the Apparent Grain Size*; ISO 643:2012; ISO: Geneva, Switzerland, 2012.
28. Smallman, R.E.; Ngan, A.H.W. *Steel Transformations. Modern Physical Metallurgy*; Elsevier: Oxford, UK, 2014; pp. 473–498.
29. Bkadeshia, H.; Honeycombe, R. *Tempering of Martensite. Steels: Microstructure and Properties*, 4th ed.; Elsevier: Oxford, UK, 2017; pp. 237–270.
30. Gulyaev, A.P. *Metal Science*; Metallurgiya: Moscow, Russia, 1986; p. 544. (In Russian)
31. Jianjun, H.; Wang, J.; Jiang, J.; Yang, X.; Hongbin, H.; Hui, L.; Guo, N. Effect of Heating Treatment on the Microstructure and Properties of Cr–Mo Duplex-Alloyed Coating Prepared by Double Glow Plasma Surface Alloying. *Coatings* **2019**, *9*, 336.
32. Aronovich, M.; Lahtin, Y. *Fundamentals of Metallurgy and Heat Treatment*; Metallurgizdat: Moscow, Russia, 1952; p. 116. (In Russian)

materials

MDPI

Article

Technology and Properties of Peripheral Laser-Welded Micro-Joints

Szymon Tofil *, Hubert Danielewski, Grzegorz Witkowski, Krystian Mulczyk and Bogdan Antoszewski

Laser Research Centre, Faculty of Mechatronics and Mechanical Engineering, Kielce University of Technology, Al. Tysiąclecia Państwa Polskiego 7, 25-314 Kielce, Poland; hdanielewski@tu.kielce.pl (H.D.); gwitkowski@tu.kielce.pl (G.W.); kmulczyk@tu.kielce.pl (K.M.); ktrba@tu.kielce.pl (B.A.)

* Correspondence: tofil@tu.kielce.pl; Tel.: +48-413424533

Abstract: This article presents the results of research on the technology and peripheral properties of laser-welded micro-couplings. The aim of this research was to determine the characteristics of properly made joints and to indicate the range of optimal parameters of the welding process. Thin-walled AISI 316L steel pipes with diameters of 1.5 and 2 mm used in medical equipment were tested. The micro-welding process was carried out on a SISMA LM-D210 Nd:YAG laser. The research methods used were macroscopic and microscopic analyses of the samples, and assessment of the distribution of elements in the weld, the distribution of microhardness and the tear strength of the joint. As a result of the tests, the following welding parameters are recommended: a pulse energy of 2.05 J, pulse duration of 4 ms and frequency of 2 Hz, beam focusing to a diameter of 0.4 mm and a rotation speed of 0.157 rad/s. In addition, the tests show good joint properties with a strength of more than 75% of the thinner pipe, uniform distribution of alloying elements and a complex dendritic structure characteristic of pulse welding.

Keywords: laser micro-welding; cryosurgical probe; butt weld; thin tube welding

Citation: Tofil, S.; Danielewski, H.; Witkowski, G.; Mulczyk, K.; Antoszewski, B. Technology and Properties of Peripheral Laser-Welded Micro-Joints. *Materials* **2021**, *14*, 3213. https://doi.org/10.3390/ma14123213

Academic Editor: Aleksander Lisiecki

Received: 12 March 2021
Accepted: 1 June 2021
Published: 10 June 2021

1. Introduction

The progressive development of technology in many fields is associated with miniaturisation. This is especially true in areas such as electrical engineering, automation, the automotive industry, aerospace and medical technology. In many solutions, there is a need for durable and reliable joints. An effective way to meet these expectations is laser micro-welding. The term is not unambiguous, but it is usually assumed that the term refers to a welding process in which the welded parts and the penetration zone have a dimension not exceeding 1 mm [1]. The literature on micro-welding addresses such topics as the following:

- Micro-welding of materials with diverse properties [2–7];
- Choice of laser type and the impact of the operating parameters on the joint properties [1,7–9];
- Micro-welding of materials with diverse shapes and sizes [2–4];
- Modelling of the micro-welding process [1,10–14];
- Evaluation of properties and metallurgical changes of laser-welded micro-joints [15–17].

Micro-scale laser welding has specific features that make the process require a special approach. Here, we are dealing with a material with a very low heat capacity, and therefore the energy supplied must be extremely precisely selected. Laser welding technology, as opposed to standard methods, is distinguished by features that make laser welding, on a micro-scale, the best possible solution next to electron-beam welding [3,4,9]. This is determined by the possibility of choosing the type of laser beam used for welding, the possibility of precise energy dosing in the form of pulses of a short duration and the possibility of focusing the beam on a very small area located in a specific place. Nd:YAG, fibre and disc lasers are the most commonly used lasers for micro-welding applications [1,18–21]. The current research interest is focused on the use of ultrashort pulse lasers [6,9]. The nanosecond and picosecond interaction of pulses causes a qualitative change in the phenomena occurring in the material and is a rewarding research area, which also enters the field of micro-welding. The above-mentioned considerations determine that each case of micro-welding, depending on the requirements set, should be treated with special care for the precise selection of all aspects of the technological process.

Macro-scale welding of stainless steels is widely reported in the literature. However, welding on a micro-scale causes significant differences. The progress of miniaturisation contributes to extensive progress in micro-welding, including laser welding of stainless steels, which is reflected in the number of articles on this subject. The authors of [22] studied the welding parameters for butt welding of 100 µm-thick foil of AISI 316 steel. Good quality joints were obtained using a pulsed Nd:YAG laser for the following process parameters: energy of 1.75 J, welding speed of 525 mm/min and pulse duration of 4 ms. According to the results of the studies in [23–31], the increase in laser power increases the depth of the weld when welding stainless steel sheets. In [31], an effective empirical model was presented in order to predict the joint properties of stainless AISI 304 and 316 steels, where an estimation error between the model and experiment equal to 15% was achieved. The laser power and scanning speed were indicated as essential factors affecting the joint strength. The grain refinement together with the formation of δ-ferrite in the weld causes a slight increase in hardness in the fusion zone. Similar results were observed in [32–34]. The authors of [7] carried out extensive research on butt welding of thin sheets (0.45 mm) of AISI 316 steel with a Nd:YAG laser. The study confirmed that the laser power is the most important parameter in the welding of thin sheets because it significantly affects the properties of mechanical and metallurgical joints. The overlap of pulses significantly affects the surface roughness and strength of welded joints. The authors of [32] presented the results of a comparative study of arc and Nd:YAG laser welding of AISI 316 steel. It was found that laser welding causes less distortion and lower residual stress compared to arc welding. The differences between laser macro- and micro-welding of stainless steels refer mainly to the narrower range of micro-laser welding parameters, which guarantees a good joint quality [22,25,29]. In addition, with micro-welding, welds have a very narrow heat-affected zone (HAZ) along the fusion line, and microstructure changes are related to grain refinement during crystallisation [10,13,30]. Therefore, hardness changes are practically negligible. A significant barrier to achieving an acceptable level of welded joints is thermal deformation, which can cause crack formation during weld formation and lead to loss of its continuity [27–30]. In the study of [35], the authors studied, in detail, the problem of gaps in lap joint welding, where, by studying the changes in the microstructure and the shape of the weld, they recommended an acceptable gap size of less than 0.1 mm. One of the main problems presented in the abovementioned publications is the problem of stress concentration and deformation of welded materials, which is particularly troublesome in circumferential lap welding [35]. As a result of deformation, the gap between the welded elements may increase, and defects may appear in the joint. For these reasons, the problem of joined element stabilisation often determines the effect of the entire operation.

While the welding of thick-walled pipes is widely described in the literature, there are no reports on the welding of thin-walled AISI 316 steel pipes with diameters on the order of millimetres.

Of special interest in recent years are medical subjects [2,6] in which laser technology is applied, such as stents, implants and dentures, probes and other medical components.

The authors of this article deal with the development of the technology of laser micro-welding of elements of a medical implement, in a design solution with basically three types of circumferential welds (Figure 1—own photos produced by the authors of this article)—lap–butt weld of a probe tip with a tube-shaped probe body; lap joint of two tubes; and fillet joint of two tubes.

The main purpose of this research team is to develop a laser welding method for thin-walled tubes which can be used as an alternative to brazing when the following requirements need to be achieved. The joints shall have a weld with no visible welding imperfections, and a regular face and back of the weld with no traces of surface scorching. Small single pores are acceptable. The weld should not degrade the strength of the probe (75% of the strength of the weaker probe component) or its corrosion resistance. The face of the weld should not extend above the parts being joined.

The developed laser micro-welding technology is innovative for the target product and can replace solutions in current use. For the implementation of the established task, it was necessary to solve the following problems: the selection of a laser device and its operating parameters, the study of mechanical properties of the joint and the assessment of metallurgical changes in the joint area.

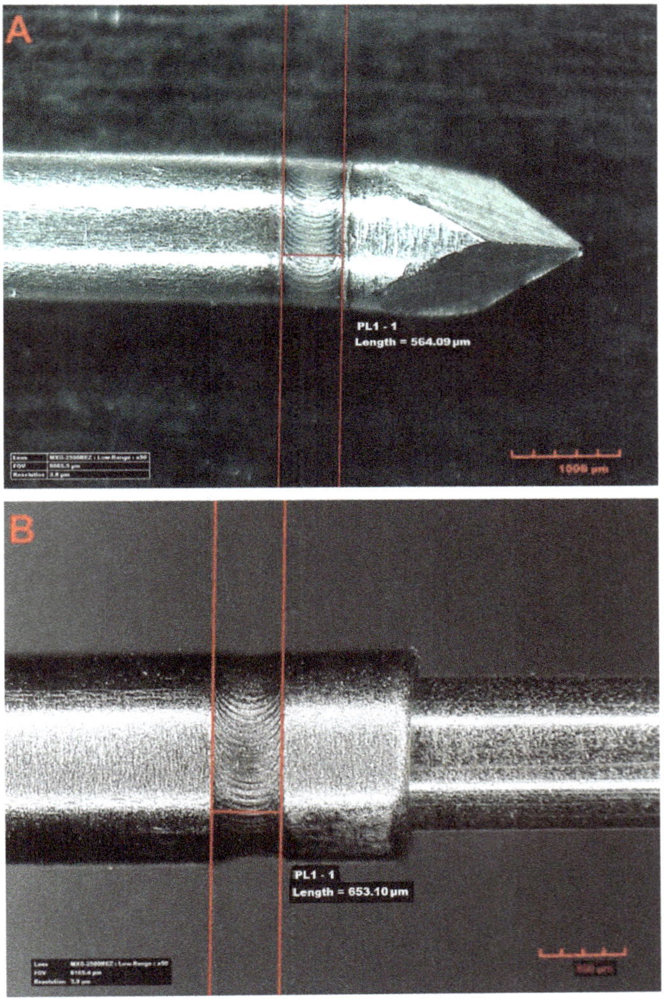

Figure 1. *Cont.*

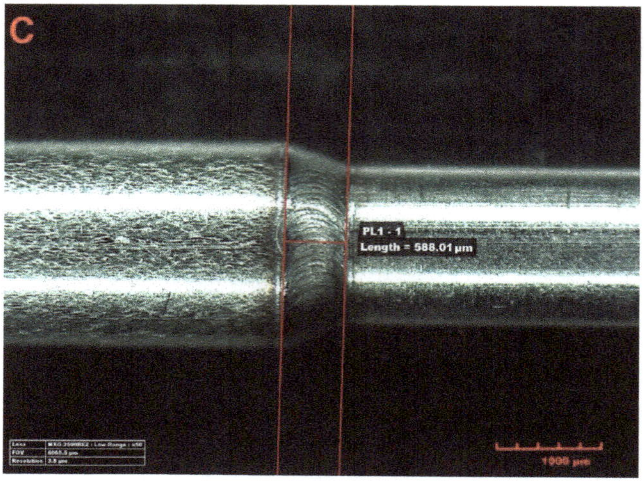

Figure 1. Example view of analysed welds: (**A**) lap–butt weld of the probe tip with a tube-shaped probe body; (**B**) lap joint of two tubes; (**C**) fillet joint of two tubes.

2. Experimental Tests

2.1. Methodology

Pulse welding was adopted as the welding method due to the small size and heat capacity of the welded parts. This provided protection to the component from excessive temperature increases that could cause structural changes and oxide formation on the surface. The micro-welding process was performed using a typical process parameter selection procedure with a professional SISMA LM-D210 laser micro-welder. In accordance with PN-EN ISO 13919-1, the correctness of the joint structure was determined, and welding imperfections were identified. Welding was performed in two variants with an argon shroud and without a gas shield. The chemical composition of the welded material is shown in Table 1. The presented paper describes a detailed study on the selection of welding process parameters and joint properties for a lap weld according to Figure 1B. The test procedure for the other welds followed the same pattern, and due to the limited volume of this article, the results for the other two types of welds are not included here.

Table 1. Chemical composition of 316 L steel according to EN10204 certified 190821-S047.

%C	%Si	%Mn	%P	%S	%Cr	%Ni	%Mo	%N
0.009	0.485	1.615	0.0307	0.0041	16.763	11.208	2.041	0.0619

The completed joints were continuously evaluated by visual tests (VT) and qualified for further testing. The planned tests included: macroscopic and microscopic tests of the weld (HIROX KH-8700 microscope), microstructure and elemental distribution with an accelerated voltage of 15 keV and a work distance of 10 mm according to ISO 22309:2011 (JOEL 7100f electron microscope and HIROX KH-8700 microscope), and mechanical property tests with a standard repeatable measurement accuracy of less than 2.5 N (INSTRON 4502 device in accordance with the recommendations of PN-EN ISO 6892-1). The strength of the resulting welded joints was determined by static tensile testing. The repeatability of the results was checked by producing two sets of samples for welding in an argon shroud and without a shielding gas. Five samples were taken for each set based on the process parameters for set 4 as shown in Table 2. The test samples were mounted in specially designed mounting fixtures to maintain the alignment of the tensile force. The strength tests were conducted at a constant tensile speed of 0.5 mm/s for the samples.

Table 2. List of welding parameters accepted after visual testing.

Weld No.	Power Setting %	Pulse Energy (J)	Pulse Power (kW)	Welding Linear Energy (J/mm])
1	7	1.36	0.34	8.69
2	8	1.59	0.4	10.13
3	9	1.82	0.45	11.58
4	10	2.05	0.51	13.03
5	11	2.27	0.57	14.47

Microhardness tests were conducted with load HV0,1 (Innovatest Nexus 4303 hardness tester in accordance with PN-EN ISO 6507-1). Measurements were recorded for samples welded in an argon shroud in the tube core material, in the centre of the penetration and at its edge.

Specimens that qualified for microscopic tests were embedded in resin and then finely ground to visualise the penetrations. The penetrations were first observed on a HIROX microscope, where the geometry of the characteristic zones was measured, as shown in the diagram (Figure 2). The width of the face (A), the depth of the penetration zone (B) and the width of the penetration at the boundary of the joined parts (C) were determined here.

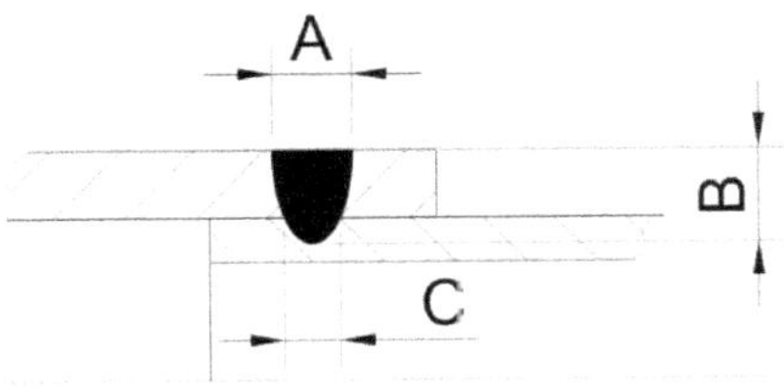

Figure 2. Schematic diagram of the analysed weld.

2.2. Technological Experiment

Drawn tubes made of AISI 316L-grade steel with nominal diameters of 1.5 × 0.15 and 2 × 0.2 were used for the experiment. The chemical composition of the welded material is shown in Table 1.

The dimensions of the workpieces to be welded allow the smaller-diameter tube to be inserted into the larger-diameter tube concentrically with a nominal clearance of 0.05 mm. Allowing for alignment errors and tube circularity, this clearance may vary locally in the range of 0 to 0.1 mm. Actual measurements of the tube series showed that the mean value of the inner diameter of the thicker tube was 1.618 mm, with a standard deviation of 0.010 mm and a spread of 0.024 mm, and the mean value of the outer diameter of the thinner tube was 1.608 mm, with a standard deviation of 0.008 mm and a spread of 0.029 mm. Such an arrangement makes it possible to perform a circumferential lap weld without having to support a second tube after the first tube is placed in the turnover fixture. At the same time, it is consistent with the recommendations available in the literature [21] according to which the allowable gap at lap joints should not exceed 25% of the spot diameter. The schematic diagram of the prepared joint is shown in Figure 2.

A test bench equipped with a Sisma LM-D210 laser micro-welder illustrated in Figure 3 was used to conduct the experiment. The device is based on a Nd:YAG pulsed laser emitting radiation of 1064 nm wavelength. The emitted beam was multi-mode. The laser was equipped with an integrated focusing and observation system with a camera and monitor. This arrangement enables a focal spot with an adjustable diameter of 0.1–2 mm. The device has digital adjustment of laser operating parameters, such as pulse duration, energy per pulse and pulse generation frequency. Pulse duration can be varied from 0.1 to

25 ms, pulse energy from 0 to 210 J and frequency from 0 to 50 Hz. The maximum power per pulse reaches 10.5 kW. A protective atmosphere in the weld area was provided by local argon flow.

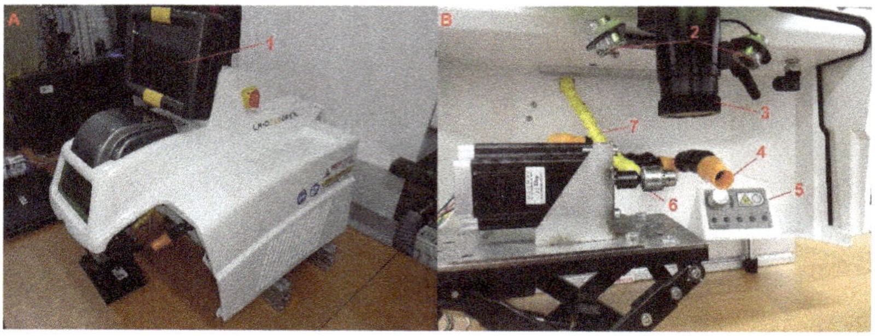

Figure 3. View of the SISMA LM-D210 laser bench. (**A**)—general view, (**B**)—workspace view. 1—front touch panel, 2—lighting spots, 3—laser head, 4—gas extractor, 5—manual parameter set panel, 6—rotary motor, 7—nozzle of gas shield.

Rotation of the workpiece around the axis during the welding process was performed using a miniature speed-controlled turnover fixture developed and made for this purpose. Rotational motion was provided by a stepping motor with microstep control. The motor controller was controlled by a PC and allows a speed range of 0.002–20 rad/s.

In order to select the optimum process parameters, preliminary experiments were conducted by performing a series of circumferential welds using different laser beam pulse energies in the range of 1.36–2.27 J. Based on the preliminary tests, the pulse duration and frequency were selected to be 4 ms and 2 Hz, respectively. The criterion adopted for the selection of the parameters was to achieve the effect of applying individual pulses at the level of 80% while maintaining the correct appearance of the weld face and the absence of clear thermal interaction in the weld area. The recommended overlap value for tight joints in the literature [21] is 80 to 90%. The focus diameter of the laser beam used in the welding process was 0.4 mm. The workpiece speed was 0.157 rad/s. The beam was quenched after the welded workpiece had rotated through an angle of 6.3 rad to ensure a continuous circumferential weld. Table 2 lists the welding parameters for which a positive evaluation was obtained from visual testing.

3. Properties of Laser Micro-Joints—Results and Discussion

3.1. Microstructural Tests

After producing metallographic sections, the images of penetrations were revealed, which are shown in Figure 4. It can be observed here that the penetration depth (Figure 2B) of welds 1, 2 and 3 (Figure 4) is less than half the thickness of the inner tube. Moreover, in the case of weld 1, there was no fusion penetration into the lower material (inner tube); therefore, we deal with the so-called adherence. At the same time, there is a clear increase in the depth of penetration with increasing pulse energy. In the case of weld 5, there is a penetration of both materials to be welded, but there was an outflow of material from the root side and a significant concavity of the weld face. Based on observations of the structure of the welds obtained, weld 4 was selected, which is characterised by a lack of buckles of the weld face, fusion penetration into the lower tube for at least half of its thickness and the absence of visible welding imperfections.

The appearance of the face as well as its width (Figure 2A) for all welds (except weld 5) is to be qualified as correct. The occurrence of a slight face concavity of welds 3 and 4 is acceptable in contrast to the occurrence of a convex face. The fusion penetration width (Figure 2C) at the boundary of the joined tubes increases with increasing pulse energy. It can be seen that the value of width A stabilises for welds 3, 4 and 5. The face width for weld 4 was 0.468 mm. The trends of changes in the A, B and C quantities characterising the penetration geometry are shown in the graph in Figure 5.

Figure 4. View of obtained welds for lap joints in cross-section.

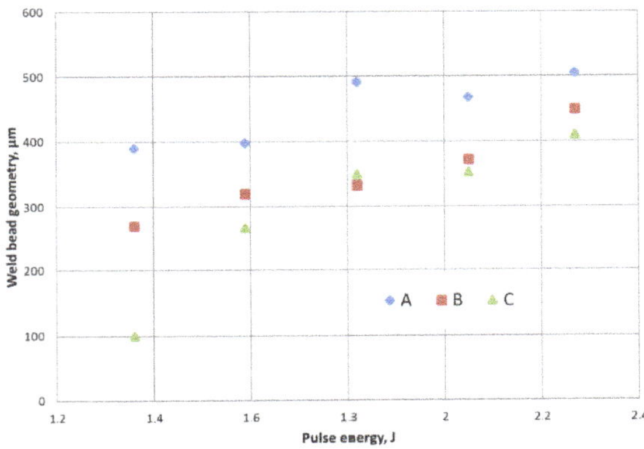

Figure 5. Dependence of characteristic dimensions of the penetration on the pulse energy value. Geometrical parameters A, B and C according to Figure 2.

The conducted tests showed that the weld width in the overlap zone (C) has the highest growth dynamics. Additionally, the weld depth (B) increases with increasing pulse energy, which is natural for the process tested. On the other hand, the weld face width (A) increases slightly due to the constant spot diameter during the performed process. The increase in the weld face width is caused by an increase in the amount of energy transferred to the surface of the material and thus the heating of the welded parts.

A joint quality level analysis was performed on the sample joints produced by pulsed laser micro-welding. Samples were taken from the circumferential lap joint and used for metallographic tests after grinding and etching processes (Figure 4).

Based on the results obtained, it was found that there is a lack of regularity in the structure of welds 1 and 5 (areas marked in red), and in the case of the first weld, the fusion penetration line is located at the boundary of the zone between the welded materials, which indicates an unstable joint in the form of incomplete fusion.

In the case of the fifth weld, we are dealing with a through penetration, but the amount of energy supplied by the laser beam was too high and a concavity of the weld face occurred due to an excessive outflow of material in the root area. Both of the abovementioned welding imperfections disqualify the resulting joints. For joints 2–4, the assumed fusion penetration into the lower material was achieved; however, due to the obtained fusion penetration depths, joint 4 was chosen for further analysis, where a fusion penetration of the lower material above half of its thickness was obtained (Figure 6).

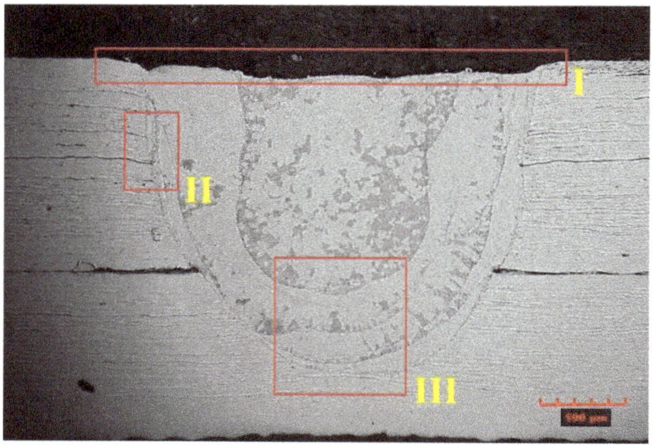

Figure 6. Structure of the weld with the characteristic areas of the joint marked. I – weld face, II and III - the fusion penetration line of the weld into the base material.

The obtained weld has a correct structure with a slight concavity of the weld face (area 1), which is 20 μm, but it does not exceed the permissible 5% of the thickness of welded materials. Area 2 shows the fusion penetration line of the weld into the base material (Figure 7).

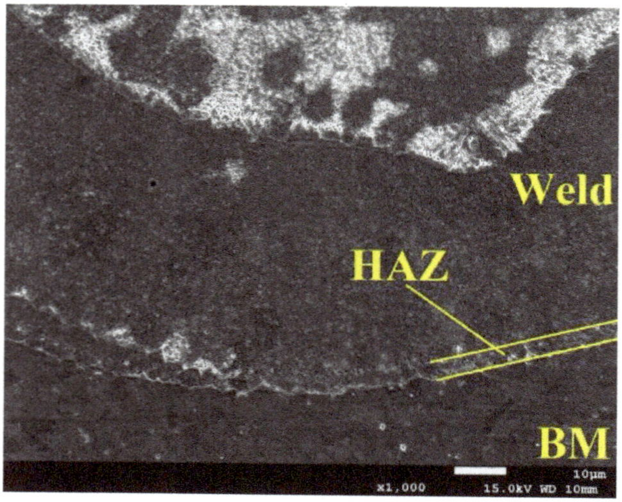

Figure 7. The fusion penetration line zone of the weld into the base material.

Within area 3, the presence of a heat-affected zone was identified, and no clear zone was found, which is typical of welds in austenitic stainless steels. In the case under consideration, no growth of austenite grains was observed, only elongated ferrite grains forming a discontinuous grid around the austenite grains. Such areas occur in steels with a base material structure consisting of austenite with ferrite δ. At high temperatures, a $\gamma \rightarrow \delta$ transformation occurs along the fusion penetration line, which begins in the existing ferrite δ grains and progresses toward the area of increased chromium concentration. Upon recooling, this area does not reach phase equilibrium, by which the proportion of ferrite δ increases, forming a narrow heat-affected zone (Figure 7). The weld root is also marked within area 3 along with the overlapping of the individual weld layers resulting from the pulsed welding mode. In pulsed welding, a series of overlapping pulses causes the material to instantaneously melt and recrystallise, producing the characteristic scaly weld bead. A dendritic structure is formed in the interaction area of the individual pulses, but for individual heat cycles, when the laser beam penetrates the lower sheet region only to a certain depth, the impact of the flow field is evident in the solidified structure and the direction of dendrite growth. In the case of partial weld penetration at the pulsed mode of the heat source, discrete growth bands are present, suggesting strong fluctuations in the flow field and affecting the grain growth process and direction.

For the weld case considered, no significant welding imperfections were found; the slight concavity of the weld face is within the acceptable range, and the penetration depth of the laser beam for joint 4 exceeds half the thickness of the lower material. No other type of welding imperfections were found; thus, the fabricated joint was classified as quality level B. Analysis of the microstructure showed a dendritic structure with visible banding of the dendrite growth resulting from the pulsed welding mode. A narrow heat-affected zone characteristic of austenitic steel was identified.

For stainless steels, especially for surgical applications, corrosion resistance is an important aspect, which is ensured by an appropriate content of alloying elements in the form of chromium and nickel as well as by the reduction in carbon in the alloy. Additionally, in the case of the considered material—austenitic stainless steel of 316L grade—molybdenum is introduced into the alloy composition, which improves the steel's resistance to corrosion, especially intercrystalline corrosion. During welding, the content of alloying elements in the weld may change, which may adversely affect the properties of the joint, including corrosion resistance. A preliminary evaluation of the joint in this regard can be conducted by performing X-ray energy dispersion spectrum analyses for the indicated elements. Analysis of the element distribution was performed for area 2 by means of which the quantitative content of selected elements was determined in defined measuring points for the joint made without a protective atmosphere (Figure 8A) and in an argon shroud (Figure 8B).

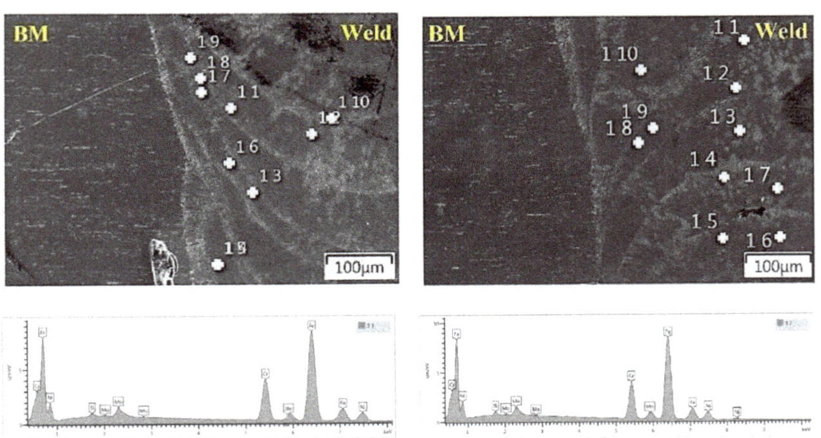

(A) (B)

Figure 8. Measurement points of the spectrum analysis and the result of the readings for point 11, for a weld made without a protective atmosphere (**A**), and for point 12, for a weld made in an argon shroud (**B**).

According to the measurement points marked above, a quantitative analysis of the chemical composition of selected alloying elements was performed for the exemplary spots, which is shown in Table 3.

Table 3. Percentage content of selected alloying elements at measurement points.

Welding Atmosphere	Measurement Point	Si %	Mn %	Cr %	Ni %	Mo %
in air	1 1	0.31	1.06	16.75	10.26	2.54
	1 2	0.32	1.07	16.76	10.49	2.42
in argon	11	0.39	1.25	16.78	10.40	2.45
	12	0.26	0.86	16.9	10.58	2.15

The quality level of a joint in terms of anti-corrosivity is defined not only by the content of selected elements within the weld but also by the difference in the content of these elements in relation to the base material. In order to determine if there are significant differences between the distribution of chromium, nickel, molybdenum and iron in the weld relative to the base material, qualitative analysis was performed on the cross-section of the joint (Figure 9).

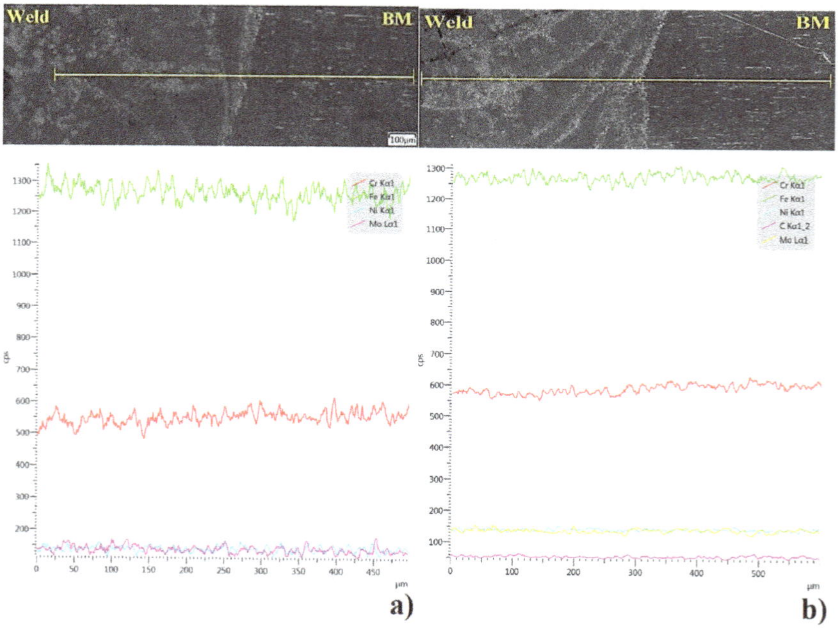

Figure 9. Linear analysis of the distribution of elements (Cr, Fe, Mo, Ni) in the weld–MR transition zone: (**a**) a weld made in air; (**b**) a weld made in an argon shroud.

The results of the analysis show a uniform distribution of the selected elements, close to linear in nature, and no significant differences between the nickel and molybdenum contents in the weld relative to the base material for both welding variants. On the other hand, a slight reduction in chromium was observed, which may be due to the migration of this element during metallurgical processes in connection with the residual content of atmospheric oxygen in the zone between welded elements, or to the burning out of a part of chromium during the welding process. A similar trend was observed for both cases. However, no separations were found in the penetration zone, especially in the chromium carbides, which may cause intercrystalline corrosion, and it was therefore assumed that the welded joint is characterised by good anti-corrosion properties. A full evaluation of the corrosion resistance of the joint will be obtained after detailed specialised corrosion testing.

3.2. Microhardness and Joint Strength Testing

Hardness measurements were performed according to the scheme shown in Figure 10. The measured microhardness values for welds (Figure 4) 3 and 4 averaged 371 HV0.1 in the tube core material (1), ranging from 220 HV0.1 in the central part of the penetration (2) to a value of 240 HV0.1 near the fusion penetration line (3). This variability is due to the different cooling conditions of the molten material. The increased microhardness value of the tube core is the result of crumple occurring during the tube drag process. The result obtained shows that there are no overheating effects on the material, which is also confirmed by the absence of discolouration of the material after the welding process. For samples without the argon shroud, there are no significant differences in the microhardness measurements recorded.

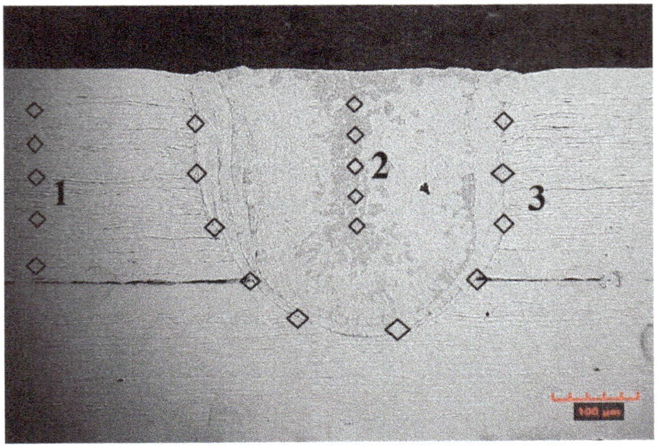

Figure 10. The scheme of microhardness measurements: the tube core material (1); the central part of the penetration (2); near the fusion penetration line (3).

3.3. Test Results for Joint Strength

The results of tensile shear test are shown in Table 4 and in Figure 11.

Table 4. Tensile shear test results of the fabricated micro-joints.

Sample No	Sisma No Shroud (N)	Sisma With Ar Shroud (N)
Sample 1	773	759
Sample 2	767	788
Sample 3	724	753
Sample 4	761	803
Sample 5	745	750
Mean value (N)	754	770.6
Standard deviation (N)	19.74	23.56

The results obtained were statistically analysed to determine the mean, minimum and maximum values and the mean dispersion of results. On the basis of the analysis performed, the mean value of the breaking force of the welded joints made was determined, which is 754 N for the weld made without the application of a shielding gas and 770.6 N for the joints made in an argon shroud. A slight increase of only 2.20% in the average strength of joints made in an argon shroud was observed. However, taking into account the dispersion of the results, it cannot be clearly stated that the shielding gas contributes to the increase in the joint strength. In view of the fact that the measured breaking strength of the thinner tube in the joint is 976 N, it must be concluded that the required strength criterion of the joint (75% of the strength of the weaker element in the joint) was met. This condition is met for both unshielded and argon-shielded welding.

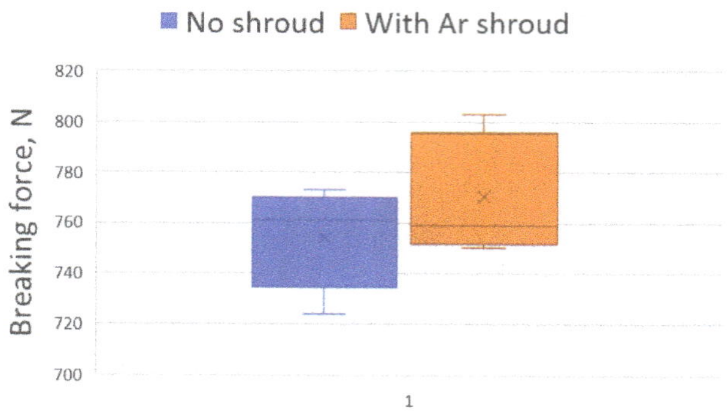

Figure 11. Results of tensile shear test.

During the static tensile test, a rupture characteristic of brittle fracture occurred in each of the welds tested. Until the moment of breaking, the samples underwent a characteristic deformation passing successively through the areas of elasticity and plasticity. No obvious plasticity growth before fracture was observed. The course of deformation in both highlighted areas should be described as typical for austenitic steel. Tensile cracking of a sample is the result of the joint interaction of tensile and shear stresses. There are two types of tensile cracking in laser welds: heat-affected zone boundary cracking and interfacial cracking. Since the hardness of the tube material is greater than the hardness of the weld, the tensile deformation is concentrated in the heat-affected zone or at its boundary. The grains in the heat-affected zone tend to grow because the heat-affected zone is kept at a high temperature for a long time; the high temperature gradient results in the transformation of the material structure. Hence, the heat-affected zone is often the weakest area of the weld. As a result, joint breaking most often occurs along the fusion penetration line or at the boundary between the heat-affected zone and the base material. The fracture characteristics observed (for sample nr 4) in the scanning microscope images are shown in Figure 12. These observations suggest that the fractures are complex in nature. On the one hand, it was noticed that fractures along the fusion penetration line were observed on some samples (Figure 12A). On the other hand, some samples, in whole or in part, were fractured across the penetrated zone (Figure 12B). In both observed mechanisms (Figure 12A,B), areas of brittle fracture (marked 1) and areas of ductile fracture (marked 2) are distinguished. At the fusion penetration line crack, traces of overlapping consecutive laser pulses are visible, on which transverse cracks are visible (Figure 12C). They are the result of accumulated stresses resulting from tensile and shear forces as well as internal stresses in the exposed fusion penetration line. Figure 12D shows the ductile fracture on the inner edge of the welded tubes. A partially developed ductile fracture is observed in the inner part of the tube outside the penetration (Figure 12E). Figure 12F shows a fracture along the fusion penetration line in the beam extinction area at the end of the circumferential weld. The authors believe that the complex nature of the fracture is the cause of the large dispersion of shear fracture strength results. This is due to the increased gap between the joined elements. It results from deformations during welding and the accuracy of welded tubes. This proves that the weld may not be uniform along the entire circumference. To avoid this, a dimensional selection of the tubes should be made before welding.

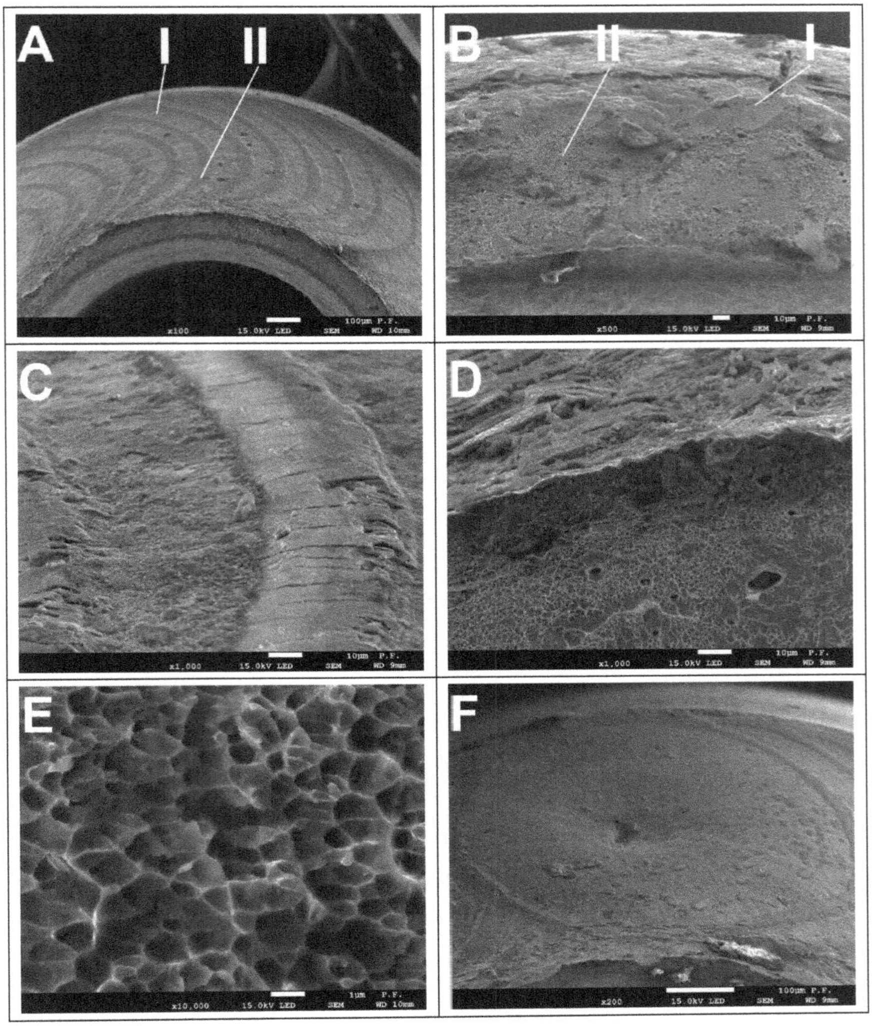

Figure 12. Morphology of the sample fracture after breaking: (**A**) 500× magnification; (**B**) 100× magnification; (**C**) 1000× magnification; (**D**) 500× magnification; (**E**) 10,000× magnification; (**F**) 200× magnification.

The shape shown in microphotograph 12f is the result of the last laser pulse and is a typical image for extinction without reducing the laser power (without power ramping). The solidification of the metal here takes place from the area of the highest temperature gradient, i.e., from the outer diameter, and the material in the liquid phase begins to crystallise at a speed higher than the flow velocity of the liquid metal; hence, the centre is only partially flooded, which results in the formation of a concave.

4. Summary

The results of the conducted tests on laser overlap welding of thin-walled tubes with nominal diameters of 1.5 × 0.15 and 2 × 0.2 mm allow for the formulation of the following conclusions.

1. The best weld properties were obtained in argon-shielded welding at a pulse energy of 2.05 J, pulse duration of 4 ms and frequency of 2 Hz, beam focusing to a diameter of 0.4 mm and a rotation speed of 0.157 rad/s.

2. The weld face is smooth with a gentle undulation corresponding to the number of pulses on the circumference, avoiding the formation of surface oxides when welding in an argon shroud.

3. The fusion penetration zone has a typical U shape, and its microstructure is characterised by the occurrence of dendrite density in the zone near the fusion penetration boundary. Moreover, the boundaries of the overlapping consecutive coagulated melting fronts of the material stand out in the fusion penetration structure.

4. The cracking of samples during static tensile testing is complex in nature. It most often occurs along the fusion penetration line, and less often by penetration. The observed fractures show areas of brittle fracture and weaker-shaped areas of ductile fracture.

5. Analysis of the chemical composition according to the defined measurement points showed no significant differences for both welding variants. No excessive burning or oxidation of alloying elements was observed in both analysed welds relative to the base material.

6. The linear nature of the chromium, nickel and molybdenum distribution between the base material and the weld shows the high level of purity of the welds obtained and the lack of reduction in the content of alloying elements. Additionally, no changes were found in the fusion penetration line area, and thus the corrosion protection capability of the material was maintained.

7. Before welding, it is recommended that a dimensional selection of the tubes should be conducted ensuring the smallest possible gap between the welded elements (maximum 0.1 mm).

Author Contributions: Conceptualization, B.A. and S.T.; methodology, B.A.; validation, B.A. and S.T.; formal analysis, H.D. and S.T.; investigation, G.W., S.T., H.D. and K.M.; data curation, G.W. and K.M.; writing—original draft preparation, S.T.; writing—review and editing, B.A. and S.T. All authors have read and agreed to the published version of the manuscript.

Funding: Supported by a grant from the National Centre for Research and Development (NCBiR), project No. POIR.04.01.04-00-0161/17.

Institutional Review Board Statement: Not applicable.

Informed Consent Statement: Not applicable.

Data Availability Statement: Not applicable.

Acknowledgments: The research reported herein was supported by a grant from the National Centre for Research and Development (NCBiR), project No. POIR.04.01.04-00-0161/17.

Conflicts of Interest: The authors declare no conflict of interest.

References

1. Naeem, M. *Developments in Laser Microwelding Technology Handbook of Laser Welding Technologies*, 1st ed.; Woodhead Publishing: Cambridge, UK, 2013.
2. Baruah, M.; Bag, S. Influence of pulsation in thermo-mechanical analysis on laser micro-welding of Ti6Al4V alloy. *Opt. Laser Technol.* **2017**, *90*, 40–51. [CrossRef]
3. Chen, X.; Brox, D.; Assadsangabi, B.; Ali, M.S.M.; Takahata, K. A stainless-steel-based implantable pressure sensor chip and its integration by microwelding. *Sens. Actuators A Phys.* **2017**, *257*, 134–144. [CrossRef]
4. Heinen, P.; Haeusler, A.; Mehlmann, B.; Olowinsky, A. Laser Beam Microwelding of Lithium-ion Battery Cells with Copper Connectors for Electrical Connections in Energy Storage Devices. *Lasers Eng.* **2017**, *36*, 24. [CrossRef]
5. Seiler, M.; Patschger, A.; Bliedtner, J. Investigations of welding instabilities and weld seam formation during laser microwelding of ultrathin metal sheets. In Proceedings of the 34th International Congress on Applications of Lasers and Electro-Optics (ICALEO), Atlanta, GA, USA, 18–22 October 2015. *J. Laser Appl.* **2016**, *28*, 022417. [CrossRef]
6. Carter, R.M.; Troughton, M.; Chen, J.; Elder, I.; Thomson, R.R.; Esser, M.J.D.; Lamb, R.A.; Hand, D.P. Towards industrial ultrafast laser microwelding: SiO_2 and BK7 to aluminum alloy. *Appl. Opt.* **2017**, *56*, 4873–4881. [CrossRef] [PubMed]

7. Jiang, X.; Chandrasekar, S.; Wang, C. A laser microwelding method for assembly of polymer based microfluidic devices. *Opt. Lasers Eng.* **2015**, *66*, 98–104. [CrossRef]
8. Mao, J.; Huang, Y.; He, P. Study on joint formation evolution in laser microwelding of Pt-10% Ir and 316 LVM SS crossed wires. In Proceedings of the International Symposium on Mechanical Engineering and Material Science (ISMEMS), Jeju Island, South Korea, 17–19 November 2016; AER-Advances in Engineering Research. Volume 93, pp. 11–16. [CrossRef]
9. Kim, S.; Park, J.; So, S.; Ahn, S.; Choi, J.; Koo, C.; Joung, Y.-H. Characteristics of an Implantable Blood Pressure Sensor Packaged by Ultrafast Laser Microwelding. *Sensors* **2019**, *19*, 1801. [CrossRef]
10. Haeusler, A.; Schuermann, A.; Schoeler, C.; Olowinsky, A.; Gillner, A.; Poprawe, R. Quality improvement of copper welds by laser microwelding with the usage of spatial power modulation. *J. Laser Appl.* **2017**, *29*, 022422. [CrossRef]
11. Danielewski, H.; Skrzypczyk, A.; Hebda, M.; Tofil, S.; Witkowski, G.; Długosz, P.; Nigrovič, R. Numerical and Metallurgical Analysis of Laser Welded, Sealed Lap Joints of S355J2 and 316L Steels under Different Configurations. *Materials* **2020**, *13*, 5819. [CrossRef] [PubMed]
12. Saha, P.; Waghmare, D. Parametric optimization for autogenous butt laser welding of sub-millimeter thick SS 316 sheets using central composite design. *Opt. Laser Technol.* **2020**, *122*, 105833. [CrossRef]
13. Wang, L.; Wei, Y.; Chen, J.; Zhao, W. Macro-micro modeling and simulation on columnar grains growth in the laser welding pool of aluminum alloy. *Int. J. Heat Mass Transf.* **2018**, *123*, 826–838. [CrossRef]
14. Wang, L.; Wang, K. Investigation on microstructural patterns and hot crack in the molten pool via integrated finite-element and phase-field modeling. *J. Manuf. Process.* **2019**, *48*, 191–198. [CrossRef]
15. Salleh, M.N.M.; Ishak, M.; Quazi, M.M.; Aiman, M.H. Microstructure, mechanical, and failure characteristics of laser-microwelded AZ31B Mg alloy optimized by response surface methodology. *Int. J. Adv. Manuf. Technol.* **2018**, *99*, 985–1001. [CrossRef]
16. Hummel, M.; Haeusler, A.; Olowinsky, A.; Gillner, A.; Poprawe, R. Comparing 1070 nm and 515 nm Wavelength Laser Beam Sources in Terms of Efficiency for Laser Microwelding Copper *Lasers Eng.* **2020**, *46*, 187–202. Available online: https://www.oldcitypublishing.com/journals/lie-home/lie-issue-contents/lie-volume-46-number-1-4-2020/lie-46-1-4-p-187-202/ (accessed on 20 January 2021).
17. Geng, S.; Jiang, P.; Guo, L.; Gao, X.; Mi, G. Multi-scale simulation of grain/sub-grain structure evolution during solidification in laser welding of aluminum alloys. *Int. J. Heat Mass Transf.* **2020**, *149*, 119252. [CrossRef]
18. Huang, S.-H.; Huang, Y.-J.; Hsieh, C.-H.; Chen, H.-Z.; Chui, H.-C. Visual-assisted laser microwelding of carbon microfiber on metal plates. *Opt. Laser Technol.* **2018**, *108*, 368–371. [CrossRef]
19. Patschger, A.; Bliedtner, J. Constraints and optimization of the laser microwelding process of thin metal foils. *J. Laser Appl.* **2017**, *29*, 22408. [CrossRef]
20. Okamoto, Y.; Nishi, N.; Nakashiba, S.; Sakagawa, T.; Okada, A. Smart laser micro-welding of difficult-to-weld materials for electronic industry. *Laser-based Micro Nanoprocessing IX* **2015**, *9351*, 935102. [CrossRef]
21. Anming, H.; Jolanta, J.-R.; Tomokazu, S. Joining Technology Innovations at the Macro, Micro, and Nano Levels. *Appl. Sci.* **2019**, *9*, 3568. [CrossRef]
22. Gao, Z.; Shao, X.; Jiang, P.; Cao, L.; Zhou, Q.; Yue, C.; Liu, Y.; Wang, C. Parameters optimization of hybrid fiber laser-arc butt welding on 316L stainless steel using Kriging model and GA. *Opt. Laser Technol.* **2016**, *83*, 153–162. [CrossRef]
23. Miyamoto, I.; Kosumi, T.; Park, S.; Uragishi, H.; Watanabe, K.; Ooie, T. Applications of single-mode fiber-lasers to novel micro welding. In Proceedings of the Fifth International Symposium on Laser Precision Microfabrication, Nara, Japan, 11–14 May 2004; Volume 5662, pp. 507–514.
24. Ventrella, V.A.; Berretta, J.R.; de Rossi, W. Pulsed Nd:YAG laser seam welding of AISI 316L stainless steel thin foils. *J. Mater. Process. Technol.* **2010**, *210*, 1838–1843. [CrossRef]
25. Landowski, M.; Świerczyńska, A.; Rogalski, G.; Fydrych, D. Autogenous fiber laser welding of 316L austenitic and 2304 lean duplex stainless steels. *Materials* **2020**, *13*, 2930. [CrossRef]
26. Lisiecki, A.; Klimpel, A. Laser welding of butt joints of austenitic stainless steel AISI 321. *J. Achieve. Mater. Manufact. Eng.* **2007**, *25*, 63–66.
27. Dontu, O.; Ganatsios, S.; Alexandrescu Nicolae, D.B. Laser micro-welding of stainless steel components used in mechatronic systems. *Mecatronica* **2006**, 40–42.
28. Kumar, A.; Gupta, M.P.; Banerjee, J.; Neogy, S.; Keskar, N.; Bhatt, R.B.; Behere, P.; Biswas, D.J. Micro-Welding of Stainless Steel and Copper Foils Using a Nano -Second Pulsed Fiber Laser. *Lasers Manuf. Mater. Process.* **2019**, *6*, 158–172. [CrossRef]
29. Liao, H.T.; Chen, Z.W. A study on fiber laser micro-spot welding of thin stainless steel using response surface methodology and simulated annealing approach. *Int. J. Adv. Manuf. Technol.* **2012**, *67*, 1015–1025. [CrossRef]
30. Türkan, M.; Karakaş, Ö. The influence of corrosion on the mechanical behavior of AISI 316L stainless steel welds. *Mechanika* **2019**, *25*, 114–118. [CrossRef]
31. Kumar, N.; Mukherjee, M.; Bandyopadhyay, A. Comparative study of pulsed Nd:YAG laser welding of AISI 304 and AISI 316 stainless steels. *Opt. Laser Technol.* **2017**, *88*, 24–39. [CrossRef]
32. Chatterjee, S.; Mahapatra, S.S.; Bharadwaj, V.; Upadhyay, B.N.; Bindra, K.S.; Thomas, J. Parametric appraisal of mechanical and metallurgical behavior of butt welded joints using pulsed Nd:YAG laser on thin sheets of AISI 316. *Opt. Laser Technol.* **2019**, *117*, 186–199. [CrossRef]

33. Sathiya, P.; Abdul Jaleel, M.Y. Measurement of the bead profile and microstructural characterization of a CO2 laser welded AISI 904 L super austenitic stainless steel. *Opt. Laser Technol.* **2010**, *42*, 960–968. [CrossRef]
34. Anawa, E.M.; Olabi, A.G. Using Taguchi method to optimize welding pool of dissimilar laser-welded components. *Opt. Laser Technol.* **2008**, *40*, 379–388. [CrossRef]
35. Wang, H.; Wang, Y.; Li, X.; Wang, W.; Yang, X. Influence of assembly gap size on the structure and properties of SUS301L stainless steel laser welded lap joint. *Materials* **2021**, *14*, 996. [CrossRef] [PubMed]

materials

MDPI

Article

Laser Dissimilar Welding of AISI 430F and AISI 304 Stainless Steels

Krzysztof Pańcikiewicz [1,*], Aleksandra Świerczyńska [2], Paulina Hućko [1,3] and Marek Tumidajewicz [4]

1 Faculty of Metals Engineering and Industrial Computer Science, AGH University of Science and Technology, al. Adama Mickiewicza 30, 30-059 Kraków, Poland; paulina.hucko@gmail.com

2 Faculty of Mechanical Engineering, Gdańsk University of Technology, Gabriela Narutowicza 11/12, 80-233 Gdańsk, Poland; aleksandra.swierczynska@pg.edu.pl

3 MOSTOSTAL KRAKÓW S.A., Ujastek 7, 30-969 Kraków, Poland

4 RaQun Sp. z o.o., Fabryczna 4, 38-300 Gorlice, Poland; m.tumidajewicz@raqun.pl

* Correspondence: krzysztof.pancikiewicz@agh.edu.pl

Received: 24 September 2020; Accepted: 8 October 2020; Published: 13 October 2020

Abstract: A dissimilar autogenous laser welded joint of AISI 430F (X12CrMoS17) martensitic stainless steel and AISI 304 (X5CrNi18-10) austenitic stainless steel was manufactured. The welded joint was examined by non-destructive visual testing and destructive testing by macro- and microscopic examination and hardness measurements. With reference to the ISO 13919-1 standard the welded joint was characterized by C level, due to the gas pores detected. Microscopic observations of AISI 430F steel revealed a mixture of ferrite and carbides with many type II sulfide inclusions. Detailed analysis showed that they were Cr-rich manganese sulfides. AISI 304 steel was characterized by the expected austenitic microstructure with banded δ-ferrite. Martensitic microstructure with fine, globular sulfide inclusions was observed in the weld metal. The hardness in the heat-affected zone was increased in the martensitic steel in relation to the base metal and decreased in the austenitic steel. The hardness range in the weld metal, caused by chemical inhomogeneity, was 184–416 HV0.3.

Keywords: laser welding; austenitic stainless steel; martensitic stainless steel; dissimilar welded joint; microstructure; sulfide inclusions

---

1. Introduction

The development of technology is aimed at reducing production and operating costs, increasing the safety of use and improving ecology in the context of the development of materials. This opens the possibility of using new materials, developing the existing ones and modifying their properties [1]. Particularly in the area of welded joints, especially dissimilar joints, there are many opportunities due to the multitude of materials, processes and additional treatments that can significantly affect the final product.

AISI 430F (X12CrMoS17) is a stainless steel with the addition of sulfur typically existing in the form of MnS inclusions, which improves the machinability of this material. Despite the high chromium content, the addition of sulfur lowers the resistance to crevice and pitting corrosion. Depending on the type of treatment, it has a ferritic or ferritic-martensitic structure [2]. This steel is not recommended for use in marine environments and highly oxidizing chemical environments. It is used in the production of screws, spindles, nuts, medical instruments, and in the automotive and transport industry. Due to its applications, this steel can be subjected to plasma nitriding or nitrocarburizing to increase hardness, corrosion, tribocorrosion and wear resistance [3–5].

AISI 304 (X5CrNi18-10) steel is a very popular grade of austenitic stainless steel, with good corrosion resistance. It is used in the construction, automotive, food, and chemical industries, for decorative purposes and kitchen equipment, as well as an element of electronic equipment [6]. There have been many studies describing the corrosion resistance, mechanical and functional properties of this steel [7,8]. Currently, many studies have focused on possible ways to improve the properties of AISI 304 steel [9–11].

The weldability of steel depends primarily on its chemical composition, structure, mechanical and physicochemical properties. Therefore, dissimilar steels usually require special care to be able to join them [12–14]. The basic difficulties include differences in the values of thermal expansion coefficients, melting points, mechanical properties, electrochemical potential as well as the formation of intermetallic compounds [15]. Additional difficulties may arise from the need to meet different heat treatment requirements for each of the welded materials. Dissimilar joints are characterized by a non-linear change of properties and often a very unfavorable segregation of elements, which may eliminate them from service. In addition, dissimilar joining requires extensive technical knowledge, appropriate equipment and careful selection of consumables. However, these inconveniences and risks are taken due to the many benefits that arise from combining two very different materials, reducing both production and operating costs and the improving mechanical properties of the joints [16,17]. In order for such a joining to be carried out without unacceptable imperfections, it is necessary to thoroughly understand the technologies and weldability of materials.

AISI 430F steel is usually regarded as difficult to weld due to, among other factors, hydrogen-induced cold cracking, lack of weld ductility, intergranular corrosion, deterioration of toughness and corrosion resistance. In order to carry out the welding it is necessary to use pre-heating at 150–200 °C, which reduces diffusible hydrogen content and welding stresses. Additionally, post-weld annealing at 790–815 °C can be performed. To reduce the risk of intergranular corrosion, chrome depletion at grain boundaries should be decreased by the use of low heat input value. Despite so many difficulties, the literature shows that there have been few effective attempts to weld this steel grade, mainly to make dissimilar joints in such applications as pressure vessels or fuel injectors. Khan et al. [18] investigated laser beam welding of dissimilar AISI 430F and AISI 440C stainless steels and the effects of laser welding parameters and heat input on weld bead geometry. They obtained qualitatively sound joints by finding appropriate welding parameters to avoid micro-crack formation. Romoli et al. [19] made a dissimilar AISI 440C with AISI 430F laser welded joint and found an empirical relationship between the shear strength of the weld and the configuration adopted during experiments. They observed that higher susceptibility to surface crack formation generated on the boundary of the weld seam with AISI 440C is connected with a higher martensitic volume fraction.

AISI 304 steel is considered to be easily weldable, with the use of specific technological procedures, such as ensuring that the interpass temperature does not exceed 200 °C. However, it should be remembered that the heat introduced into the material during welding makes it more sensitive to intergranular corrosion and solidification cracking. Pankaj et al. [20] investigated the possibility of laser welding of AISI 304 steel with low carbon steel and found a correlation between heat input and the properties of the joints. Rogalski et al. [21] welded 304L steel with Incoloy 800HT and conducted an analysis of the properties and structure of the joint, confirming that the TIG method can be used to join these materials even with an unfavorable design resulting from the working conditions of the joint. On the other hand, Pańcikiewicz et al. [22] indicated in their work that although austenitic steels are treated as easily weldable, they cannot be ignored and potentially dangerous imperfections should be identified using both traditional and new test methods. At the same time, Kurc-Lisiecka et al. [23] obtained the sound joints during laser welding of AISI 304 steel in a wide range of welding parameters, confirming the good weldability of this material, even when using a concentrated heat source. Welding of AISI 304 austenitic steel with AISI 430 ferritic steel was performed by Wang et al. [24]. They used the gas tungsten arc welding process, assessed the structure and corrosion resistance of welded joints and proved that it is possible to obtain the correct joint between these steels.

Laser welding by concentrating the heat source provides many opportunities compared to the conventional arc welding: simplicity of automation, very narrow HAZ, slight material deformation, deep penetration without the need for beveling and the use of filler metal [25,26]. These features make laser technologies are constantly evolving so that the expensive equipment, which is necessary in this method, is becoming more accessible and at the same time the process simulation programs are more and more accurate [27,28]. In addition, laser welding was initially considered unsuitable for difficult-to-weld materials and carried a high risk due to the concentrated heat source. A lot of research has been done on the possibility to use laser welding for joining various materials and show that this method makes it possible to obtain high quality welded joints using the wide range of conditions and parameters [29–31].

Each of the previously discussed steel grades can cause some problems during and after welding. According to the literature, each of the steel grades was subjected to laser welding, but there are no data on the possibility of laser beam welding of AISI 430F steel with AISI 304 steel. The purpose of this work was to analyze the possibility of making the overlap autogenous welded joint with the use of laser beam welding and to show the mechanism of structure formation in this welded joint. This goal was achieved by metallographic macro- and microscopic observations, EDS analysis, simulations and hardness measurements.

2. Materials and Methods

The overlap joint of AISI 430F drill rod and AISI 304 tube (Ø25 × 2.5 mm) was welded by the 521 laser beam welding (LBW) process without a filler metal. A general view and the design of the welded joint are presented in Figure 1. This joint, a part of a specialized drill produced by RaQun Sp. z o.o. (Gorlice, Poland) for the mining industry, was made during the start of development and first testing of the welding procedure qualification.

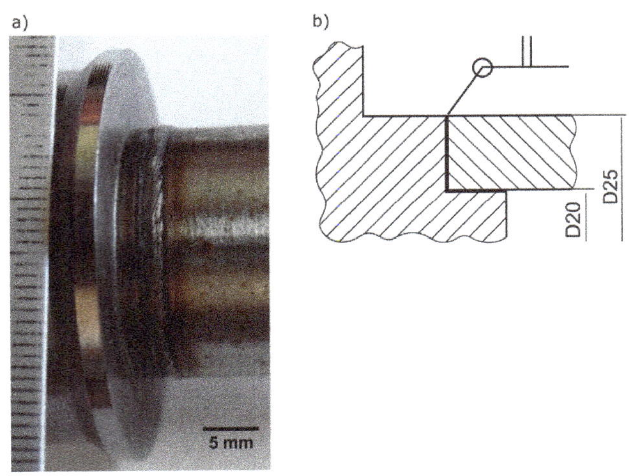

Figure 1. The welded joint: (**a**) general view, (**b**) design.

Visual testing was performed according to European Standards ISO 17637 and EN 13018 by direct method in intense lighting ~600 lx with universal welding gauge, with resolution 0.05 mm. Chemical composition of base metals was measured by optical emission spectroscopy (OES) using a Foundry Master-WAS Spectrometer (Hitachi, Tokyo, Japan). Macro- and microscopic examinations were performed on two cross sections after mechanical grinding, polishing and two-step etching: chemical in Kalling's reagent (5 g $CuCl_2$ + 100 mL HCl + 100 mL C_2H_5OH) and electrolytic in 10% CrO_3 water solution. Base metals, heat-affected zones and weld metal were examined. Microstructure was characterized using a light microscope (Leica, Wetzlar, Germany) and scanning electron microscope (SEM) (Japan Electron Optics Laboratory Co., Ltd., Tokyo, Japan) with X-ray energy-dispersive spectroscopy (EDS) (Japan Electron Optics Laboratory Co., Ltd., Tokyo, Japan). The depth of EDS analysis was assessed by CASINO v3.3.0.4 software (Université de Sherbrooke, Sherbrooke, QC, Canada). Thermodynamic analyses were performed by Thermo-Calc 2020a (Thermo-Calc Software AB, Solna, Sweden) with the TC-FE7 database (Thermo-Calc Software AB, Solna, Sweden). Hardness measurements were performed by the Vickers method (2.942 N) on a Tukon 2500 universal hardness tester (Wilson, NY, USA) in 8 rows of 25 indentations on the cross section of the joint. The distance between the indentations was 0.2 mm and indentation time was 5 s.

3. Results and Discussion

Chemical compositions of base metals, verified by the OES, and weld metal, verified by the EDS, are presented in Table 1. The compositions comply with the requirements of product standards. The main alloying elements in AISI 430F steel were 15.5% chromium for corrosion resistance and almost 0.5% of sulfur for machinability in automatic machining processes. AISI 304 steel contains 18% chromium and 8% nickel to provide an austenitic structure with δ-ferrite at ambient temperature.

Table 1. Chemical composition of base metals (OES) and weld metal (EDS).

Steel	Chemical Composition, Mass %									
	C	Si	Mn	P	S	Cr	Mo	Ni	Cu	V
AISI 430F (X12CrMoS17)	0.17	0.48	0.80	0.028	0.497	15.50	0.26	0.33	0.12	0.05
AISI 304 (X5CrNi18-10)	0.05	0.59	1.95	0.009	0.003	18.02	0.04	8.02	0.05	-
Weld metal	-	0.7	1.9	-	0.2	16.8	0.4	4.5	-	-

The butt welded joint was B quality level (acc. to EN ISO 13919-1) in visual testing (VT) and was C quality level in macroscopic examination. Full penetration and a few gas pores were observed (Figure 2). The largest pore was 0.16 mm in diameter and all the pores occupied 1.2% of the weld area (Table 2). For C quality level, the maximum dimension of the pore (with a tube thickness of 2.5 mm) was 1 mm and the maximum dimension of the area of the imperfection related to the projected area was 2%.

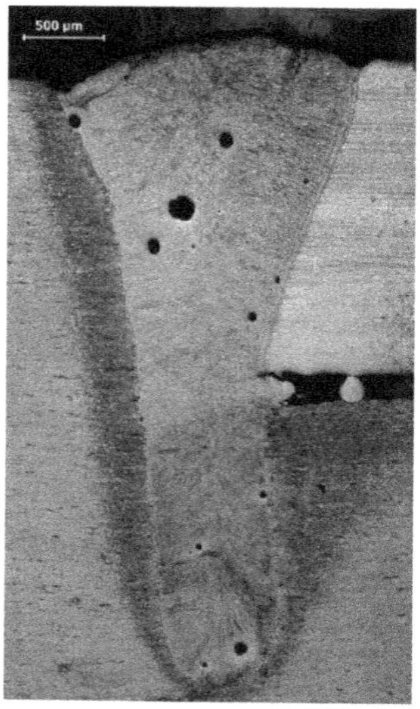

Figure 2. Macrostructure of the welded joint.

Table 2. The results of the weld and pores surface area measurements.

Zone		Area, μm^2
Weld Metal		4,206,827
Pore	1	554
	2	5453
	3	5550
	4	893
	5	19,691
	6	236
	7	318
	8	80
	9	6326
	10	788
	11	3923
	12	192
	13	536
	14	1104
	15	796
	16	144
	17	3987
	18	483
	19	75
	sum	51,129

Figure 3 presents the microstructure on the AISI 430F steel side of the joint. Base metal (Figure 3d) was characterized by a mixture of ferrite and carbides, which indicates a soft annealing or a high tempering delivery condition. This is a particularly favorable condition for machining due to the avoidance of a martensitic structure and low hardness. Gray inclusions elongated in the rolling direction, characteristic of manganese sulfides, were also observed (Figure 3d). The shape of the inclusions suggests that they are type II sulfides, acc. to Sims's classification [32]. EDS analysis confirmed the high content of manganese and sulfur, and additionally the inclusion was enriched by chromium (Figure 4), as indicated by a similar chromium content in the inclusion and the matrix. In high chromium steels it is possible to enrich the inclusions with this element. Considering also the small amount of iron, pyrrhotite with the formula (Cr, Fe, Mn)S can be formed [33–35]. Atomic concentration analysis indicates a 3:1 ratio between iron+manganese and chromium, and a 1:1 ratio between metallic elements and sulfur; thus, the observed inclusions were $[(Fe,Mn)_3Cr]S$.

In the heat-affected zone, Figure 3b, intercritical HAZ (ICHAZ, heated to the $\delta + \gamma$ + MnS range) and coarse-grain HAZ (CGHAZ, heated much above A3) were observed. In ICHAZ an incomplete normalized microstructure with fine and medium grains and more carbides than base metal (Figure 3e) and in CGHAZ an overheated microstructure with coarser grains than in other zones were observed (Figure 3c). Due to the presence of a higher content of ferrite stabilizers (particularly chromium), martensite with δ-ferrite was observed, clearly visible in CGHAZ (Figure 3c). With increasing chromium content, the γ-field on the phase diagram contracts to a small region, and with increasing carbon content, initially two-phase ($\delta + \gamma$)-field expands. Accelerated cooling in the two-phase field, in the absence or narrow range of homogeneous austenite, leads to the transformation of the austenite into martensite and some δ-ferrite remaining (Figure 3a). Unfortunately, a negative effect on mechanical properties is observed – ductile-brittle transition temperature increases and tensile strength decreases with increase δ-ferrite in volume fraction [36,37].

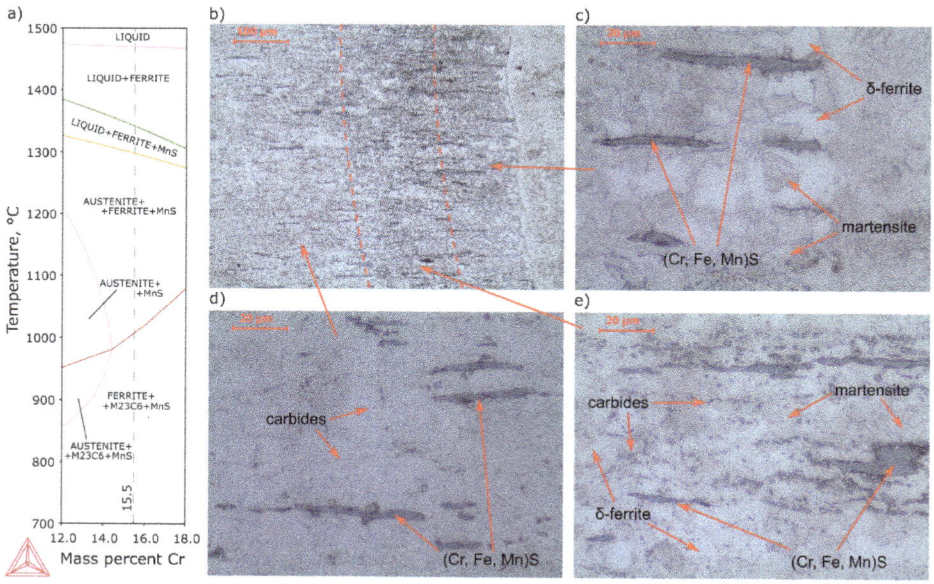

Figure 3. Microstructure of AISI 430F (X12CrMoS17) steel heat-affected zone: (**a**) pseudobinary phase diagram of 0.17C-0.48Si-0.8Mn-0.028P-0.497S-0.26Mo-0.33Ni-0.12Cu-0.05V steel, (**b**) general view, (**c**) coarse-grain heat-affected zone, (**d**) base metal, (**e**) intercritical heat-affected zone (Kalling's and CrO_3 etched).

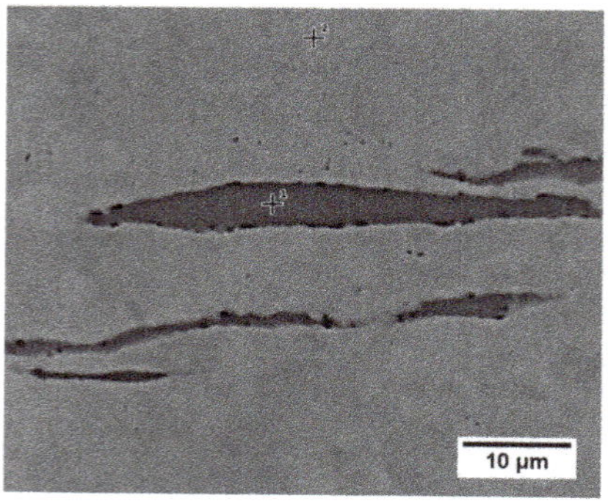

Element Symbol	Weight Concentration		Atomic Concentration	
	Point 1	Point 2	Point 1	Point 2
Mn	46.1	0.9	36.1	0.9
S	36.6	0.3	49.2	0.6
Cr	13.1	14.6	10.8	15.3
Fe	3.4	83.2	2.6	81.5
Si	0.8	0.8	1.3	1.6
Mo	-	0.2	-	0.1

Figure 4. SEM microstructure and EDS point analysis of AISI 430F (X12CrMoS17) base metal (polished).

Near the fusion line, on both sides, the shape of the inclusions was changed from elongated to spherical, which was caused by high temperature from laser beam welding (Figure 5). The mechanism is shown in Figure 6.

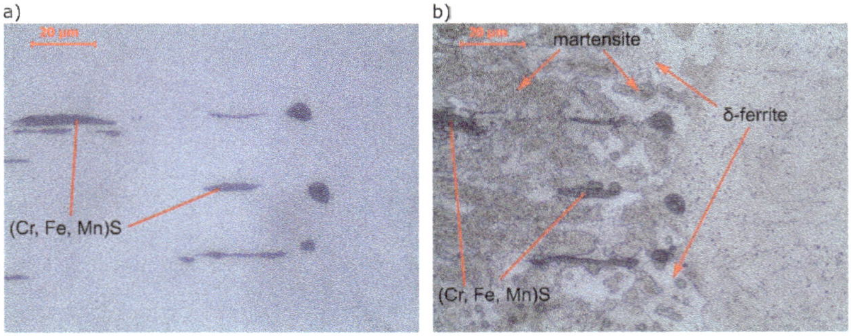

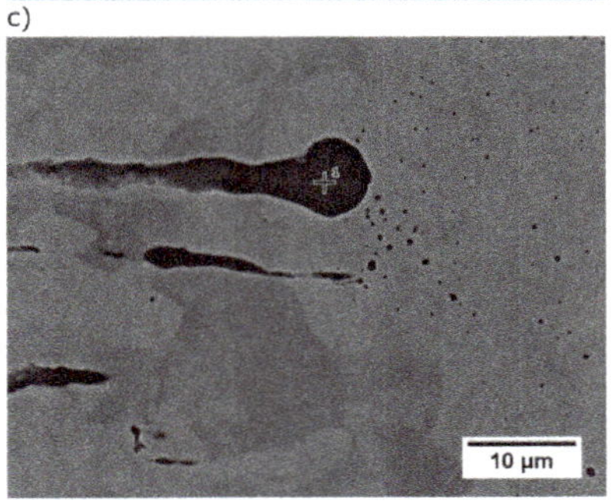

Element Symbol	Weight Concentration Point 1
Mn	45.2
S	35.9
Cr	14.3
Fe	4.6

Figure 5. Microstructure near fusion line on the AISI 430F (X12CrMoS17) base metal side: (**a**) microstructure (polished), (**b**) microstructure (Kalling's etched), (**c**) SEM microstructure and EDS point analysis (polished).

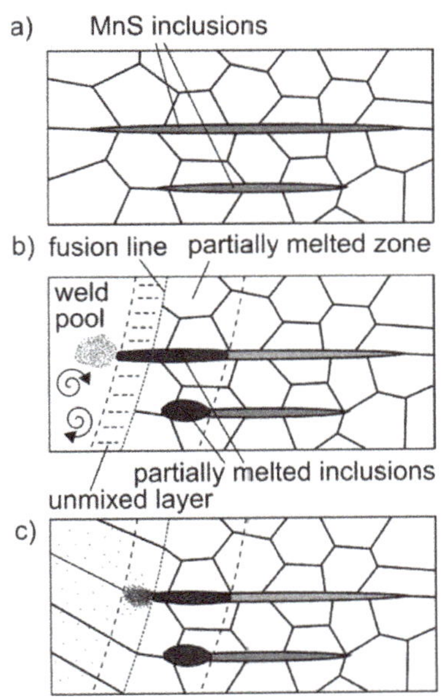

Figure 6. Constitutional liquation mechanism of sulfide in partially melted zone. (**a**) before welding; (**b**) during welding; (**c**) after welding.

Sulfides, which were in the high-temperature HAZ, melted with the formation of a liquid film as a result of constitutional liquation. Part of the sulfur-rich liquid was located at grain boundaries in the partially melted zone, while part was located in the weld. At the fusion line, sulfur-rich liquid was not mixed with the weld metal due to the presence of an unmixed zone near the fusion line (Nernst layer). During cooling, the areas with a higher melting point (poorer in impurity) were first crystallized. Figure 7 shows a Scheil Simulation for AISI 430F steel solidification, when the chemical composition corresponds to the tested steel (sulfur mass content was 0.497%) and when the sulfur content was 0%. The solidification temperature of steel with 0.497%S was lower than that of the sulfur-free steel by 394.2 deg. Sulfur-rich liquid is trapped at the grain boundaries and crystallizes in the form of eutectics.

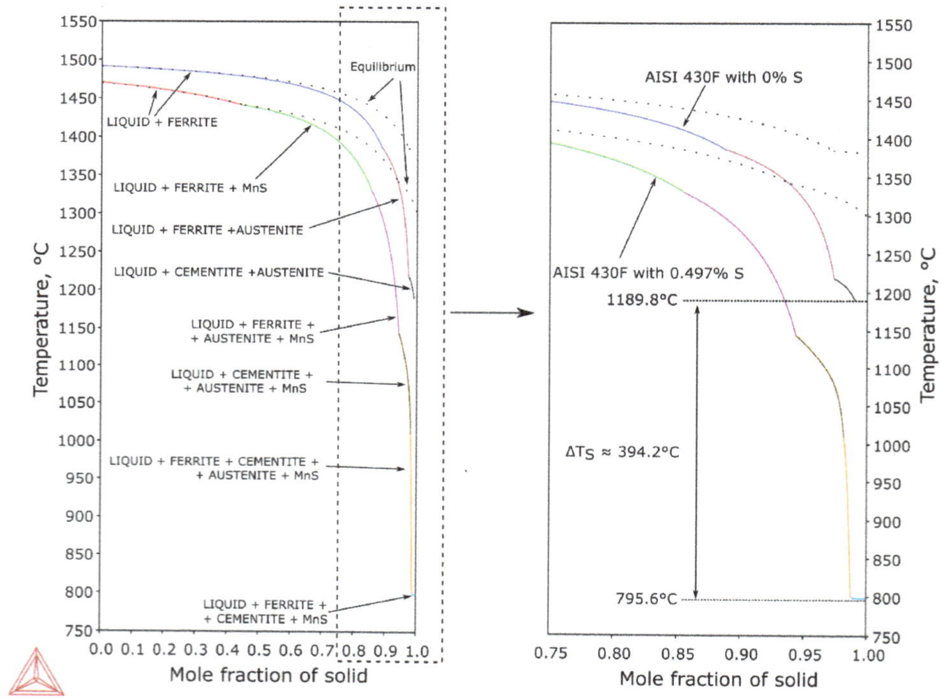

Figure 7. AISI 430F (X12CrMoS17) steel solidification (Scheil Simulation).

In the weld metal, martensite with many spheroidal inclusions was observed (Figure 8). δ-ferrite was not observed in microstructure, which is caused by the presence of nickel, derived from partially melted AISI 304 base metal. With increasing nickel content, the γ-field on the phase diagram expands. Accelerated cooling from homogeneous austenite transforms the austenite into martensite. EDS analysis showed enrichment of inclusions with manganese and sulfur and a similar chromium content to the matrix, which indicated that they are (Cr, Fe, Mn)S inclusions. The small size of the inclusions is responsible for the high iron content, which comes from the matrix. An example of the effective depth of electron penetration in the chemical analysis of small-size MnS inclusion, performed by Monte Carlo Simulation, is presented in Figure 9. Characteristic X-rays were emitted approximately equally from the inclusion and the matrix.

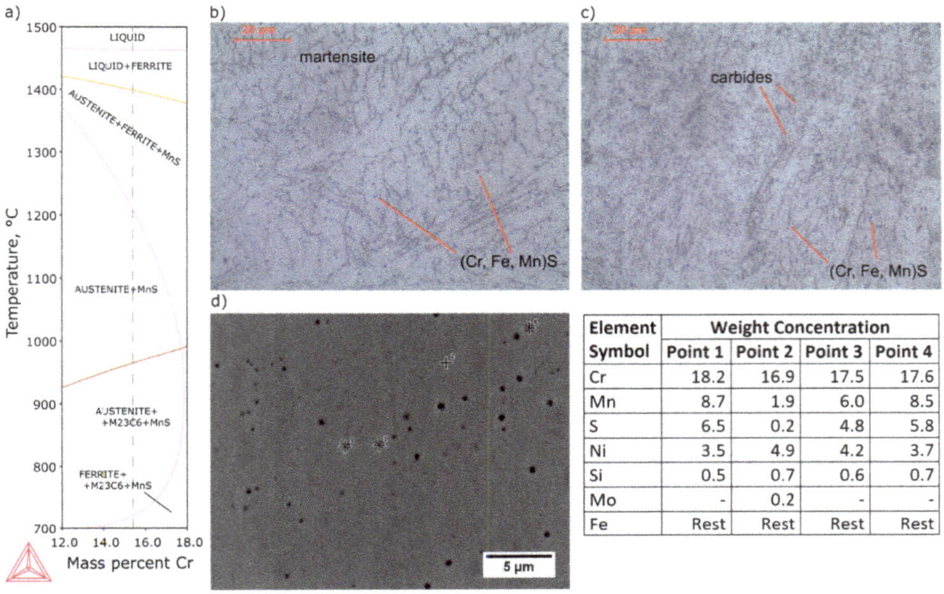

Element Symbol	Weight Concentration			
	Point 1	Point 2	Point 3	Point 4
Cr	18.2	16.9	17.5	17.6
Mn	8.7	1.9	6.0	8.5
S	6.5	0.2	4.8	5.8
Ni	3.5	4.9	4.2	3.7
Si	0.5	0.7	0.6	0.7
Mo	-	0.2	-	-
Fe	Rest	Rest	Rest	Rest

Figure 8. Microstructure and EDS point analysis of weld metal: (**a**) pseudobinary phase diagram of 0.11C-0.7Si-1.9Mn-0.2S-0.4Mo-4.5Ni steel, (**b**,**c**) microstructure (Kalling's + CrO_3 etched), (**d**) SEM microstructure and EDS points analysis (polished).

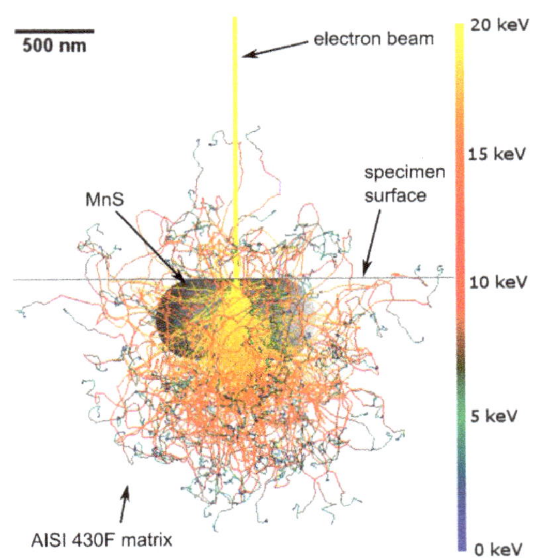

Figure 9. Monte Carlo Simulation of electron trajectory in MnS inclusion and AISI 430F (X12CrMoS17) matrix.

Base metal of AISI 304 steel (Figure 10a,b) was characterized by austenitic microstructure with banded δ-ferrite. The orientation of the δ-ferrite and a large content of twins indicate a mechanical worked delivery condition. In high-temperature HAZ (Figure 10c,d), disappearance of twins due to the recrystallization was observed, which significantly reduces strength properties [38].

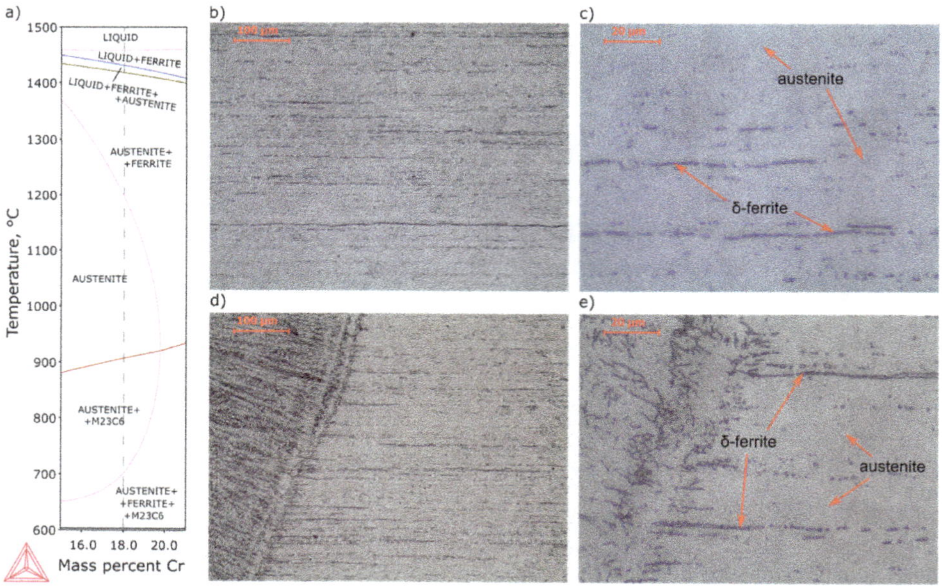

Figure 10. Microstructure of AISI 304 (X5CrNi18-10) steel heat-affected zone: (**a**) pseudobinary phase diagram of 0.05C-0.59Si-1.95Mn-0.009P-0.003S-0.04Mo-8.02Ni-0.05Cu steel, (**b**,**c**) base metal, (**d**,**e**) heat-affected zone (Kalling's + CrO_3 etched).

Comparison of the expected weld structure with the obtained results of the EDS analysis was performed with the Schaeffler diagram (Figure 11). OES analysis results of base metals, average values determined assuming of 50% dilution of both material and EDS analysis results of weld metal are presented in Table 3. In the case of AISI 430F steel, the predicted structure is martensite with approx. 24% ferrite, which essentially coincides with the obtained microstructure in the HAZ. Base metal was heat treated before welding; therefore, the structure does not correspond to that of the Schaeffler diagram. The structure of the AISI 304 steel according to the diagram is composed of austenite with approx. 8% ferrite and a small amount of martensite, which correspond to the observed microstructure. In the case of the weld, a difference between the average and measured chemical composition was identified (Table 3). Chemical composition of the weld obtained in the EDS analysis indicates that dilution of AISI 304 steel in the weld was 65% and AISI 430F was 35%. Non-axial arrangement of the laser beam in relation to the contact point of both welded materials and the difference in melting temperature and thermal conductivity coefficient values were the reasons for uneven melting. Regardless of the degree of dilution, the microstructure should consist of martensite, residual austenite and approx. 13% ferrite.

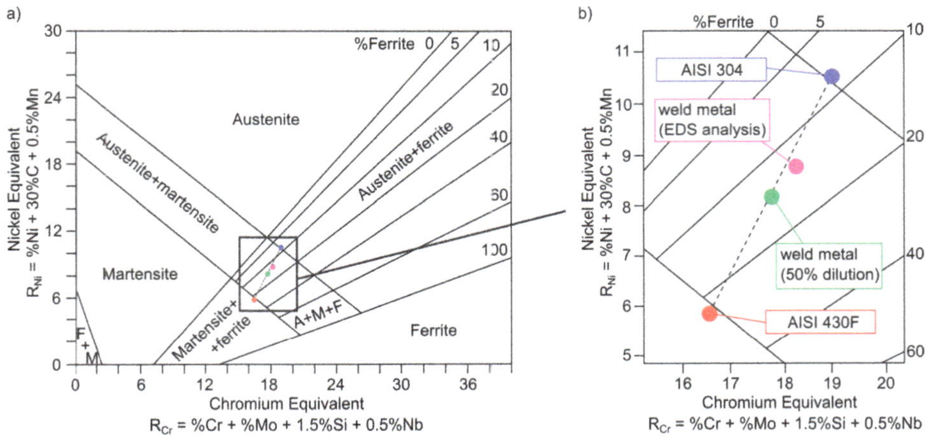

Figure 11. Schaeffler constitutional diagram: (**a**) general view, (**b**) enlarged view.

Table 3. Chemical composition of base metals (OES) and weld metal (arithmetic mean of the chemical composition of base metals and EDS analysis).

Steel	Chemical Composition, Mass %								
	C	Si	Mn	Cr	Mo	Ni	S	R_{Cr}	R_{Ni}
AISI 430F	0.17	0.48	0.80	15.50	0.26	0.33	0.497	16.48	5.83
AISI 304	0.05	0.59	1.95	18.02	0.04	8.02	0.003	18.95	10.50
Weld metal [1]	0.11	0.54	1.38	16.76	0.15	4.18	0.25	17.72	8.17
Weld metal [2]	0.11 [1]	0.7	1.9	16.8	0.4	4.5	0.2	18.25	8.75

[1] arithmetic mean of the chemical composition of base metals, [2] EDS analysis.

The hardness distribution map on the cross section of the joint is shown in Figure 12a. Hardness of the AISI 430F base metal was the lowest in the welded joint and amounted to 178–205 HV0.3. In the HAZ an increase in hardness to 417 HV0.3 was noted. On the AISI 304 steel side, the hardness was 259–333 HV0.3, which decreased to 191 HV0.3 in HAZ, reaching the hardness of the AISI 430F steel in the delivery condition. In the weld metal, hardness ranged from 184 to 416 HV0.3, which corresponds to the lowest and highest hardness in the other zones of the joint. The large difference in the hardness of the weld was caused by chemical segregation, shown in Figure 12b. EDS analyses were performed in the areas of 192 × 192 µm (0.037 mm^2), where the centers of indentation and analyzed areas were a common point. The zones with the highest hardness coincide with the areas with decreased chromium and nickel contents, i.e., the main components of chromium and nickel equivalents (R_{Cr} and R_{Ni}) decreased in these areas (Figure 12c). The determination of distribution of carbon in the weld metal using EDS analysis was not possible. When this distribution was simplified to be homogeneous, as the nickel equivalent value was decreased, the volume fractions of martensite and ferrite increased at the cost of austenite, and the hardness in the weld increased (Figure 12d).

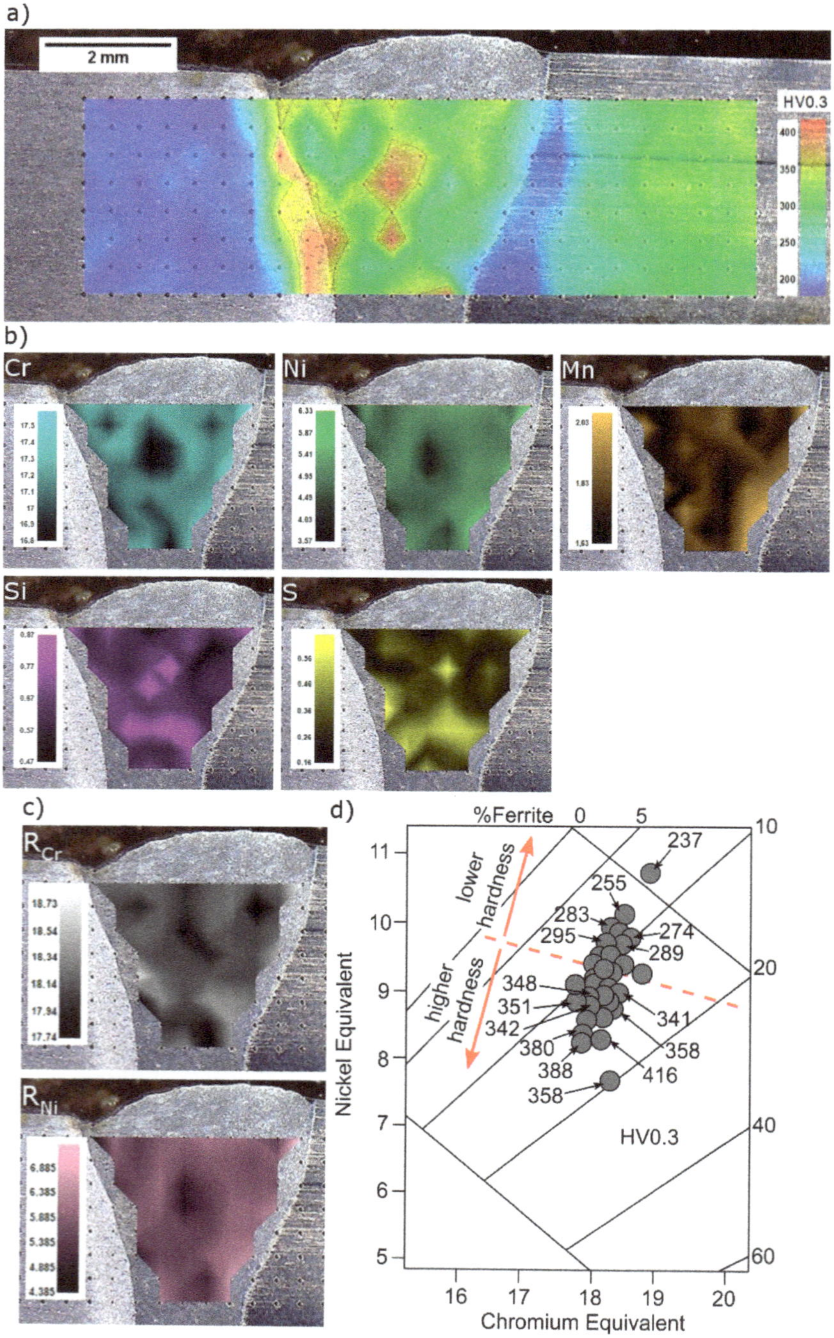

Figure 12. Hardness map of the joint and chemical composition distribution in weld metal. (**a**) hardness map; (**b**) chemical composition map; (**c**) chromium and nickel equivalents map; (**d**) hardness distribution depending on the chromium and nickel equivalents.

4. Conclusions

1. Laser welded joint of AISI 430F martensitic stainless steel and AISI 304 austenitic stainless steel was characterized as C level, with reference to the ISO 13919-1 standard, due to the detection of gas pores. The final welding procedure qualification of the product has been refined on the basis of the present findings.
2. Heat input during welding and a rapid cooling cycle after welding, characteristic of laser beam welding, causes hardening of the AISI 430F steel HAZ to a mixture of martensite and ferrite, and softening of the AISI 304 steel HAZ due to recrystallization. The inclusions observed in the AISI 430F steel were Cr-rich manganese sulfides of the (Cr, Fe, Mn)S pyrrhotite type.
3. The weld metal, formed by the melting of two base metals, was characterized by the high-alloy martensite structure, in which the hardness corresponds to local changes in the chemical composition, occurring as a result of the chemical inhomogeneity of the weld. Fine, globular sulfide inclusions were also observed.

Author Contributions: Conceptualization, K.P., P.H. and M.T.; formal analysis, K.P., A.Ś., P.H.; investigation, K.P. and P.H.; methodology, K.P. and P.H.; writing–original draft, K.P., A.Ś.; writing–review and editing, K.P., A.Ś., P.H. and M.T. All authors have read and agreed to the published version of the manuscript.

Funding: This research was supported by the Polish Ministry of Science and Higher Education (Poland), grant number 16.16.110.663 (AGH University of Science and Technology).

Acknowledgments: The authors are grateful to Krzysztof Wieczerzak and Piotr Bała (AGH University of Science and Technology in Krakow-Academic Centre for Materials and Nanotechnology, Poland) for the hardness analysis. Appreciation is also expressed to A.S. Wronski (University of Bradford, UK) for editing the manuscript.

Conflicts of Interest: The authors declare no conflict of interest.

References

1. Szala, M.; Łatka, L.; Walczak, M.; Winnicki, M. Comparative study on the cavitation erosion and sliding wear of cold-sprayed Al/Al_2O_3 and Cu/Al_2O_3 coatings, and stainless steel, aluminium alloy, copper and brass. *Metals* **2020**, *10*, 856.
2. Kim, D.K.; Lee, D.G.; Lee, S. Correlation of microstructure and surface roughness of disc drums fabricated by hot forging of an AISI 430F stainless steel. *Met. Mater. Trans. A* **2001**, *32*, 1111–1118.
3. Jashari, N.; Atapek, H.; Polat, Ş.; Çelik, G.A. Ion plasma nitriding of ferritic steel AISI 430 F. *Mater. Sci. Nonequilib. Phase Transform.* **2018**, *4*, 138–141.
4. Alphonsa, J.; Mukherjee, S.; Raja, V.S. Study of plasma nitriding and nitrocarburising of AISI 430F stainless steel for high hardness and corrosion resistance. *Corros. Eng. Sci. Technol.* **2018**, *53* (Suppl. 1), 51–58.
5. Dalbert, V.; Mary, N.; Normand, B.; Verdu, C.; Douillard, T.; Saedlou, S. The effects of microstructures and repassivation kinetics on the tribocorrosion resistance of ferrite and ferrite-martensite stainless steels. *Wear* **2019**, *420*, 245–256.
6. Pańcikiewicz, K.; Tuz, L.; Zielińska-Lipiec, A. Zinc contamination cracking in stainless steel after welding. *Eng. Fail. Anal.* **2014**, *39*, 149–154.
7. Shtefan, V.; Kanunnikova, N.; Pilipenko, A.; Pancheva, H. Corrosion behavior of AISI 304 steel in acid solutions. *Mater. Today Proc.* **2019**, *6*, 150–157.
8. Michler, T.; Bruder, E. Local strains in 1.4301 austenitic stainless steel with internal hydrogen. *Mater. Sci. Eng. A* **2018**, *725*, 447–455.
9. Naeem, M.; Iqbal, J.; Zakaullah, M.; Shafiq, M.; Mujahid, Z.I.; Díaz-Guillén, J.C.; Lopez-Badillo, C.M.; Sousa, R.R.M.; Khan, M.A. Enhanced wear and corrosion resistance of AISI-304 steel by duplex cathodic cage plasma treatment. *Surf. Coat. Technol.* **2019**, *375*, 34–45.
10. Maleki, E.; Unal, O. Fatigue limit prediction and analysis of nano-structured AISI 304 steel by severe shot peening via ANN. *Eng. Comput.* **2020**, *36*, 1–16.
11. Amininejad, A.; Jamaati, R.; Hosseinipour, S.J. Achieving superior strength and high ductility in AISI 304 austenitic stainless steel via asymmetric cold rolling. *Mater. Sci. Eng. A* **2019**, *767*, 138433. [CrossRef]

12. Tomków, J.; Fydrych, D.; Rogalski, G. Dissimilar underwater wet welding of HSLA steels. *Int. J. Adv. Manuf. Technol.* **2020**, *109*, 717–725.
13. Ramkumar, K.D.; Sidharth, D.; Prabhakar, K.V.P.; Rajendran, R.; Mugundan, K.G.; Narayanan, S. Microstructure and properties of Inconel 718 and AISI 416 laser welded joints. *J. Mater. Proc. Technol.* **2019**, *266*, 52–62.
14. Zhao, X.; Shi, Z.; Deng, C.; Liu, Y.; Li, X. The effect of laser offset welding on microstructure and mechanical properties of 301L to TA2 with and without Cu intermediate layer. *Metals* **2020**, *10*, 1138.
15. Shamsolhodaei, A.; Oliveira, J.P.; Schell, N.; Maawad, E.; Panton, B.; Zhou, Y.N. Controlling intermetallic compounds formation during laser welding of NiTi to 316L stainless steel. *Intermetallics* **2020**, *116*, 106656.
16. Landowski, M.; Świerczyńska, A.; Rogalski, G.; Fydrych, D. Autogenous fiber laser welding of 316L austenitic and 2304 lean duplex stainless steels. *Materials* **2020**, *13*, 2930.
17. Shreyas, P.; Panda, B.; Kumar, R. Mechanical properties and microstructure of 316L-galvanized steel weld. *Mater. Today Proc.* **2020**, *23*, 600–607. [CrossRef]
18. Khan, M.M.A.; Romoli, L.; Dini, G. Laser beam welding of dissimilar ferritic/martensitic stainless steels in a butt joint configuration. *Opt. Laser Technol.* **2013**, *49*, 125–136.
19. Romoli, L.; Rashed, C.A.A. The influence of laser welding configuration on the properties of dissimilar stainless steel welds. *Int. J. Adv. Manuf. Technol.* **2015**, *81*, 563–576.
20. Pankaj, P.; Dhara, L.N.; Tiwari, A.; Biswas, P. Experimental study on monitoring of heat input in continuous CO_2 laser butt welding of dissimilar thin steels. *Lasers Manuf. Mater. Process.* **2020**, *7*, 1–27. [CrossRef]
21. Rogalski, G.; Świerczyńska, A.; Landowski, M.; Fydrych, D. Mechanical and microstructural characterization of TIG welded dissimilar joints between 304L austenitic stainless steel and Incoloy 800HT nickel alloy. *Metals* **2020**, *10*, 559.
22. Pańcikiewicz, K.; Radomski, W. Lack of tightness analysis of concealed welded radiators. *Eng. Fail. Anal.* **2020**, *114*, 104579.
23. Kurc-Lisiecka, A.; Lisiecki, A. Laser welding of stainless steel. *J. Achiev. Mater. Manuf. Eng.* **2020**, *98*, 32–40. [CrossRef]
24. Wang, C.; Yu, Y.; Yu, J.; Zhang, Y.; Zhao, Y.; Yuan, Q. Microstructure evolution and corrosion behavior of dissimilar 304/430 stainless steel welded joints. *J. Manuf. Proc.* **2020**, *50*, 183–191. [CrossRef]
25. Tuz, L. Evaluation of microstructure and selected mechanical properties of laser beam welded S690QL high-strength steel. *Adv. Mater. Sci.* **2018**, *18*, 34–42. [CrossRef]
26. Xu, C.; Zhang, Y.; Liu, W.; Jin, Y.; Wen, L.; Sun, D. Influence of laser-welding on microstructure and corrosion properties of TWIP steel. *Materials* **2020**, *13*, 4315.
27. Kik, T. Heat source models in numerical simulations of laser welding. *Materials* **2020**, *13*, 2653. [CrossRef]
28. Majkowska-Marzec, B.; Tęczar, P.; Bartmański, M.; Bartosewicz, B.; Jankiewicz, B.J. Mechanical and corrosion properties of laser surface-treated Ti13Nb13Zr alloy with MWCNTs coatings. *Materials* **2020**, *13*, 3991.
29. Landowski, M. Influence of parameters of laser beam welding on structure of 2205 duplex stainless steel. *Adv. Mater. Sci.* **2019**, *19*, 21–31. [CrossRef]
30. Górka, J. Assessment of the weldability of T-welded joints in 10 mm thick TMCP steel using laser beam. *Materials* **2018**, *11*, 1192.
31. Quazi, M.M.; Ishak, M.; Fazal, M.A.; Arslan, A.; Rubaiee, S.; Qaban, A.; Aiman, M.H.; Sultan, T.; Ali, M.M.; Manladan, S.M. Current research and development status of dissimilar materials laser welding of titanium and its alloys. *Opt. Laser Technol.* **2020**, *126*, 106090. [CrossRef]
32. Sims, C.E. The nonmetallic constituents of steel. *Trans. Am. Inst. Min. Met. Eng.* **1959**, *215*, 367–393.
33. Dilner, D.; Mao, H.; Selleby, M. Thermodynamic assessment of the Mn–S and Fe–Mn–S systems. *CALPHAD* **2015**, *48*, 95–105. [CrossRef]
34. Hack, K.; Jantzen, T.; Yazhenskikh, E.; Müller, M. Technologies sulfide database: Evaluation of thermodynamic data and phase equilibria. In Proceedings of the GTT-Workshop, Herzogenrath, Germany, 1–3 July 2015.
35. Waldner, P. Thermodynamic modeling of the Cr-Fe-S system. *Met. Mater. Trans. A* **2014**, *45A*, 798–814. [CrossRef]
36. Pandey, C.; Mahapatra, M.M.; Kumar, P.; Saini, N.; Thakre, J.G.; Vidyarthy, R.S.; Narang, H.K. A brief study on δ-ferrite evolution in dissimilar P91 and P92 steel weld joint and their effect on mechanical properties. *Arch. Civ. Mech. Eng.* **2018**, *18*, 713–722. [CrossRef]

37. Zala, A.B.; Jamnapara, N.I.; Badheka, V.J.; Sasmal, C., Sam, S.; Ranjan, M. Delta (δ) ferrite formation in the welds of aluminized 9Cr-1Mo steels. *Met. Microstruct. Anal.* **2019**, *8*, 256–262. [CrossRef]
38. Li, X.; Wei, Y.; Wei, Z.; Zhou, J. Effect of cold rolling on microstructure and mechanical properties of AISI 304N stainless steel. *IOP Conf. Ser. Earth Environ. Sci.* **2019**, *252*, 1–9. [CrossRef]

materials

MDPI

Article

Autogenous Fiber Laser Welding of 316L Austenitic and 2304 Lean Duplex Stainless Steels

Michał Landowski *, Aleksandra Świerczyńska, Grzegorz Rogalski and Dariusz Fydrych

Faculty of Mechanical Engineering, Gdańsk University of Technology, 80-233 Gdańsk, Poland; aleksandra.swierczynska@pg.edu.pl (A.Ś.); grzegorz.rogalski@pg.edu.pl (G.R.); dariusz.fydrych@pg.edu.pl (D.F.)

* Correspondence: michal.landowski@pg.edu.pl; Tel.: +48-583-471-291

Received: 1 June 2020; Accepted: 28 June 2020; Published: 30 June 2020

Abstract: This study presents results of experimental tests on quality of dissimilar welded joints between 316L austenitic and 2304 lean duplex stainless steels, welded without ceramic backing. Fiber laser welded butt joints at a thickness of 8 mm were subjected to non-destructive testing (visual and penetrant), destructive testing (static tensile test, bending test, and microhardness measurements) and structure observations (macro- and microscopic examinations, SEM, element distribution characteristics, and ferrite content measurements). Non-destructive tests and metallographic examinations showed that the welded joints meet the acceptance criteria for B level in accordance with EN ISO 13919–1 standard. Also the results of the destructive tests confirmed the high quality of the joints: specimens were fractured in base material with lower strength—316L austenitic stainless steel and a 180° bending angle was obtained confirming the high plasticity of the joints. Microscopic examination, SEM and EDS analysis showed the distribution of alloying elements in joints. The microhardness of the autogenous weld metal was higher by about 20 HV0.2 than that of the lean duplex steel. Ferrite content in the root was about 37% higher than in the face of the weld. The Schaeffler phase diagram was used to predict the phase composition of the welded joints and sufficient compliance with the magnetic method was found. The presented procedure can be used for welding of 316L–2304 stainless steels dissimilar welded joints of 8 mm thickness without ceramic backing.

Keywords: laser welding; austenitic stainless steel; lean duplex stainless steel; dissimilar welded joints; mechanical properties; microstructure

1. Introduction

Lasers as a heat source have been widely applied in industrial processes [1,2]. Currently, they are used for welding, cutting, remelting, and various processes with surface modification of materials [3–6]. Laser sources are characterized by: high energy density, high efficiency, large temperature gradients, high repeatability of the process, and formation of a narrow heat affected zone (HAZ). Many authors report beneficial changes in the properties of various materials subjected to laser treatment in such sectors as: medical, energy, ocean engineering, etc. [7–9]. The area of application of various technological laser solutions includes processing of many types of metals and their alloys, ceramics, polymers, and composites [1,5,10,11], including underwater local dry welding and wet welding [12,13]. Limiting the disadvantageous effects of technological use of lasers in some applications, e.g., the small width of weld, can be obtained by combining it with other heat sources to create hybrid processes. The most popular solutions include laser-metal active gas arc welding (MAG) and laser- tungsten inert gas arc welding (TIG) processes [14–16].

Often, high-alloy steels, especially stainless steels, are subjected to laser treatment. Stainless steels occupy an irreplaceable position in many branches of industry [17]. These steels are classified, among others according to the criteria of chemical composition and structure into: ferritic, martensitic, austenitic, and duplex (ferritic-austenitic) steels [18–20]. Austenitic steels are the most popular group of high-alloy stainless steels. The content of chromium (above 18%) and nickel (minimum 8%) provides the structure stability and the corrosion resistance [21–23]. In order to improve the properties of these steels, alloying elements: molybdenum, titanium, and niobium are introduced. The last two are excellent stabilizers and reduce the risk of intergranular corrosion. Austenitic steels are characterized by high corrosion resistance, good strength, and plastic parameters, hence their wide application. Improvement of mechanical properties can be achieved through the use of cold forming, and increase in corrosion resistance through the use of heat treatment. The group of austenitic stainless steels is considered well weldable by using many processes and maintaining an appropriate technological regime that includes all treatments related to the welding process, e.g., the prefabrication of the elements to be welded by using appropriate tools, the use of the consumables with an appropriate chemical composition, the control of the heat input and others. The main risks directly associated with the welding process are the formation of hot cracks and intergranular corrosion. General guidelines for welding this group of material can be found, among others, in many scientific reports [17,18,24], regulations and standards, e.g., EN 1011–3.

Duplex ferritic-austenitic steels are becoming increasingly popular due to their specific properties. They combine the advantages of other groups of high-alloy steels and the most important of them are high strength with good plasticity and corrosion resistance [25–27]. The two-phase structure is obtained by appropriate selection of the ratio of ferrite-forming elements (mainly Cr—21–28%) and austenite-forming elements (Ni—1.5–8%, N—0.05–0.3%). Depending on the content of these elements, duplex steels are divided into lean duplex, duplex, super duplex, and hyper duplex. Lean duplex steels represent a reasonable compromise between the need to provide the appropriate strength properties, desired corrosion resistance, and the requirement of low material costs [28]. Compared to austenitic stainless steels, they have higher strength parameters besides plasticity and higher resistance to stress corrosion. Steels from this group are characterized by good weldability, provided that the technological regime is maintained: the control of the value of heat input, the use of appropriate welding consumables dedicated to welded grade, the dilution rate, the suitable dimensions of the welding groove, the use of the appropriate type of shielding gas on the face and root side, maintaining high purity of welded components and other [29–31].

Duplex steels are sensitive to structural changes resulting from the welding thermal cycle. This can reduce the mechanical properties of the joints as well as their corrosion resistance. High cooling rate of duplex steel joints may result in higher ferrite content in HAZ and in the weld [31]. According to ASTM E562 standard, no more than 70% of ferrite is recommended, and according to Norskom M-601 it should be in the range of 30–70%.

Despite the fact that duplex steels are considered sensitive to high cooling rate, they can be successfully welded in conditions of large negative temperature gradients, e.g., under water and using concentrated heat sources [32–34]. In addition, a significant problem during welding is the risk of chromium nitride precipitation, as well as the occurrence of micro–areas depleted in Cr and Ni. Guidelines for welding duplex steels can be found in scientific reports [17,19], regulations and standards, e.g., EN 1011–3, Annex C.

In industrial applications (for example in: power generation, nuclear, petrochemical, aerospace, and shipbuilding sector) there is often a need to perform various variants of dissimilar joints [28,35,36]. A particular group of dissimilar joints are connections between steels from different groups of stainless steels. Of these, joints of the austenitic steel type with ferritic, martensitic and duplex steel grades are most often made. For joining stainless steels with other metals in addition to laser welding also other methods can be used: metal active gas arc welding (MAG), tungsten inert gas arc welding (TIG), shielded metal arc welding, arc stud welding, friction welding, friction stir welding, diffusion welding, explosive welding, and furnace brazing. For welding different grades of austenitic and duplex steels, many variants of technologies were used, which led to obtain joints with various morphology, mechanical properties, and corrosion resistance [37–44].

Welding of dissimilar stainless steel joints can be done between two main types of filler metals: austenitic (e.g., 309L) or duplex (e.g., 2209). Vincente et al. stated that the higher corrosion resistance of joints made by the MAG process is favored by the use of duplex steel consumable [45]. Similarly, Rahmani et al. recommend using duplex instead of austenitic consumable when using the TIG process [46]. However, among the various austenitic consumables, the most beneficial, both in terms of strength properties, as well as morphology and corrosion resistance, is the use of 309L [47,48] or super austenitic 904L filler metal [49]. The literature also describes attempts to weld such a material combination without filler material: using different variants of the TIG process [50,51] or concentrated welding sources [33,52,53]. Ridha Mohammed et al. stated that the mechanical properties of the duplex-austenitic steel fiber laser joints were better compared with the base materials (BM) because of the small HAZ [52]. In the same work, the authors found the presence in welded joints the regions with a completely austenitic solidification mode which were susceptible to solidification cracking. An important observation is that when laser welding is used for super duplex and austenitic steels, the ferrite/austenite phase balance is not significantly changed by different heat input values [39]. Existence of an unmixed zone that originated from each base material was stated by Chun et al. [53]. The approximately 50:50 ferrite/austenite microstructural balance can be reached with solution annealing in the range of 1050–1100 °C, and it is especially important in autogenous welds [33]. Saravanan et al. stated that the higher microhardness of the weld zone is attributed by the formation of finer and uniform grains following high cooling rate [54].

Although researches on laser welding of dissimilar stainless steels have been reported, the study on structure and properties of laser beam welded austenitic/lean duplex steel joints is still relatively rarely mentioned. Therefore, the present work aims to show the ability of making 316L austenitic stainless steel–2304 lean duplex stainless steel dissimilar joints using the autogenous fiber laser welding process.

2. Materials and Methods

The test pieces were a flat plates with thickness of 8 mm made of 316L austenitic stainless steel (1.4404, UNS S31603) and 2304 lean duplex stainless steel (1.4362, UNS S32304) in delivery condition. Both materials were after heat treatment: 316L steel after solution annealing from 1050 °C and 2304 steel was hot rolled and solution annealed. The chemical composition of the used materials in accordance with the inspection certificate (spectral analysis) and the requirements of the EN 10088-2:2014 standard (minimum and maximum wt. %) are given in Table 1. The mechanical properties of the materials in accordance with the requirements of the EN 10088–2–2014 standard are presented in Table 2. Chromium and nickel equivalents (Creq and Nieq) were calculated according to [55].

Table 1. Chemical composition of base materials, wt.%.

Material	Value	C	Si	Mn	Ni	Cr	Mo	Cu	N	S	P	Creq	Nieq
316L	min	0.000	0.00	0.0	10.00	16.50	2.00	-	0.000	0.000	0.000		
	max	0.030	0.75	2.0	13.00	18.00	2.50	-	0.100	0.015	0.045		
	analysis	0.024	0.43	1.3	10.02	16.74	2.04	-	0.026	0.003	0.028	19.43	11.39
2304	min	0.000	0.0	0.0	3.5	22.0	0.10	0.10	0.05	0.000	0.000		
	max	0.030	1.0	2.0	5.5	24.0	0.60	0.60	0.20	0.015	0.035		
	analysis	0.027	0.4	1.4	4.9	23.4	0.39	0.32	0.14	0.001	0.029	24.39	6.41

Table 2. Mechanical and physical properties of base materials.

Material	Value	Rp0.2 (MPa)	Rm (MPa)	A50 (%)	λ (W/(m·K))
316L	min max	220	530 680	40	16.2
2304	min max	400	630 800	25	14.3

A high power continuous wave ytterbium fiber laser IPG Photonics YLS–6000 (IPG Photonic, Oxford, MS, USA) with a maximum power of 6 kW and a wavelength of 1070 nm was used (Figure 1). The joints were made at 6 kW laser output power. The laser beam was delivered through the optical feeding fiber (transmitting fiber) of 300 µm core diameter. A welding head having a focal length of 250 mm and a focus collimator lens 150 mm was used. Focusing position was set on the surface of the plate. Welding speed was set to 25 mm/s (1.5 m/min). Welding process was autogenous—without filler metal. Shielding gas was supplied by a gas nozzle mounted at laser head to avoid the oxidation of weld beads. The flow rate of shielding gas—argon 5.0 (I1 in accordance with ISO 14175)—was set at 16 L/min. Test pieces were welded without using a ceramic backing or forming gas.

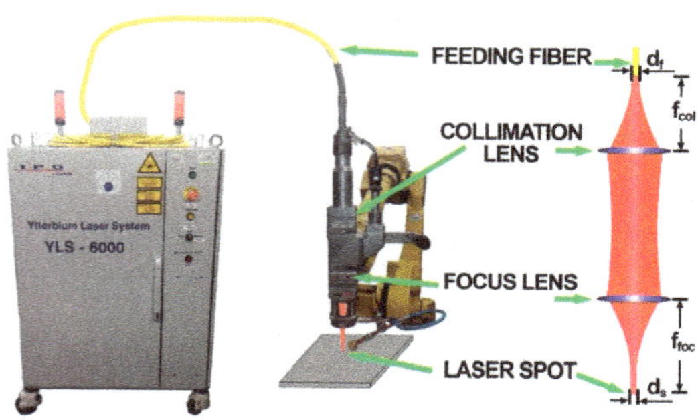

Figure 1. Experimental set-up for fiber laser welding [56].

In order to identify and then eliminate any troubles with welding process the first stage of research consisted of making preliminary test joints. Initial joints were made to determine the values of significant variables, such as focal length, laser power and welding speed. The parameters of laser welding for austenitic stainless steel and lean duplex stainless steel differ, therefore, after initial experiments with similar materials, parameters matching both grades were selected. Before welding, the elements were cleaned with abrasive paper and degreased with acetone. Preliminary tests on stainless steels showed the presence of spatter if the surface was not cleaned properly just before welding.

To assess the quality of the welded joints, non-destructive tests (NDT) were performed: visual testing (VT)—according to the EN ISO 17637 standard and penetrant testing (PT)—according to the EN ISO 571–1 standard. Tensile and bending tests were conducted at ambient temperature of 20 °C using Instron 1195 (INSTRON, Norwood, MA, USA) universal testing machine in accordance with EN ISO 6892–1 and EN ISO 4136 standards. For bending test according to EN ISO 15614–11 bend former with a diameter of $\phi = 27$ mm was used (the bend former diameter for materials with elongation above 25% is four times the specimen thickness). The acceptance criterion was defined as a bend angle $\alpha = 180°$. Specimens in the face and root areas were slightly grinded to remove notches before performing tensile and three-point bending tests.

In order to prevent changes in the material structure under the influence of temperature during cutting, the specimens for metallographic examinations (according to EN ISO 17639) were cut mechanically in a direction transverse to the welding axis at the cutting machine with intensive cooling.

Then it was grinded on abrasive papers with gradation 600–2400 and polishing on a polishing cloth using an aqueous suspension of 3 µm diamonds. This preparation of the specimen was followed by two-stage etching. The first stage was etching to reveal the austenite grain boundaries for which a mixture of 50 mL of boiling water, 3 mL of HNO_3, and 1 mL of HF was used. The etching was performed by a two-minute immersion of specimens in an 80 °C temperature solution. Then the specimen was thoroughly rinsed and cooled under water. The next stage was etching in Beraha-type reagent (85 mL of water, 15 mL of HCl, and 1 g $K_2S_2O_5$). Etching was carried out by about 1-3 min immersion in 20 °C temperature solution (until the corresponding colored structure was obtained on the specimen). At the end the specimen was thoroughly rinsed and dried with a stream of compressed air. Macroscopic metallographic observations were performed using DSLR Nikon d7000 with Tamron 90 mm f/2.8 macro lens (Nikon Corporation, Tokyo, Japan), while metallographic microscopy tests were done on a light microscope (LM) Olympus BX51 (Olympus, Tokyo, Japan). It offers imaging in a bright field, dark field and polarized light. The tests were also carried out using the scanning electron microscope JOEL JSM-7800F (SEM) with the EDAX adapter (Japan Electronics Corporation, Tokyo, Japan) enabling EDS analysis.

Microhardness measurements—HV0.2 were carried out using FM-800 tester with a load of F = 1.9614 N (Future-Tech, Tokyo, Japan).

The ferrite number was determined using a Fischer Feritscope FMP30 for both base materials and weld (Helmut Fischer GmbH Institut für Elektronik und Messtechnik 71069 Sindelfingen, Germany). The Feritscope was calibrated on calibration standards prior to measurements. The ferrite content measurements were carried out in accordance with ISO 17655, at 6 measuring points for each place of measure.

3. Results and Discussion

3.1. Non-Destructive Tests

To detect surface imperfections VT and PT were conducted. Positive results of those tests were found for all specimens. Figure 2 shows face and root side of one of the welded joints.

(b)

Figure 2. View of the welded joint: (**a**) face side and (**b**) root side.

Figure 2a shows the view of the face of welded joint. Underfilling of the weld face can be seen, which in the case of laser welding of large thicknesses without consumable is expected. The welded joints met the assumed acceptance criterion of quality level B in accordance with the EN ISO 13919–1 standard. Therefore, on a basis of NDT results it was determined that the next part of the research, consisting of destructive tests, can be carried out.

3.2. Destructive Tests

3.2.1. Static Tensile Test

Figure 3 shows the view of two specimens (signed A and B) which were subjected to transverse tensile tests. The results of the test are presented in Table 3. Minimum tensile strength required for the 316L austenitic stainless steel should be 530 MPa, and minimum tensile strength required for the 2304 duplex stainless steel should be 630 MPa (Table 2). For this test, the acceptance criterion was to exceed the minimum tensile strength of 316L austenitic stainless steel.

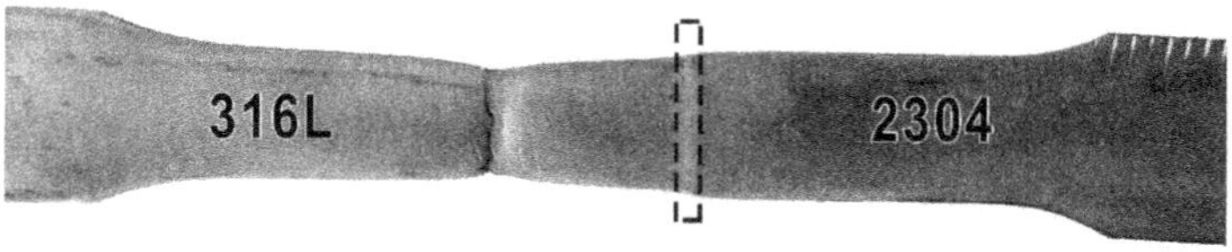

(a)

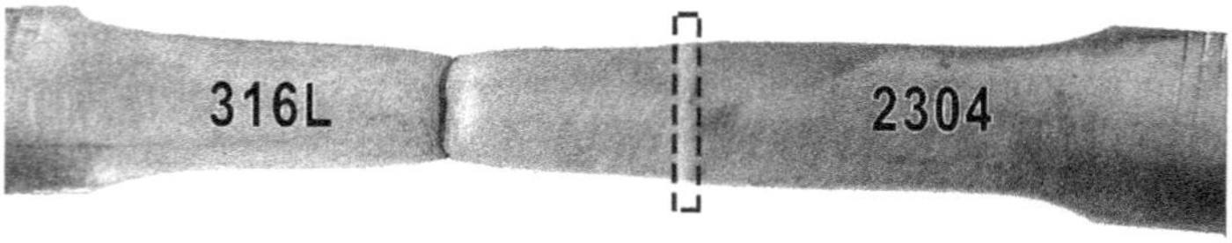

(b)

Figure 3. View of the specimens after a static tensile test: (**a**) specimen A and (**b**) specimen B. Welds marked by dashed contour lines.

Table 3. Tensile test results of welded joints.

No Specimen	Cross-Section Area (mm^2)	Maximum Loading Force (kN)	Tensile Strength, R_m (MPa)	Place of Fracture
1	170.7	104	609	Base material 316L
2	170.7	101	592	Base material 316L

The obtained results meet the acceptance criterion and the obtained values are 10–15% higher compared to the requirements. The fracture was ductile and occurred in the 316L base material. It proves the correctness of welding parameters in terms of laser beam power, focus position and welding speed. The identified welding imperfections in the VT and PT tests did not affect the tensile strength. Tests showed that no structural changes reducing mechanical properties occurred under the influence of the laser welding thermal cycle. The absence of structural changes will be verified by metallographic examinations (Section 3.2.3).

3.2.2. Bending Test

The bending tests were carried out on four specimens—in two the tensiled side was the face of the weld and in two the tensiled side was the root of the weld. The aim of this test was to investigate the plastic properties of the welded joint. The bending tests are also carried out to reveal welding imperfections, e.g., incomplete fusion, lack of penetration, pores, and others. Specimens after the bending test are presented in Figure 4.

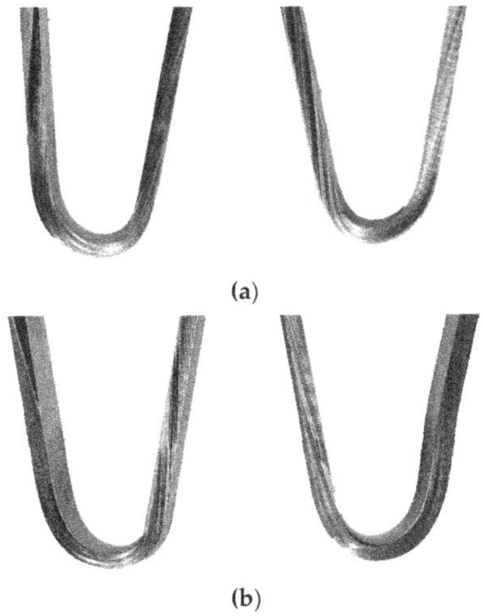

Figure 4. Specimens after bending tests: (**a**) bending from the face and (**b**) bending from the root.

The bending test was executed until an angle of 180° was reached. The welds were subjected to significant plastic deformation and no surface cracks were visible on the root or face side, which indicates that the laser welded dissimilar joints exhibited good ductility and adequate bending strength. Such a test result proves very good plastic properties, but also a lack of welding imperfections in the tested specimens. As in the tensile test, underfill did not lead to cracks on the tensiled surfaces during the bending test.

3.2.3. Macro- and Microscopic Examinations

Figure 5 presents the macrostructure of the welded joint observed on the cross-section of the weld axis. The joint has a regular symmetrical shape with visible underfilling of the weld face, without visible pores or excess penetration from the root side. Typical geometry of laser welded joint was observed. Keyhole laser welding forms a 'chalice' shaped weld bead profile. Examined welded joints were made without welding imperfections such as porosity and humping beads. Full penetration was achieved for investigated welded joints without changing beam focusing position.

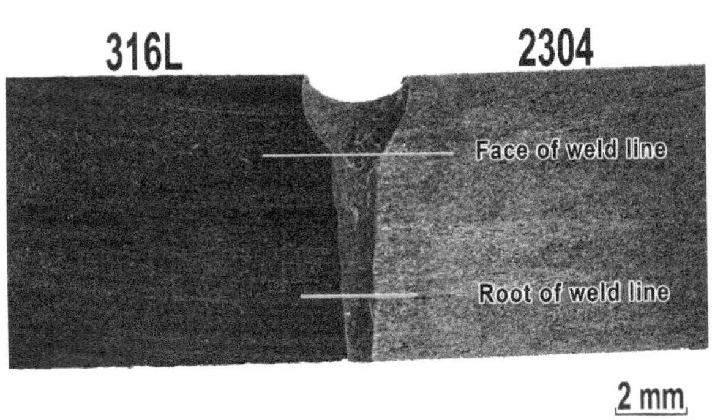

Figure 5. Cross-section of the 316L–2304 stainless steel welded joint. The arrangement of microhardness measurement points on the specimen is marked by lines.

Figure 6 shows the structure of base materials: 316L austenitic stainless steel and 2304 lean duplex stainless steel. Figure 6a presents a microstructure of 316L including twins, slip bands, and δ ferrite (darker due to Beraha reagent etching). Austenite grains are quite fine, elongated in the direction of forming and between them big amount of darker δ ferrite precipitates. As demonstrated in Figure 6b base material of 2304 lean duplex steel is characterized by a dark continuous matrix of the ferrite phase (δ) and white island of austenitic (γ) phase with characteristic twins made visible through two-stage etching. Visible directionality of the structure is related to the rolling process.

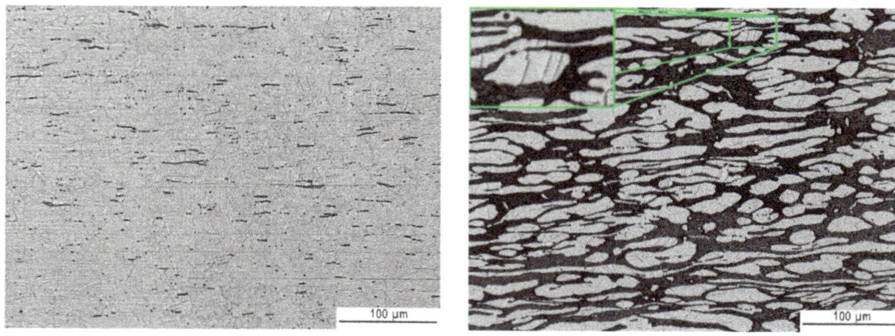

(**a**) (**b**)

Figure 6. Metallographic structure (light microscope (LM)) of: (**a**) 316L austenitic stainless steel and (**b**) 2304 duplex stainless steel.

Figures 7 and 8 show the transition areas from base material through the HAZ to the weld metal (WM). The austenitic structure of the weld consists of equiaxed coarse and fine dendrites. In addition, no eutectics or microcracks were found in the weld structure. During cooling the liquid metal in welding pool firstly solidify as ferrite. Further cooling cause partial transformation of ferrite to austenite. Austenite initially form as grain boundary allotriomorphic austenite (GBA), then as Widmanstätten side plates of austenite (WA) and finally as intragranular precipitates of austenite (IGA). In general, the first two types of austenite (GBA and WA) require less driving force than intragranular needle austenite, which means that the grain boundary and Widmanstätten austenite formed earlier at higher temperatures, whereas intragranular austenite particles precipitated on further cooling at a lower temperature [55,56]. GBA morphology resembles a coherent island arranged along the border of ferrite. WA morphology can be described as small needles in a ferrite matrix. While the IGA morphology is like square islands in a ferrite matrix.

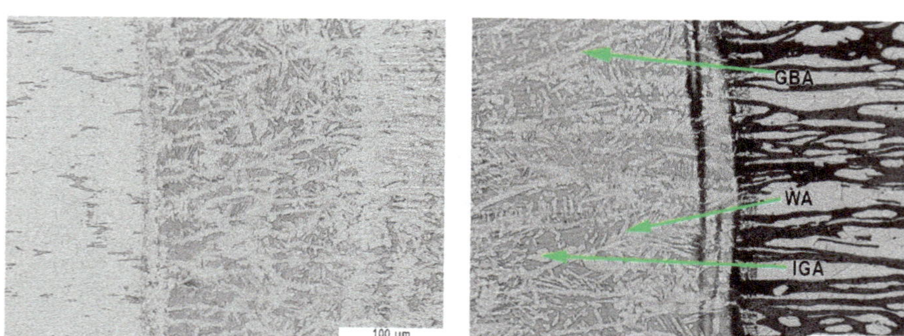

(**a**) (**b**)

Figure 7. Metallographic structure (LM) of: (**a**) 316L austenitic stainless steel→HAZ→WM and (**b**) WM→HAZ→2304 duplex stainless steel.

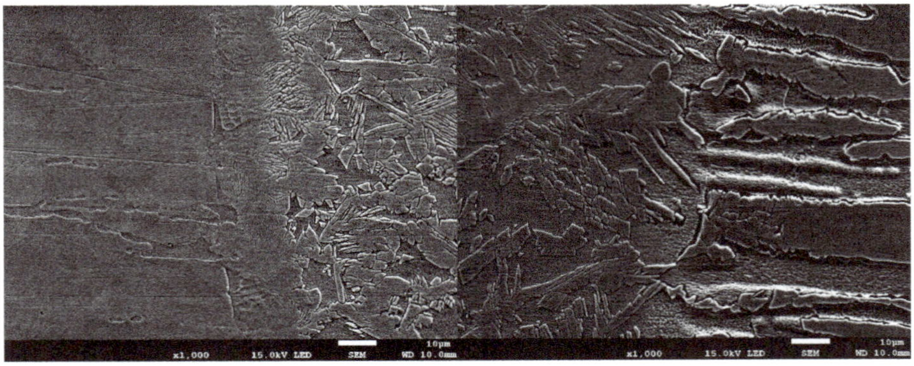

(**a**) (**b**)

Figure 8. Metallographic structure (SEM) of: (**a**) 316L austenitic stainless steel→HAZ→WM and (**b**) WM→HAZ→2304 duplex stainless steel.

As presented in Figure 7 heat affected zones were barely visible. HAZ of laser welded stainless steel joints is very narrow, in this case it was about 20 μm for 316L steel and 50 μm for 2304 steel as can be seen on Figure 8a and b, respectively. Both for lower (LM—Figure 7) and higher magnifications (SEM—Figure 8), a difference in the austenite-ferrite ratio in the weld close to HAZ is noticeable. As expected, from the 316L steel side there is a larger amount of austenite in the weld metal than from the 2304 steel side.

Linear EDS analysis was carried out through all welded joint: from 316L steel, first HAZ, weld metal, second HAZ ending at 2304 steel (Figure 9). The distribution of the analyzed elements along the measuring line is visible on individual curves. Level of the line does not show the content of a given element in the analyzed alloy, but only show the variability of its content along the measuring line. As could be predicted, the weld metal has a composition resulting from mixing of both materials—due to autogenous laser welding. The content of nickel which is stabilizing austenite element increases towards 316L steel. The heterogeneous composition in the weld metal can be the reason for the different austenite-ferrite ratio on both sides of the weld axis. Due to diffusion, differences in elements content occur in HAZ. The nickel content in HAZ next to 2304 steel decreases compared to the weld. The chromium content increases in the HAZ on the 2304 steel side. This increase may be due to a visible larger ferrite grains in this area and chromium is ferrite stabilizer. This can be seen in the photo below the graph (Figure 9), and is the effect of heat affecting the structure.

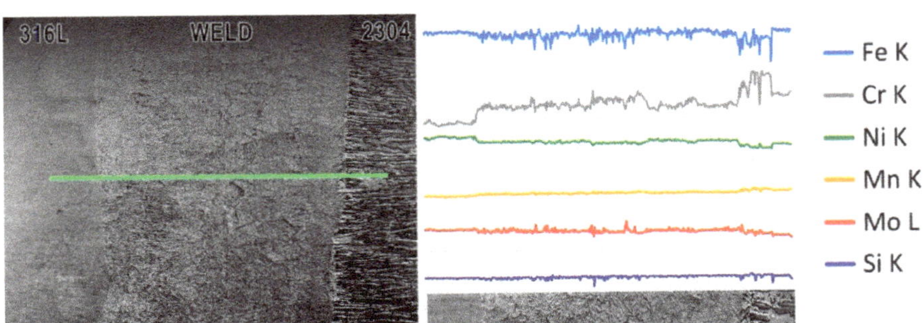

Figure 9. Results of EDS linear analysis through the welded joint from 316L austenitic stainless steel (beginning of the line) to 2304 duplex stainless steel (end of the line). The analysis was carried out along the line indicated in the SEM image on the left.

The low chromium content in the welding pool caused the formation of a structure with a higher content of austenite than is observed in duplex laser-welded joints with the same parameters and geometry [56]. As a result, the austenite to ferrite ratio is closer to 50:50. The obtained volume fraction of austenite may also be induced by the difference in the thermal conductivity coefficient of the austenitic steel in relation to the duplex steel, which resulted in an increase in the cooling time and allowed the transformation of ferrite into austenite [31].

Metallographic microscopic examination—SEM and EDS analysis—did not show the segregation of alloying elements between the ferritic and austenitic phases. This indicates that the areas depleted in Cr and Ni, which can cause degradation phenomena, e.g., corrosion, were not formed.

3.2.4. Microhardness Measurements

Microhardness measurements were made in base materials, weld metal and HAZs in two lines as can be seen in Figure 5. Figure 10 shows the results of the measurements.

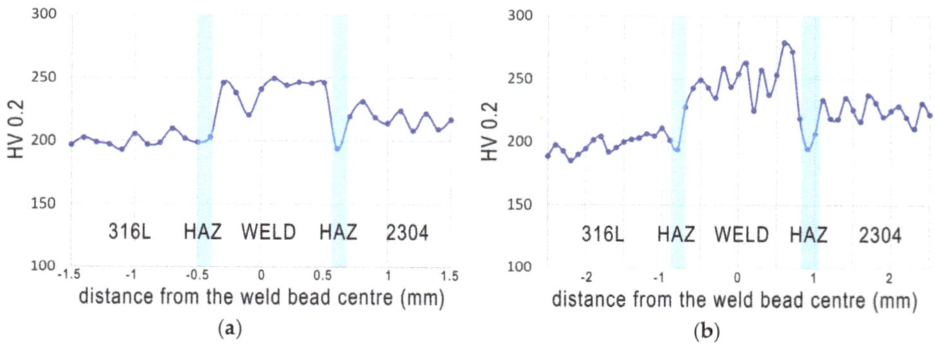

Figure 10. Microhardness (HV0.2) distribution across the welded joint: (**a**) 2 mm below the weld face and (**b**) 2 mm above the weld root.

For 316L austenitic stainless steel microhardness value was in a range of 186–209 HV0.2, which corresponds well to the literature [38,50]. The microhardness fluctuations in the 316L steel are due to the presence of a large number of delta ferrite precipitates in the austenitic structure. In the HAZ (on 316L steel side) microhardness values are lower (average 196 HV0.2), which confirms that both heat input and cooling rate were appropriate for this process. The average HV0.2 value for weld metal is 254 HV0.2, with higher values on weld face than in the weld root. Differences in hardness values between the line passing through the face of the weld in relation to the line close to the root can be explained by differences in the amount of heat accumulated in each area. The average value of hardness for 2304 steel was 233 HV0.2—typical for this grade. However, its decrease in the HAZ on 2304 steel side (average value 197 HV0.2) was observed, which is a result of the structural changes shown in the microscopic studies. Due to the different morphology and austenite arrangement between weld metal and lean duplex, the microhardness of the autogenous weld metal was higher (about 20 HV0.2) than microhardness of the lean duplex steel [33]. The results are distinctive for dissimilar austenitic—duplex stainless steel welded joints—which was also demonstrated by other authors [33,52,54].

3.2.5. Ferrite Content Measurements and Calculations

Points for ferrite measurement were chosen within the areas of 316L austenitic stainless steel and 2304 lean duplex stainless steel on the welded joint surface. Similarly, points were selected along the weld from the face and root of the weld. Table 4 shows the results of ferrite content measurements.

Table 4. Experimental results of ferrite content.

Place of Measure	Ferrite Content (%)						Average Value	Standard Deviation
316L	0.23	0.24	0.23	0.26	0.25	0.24	0.24	0.01
2304	43.4	45.6	45.4	44.3	45.3	49.7	45.62	2.17
Weld face	13.9	12.6	13.0	14.9	11.3	12.0	12.95	1.30
Weld root	19.1	18.3	23.5	19.4	23.2	22.0	20.92	2.26

The results of the delta ferrite measurements showed its correct, expected content in base materials. However, differences in the ferrite content between the face and the root of the weld were observed. Higher (by about 37%) ferrite content in the weld root is a consequence of different heat distribution between the border surfaces during and after the welding process. This distribution is also associated with various values of the thermal conductivity coefficient of base materials and the intensity of the laser heat source. Laser welding is characterized by high power density, which means that the energy distribution is constant over the entire depth (that's why a cylindrical model of a heat source is usually used for numerical simulations).

Geometry of the weld is not constant throughout the thickness of the specimen, but it changes significantly in the area of the weld face (Figure 5). Forming a 'chalice' shaped weld bead is characteristic for the keyhole laser welding of thick materials. The change in the fusion line shape also indicates different heat distribution conditions in the welding area. This causes a change in the cooling rate of the joint on the upper and lower boundary surface (face/root). This phenomenon and different thermal properties can be explained by the increase in ferrite content in the weld root, as this leads to an increase in the cooling rate of the joint.

The results of ferrite content measurements were compared with the results obtained from the Schaeffler diagram—a graphic method for assessing the weldability of high alloy steels and dissimilar joints. Schaeffler diagram is a good preliminary prediction method for weldability of steels subjected to fusion welding [54]. It was assumed, according to the welding conditions presented in Chapter 2, that a dilution rate was 50%, so the values of Creq and Nieq for the weld metal are 21.9% and 8.9% respectively, and this corresponds to about 30% of the amount of ferrite (Figure 11). Comparative analysis of the results of measurements by the magnetic method and the Schaeffler diagram confirm that Creq and Nieq in this form does not fully describe the ferrite forming tendency during laser welding. Obtained results of prediction of ferrite content were also confirmed using WRC-92 diagram.

Solidification mode of stainless steels can be predicted by the Fe–Ni–Cr pseudo-binary phase diagram shown in Figure 12 based on Creq/Nieq ratio [50]. As a result of welding of 316L steel without filler metal (Creq/Nieq = 1.71) the weld solidification mode is ferrite–austenite (FA). According to the Figure 12, Creq/Nieq ratio of 2304 steel equal to 3.80 during rapid cooling is enough to cause δ-ferrite solidification and then ferrite and austenite transformation (ferrite mode—F). Chemical composition of the weld metal, which is an alloy of 50% dilution rate, described by the value of Creq/Nieq = 2.46 also causes solidification in the F mode. The results of these analyzes are generally consistent with the results of metallographic examinations and ferrite content measurements.

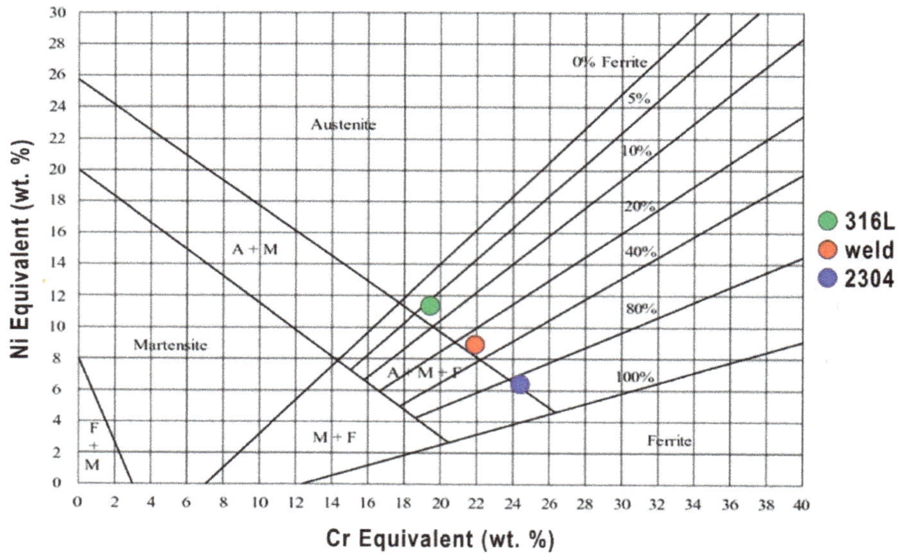

Figure 11. Schaeffler constitutional diagram.

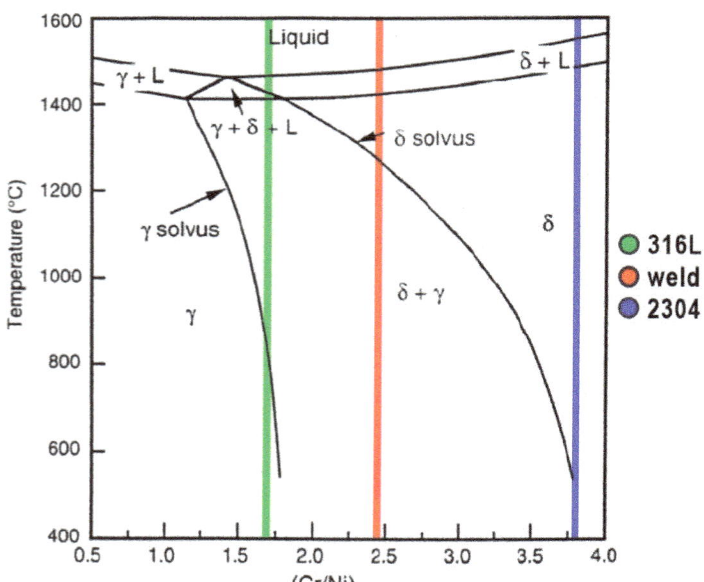

Figure 12. Fe-Ni-Cr pseudo-binary phase diagram.

4. Conclusions

Tests regarding dissimilar laser welding of butt joints made of 316L austenitic stainless steel and 2304 lean duplex stainless steel showed that the use of IPG YLS–6000 fiber laser with a maximum power of 6 kW allowed to obtain sound butt joints meeting the requirements of quality level B in accordance with EN ISO 13919–1 standard. During welding, the only imperfection detected by VT and

PT was the underfilling of the face, while microscopic examinations did not show the presence of any other welding imperfections. Due to the very low (as for laser welding) welding speed, it was possible to achieve full penetration of 8 mm thick plates with one-sided course of the laser beam.

Obtained joints have good strength properties (higher than the minimum values required by the standard), which were confirmed in the static tensile tests (average R_m = 600 MPa). All specimens fractured in 316L austenitic stainless steel BM. The bending tests showed that plastic properties of base materials were not deteriorated and confirmed the absence of welding imperfections.

The microstructure of the dissimilar welded joint has a better austenite to ferrite ratio (closer to 50:50), compared to the microstructure observed on similar duplex steel welded joints [56]. This is caused by mixing of 316L steel and 2304 steel as well as by lower thermal conductivity of austenitic steel, which extends the cooling process and the time for austenite formation. Within the weld face, the weld width is greater, which affects a different amount of heat to be dissipated by the base material than in the case of weld root. The differences in thermal cycles, in these areas, cause changes in weld microhardness in the face and in the root of the weld. The microhardness of the autogenous weld metal (average 254 HV0.2) is higher than microhardness of the 2304 lean duplex stainless steel (average 233 HV0.2). The solidification microstructures of laser welds are generally consistent with the prediction from the Schaeffler diagram.

Based on the results of present study, autogenous fiber laser welding of 316L austenitic and 2304 lean duplex stainless steels without using a ceramic backing can be recommended as a suitable procedure for sound welded joints.

Author Contributions: Conceptualization, M.L. and D.F.; formal analysis, G.R., A.Ś., M.L., and D.F.; investigation, M.L.; methodology, G.R., D.F., and M.L.; writing–original draft, M.L., G.R., A.Ś., and D.F.; writing–review & editing, D.F., A.Ś., M.L., and G.R. All authors have read and agreed to the published version of the manuscript.

Funding: This research received no external funding.

Acknowledgments: Authors want to thank CRIST S.A. from Gdynia for providing materials for research.

Conflicts of Interest: The authors declare no conflict of interest.

References

1. Quazi, M.M.; Ishak, M.; Fazal, M.A.; Arslan, A.; Rubaiee, S.; Aiman, M.H.; Qaban, A.; Yusof, F.; Sultan, T.; Ali, M.M.; et al. A comprehensive assessment of laser welding of biomedical devices and implant materials: Recent research, development and applications. *Crit. Rev. Solid. State* **2020**, *4*, 1–43. [CrossRef]
2. Kik, T. Computational techniques in numerical simulations of arc and laser welding processes. *Materials* **2020**, *13*, 608. [CrossRef]
3. Górka, J. Assessment of the effect of laser welding on the properties and structure of TMCP steel butt joints. *Materials* **2020**, *13*, 1312. [CrossRef]
4. Lisiecki, A. Study of optical properties of surface layers produced by laser surface melting and laser surface nitriding of titanium alloy. *Materials* **2019**, *12*, 3112. [CrossRef]
5. Janicki, D.; Górka, J.; Kwaśny, W.; Pakieła, W.; Matus, K. Influence of solidification conditions on the microstructure of laser-surface-melted ductile cast iron. *Materials* **2020**, *13*, 1174. [CrossRef] [PubMed]
6. Sroka, M.; Jonda, E.; Pakieła, W. Laser surface modification of aluminium alloy AlMg9 with B4C powder. *Materials* **2020**, *13*, 402. [CrossRef]
7. Wang, G.; Wang, J.; Yin, L.; Hu, H.; Yao, Z. Quantitative correlation between thermal cycling and the microstructures of X100 pipeline steel laser-welded joints. *Materials* **2020**, *13*, 121. [CrossRef] [PubMed]
8. Lisiecki, A.; Kurc-Lisiecka, A. Automated laser welding of AISI 304 stainless steel by disk laser. *Arch. Metall. Mater.* **2018**, *63*, 1663–1672.
9. Tęczar, P.; Majkowska-Marzec, B.; Bartmański, M. The influence of laser alloying of Ti13Nb13Zr on surface topography and properties. *Adv. Mater. Sci.* **2019**, *19*, 44–56. [CrossRef]
10. Herthoge, M.; De Pelsmaeker, J.; Boone, M.; De Baere, I.; Van Paepegem, W.; Van Vlierberghe, S. Laser welding of carbon fiber filled polytetrafluoroethylene. *J. Mater. Proc. Technol.* **2020**, *282*, 116681. [CrossRef]

11. Liao, H.; Zhu, J.; Chang, S.; Xue, G.; Zhu, H.; Chen, B. Al_2O_3 loss prediction model of selective laser melting Al_2O_3–Al composite. *Ceram. Int.* **2020**, *46*, 13414–13423. [CrossRef]
12. Wen, X.; Jin, G.; Cui, X.; Feng, X.; Lu, B.; Cai, Z.; Zhao, Y.; Fang, Y. Underwater wet laser cladding on 316L stainless steel: A protective material assisted method. *Opt. Laser Technol.* **2019**, *111*, 814–824. [CrossRef]
13. Fu, Y.; Guo, N.; Cheng, Q.; Zhang, D.; Feng, J. In-situ formation of laser-cladded layer on Ti-6Al-4 V titanium alloy in underwater environment. *Opt. Lasers Eng.* **2020**, *131*, 106104. [CrossRef]
14. Kik, T.; Górka, J. Numerical simulations of laser and hybrid S700MC T-joint welding. *Materials* **2019**, *12*, 516. [CrossRef] [PubMed]
15. Wang, H.; Liu, X.; Liu, L. Research on laser-TIG hybrid welding of 6061-T6 aluminum alloys joint and post heat treatment. *Metals* **2020**, *10*, 130. [CrossRef]
16. Yazdian, N.; Mohammadpour, M.; Razavi, R.; Kovacevic, R. Hybrid laser/arc welding of 304L stainless steel tubes, part 2–Effect of filler wires on microstructure and corrosion behavior. *Int. J. Press. Vessel. Pip.* **2018**, *163*, 45–54. [CrossRef]
17. Mohammed, G.R.; Ishak, M.; Aqida, S.N.; Abdulhadi, H.A. Effects of heat input on microstructure, corrosion and mechanical characteristics of welded austenitic and duplex stainless steels: A review. *Metals* **2017**, *7*, 39. [CrossRef]
18. Kuryntsev, S.V.; Gilmutdinov, A.K. Welding of stainless steel using defocused laser beam. *J. Constr. Steel Res.* **2015**, *114*, 305–313. [CrossRef]
19. Verma, J.; Taiwade, R.V. Effect of welding processes and conditions on the microstructure, mechanical properties and corrosion resistance of duplex stainless steel weldments—A review. *J. Manuf. Process.* **2017**, *25*, 134–152. [CrossRef]
20. Prijanovič, U.; Prijanovič Tonkovič, M.; Trdan, U.; Pleterski, M.; Jezeršek, M.; Klobčar, D. Remote fibre laser welding of advanced high strength martensitic steel. *Metals* **2020**, *10*, 533. [CrossRef]
21. Pańcikiewicz, K.; Radomski, W. Lack of tightness analysis of concealed welded radiators. *Eng. Fail. Anal.* **2020**, *114*, 104579. [CrossRef]
22. Skowrońska, B.; Chmielewski, T.; Pachla, W.; Kulczyk, M.; Skiba, J.; Presz, W. Friction weldability of UFG 316L stainless steel. *Arch. Metall. Mater.* **2019**, *64*, 1051–1058.
23. Krella, A.K.; Krupa, A. Effect of cavitation intensity on degradation of X6CrNiTi18-10 stainless steel. *Wear* **2018**, *408*, 180–189. [CrossRef]
24. Kuryntsev, S.V. Effect of heat treatment on the phase composition and corrosion resistance of 321 SS welded joints produced by a defocused laser beam. *Materials* **2019**, *12*, 3720. [CrossRef]
25. Świerczyńska, A.; Fydrych, D.; Landowski, M.; Rogalski, G.; Łabanowski, J. Hydrogen embrittlement of X2CRNiMoCuN25-6-2 super duplex stainless steel welded joints under cathodic protection. *Constr. Build. Mater.* **2020**, *238*, 117697. [CrossRef]
26. Varbai, B.; Májlinger, K. Optimal etching sequence for austenite to ferrite ratio evaluation of two lean duplex stainless steel weldments. *Measurement* **2019**, *147*, 106832. [CrossRef]
27. Ouali, N.; Khenfer, K.; Belkessa, B.; Fajoui, J.; Cheniti, B.; Idir, B.; Branchu, S. Effect of heat input on microstructure, residual stress, and corrosion resistance of UNS 32101 lean duplex stainless steel weld joints. *J. Mater. Eng. Perform.* **2019**, *28*, 4252–4264. [CrossRef]
28. Verma, J.; Taiwade, R.V. Dissimilar welding behavior of 22% Cr series stainless steel with 316L and its corrosion resistance in modified aggressive environment. *J. Manuf. Proc.* **2016**, *24*, 1–10. [CrossRef]
29. Świerczyńska, A.; Łabanowski, J.; Michalska, J.; Fydrych, D. Corrosion behavior of hydrogen charged super duplex stainless steel welded joints. *Mater. Corros.* **2017**, *68*, 1037–1045. [CrossRef]
30. Varbai, B.; Májlinger, K. Physical and theoretical modeling of the nitrogen content of duplex stainless steel weld metal: Shielding gas composition and heat input effects. *Metals* **2019**, *9*, 762. [CrossRef]
31. Yang, Y.; Guo, Y.; Liu, Y.; Li, J.; Jiang, Y. The microstructure and pitting resistance of 2002 lean duplex stainless steel after the simulated welding thermal cycle process. *Materials* **2019**, *12*, 70. [CrossRef]
32. Łabanowski, J.; Fydrych, D.; Rogalski, G.; Samson, K. Underwater welding of duplex stainless steel. *Solid State Phenom.* **2012**, *183*, 101–106. [CrossRef]
33. Silva Leite, C.G.; da Cruz Junior, E.J.; Lago, M.; Zambon, A.; Calliari, I.; Ventrella, V.A. Nd: YAG pulsed laser dissimilar welding of UNS S32750 duplex with 316L austenitic stainless steel. *Materials* **2019**, *12*, 2906. [CrossRef] [PubMed]

34. Fydrych, D.; Łabanowski, J.; Tomków, J.; Rogalski, G. Cold cracking of underwater wet welded S355G10+N high strength steel. *Adv. Mater. Sci.* **2015**, *15*, 48–56. [CrossRef]
35. Rogalski, G.; Świerczyńska, A.; Landowski, M.; Fydrych, D. Mechanical and microstructural characterization of TIG welded dissimilar joints between 304L austenitic stainless steel and Incoloy 800HT nickel alloy. *Metals* **2020**, *10*, 559. [CrossRef]
36. Tomków, J.; Fydrych, D.; Rogalski, G. Dissimilar underwater wet welding of HSLA steels. *Int. J. Adv. Manuf. Tech.* **2020**, in press.
37. Wang, W.; Hu, Y.; Zhang, M.; Zhao, H. Microstructure and mechanical properties of dissimilar friction stir welds in austenitic-duplex stainless steels. *Mater. Sci. Eng. A* **2020**, *787*, 139499. [CrossRef]
38. Ramkumar, K.D.; Singh, A.; Raghuvanshi, S.; Bajpai, A.; Solanki, T.; Arivarasu, M.; Arivazhagan, N.; Narayanan, S. Metallurgical and mechanical characterization of dissimilar welds of austenitic stainless steel and super-duplex stainless steel–a comparative study. *J. Manuf. Process.* **2015**, *19*, 212–232. [CrossRef]
39. Yıldızlı, K. Investigation on the microstructure and toughness properties of austenitic and duplex stainless steels weldments under cryogenic conditions. *Mater. Des.* **2015**, *201577*, 83–94. [CrossRef]
40. Başyiğit, A.B.; Kurt, A. Investigation of the weld properties of dissimilar S32205 duplex stainless steel with AISI 304 steel joints produced by arc stud welding. *Metals* **2017**, *7*, 77. [CrossRef]
41. Abdollahi, A.; Shamanian, M.; Golozar, M.A. Comparison of pulsed and continuous current gas tungsten arc welding in dissimilar welding between UNS S32750 and AISI 321 in optimized condition. *Int. J. Adv. Manuf. Tech.* **2018**, *97*, 687–696. [CrossRef]
42. Theodoro, M.C.; Pereira, V.F.; Mei, P.R.; Ramirez, A.J. Dissimilar friction stir welding between UNS S31603 austenitic stainless steel and UNS S32750 superduplex stainless steel. *Metall. Mater. Trans. B* **2015**, *46*, 1440–1447. [CrossRef]
43. Neissi, R.; Shamanian, M.; Hajihashemi, M. The effect of constant and pulsed current gas tungsten arc welding on joint properties of 2205 duplex stainless steel to 316L austenitic stainless steel. *J. Mater. Eng. Perform.* **2016**, *25*, 2017–2028. [CrossRef]
44. Szala, M.; Łukasik, D. Pitting corrosion of the resistance welding joints of stainless steel ventilation grille operated in swimming pool environment. *Int. J. Corros.* **2018**, *2018*, 1–7. [CrossRef]
45. Vicente, T.A.; Oliveira, L.A.; Correa, E.O.; Barbosa, R.P.; Macanhan, V.B.P.; Alcântara, N.G. Stress corrosion cracking behaviour of dissimilar welding of AISI 310S austenitic stainless steel to 2304 duplex stainless steel. *Metals* **2018**, *8*, 195.
46. Rahmani, M.; Eghlimi, A.; Shamanian, M. Evaluation of microstructure and mechanical properties in dissimilar austenitic/super duplex stainless steel joint. *J. Mater. Eng. Perform.* **2014**, *23*, 3745–3753. [CrossRef]
47. Moteshakker, A.; Danaee, I. Microstructure and corrosion resistance of dissimilar weld-joints between duplex stainless steel 2205 and austenitic stainless steel 316L. *J. Mater. Sci. Technol.* **2016**, *32*, 282–290. [CrossRef]
48. Moteshakker, A.; Danaee, I.; Moeinifar, S.; Ashrafi, A. Hardness and tensile properties of dissimilar welds joints between SAF 2205 and AISI 316L. *Sci. Technol. Weld. Join.* **2016**, *21*, 1–10. [CrossRef]
49. Taheri, A.; Beidokhti, B.; Boroujeny, B.S.; Valizadeh, A. Characterizations of dissimilar S32205/316L welds using austenitic, super-austenitic and super-duplex filler metals. *Int. J. Min. Metall. Mater.* **2020**, *27*, 119–127. [CrossRef]
50. Fei, Z.; Pan, Z.; Cuiuri, D.; Li, H.; Van Duin, S.; Yu, Z. Microstructural characterization and mechanical properties of K-TIG welded SAF2205/AISI316L dissimilar joint. *J. Manuf. Process.* **2019**, *45*, 340–355. [CrossRef]
51. Ramkumar, K.D.; Bajpai, A.; Raghuvanshi, S.; Singh, A.; Chandrasekhar, A.; Arivarasu, M.; Arivazhagan, N. Investigations on structure–property relationships of activated flux TIG weldments of super-duplex/austenitic stainless steels. *Mater. Sci. Eng. A* **2015**, *638*, 60–68. [CrossRef]
52. Ridha Mohammed, G.; Ishak, M.; Ahmad, S.N.A.S.; Abdulhadi, H.A. Fiber laser welding of dissimilar 2205/304 stainless steel plates. *Metals* **2017**, *7*, 546. [CrossRef]
53. Chun, E.J.; Lee, J.H.; Kang, N. Unmixing behaviour in dissimilar laser welds for duplex and austenitic stainless steels. *Sci. Technol. Weld. Join.* **2019**, *24*, 263–275. [CrossRef]
54. Saravanan, S.; Raghukandan, K.; Sivagurumanikandan, N. Studies on metallurgical and mechanical properties of laser welded dissimilar grade steels. *J. Braz. Soc. Mech. Sci. Eng.* **2017**, *39*, 3491–3498. [CrossRef]

55. Cao, F.; Zhang, Y.; Shen, Y.; Jin, Y.; Li, J.; Hou, W. Effects of beam offset on the macro defects, microstructure and mechanical behaviors in dissimilar laser beam welds of SDSS2507 and Q235. *J. Manuf. Process.* **2020**, *55*, 335–347. [CrossRef]
56. Landowski, M. Influence of parameters of laser beam welding on structure of 2205 duplex stainless steel. *Adv. Mater. Sci.* **2019**, *19*, 21–31. [CrossRef]

materials

MDPI

Article

Assessment of the Effect of Laser Welding on the Properties and Structure of TMCP Steel Butt Joints

Jacek Górka

Department of Welding Engineering, Silesian University of Technology, Konarskiego 18A, 44-100 Gliwice, Poland; jacek.gorka@polsl.pl; Tel.: +48-32-237-14-45; Fax: +48-32-237-1232

Received: 24 February 2020; Accepted: 12 March 2020; Published: 13 March 2020

Abstract: The research work and related tests aimed to identify the effect of filler metal-free laser beam welding on the structure and properties of butt joints made of steel 700MC subjected to the TMCP (thermo-mechanically controlled processed) process. The tests involved 10-mm thick welded joints and a welding linear energy of 4 kJ/mm and 5 kJ/mm. The inert gas shielded welding process was performed in the flat position (PA) and horizontal position (PC). Non-destructive testing enabled classification of the tested welded joints as representing the quality level B in accordance with the requirements set out in standard 13919-1. Destructive tests revealed that the tensile strength of the joints was 5% lower than S700MC steel. The results of tensile tests and changes in structure were referred to joints made using the MAG (Metal Active Gas) method. The tests of thin films performed using a high-resolution scanning transmission electron microscope revealed that, during laser beam welding, an increase in dilution was accompanied by an increase in the content of alloying microadditions titanium and niobium, particularly in the fusion area. A significant content of hardening phases in the welded joint during cooling led to significant precipitation hardening by fine-dispersive (Ti,Nb)(C,N) type precipitates being of several nanometres in size, which, in turn, resulted in the reduction of plastic properties. An increase in the concentration of elements responsible for steel hardening, i.e., Ti and Nb, also contributed to reducing the weld toughness below the acceptable value, which amounts to 25 J/cm^2. In cases of S700MC, the analysis of the phase transformation of austenite exposed to welding thermal cycles and the value of carbon equivalent cannot be the only factors taken into consideration when assessing weldability.

Keywords: laser beam; weldability; thermo-mechanically controlled processed; S700MC TMCP steel; MAG

---

1. Introduction

Laser welding involves the melting of the interface of elements being welded by means of heat supplied to the previously mentioned area and result from the use of a highly concentrated beam of coherent light characterised by very high power density (i.e., restricted within the range of approximately 10^2 to 10^{11} W/mm^2) [1–3]. Processes can be performed using the melt-in welding technique (as in classical arc welding) or with full joint penetration in one run or with many layers, with or without the filler metal, i.e., using the key-hole welding technique. Very high laser beam power density is responsible for the fact that welding linear energy is at the level of the minimum energy required to melt the joint. Therefore, the heat affected zone (HAZ) and the fusion zone are very narrow. At the same time, deformations of joints are so insignificant that welded elements, once finished, do not require additional mechanical treatment [4–9]. Until recently, laser beam welding has seldom been implemented in the industry, mainly due to high investment costs and some technological problems. Presently, investing in laser beam welding no longer poses such a risk since, only a few years ago, everything indicates that the future market will be dominated by companies using laser technologies within a wide range. The foregoing results from the fact that many welding companies are increasingly interested in the application of technological lasers in production processes because it enables the introduction of such lasers for these companies to modernise during their production processes, start the production of technologically-advanced products, or manufacture new product generations. Very often, this process is used to join plates (of thicknesses exceeding 4 mm) since the application of the laser beam makes it possible to obtain high-quality joints without using the filler metal, which reduces costs and restricts the formation of welding strains [10–16]. Now, laser welding is used in the ship-building or power generation industries, where it is necessary to make thick joints in high-strength steels, e.g., TMCP steels. The implementation of TMCP steels significantly decreases the time of welding works by reducing the preheating temperature or even eliminating the preheating process as such. In addition, the use of TMCP steel makes it possible to reduce the cross-sections of structural elements. As a result, welded structures with the same load capacity are now thinner and lighter. Furthermore, the use of TMCP steels reduces welding costs by decreasing the consumption of filler metals, shortening the time of welding processes, and reducing costs related to the straightening of structures and the testing of welds [17–22]. It should also be noted that the welding thermal cycle itself significantly differs from the classical thermomechanical treatment cycle (by being considerably more intense). As a result of the thermal welding cycle, carbide and nitride deposits are partially dissolved in austenite. In addition, rapid cooling leads to supersaturation of the α solution with micro-additives, i.e., C and N and/or their uncontrolled precipitation. During the welding of TMCP steels, the weld is provided with micro-additions such as niobium, vanadium, and titanium. During cooling, the previously mentioned micro-additions precipitate in the form of carbides and carbonitrides. The amount of precipitates depends on the rate of cooling. An increase in the rate of cooling is accompanied by an increase in the concentration of micro-additions in the solution. A similar situation can be noticed in the heat affected zone. The amount of micro-additions left in the solution significantly affects phase transformations during cooling as well as changes of properties following the heat treatment [23–31]. This, in turn, results in an increase in the content of the products of the diffusionless and intermediate (bainitic) transformation. To a significant extent, the previously mentioned structures are responsible for the reduction of toughness. In cases of pure chemical elements, the stability of phases responsible for hardening can be presented on the basis of Gibbs-free enthalpy [32] (Figure 1). The analysis of the stability of hardening phases during the production of steel is impeded since the above-named phases are composed of constituents dissolved in the solution. During welding, the short time of heating, the size of the liquid metal pool, and high cooling rates impede the analysis of decomposition processes and the re-precipitation of hardening phases in the weld and in the HAZ (Heat Affectet Zone) area.

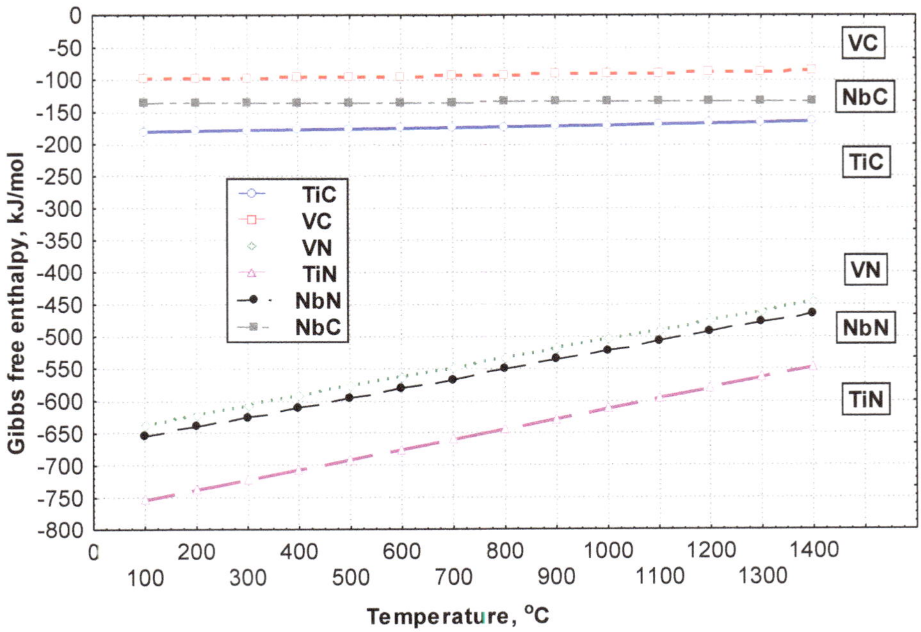

Figure 1. Free enthalpy of the formation of carbides and nitrides, calculated per one atom of carbon or nitride [32].

In the process of hardening, a significant role is played by the diffusion coefficient [33,34]. High diffusion coefficient values favour the fast formation of carbides and nitrides at high temperatures. The highest values of the diffusion coefficient in solution α are characteristic of interstitial elements such as carbon, nitrogen, and boron (D = 10^{-7} cm^2/s) and, among substitutional elements, titanium (D = 10^{-8} cm^2/s). Vanadium is characterised by a lower diffusion coefficient value than that of interstitial elements (D = 10^{-12} cm^2/s) [35–37]. The microstructure of the tested S700MC steel is obtained by rolling with a controlled degree of deformation, deformation speed, and the appropriate sequence of precipitation processes. The welding process (method, linear energy) can naturally disturb this balance, which results in a significant deterioration of the plastic properties in both the HAZ and weld area. Different durability of Nb, V, and Ti nitrides and carbides is synonymous with their different ability to limit the growth of austenite grain in HAZ and a different impact on weld properties. Lack of control over the decay and re-separation of strengthening phases can affect the deterioration of the mechanical and plastic properties of the joints. A welding thermal cycle will have a significant impact on structural and phase changes as well as the durability of strengthening phases in the area of HAZ and weld. Therefore, the aim of the work was to explain the phenomenas occurring in the HAZ and welds, caused by the thermal cycle of the laser beam welding.

2. Own Research

The aim of the study was to assess the impact of laser beam welding without additional material on the structural and strength properties of S700MC thermo-heat rolled steel with a thickness of 10 mm. The actual chemical composition and properties of steel are shown in Table 1.

Table 1. Actual chemical composition of S700MC steel.

<table>
<tr><th colspan="11">Chemical Composition, % by Weight</th></tr>
<tr><th>C</th><th>Mn</th><th>Si</th><th>S</th><th>P</th><th>Al</th><th>Nb</th><th>Ti</th><th>V</th><th>N</th><th>C_e</th></tr>
<tr><td>0.056</td><td>1.68</td><td>0.16</td><td>0.005</td><td>0.01</td><td>0.027</td><td>0.044</td><td>0.12</td><td>0.006</td><td>72</td><td>0.33</td></tr>
<tr><td colspan="11">Mechanical properties</td></tr>
<tr><td colspan="3">Tensile strength R_m, MPa</td><td colspan="3">Yield point R_e, MPa</td><td colspan="2">Elongation A_5, %</td><td colspan="3">Toughness, J/cm^2 (−30 °C)</td></tr>
<tr><td colspan="3">820</td><td colspan="3">700</td><td colspan="2">17</td><td colspan="3">50</td></tr>
<tr><td colspan="11">$C_e = C + \frac{Mn}{6} + \frac{Ni + Cu}{15} + \frac{Cr + Mo + V}{5}$, (%)</td></tr>
</table>

The structure of the test steel was bainitic-ferritic. The thermomechanical treatment of steel S700MC leads to the refinement and defecting of its structure (Figure 2) and supersaturation of the structure with hardening components. The thermomechanical rolling process leads to selective plastic deformation of the grain. The steel was subjected to precipitation hardening, solid solution hardening, and strain hardening. Structural changes occurring in the welded joint during cooling are described in the CCT (Continuous Cooling Transformation) diagram (Figure 3).

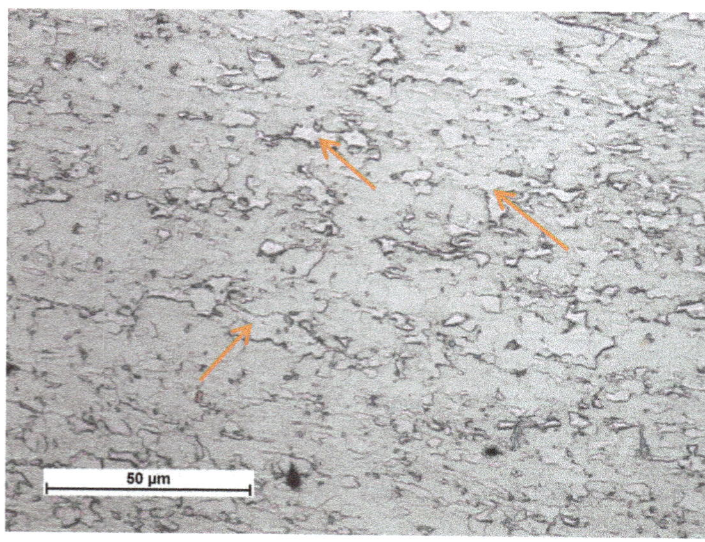

Figure 2. Structure of bainitic-ferritic steel S700MC (plastic deformed areas are marked with an arrow).

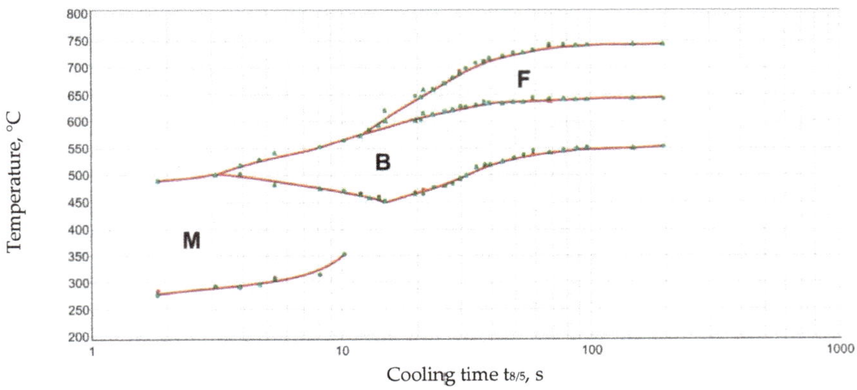

Figure 3. CCT diagram of the S700MC steel.

2.1. *Welding Process*

The tests involved 10-mm thick joints made of steel S700MC using a TruDisk 12002 disc laser (Trumpf, Ditzingen, Germany), being a component element of a robotic welding station. The identification of welding parameters required the melting of the plates using variable process parameters (Table 2). Exemplary penetrations are presented in Figure 4.

Table 2. Parameters of the remelting process.

Specimen Test No.	Beam Power, W	Melting Rate, mm/min	Linear Energy, J/mm	Position of Focus in Relation to the Plate Surface, mm
1	4000	2000	120	0
2	4000	2000	120	−3
3	4000	1000	240	0
4	4000	1000	240	−3
5	6000	2000	180	0
6	6000	2000	180	−3
7	6000	1000	360	0
8	6000	1000	360	−3
9	8000	2000	240	0
10	8000	2000	240	−3
11	8000	1000	480	0
12	8000	1000	480	−3
13	10,000	2000	300	0
14	10,000	2000	300	−3
15	10,000	1000	600	0
16	10,000	1000	600	−3

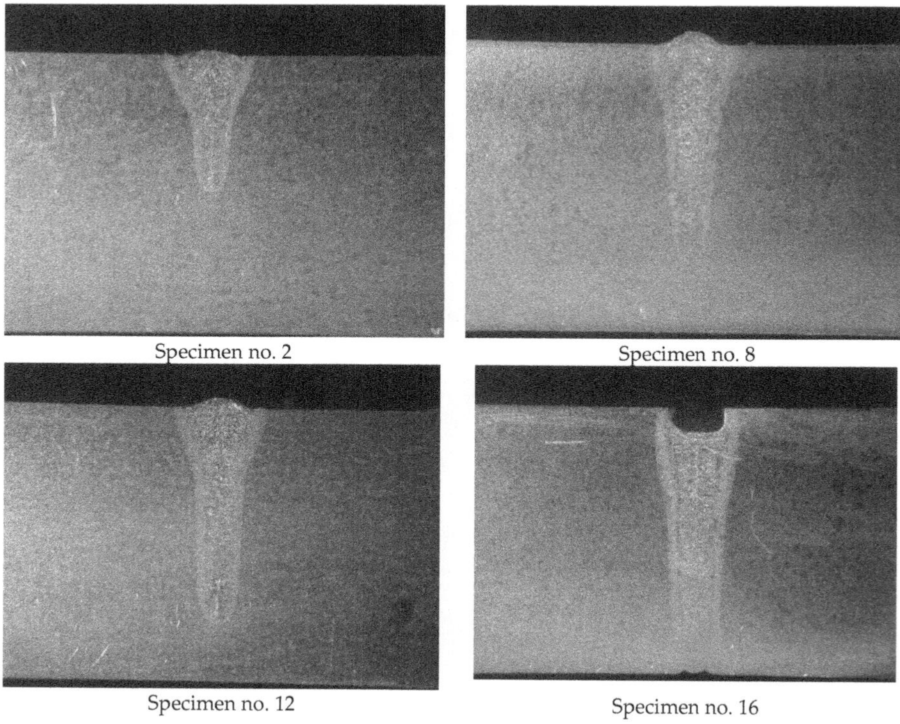

Figure 4. Macrostructures of the remelted specimens.

Based on melting tests and assessments of penetration depths, it was possible to determine the range of welding parameters (Table 3). The welding process was performed in the flat position (PA) and horizontal position (PC) with inert shielding gas (helium fed at a flow rate of 20 dm^3/min). The horizontal position made it possible to prevent the outflow of a significant volume of the liquid weld metal formed during the process, which could take place during welding in the flat position. Before welding, the edges of the plates were dried by being heated at a temperature of 65 °C. The results of strength tests and changes in structure were referred to the joints made using the MAG method [38]. Additional material used solid wire G Mn4Ni1.5CrMo. Linear welding energy was 8 kJ/cm.

Table 3. Parameters welding S700MC a thickness of 10 mm with a laser beam.

Preparing Metal Welding	The Stacking Order of the Bead
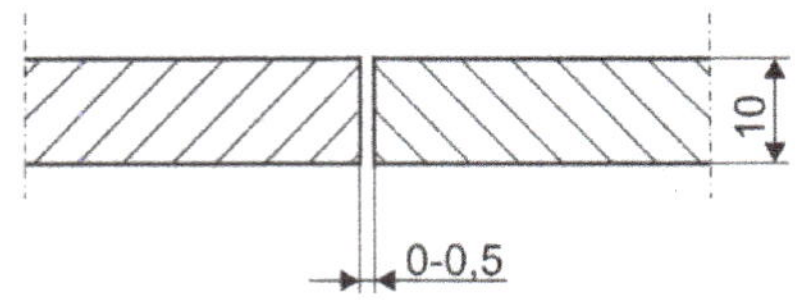	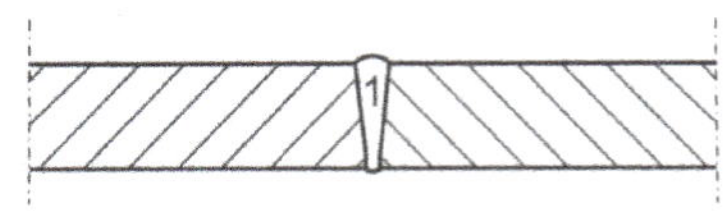

Bead	Beam Power, W	Focus Position Relative to the Surface of the Plate, mm	Welding Position	Welding Speed v, m/min	Energy Linear Welding E, kJ/cm
Joint 1					
2	3500	0	PC	1	4
Joint 2					
1	8400	0	PC	1	5
Joint 3					
1	8400	0	PA	1	5
Helium with a flow rate of 20 dm^3/min was used as the shielding gas. The joints were heated to a temperature before welding at 65 °C.					

The welding of an exemplary joint is presented in Figure 5.

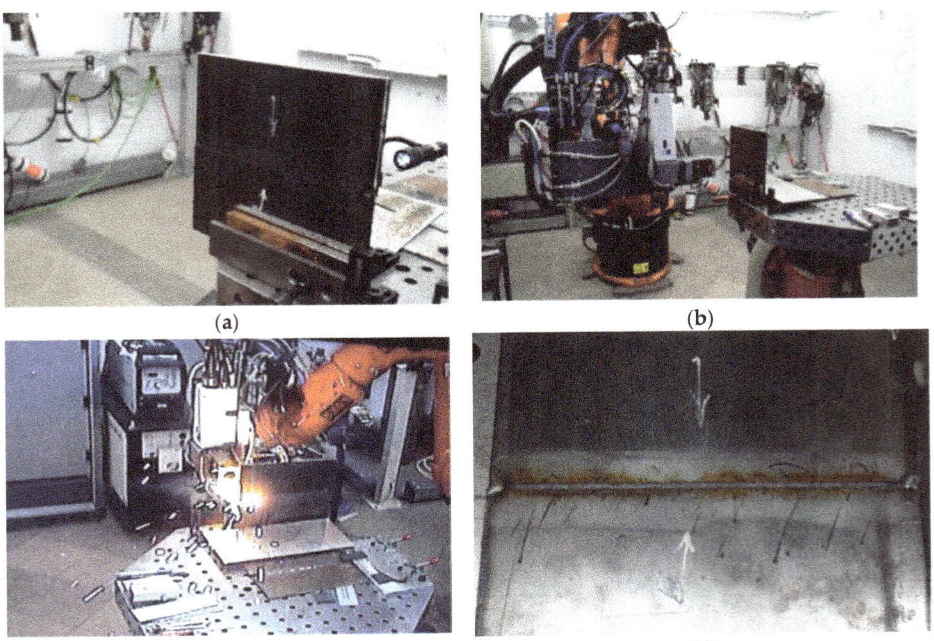

(a) (b) (c) (d)

Figure 5. Laser welding of the joint in the PC position with 5 kJ/cm linear energy: (**a**) tacked plates fixed vertically, (**b**) determination of the weld axis, (**c**) course of the welding process, and (**d**) finished joint.

2.2. The Study of Welded Joints

Obtained welded joints were subjected to non-destructive testing:

- visual tests,
- magnetic-particle tests,
- radiographic tests.

After the non-destructive tests, the welded joints were subjected to the following destructive tests:

- static tensile strength test,
- technological face bend test,
- impact tests (tests were carried out at a temperature of −30 °C),
- macro and microscopic metallographic tests, ent,
- Vickers hardness tests,
- analysis of the chemical composition using a Magellan Q8 Bruker (Bruker Austria GmbH, Vienna, Austria) and optical emission spectroscopy,
- microanalysis of the chemical composition, performed using a JOEL JSM-5800LV EDX Penta FETx3 scanning microscope (JEOL Ltd., Akishima, Japan) and an accelerating voltage of 13 kV and 20 kV,
- tests of the chemical composition in the micro-areas of the welded joints performed using a SUPRA 35 scanning electron microscope (Zeiss, Jena, Germany) equipped with an EDS attachment (EDAX) and the back-scattered electron observation technique (BSE),
- quantitative composition measured using a JXA-8230 electron probe microanalyser (JEOL Ltd., Akishima, Japan) and wavelength dispersion spectroscopy. In relation to the above-named method, the limit of detectability of heavy elements amounts to approximately 100 ppm. Images of the tested areas (with marked areas subjected to analysis) were made in light of back-scattered electrons (composition-contrast depending on the average atomic number),
- tests of thin films performed using a Titan 80–300 kV high-resolution scanning transmission electron microscope (HR S/TEM, Thermo Fisher Scientific, Waltham, MA, USA), the diffraction solution was based on the publication [39],
- X-ray phase analysis performed using an X'Pert PRO diffractometer (PANalytical, Almelo, The Netherlands) an X'Celerator strip detector (PANalytical, Almelo, The Netherlands), and a lamp equipped with a cobalt anode.

3. Results and Discussion

3.1. Analysis of the Base Material

The analysis of the chemical composition of tested steel confirmed the compliance of the chemical composition of steel with standard. The steel was characterised by a carbon content of 0.056% by weight as well as by the following contents of hardening micro-additions: titanium–0.12%, niobium–0.044%, and vanadium–0.006%. The nitrogen content was about 70 ppm, which was in accordance with the material compliance certificate. The above-named steel is characterised by very low carbon content and relatively high titanium content. Because of the high reactivity of titanium with nitrogen and carbon, titanium was bonded in stable precipitates of TiN and TiC as well as in complex precipitates of (Ti,Nb) (C,N) (Figure 6). Taking into consideration the atomic masses of the previously mentioned chemical elements, it was calculated that titanium and niobium bonded above 0.02% of carbon. As a result, the content of free carbon involved in phase curing and structural transformations was very low (approximately 0.03%).

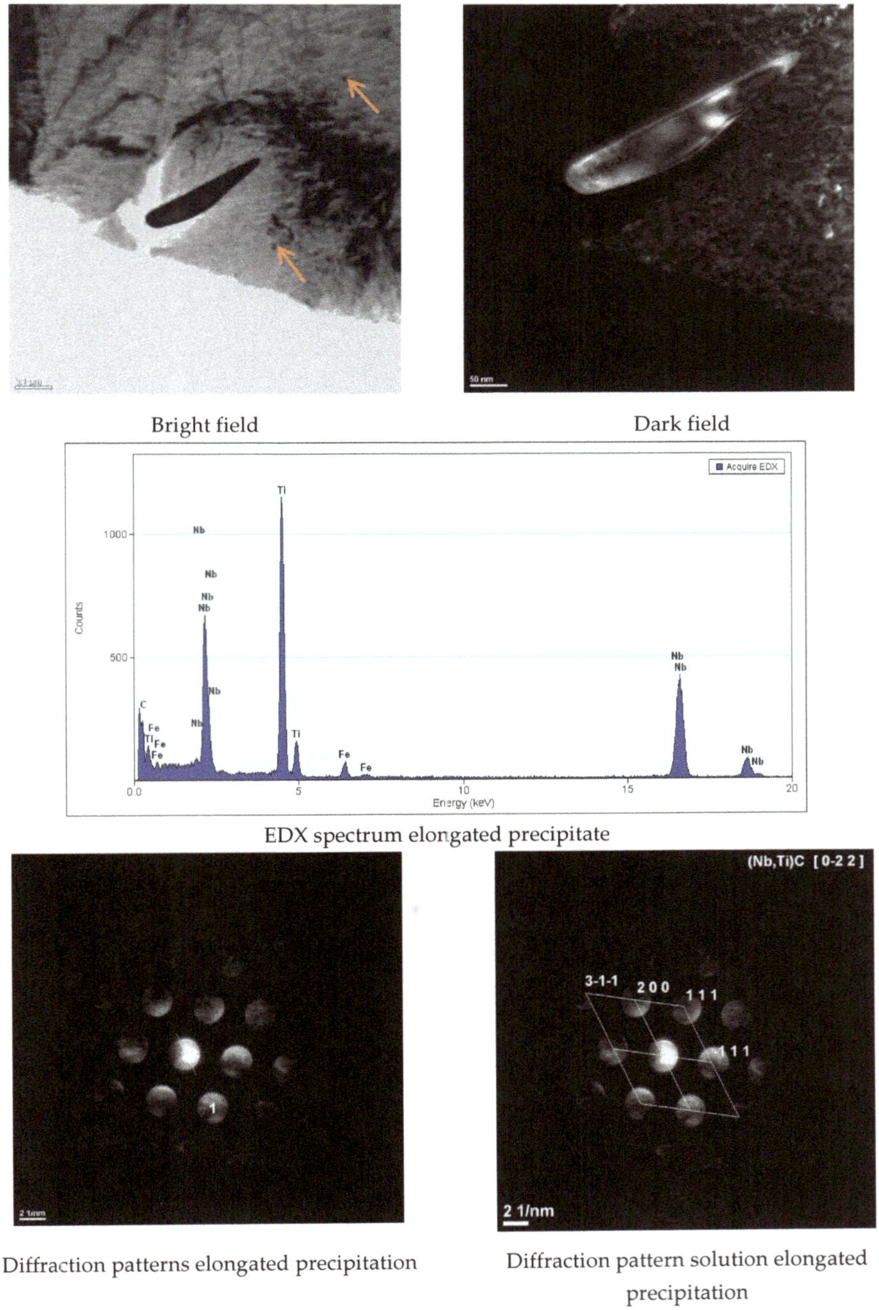

Bright field

Dark field

EDX spectrum elongated precipitate

Diffraction patterns elongated precipitation

Diffraction pattern solution elongated precipitation

Figure 6. Carbonitride precipitate elongated precipitation (Nb,Ti)C with single fine-dispersive precipitates responsible for steel hardening (marked with an arrow).

Microscopic tests performed using optical microscopy revealed that the steel tested contained large precipitates (of up to tens of μm) of characteristic sharp shapes, i.e., most likely Ti carbonitride precipitates that crystallised on impurities in the steel (Figures 7 and 8). The static tensile test revealed that the S700MC steel was characterised by a tensile strength R_m of 820 MPa, a yield point R_e of approximately 700 MPa, and an elongation A_5 of 17%. The impact strength test of the base material performed at a temperature of −30 °C revealed a toughness of approximately 50 J/cm^2.

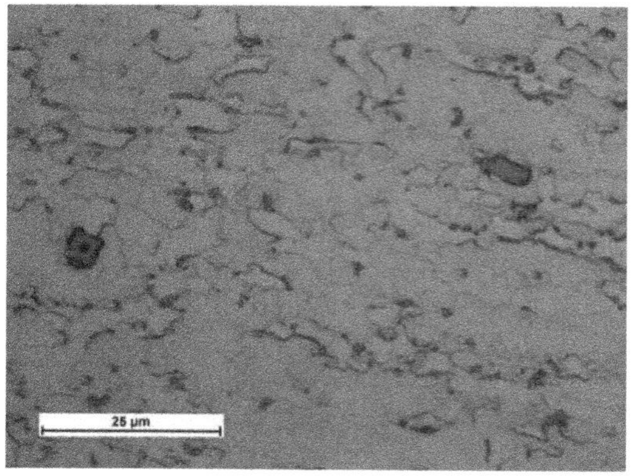

Figure 7. Characteristic sharp-shaped precipitate crystallised on impurity.

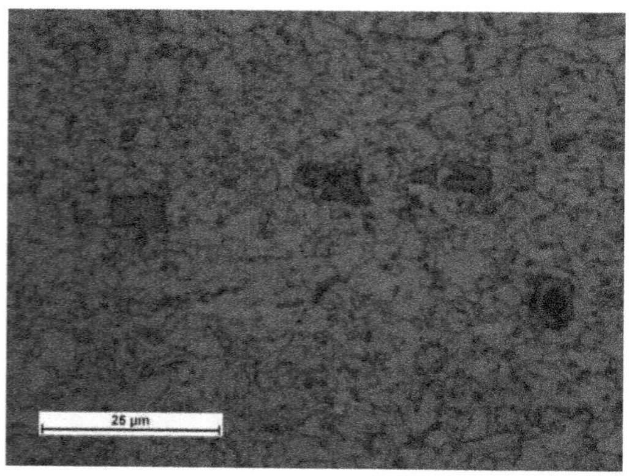

Figure 8. Clusters of characteristic sharp-shaped precipitates crystallised on impurity.

3.2. *Analysis of the Welded Joints*

The non-destructive tests carried out did not show welding defects. Based on the non-destructive tests carried out, the welded joints were classified as meeting the quality level B in accordance with ISO 13919-1 [40]. Additionally, macroscopic metallographic tests did not show the presence of welding defects in the weld and HAZ areas (Figure 9). Additionally, the MAG joint was free from welding incompatibilities. However, it has a much larger area of the cross-section (Figure 9). Microscopic tests revealed the bainitic-ferritic structure in the weld area. The grains in the HAZ area of individual joints did not differ significantly in size, which can be attributed to the similar linear welding energy (Figure 10). MAG microscopic tests of joints showed changes in the microstructure of the weld and HAZ areas compared to the base material. Both the weld and HAZ eliminated the effect of plastic deformation in the form of grains elongated in the direction of rolling, obtained during the production of plates. The weld area structure was dendritic and consisted of bainite and ferrite lamellas formed from retained austenite grains. HAZ contained a fine-grained microstructure clearly dominated by ferrite (Figure 10).

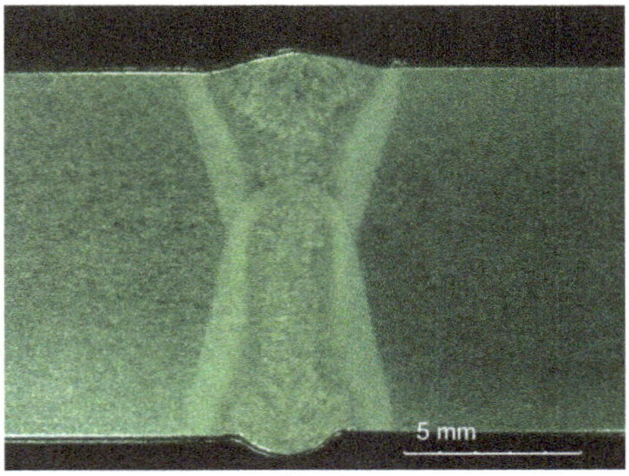

Joint 1

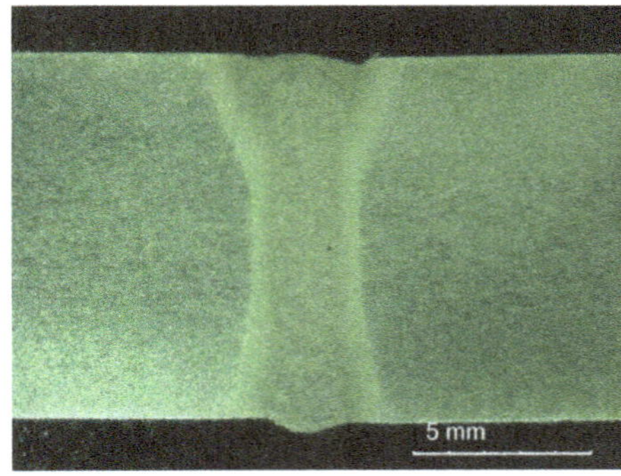

Joint 2

Joint 3

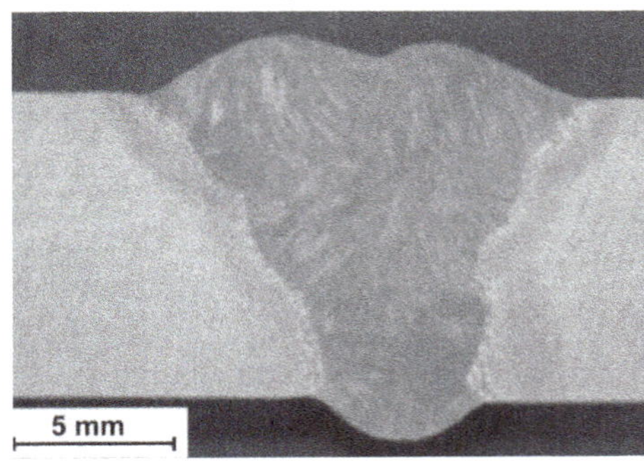

Joint 4

Figure 9. View of the macrostructures of the laser and MAG S700MC steel welded joints.

Weld	HAZ
Joint 1	
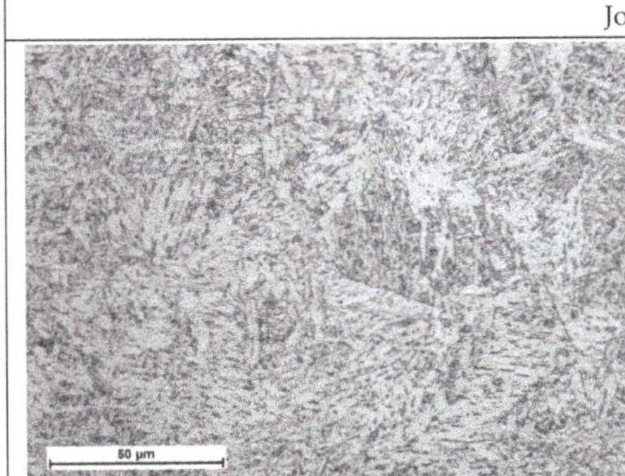	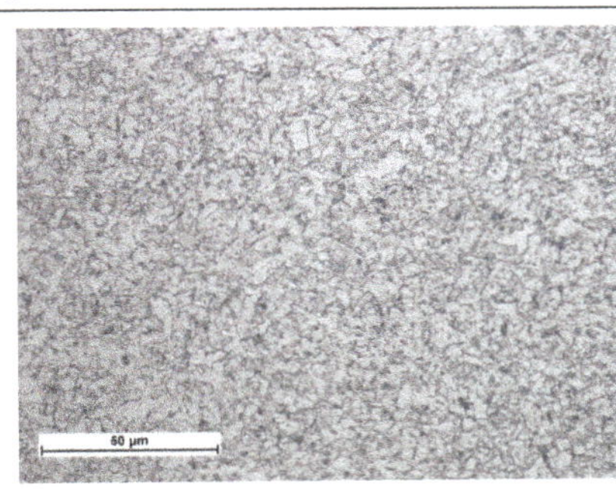

Figure 10. *Cont.*

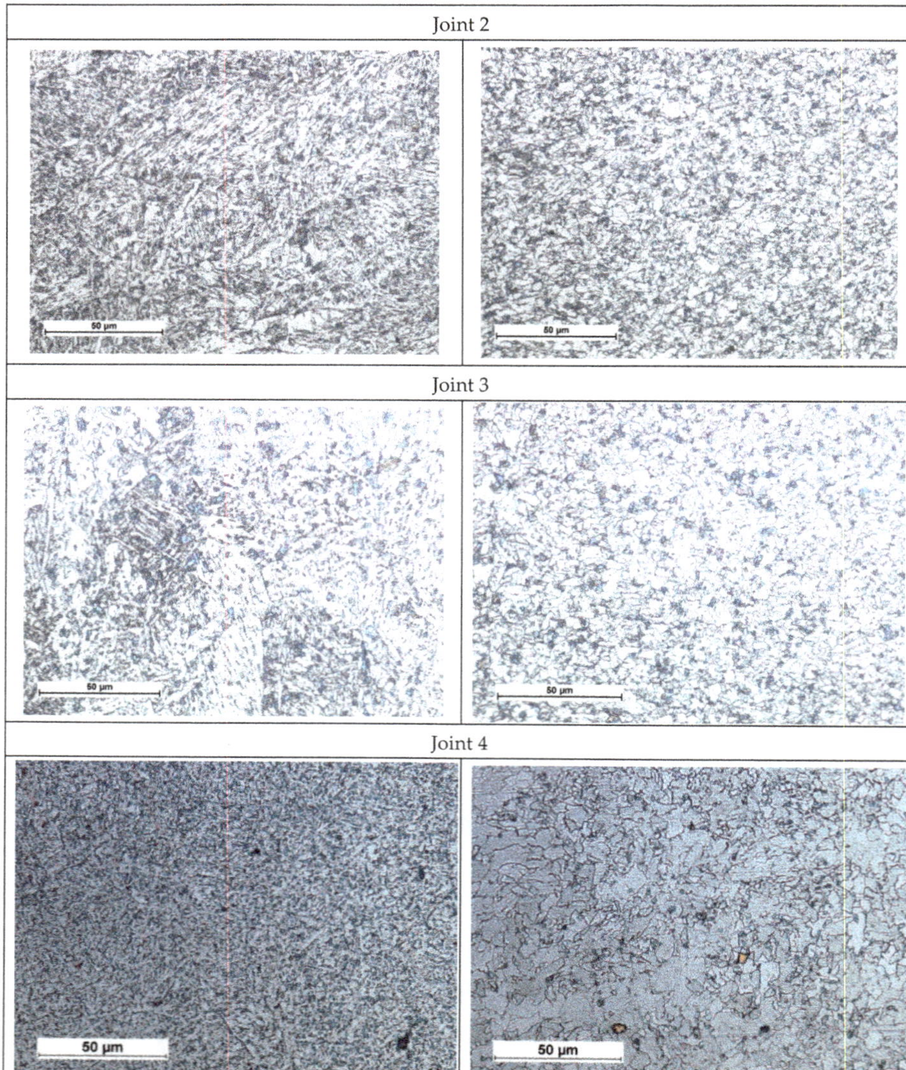

Figure 10. Middle part microstructure of S700MC steel laser and MAG welded joints.

The influence of the welding method on the strength and plastic properties of the joint is presented in Table 4.

Table 4. Strength and ductility of the S700MC steel laser beam welded joints.

Designation of the Welded Joint	Tensile Strength		Bending, Bending Angle, °		Impact Strength KCV, J/cm² (Test Temperature −30 °C)					
					Weld		FL		HAZ	
	R_m, MPa	Place Breaking	Place of Rupture	Face of Weld	KCV, J/cm²	Fracture	KCV, J/cm²	Fracture	KCV, J/cm²	Fracture
Joint 1	785	FL	180	180	20	fragile	23	fragile	41	mixed
Joint 2	790	Weld	180	180	25	fragile	22	fragile	38	mixed
Joint 3	792	Weld	180	180	22	fragile	27	fragile	39	mixed
Joint 4 *	810	BM	180	180	94	mixed	82	mixed	86	mixed

* MAG welded joint [38].

Laser welding led to a decrease in tensile strength to approximately 790 MPa in relation to that of the base material (820 MPa). Similar results were also obtained for MAG welded joints. The rupture took place in the fusion line area and was accompanied by the formation of a structural notch. The reduction of tensile strength was connected with the loss of properties obtained by steel S700MC in the TMCP process. The angle obtained in a bend test amounted to 180°. The above-named angle was obtained in relation to the tension affecting the face side as well as the root side. An impact strength test performed at a temperature of −30 °C revealed very low toughness values. The toughness in the weld area amounted to 23 J/cm^2, i.e., significantly below an acceptable level of 27 J/cm^2. The toughness in the fusion area was similar to that of the weld. In the HAZ area, toughness amounted to 40 J/cm^2. Fractographic images made after the impact strength test revealed that the material cracked without any noticeable plastic strain. The fractures of the specimens in the weld area were brittle and glossy with slightly visible drainage areas and single cavities of material extraction, which indicated the deterioration of plastic properties in the weld exposed to very low temperatures. The specimens fractured (using the Charpy pendulum machine) in other areas, i.e., in the fusion line, were also brittle. However, in the HAZ area, the fracture of the previously mentioned specimens was delaminated and predominantly matt with only a slight gloss content (Figure 11). The previously presented type and shape of the fracture may indicate high material anisotropy, which likely results from the significant plastic strain of the material or from the presence of very small precipitates or impurities. MAG welded joints have significantly higher impact values, especially in the weld area. This is due to the reduced concentration of micro-additives as a result of mixing the base material with the filler material.

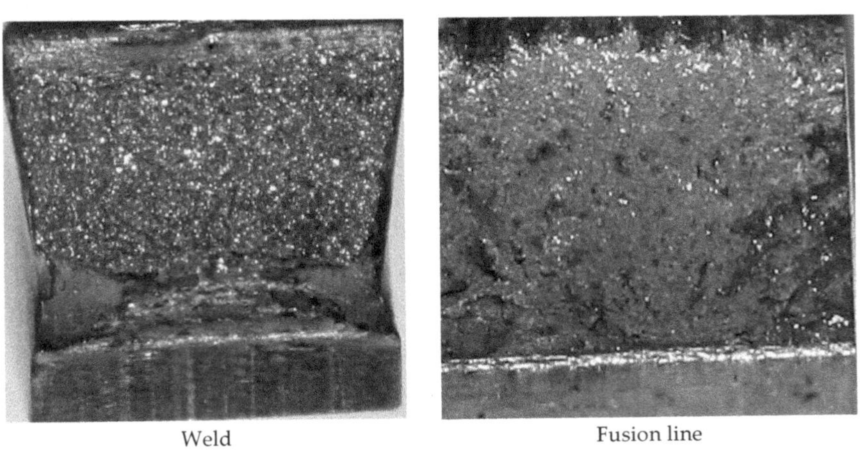

Figure 11. *Cont.*

Figure 11. Fractures after the impact test of the S700MC steel laser welded joints in the PC position.

The hardness tests of the welded joint revealed that the weld hardness was lower by approximately 40 HV1 in relation to that of the base material (280 HV1). The HAZ hardness was restricted between the hardness of the weld and that of the base material (Figure 12). In the case of MAG welding, hardness increases in the weld area, which is related to the introduction of elements increasing the chart (Ni,Cr,Mo) to the weld. In the HAZ area, there is a decrease in hardness as a result of grain growth and partial recrystallization. The X-ray phase analysis revealed that the weld made using the laser beam was entirely composed of the Feα (Figure 13). The weld using a MAG method contained phase Feα and a slight amount of phase Feγ (Figure 14). The presence of the Feγ phase can be attributed to the presence of austenitic alloying elements, e.g., Ni or C in the weld bed. During laser welding, toughness was very low, i.e., below 20 J/cm^2. In laser welds made without filler metal, the content of titanium and niobium was much higher than in welded joints made using an arc (filler metal does not contain titanium and niobium). The higher content of hardening components was responsible for the lower toughness of the weld (if compared with that of the base material). The increased content of hardening components was particularly noticeable near the melting area (which was confirmed by a detailed analysis of the chemical composition made using a microanalyzer with an electron probe) (Figures 15 and 16).

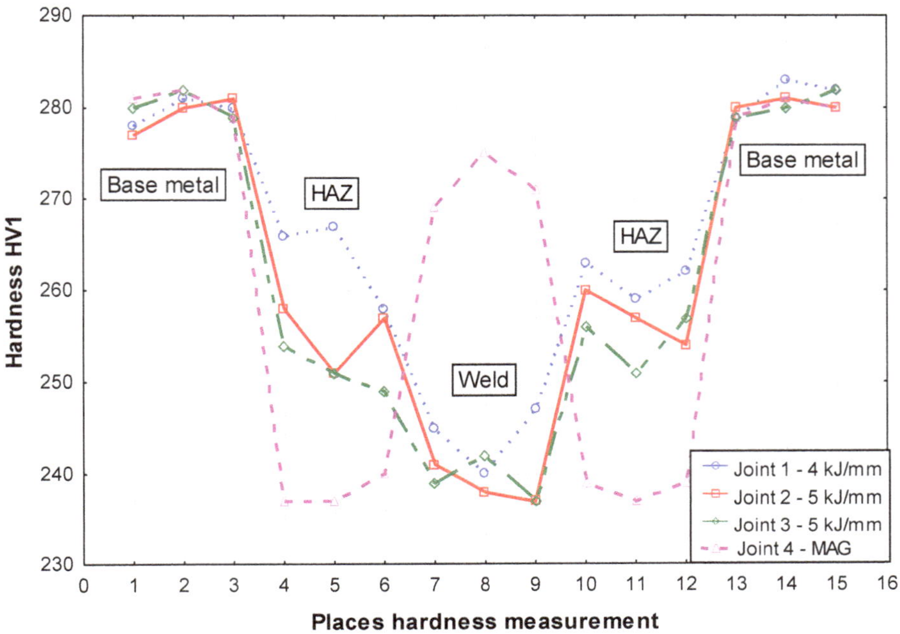

Figure 12. Hardness distribution in tested welded joints.

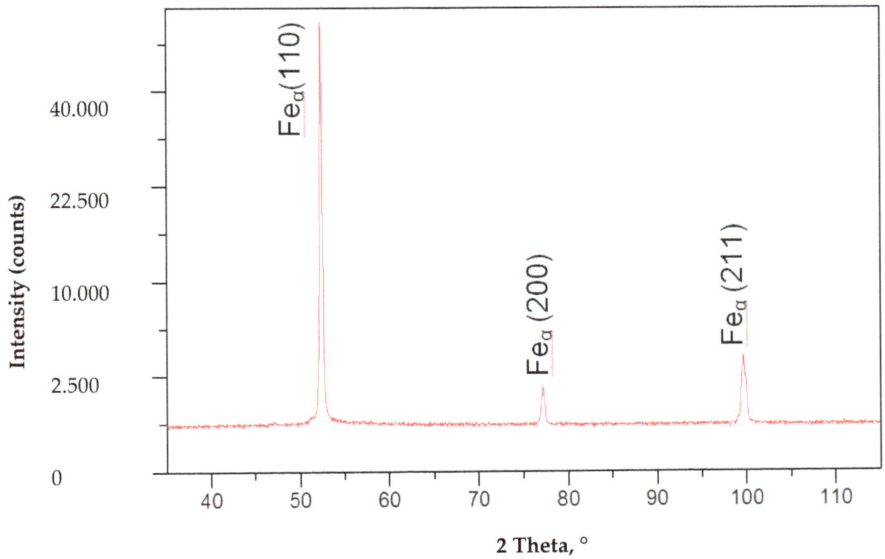

Figure 13. X-ray diffraction of a laser S700MC steel butt joint welded in the PC position with 5 kJ/cm linear energy.

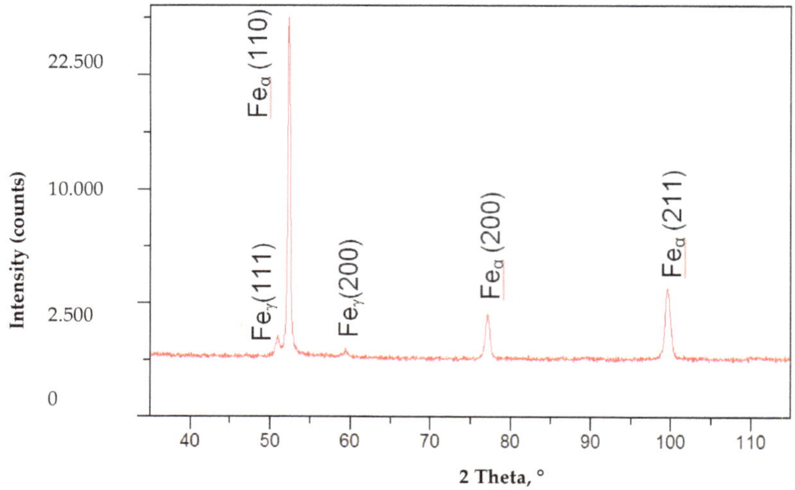

Figure 14. X-ray diffraction of a MAG S700MC steel butt joint welded with 8 kJ/cm linear energy.

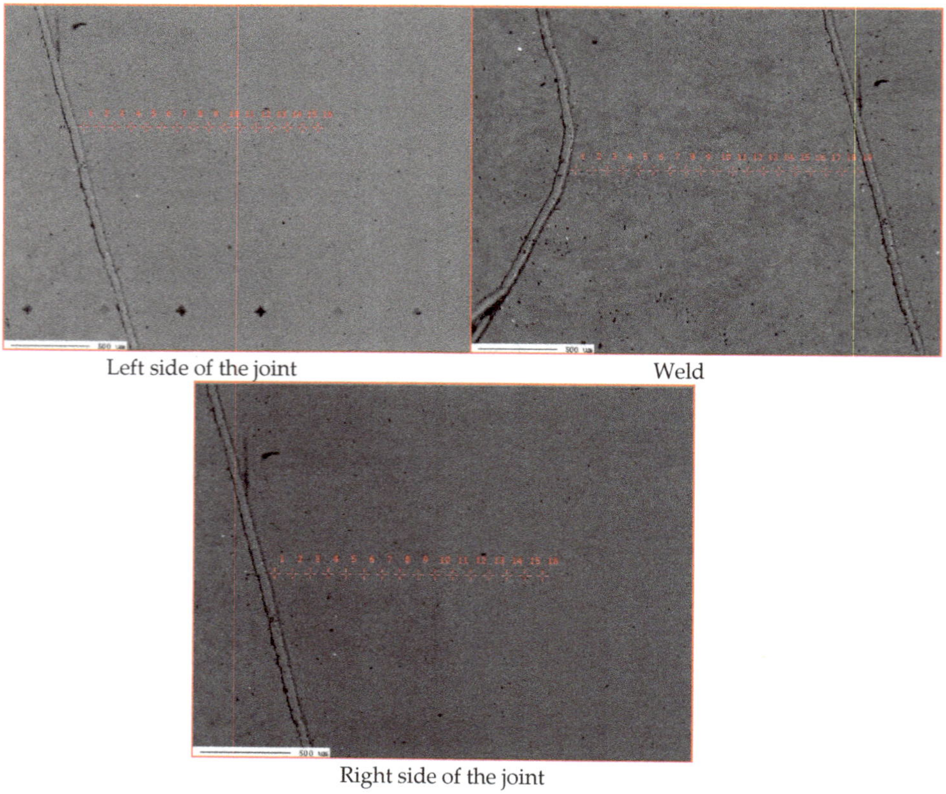

Figure 15. Points of quantitative composition measurements performed using the X-ray microanalyser and wavelength dispersion spectroscopy (WDS) for a laser welded joint in the PC position with 5 kJ/cm linear energy.

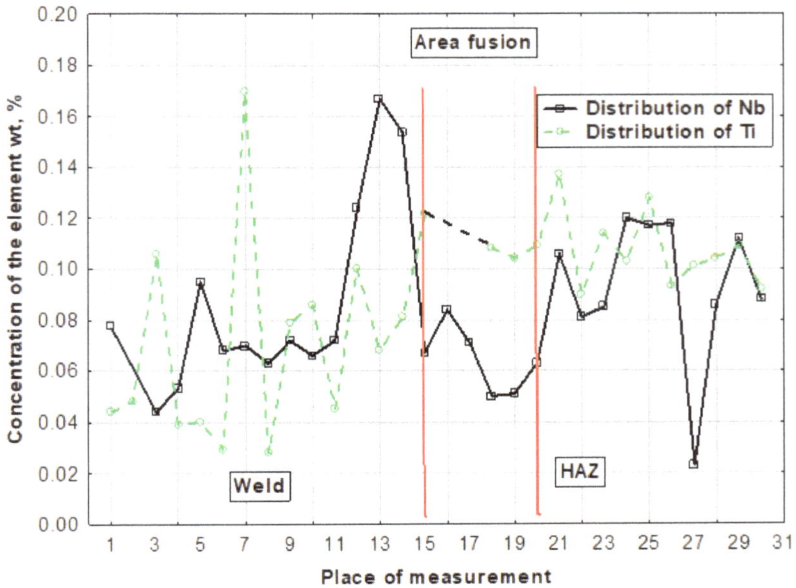

Figure 16. Distribution of Ti and Nb in the area of the fusion zone of the S700MC steel laser welded joint in the PC position with 5kJ/cm linear energy (weld and right side of the joint).

Some areas of the weld, near the fusion area, contained very high contents of Ti and Nb, which points to the presence of clusters of carbonitrides that did not dissolve completely in the liquid in the weld pool. The excessive concentration of hardening phases in the fusion area may adversely affect the plastic properties of the weld.

A significant content of hardening phases in the weld pool triggered intense precipitation hardening through fine-dispersive precipitates being several nm in size and precipitated next to larger (Ti,Nb)N and TiC particles with a size of 100 nm (Figures 17 and 18). The result of the previously mentioned situation was the reduction of plastic properties.

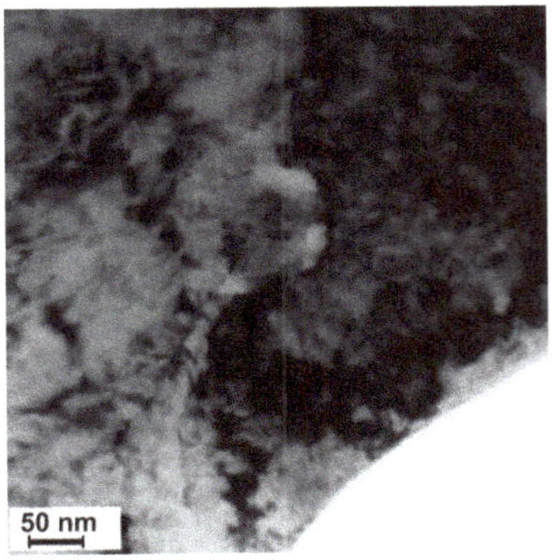

Figure 17. Precipitation of nitride (Ti,Nb)N in the S700MC steel (marked with an arrow) laser welded joint in the PC position with 5 kJ/cm linear energy.

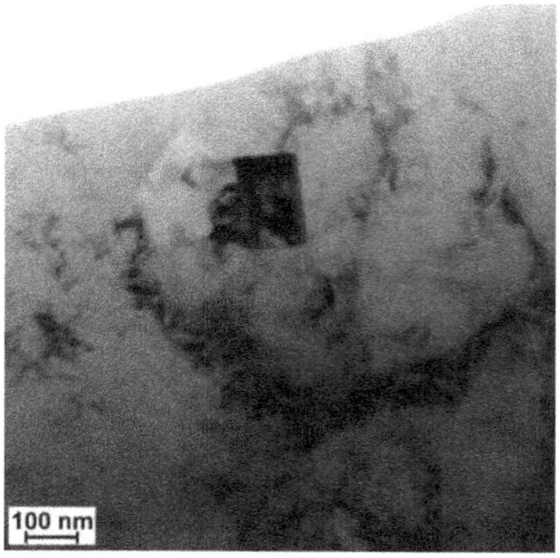

Figure 18. Precipitation of TiC carbide in the S700MC steel laser welded joint in the PC position with 5 kJ/cm linear energy.

4. Conclusions

The analysis of reference publications [1,5,7,10,18,22] and individual research [4,9,17,38] revealed that, in cases of S700MC steel, the analysis of the phase transformations of austenite was affected by welding thermal cycles and the value of carbon equivalent cannot constitute the only factors taken into consideration during the assessment of weldability. The tested steel has a low carbon content (0.056% by mass), limited primarily by hardening elements (titanium and niobium), which reduces its role in hardening by supersaturation of ferrite and by limiting its action during γ-α transformation. The short cooling time creates a martensitic phase. Martensite formed after cooling is a low-carbon variety (does not reduce the plasticity of steel). The properties of welded joints made of the previously mentioned steel are primarily influenced by their structure and stability of the hardening phases and changes in their dispersion as well as aging processes. In precipitation-hardened S700MC steels, the process of manufacturing results in obtaining quasi-equilibrium between mechanical and plastic properties of such steels. The primary components of these steels include carbon, manganese, and microadditions of V, Nb, Ti, and N, which, on one hand, reduce weldability, whereas, on the other hand, provide high mechanical properties. To maintain a desirable balance between the previously mentioned properties, it is necessary to maintain a compromise between the general content of alloying components, their quantitative ratio, and good weldability. The welding process can significantly disturb the mentioned balance, which, in turn, may lead to considerably deteriorated plastic properties both in the HAZ and in the weld. The varying stability of Nb, V, and Ti nitride precipitates and carbide precipitates is tantamount to their varying ability to restrict the growth of austenite grains in the HAZ to a varied effect on weld properties [33–37]. The lack of control over the decomposition and repeated precipitation of hardening phases may worsen both plastic and mechanical properties of joints. The welded joints represented quality level B in accordance with the ISO 13919-1 standard. The horizontal position made it possible to prevent the outflow of a significant volume of the liquid weld metal formed during the process, which could take place during welding in the flat position. Laser welding led to a decrease in tensile strength to approximately 790 MPa in relation to that of the base material (820 MPa). Similar results were also obtained for MAG welded joints. The hardness tests of the welded joint revealed that the weld hardness was lower by approximately 40 HV1 in relation to that of the base material (280 HV1). In the case of MAG welding, hardness increases in the weld area, which is related to the introduction of elements increasing the chart (Ni,Cr,Mo) to the weld. MAG welded joints have significantly higher impact values, especially in the weld area. This is due to the reduced concentration of micro-additives as a result of mixing the base material with the filler material. During laser welding of the test steel additive material, the increase in dilution is accompanied by a local increase in the content of micro titanium and niobium alloy, especially near the melting area. A significant amount of precipitates of hardening elements has a very adverse impact on the plastic properties of the weld. For welded joints made with a laser beam, despite the very low linear energy (5 kJ/cm) and low carbon equivalent (0.33%), toughness is unsatisfactory (i.e., below 20 J/cm^2). The concentration of microagents in the weld made using the laser beam without filler material is significantly higher than that in welds made using arc, which, in turn, leads to a higher amount of dispersive precipitates [38]. The high content of hardening phases in the joint during cooling leads to intensive precipitation hardening by fine dispersion deposits (Ti,Nb) (C,N) with a size of several nm, which consequently reduces plastic properties. In order to increase weld toughness, it seems advisable to use hybrid welding or to perform the welding process using the filler metal, which reduces the content of titanium and niobium alloying micro-agents in the weld. This is confirmed by the results of MAG welded joints testing.

Funding: The research Rector's pro-quality grant, 2019; Silesian University of Technology.

Conflicts of Interest: I declare no conflict of interest.

References

1. Rajashekhar, S.; Pal, M. Weldability of advanced high strength steels using an Yb:YAG disk laser. *J. Mater. Process. Technol.* **2011**, *211*, 1888–1897.
2. Gao, M.; Zeng, X.; Hu, Q.; Yan, J. Laser-TIG hybrid welding of ultra-fine grained steel. *J. Mater. Process. Technol.* **2009**, *209*, 785–791. [CrossRef]
3. Grajcar, A.; Różański, M.; Stano, S.; Kowalski, A.; Grzegorczyk, B. Effect of heat input on microstructure and hardness distribution of laser welded Si-Al TRIP-Type steel. *Adv. Mater. Sci. Eng.* **2014**, *2014*, 1–8. [CrossRef]
4. Górka, J. Microstructure and properties of the high-temperature (HAZ) of thermo-mechanically treated S700MC high-yield-strength steel. *Mater. technol.* **2016**, *50*, 616–621. [CrossRef]
5. Bang, K.S.; Kim, W.Y. Estimation and prediction of HAZ softening in thermomechanically controlled–rolled and accelerated–cooled steel. *Weld. J. N. Y.* **2002**, *81*, 174-S.
6. Żuk, M.; Górka, J.; Czupryński, A.; Adamiak, M. Properties and structure of the weld joints of quench and tempered 4330V steel. *Metalurgija* **2016**, *55*, 613–616.
7. Hildebrand, J.; Werner, F. Change of structural condition of welded joints between high-strength fine-grained and structural steels. *J. Civ. Eng. Manag.* **2004**, *10*, 87–95. [CrossRef]
8. Porter, D.; Laukkanen, A.; Nevasmaa, P.; Rahka, K.; Wallin, K. Performance of TMCP steel with respect to mechanical properties after cold forming and post-forming heat treatment. *Int. J. Press. Vessels Pip.* **2004**, *81*, 867–877. [CrossRef]
9. Górka, J.; Janicki, D.; Fidali, M.; Jamrozik, W. Thermographic Assessment of the HAZ Properties and Structure of Thermomechanically Treated Steel. *Int. J. Thermophys.* **2017**, *38*, 183. [CrossRef]
10. Fydrych, D.; Labanowski, J.; Rogalski, G. Weldability of high strength steels in wet welding conditions. *Pol. Marit. Res.* **2013**, *20*, 67–73. [CrossRef]
11. Kurc-Lisiecka, A.; Lisiecki, A. Laser welding of the new grade of advanced high-strength steel Domex 960. *Mater. Technol.* **2017**, *51*, 199–204. [CrossRef]
12. Janicki, D. "Disk laser welding of armor steel". *Arch. Metall. Mater.* **2014**, *59*, 1641–1646. [CrossRef]
13. Grajcar, A.; Różański, M.; Kamińska, M.; Grzegorczyk, B. Effect of gas atmosphere on the non-metallic inclusions in laser-welded TRIP steel with Al and Si additions. *Mater. Tehnol.* **2016**, *50*, 945–950. [CrossRef]
14. Suder, W.; Ganguly, S.; Williams, S.; Yudodibroto, B. Root stability in hybrid laser welding. *J. Laser Appl.* **2017**, *29*, 022410. [CrossRef]
15. Janicki, D. Fiber laser welding of nickel based superalloy Rene 77. In Proceedings of the Laser Technology 2012: Applications of Lasers, Szczecin, Poland, 22 January 2013; Volume 8703, p. 87030Q.
16. Węgrzyn, T.; Piwnik, J.; Hadryś, D.; Wszołek, Ł. Low alloy steel structures after welding with micro-jet cooling. *Arch. Metall. Mater.* **2017**, *62*, 115–118. [CrossRef]
17. Górka, J. Welding Thermal Cycle-Triggered Precipitation Processes in Steel S700MC Subjected to the Thermo-Mechanical Control Processing. *Arch. Metall. Mater.* **2017**, *62*, 331–336. [CrossRef]
18. Yurioka, M. TMCP steel and their welding. *Weld. World* **1995**, *35*, 375–390.

19. Bajić, N.; Šijački-Žeravčić, V.; Rakin, M.; Kovačević, K. Effect of the welding mode and filler content on the structure of microalloyed Nb/Ti steel weldments. *Mater. Sci.* **2010**, *46*, 124–133. [CrossRef]
20. Winczek, J.; Modrzycka, A.; Gawrońska, E. Analytical description of the temperature field induced by laser heat source with any trajectory. *Procedia Eng.* **2016**, *149*, 553–558. [CrossRef]
21. Kik, T.; Górka, J. Numerical Simulations of Laser and Hybrid S700MC T-Joint Welding. *Materials* **2019**, *12*, 516. [CrossRef]
22. Skowrońska, B.; Szulc, J.; Chmielewski, T.; Sałaciński, T.; Swiercz, R. Properties and microstructure of hybride Plasma + MAG welded joints of thermomechanically treated S700MC steel. In Proceedings of the 27th Anniversary International Conference on Metallurgy and Materials (METAL), Brno, Czech Republik, 23–25 May 2018.
23. Shin, Y.T.; Kang, S.W.; Lee, H.W. Fracture characteristics of TMCP and QT steel weldments with respect to crack length. *Mater. Sci. Eng. A* **2006**, *434*, 365–371. [CrossRef]
24. Kim, Y.; Kwon, J.; Lee, H.; Jang, W. Effect of microstructure on fatigue crack propagation and S–N fatigue behaviours of TMCP steels with yield strengths of approximately 450 MPa. *Metall. Mater. Trans.* **2011**, *42*, 986–999. [CrossRef]
25. Altuna, M.A.; Iza-Mendia, A.; Guitierrez, I. Precipitation strengthening by formationin ferrite of Nb carbides. In Proceedings of the 3rd International Conference on Thermomechanical Processing of Steels, Padua, Italy, 10–12 September 2008; pp. 1–12.
26. Kik, T.; Moravec, J.; Novakova, I. New method of processing heat treatment experiments with numerical simulation support. In *IOP Conference Series-Materials Science and Engineering*; IOP Publishing: Bristol, UK, 2017; Volume 227.
27. Hashimoto, S.; Ikeda, S.; Sugimoto, K.; Miyake, S. Effects of Nb and Mo addition to 0.2%C–1.5%Mn steel on mechanical properties of hot rolled TRIP-aided steel sheets. *ISIJ Int.* **2004**, *44*, 1590–1598. [CrossRef]
28. Wang, X.D.; Huang, B.X.; Wang, L.; Rong, Y.H. Microstructure and mechanical properties of microalloyed high-strength transformation-induced plasticity steels. *Metall. Mater. Trans. A* **2008**, *39*, 1–7. [CrossRef]
29. Yasuharu, S.; Osamu, M.; Hiroshi, T. Mechanical properties and retained austenite in intercritically heat-treated bainite-transformed steel and their variation with Si and Mn additions. *Metall. Trans. A* **1991**, *22*, 489–498.
30. Wang, S.H.; Chiang, C.C.; Chan, L.I. Effect of initial microstructure on the creep behaviour of TMCP EH36 and S490MC steels. *Mater. Sci. Eng. A* **2003**, *344*, 288–295. [CrossRef]
31. Timokhaina, I.B.; Hodgson, P.D.; Pereloma, E.V. Effect of microstructure on the stability of retained austenite in transformation-induced-plasticity steels. *Metall. Mater. Trans. A* **2004**, *35*, 2331–2341. [CrossRef]
32. Barin, I. *Thermochemical Data of Pure Substances*; VCH: Weinheim, Germany, 1995.
33. Bhadeshia, H.K. *Bainite in Steels*, 2nd ed.; Institute of Materials, Minerals and Mining: London, UK, 2001; pp. 265–268.
34. Opiela, M. Thermodynamic analysis of the precipitation of carbonitrides in microalloyed steels. *Mater. Technol.* **2015**, *49*, 395–401. [CrossRef]
35. Xue, E.; Shan, Y.; Zheng, S.; Lou, S. Microstructural characteristic of low carbon microalloyed steels produced by termomechanical controlled process. *Mater. Sci. Eng.* **2006**, *438*, 285–287. [CrossRef]
36. Medina, S.F.; Gomez, M.; Gomez, P. Effects of V and Nb on static recrystallization of austenite and precipitate size in microalloyed steels. *J. Mater. Sci.* **2010**, *45*, 5553–5557. [CrossRef]
37. Nishioka, K.; Ichikawa, K. Progress in termomechanical control of steel plates and their commercialization. *Sci. Technol. Adv. Mater.* **2012**, *13*, 023001. [CrossRef] [PubMed]
38. Górka, J. Assessment of Steel Subjected to the Thermomechanical Control Process with Respect to Weldability. *Metals* **2018**, *8*, 169. [CrossRef]
39. Hirsch, P.B.; Howie, B.; Nicholson, A.M.; Pasheley, D.W.; Whelan, M.J. *Electron Microscopy of Thin Crystals*; Butterworths: London, UK, 1965.
40. *ISO 13919-1:2002 Welding-Electrons and Laser Beam Welded Joints-Guidance on Quality Levels for Imperfections-Part 1: Steel*; ISO: Geneva, Switzerland, 2002.

materials

MDPI

Article

Quantitative Correlation between Thermal Cycling and the Microstructures of X100 Pipeline Steel Laser-Welded Joints

Gang Wang [1], Jinzhao Wang [2], Limeng Yin [1,2,*], Huiqin Hu [3] and Zongxiang Yao [1]

1 School of Metallurgy and Materials Engineering, Chongqing University of Science and Technology, Chongqing 401331, China; wwg_16@163.com (G.W.); yaozongx@163.com (Z.Y.)

2 Guangdong Provincial Key Laboratory of Advanced Welding Technology, Guangdong Welding Institute (China-Ukraine E.O. Paton Institute of Welding), Guangzhou 510650, China; jinzhao_wang@foxmail.com

3 School of Natural and Applied Sciences, Northwestern Polytechnical University, Xian 710129, China; huiqin_hu@foxmail.com

* Correspondence: yeenlm@cqust.edu.cn; Tel.: +86-150-2373-0501

Received: 2 December 2019; Accepted: 23 December 2019; Published: 26 December 2019

Abstract: Due to the limitations of the energy density and penetration ability of arc welding technology for long-distance pipelines, the deterioration of the microstructures in the coarse-grained heat-affected zone (HAZ) in welded joints in large-diameter, thick-walled pipeline steel leads to insufficient strength and toughness in these joints, which strongly affect the service reliability and durability of oil and gas pipelines. Therefore, high-energy-beam welding is introduced for pipeline steel welding to reduce pipeline construction costs and improve the efficiency and safety of oil and gas transportation. In the present work, two pieces of X100 pipeline steel plates with thicknesses of 12.8 mm were welded by a high-power robot laser-welding platform. The quantitative correlation between thermal cycling and the microstructure of the welded joint was studied using numerical simulation of the welding temperature field, optical microscopy (OM), and scanning electron microscopy (SEM) with energy-dispersive spectroscopy (EDS). The results show that the heat-source model of a Gaussian-distributed rotating body and the austenitization degree parameters are highly accurate in simulating the welding temperature field and characterizing the austenitization degree. The effects of austenitization are more significant than those of the cooling rate on the final microstructures of the laser-welded joint. The microstructure of the X100 pipeline steel in the HAZ is mainly composed of acicular ferrite (AF), granular bainite (GB), and bainitic ferrite (BF). However, small amounts of lath martensite (LM), upper bainite (UB), and the bulk microstructure are found in the columnar zone of the weld. The aim of this paper is to provide scientific guidance and a reference for the simulation of the temperature field during high-energy-beam laser welding and to study and formulate the laser-welding process for X100 pipeline steel.

Keywords: laser welding; numerical simulation; X100 pipeline steel; welding thermal cycle; microstructure

1. Introduction

As a result of the immense consumption of oil and gas energy driven by the rapid development of the world economy, the scope of energy exploration is constantly expanding, and oil gas fields are usually far from the end of the consumer market. The use of high-grade steel pipelines with large pipe diameters, thick walls, and high strength and toughness substantially improves the cost-effectiveness of construction to ensure transport efficiency and safety [1–3].

The efficiency and quality requirements for welding in the installation of large-diameter, thick-walled pipes in harsh environments and over long distances will become more stringent. Therefore, many advanced welding methods are used for welding high-grade pipeline steel in recent years, such as laser welding [4], laser-arc hybrid welding [5,6], friction stir welding [7], and more. Laser welding has good application prospects due to its advantages such as fast welding speed, large penetration depth, and small deformation, and has been reported in many welding fields. Guo et al. [8] reported the microstructure and mechanical properties of underwater laser welding of Titanium alloy. Xu et al. [9] as well as Xin et al. [10] investigated the characteristics and process mechanism of laser welding or laser-arc hybrid welding of aluminum alloys. Zhou et al. [11], Chen et al. [12], and Casalino et al. [13] studied the laser welding-brazing process of Titanium-Aluminum dissimilar metals. Silva Leite et al. [14] and Hipp et al. [15] discussed the laser welding process and mechanism of various stainless steels.

However, the final microstructure of high-grade pipeline steel in a laser-welded joint depends on the temperature field formed by the laser heat source and on the influence of thermal cycling during the welding process, which directly determines the strength, toughness, and service reliability of a welded joint [16–19]. Therefore, to accurately control the strength and toughness of a laser-welded joint in a high-grade long-distance pipeline and to improve the quality, efficiency, and cost benefits of pipeline installation, building the quantitative correlation between the thermal cycling and the microstructure of high-grade pipeline steel within laser-welded joints is important [2,16,20]. Different thermal cycling leads to different microstructures within welded joints and, accordingly, to different strengths and toughness in these welded joints [17,21–23].

In this research, based on the actual working conditions of an X100 pipeline steel plate with a thickness of 12.8 mm, reasonable and feasible process parameters are chosen to obtain high-quality welded joints. The microstructures of the metallographic specimens obtained from the laser-welded joints were analyzed by optical microscopy (OM) and scanning electron microscopy (SEM) in conjunction with energy-dispersive X-ray spectrometry (EDS). In addition, a finite-element temperature field simulation was carried out using the parameters of the thermal physical properties of X100 pipeline steel, an accurate self-developed laser-welding heat-source model, a reasonably simplified finite-element mesh, and welding temperature field initialization conditions. Ultimately, to reveal the quantitative correlation between thermal cycling and the microstructures of X100 pipeline steel laser-welded joints, the intrinsic correlation between the temperature field of the laser welding and the microstructure within the welded joint was analyzed and discussed on the basis of metallurgical and phase-change theories.

2. Experimental Procedures for X100 Laser Welding

2.1. Experimental Materials

The composition of the X100 pipeline steel plate with a thickness of 12.8 mm that conformed to the API-5L standard is shown in Table 1. The production of X100 pipeline steel is based on alloying technology and the thermomechanical control process (TMCP), which compensates for the loss of strength caused by the reduction of the carbon content from adding alloying elements and improves the comprehensive properties of the steel via alloy phase-transformation strengthening, precipitation strengthening, and fine-grain strengthening [24–26]. The original microstructure of the X100 pipeline steel is mainly bainite, including granular bainite (GB), acicular ferrite (AF), and martensite-austenite (M-A) constituent. A large amount of the M-A constituent is distributed between or within grains, and the dislocation density of the lattice distortions increases due to the martensitic transformation. Transformed martensite is also the location of microcracks [27,28]. Therefore, controlling the shape and distribution of the M-A constituent in laser-welded joints of X100 pipeline steel is important [25,27,29,30].

Table 1. Chemical composition of X100 pipeline steel (wt.%).

AL	C	CO	CR	CU	FE	MN	MO	N	NB	NI	P	SI	TI	V
0.012	0.064	0.003	0.023	0.28	96.90	1.87	0.003	0.017	0.017	0.47	0.009	0.099	0.017	0.002

2.2. Experimental Methods

The processing equipment consisted of a laser-welding head with TRUMPF BEO D70 90° focusing optics (200-mm collimation focal length, 200-mm focus length, and 200-μm transmission fiber) mounted on a precision six-axis robotic arm (KUKA KR 60 HA). The robotic arm was configured to perform the necessary motions required for welding as per the welding procedure specification (WPS) through laser-welding control software. The high-energy laser beam was supplied by a disc laser (TRUMPF TruDisk 10002, TRUMPF, Ditzingen, Germany) with a maximum output power of 10 kW and was protected by a gas protection device (Figure 1) during welding.

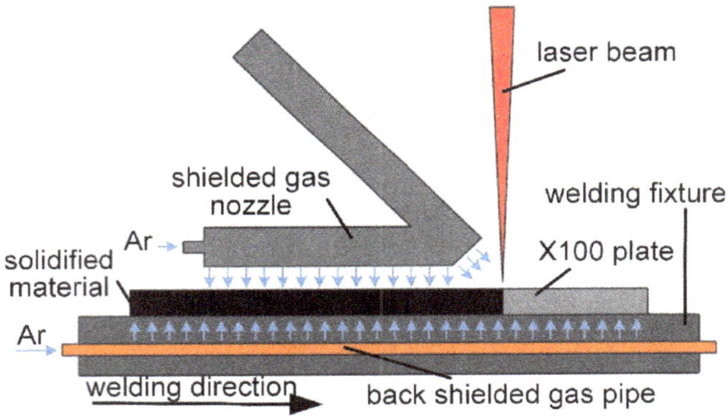

Figure 1. Schematic of the X100 laser-welding device.

The welding specimens were thoroughly cleaned prior to welding. The welding was performed in a flat position (PA) on both sides. The process parameters employed for the laser welding are given in Table 2. The laser power is identified based on the penetration capacity of the laser welding. The other parameters are determined by the orthogonal experiments. The welding of the X100 steel specimens was performed in the I-groove joint configuration. The metallographic samples (30 mm × 20 mm × 12.8 mm) for the microstructural examinations were machined by wire electric discharge machining (WEDM). The machined microstructural samples were then polished following the standard microstructural examination procedure and were etched using a 4% Nital solution (for 3–8 s). The bead shape measurements and the microstructural examinations were performed using OM (Axio Imager M2m, ZEISS, Oberkochen, Germany) and SEM (Nova NanoSEM 430, FEI, Hillsboro, OR, USA). OM was performed on a metallurgical microscope. SEM was performed on a scanning electron microscope (Hitachi S-3700N). The quantitative analyses of the micrographic images were performed using image analysis software (Image-Pro Plus).

Table 2. Parameters for the laser welding of the pipeline steel.

Material (mm)	Laser Power (kw)	Welding Speed (m/s)	Amount of Defocusing (mm)	Frontal Protective Gas Flow (L/min)	Back Protective Gas Flow (L/min)
X100 (12.8)	10	0.035	−4	15	25
	8	0.035	−4	15	25

3. Numerical Calculation of the Laser-Welding Temperature Field

3.1. Establishment and Solution of the Finite-Element Model

The Multiphysics Object-Oriented Simulation Environment (MOOSE), which is an open-source finite-element analysis software framework developed and maintained by Idaho National Laboratory, was used for the X100 pipeline steel laser-welding temperature field simulation. To compare the laser-welding temperature field simulation results with the actual working conditions of the welding process, a two-dimensional finite-element model consistent with the actual welding joint size was established through reasonable model abstraction and reliable data for the material thermophysical parameters that were calculated by JmatPro [31–35]. The material thermophysical parameters are shown in Table 3.

Table 3. Thermophysical parameters of X100 high-grade pipeline steel.

Temperature (K)	293	373	473	673	873	1073	1273	1473	1773
Density (kg/m^3)	7810	7790	7770	7720	7650	7610	7560	7500	7500
Thermal Conductivity [W/(m·K)]	54.42	54.01	52.75	52.75	34.67	27.55	23.18	21.54	21.54
Specific Heat [J/(kg·K)]	423	493	536	662	827	614	565	516	516

The model size was 10 mm × 12.8 mm, and the mesh size was 0.05 mm × 0.05 mm. According to the shape parameters of the weld seam and the welding heat-affected zone (HAZ), the heat-source model for a Gaussian-distributed rotating body is proposed (Equations (1)–(3)). Heat convection, heat conduction, and heat radiation are taken into account in the boundary conditions (Equations (4) and (5)) [32,34,36–40]. The model was solved by the Preconditioned Jacobian-Free Newton-Krylov algorithm (PJFNK). The linear tolerance was 1×10^{-2}, the nonlinear absolute tolerance was 1×10^{-8}, and the time step was 0.01 s.

Heat-source model

$$Q(x,y,z) = \frac{\eta_L P}{V_{hs}} exp(-\frac{3((x-vt)^2+y^2)}{R^2}) \tag{1}$$

$$R = R_0\left(0.2sin\left(\frac{4\pi z}{H}\right) + \frac{z^{\frac{2}{5}}}{H}\right) \tag{2}$$

$$V_{hs} = \pi \int_0^H R^2 dz \tag{3}$$

Governing equation

$$\left\{k(T)(\frac{\partial^2 T}{\partial x^2} + \frac{\partial^2 T}{\partial y^2} + \frac{\partial^2 T}{\partial z^2})\right\} + Q = \rho(T)C_p(T)(\frac{\partial T}{\partial t}) \tag{4}$$

Boundary condition

$$k\frac{\partial T}{\partial r} = -h_A(T - T_{ref}) - \sigma_s \varepsilon_r (T^4 - T^4_{ref}) - q_{con} \tag{5}$$

where $Q(x, y, z)$ is the heat flux density at point (x, y, z), η_L is the thermal efficiency of laser welding, P is the laser power (W), v is welding speed (mm/s), t is welding time (s), R_0 is the radius of the laser welding keyhole (mm), H is the height of the heat source (mm), k is the thermal conductivity of X100 pipeline steel, ρ is the density, C_p is the specific heat, T is the temperature (K), h_A is the convective heat transfer coefficient on the test plate surface, T_{ref} is the ambient temperature (K), σ_s is the Stefan-Boltzmann constant, ε_r is the emissivity of the blackbody, and q_{con} is the energy density of heat conduction. In order to improve the numerical calculation efficiency of the welding temperature field, the three heat transfer effects are simplified as a parameter of the comprehensive convection heat transfer coefficient. For the numerical calculation of laser welding temperature of X100 pipeline steel

$\eta_L = 0.85$, $v = 35$ mm/s, $P = 10{,}000$ W, $H = 10$ mm, $R_0 = 1.12$ mm, and $T_{ref} = 300$ K. In addition, the comprehensive convection heat transfer coefficient was 0.2 W/(m^2·K) [37–39,41,42].

3.2. Post-Processing of Welding Temperature Field Data

The results calculated for the laser-welding temperature field were output in the EXODUS II format. The welding-temperature field contour, the welding thermal cycle curve, the t_{85} field contour, and the weld-pool section profile were obtained from the temperature field data through data visualization software (Paraview, Kitware, New York, NY, USA) and MATLAB (MathWorks, Natick, MA, USA)). The simulated results for the laser-welding temperature field are in good agreement with the experimental results, as shown in Figure 2b.

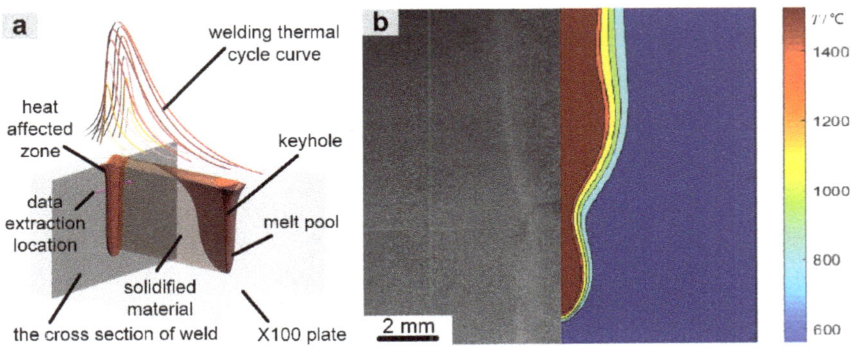

Figure 2. Principle of laser welding (**a**) and comparison of the temperature field simulation and experimental results (**b**) for X100 pipeline steel.

According to the symmetry of the microstructural state of the laser-welded joint, the position 2 mm from the upper surface of the welding plate on the cross section of the weld is taken as the focus of attention. As shown in the diagram in Figure 2a, we extract the welding heat cycle curve in the grid node from the base material of the weld and calculate the $t_{8/5}$ value, which is used for analysis of the microstructural distribution in the laser-welded head.

4. Results and Discussion

4.1. Extraction of the Characteristic Parameters of the Welding Thermal Cycle

Because the thermal cycling process of a welded joint determines its microstructure, deep analysis of the thermal cycling process of the X100 pipeline steel is beneficial for the determination and analysis of the microstructure. The laser-welded joint of X100 pipeline steel consists of the base material, a partial-phase-transition zone, a normalized zone, an overheated zone, and a melted zone from the base material to the weld, according to the process of thermal cycling. The corresponding microstructures are the base material, a banded-microstructural zone, a fine-grain zone, a coarse grain zone, and a columnar zone.

The maximum temperature field (Figure 3a) and the $t_{8/5}$ field (Figure 3b) are obtained by post-processing the numerical results of the laser-welding temperature field. The $t_{8/5}$ field, just like the temperature field, is a general term for the distribution of $t_{8/5}$ at all points in the welded joint. The welding HAZ exhibits a clear temperature gradient, even though the change in $t_{8/5}$ is relatively small. However, the change in the maximum temperatures of the weld is relatively small, and the gradient of change in $t_{8/5}$ is clear.

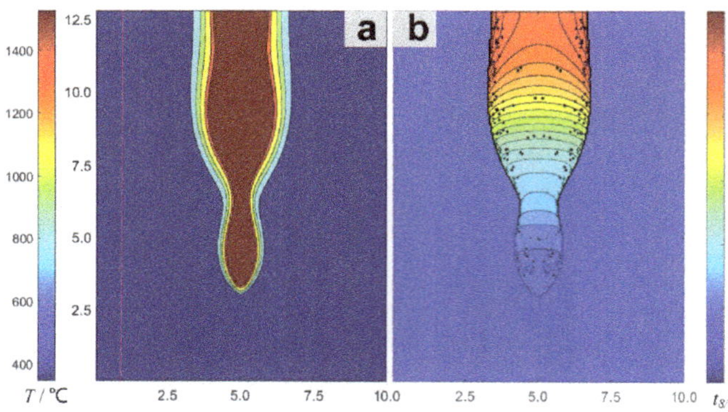

Figure 3. Maximum temperature field (**a**) and $t_{8/5}$ field (**b**) of the weld section.

The maximum temperature ranges are observed in each microstructure zone, and the variation in the welding thermal cycle for each zone is relatively small. Therefore, the average value curve is taken as the characteristic thermal cycle curve of the zone. As shown in Figure 4, the characteristic thermal cycle curve is surrounded by a heated region with room temperature T_0 and peak temperature T_{max}. The heating time is T_{max}–T_0. The high-temperature zone is surrounded by the T_{800} curve, and the high-temperature residence time is $T_{800}{}^{max}$–$T_{800}{}^{min}$. The cooling zone is surrounded by t_{800} and T_{500}, and the cooling time is t_{500}-t_{800}.

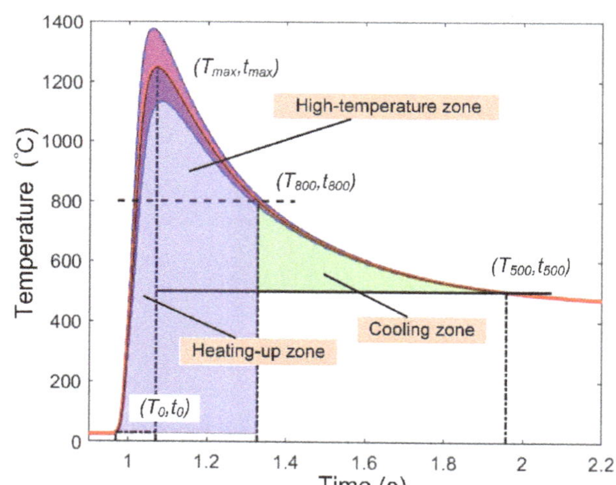

Figure 4. Thermal cycle curve and mean value curve of the laser-welding temperature field in the coarse-grained heat-affected zone.

The original microstructure of the X100 pipeline steel underwent austenitization and continuous cooling of the undercooled austenite during the welding thermal cycling. The degree of austenitization is determined by the heating rate, the peak temperature, and the residence time at a high temperature. The parameterization can be defined as what corresponds to the region enclosed by the characteristic thermal cycling curve T_0 and t_{800} in Figure 5.

$$D_A = \int_{t_0}^{t_{800}} T(x, y, z, t) - T_0 dt \tag{6}$$

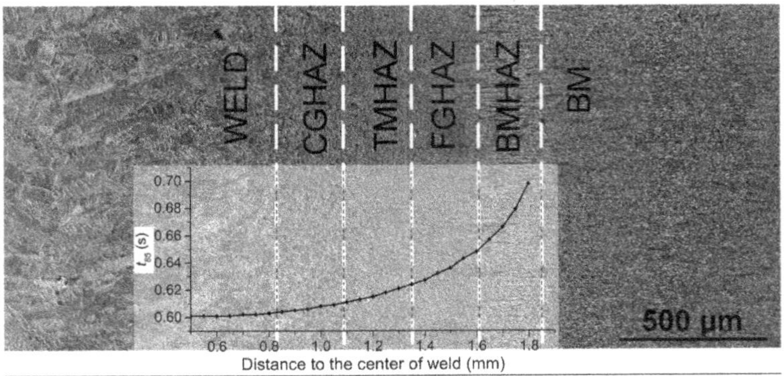

Figure 5. Microstructural zones and the trend of $t_{8/5}$ in the X100 laser-welded joint.

The function $T(x, y, z, t)$ of temperature (T) with time (t) at point (x, y, z) in a welded joint is called the welding thermal cycle curve. Since $t_{8/5}$ determines the cooling transformation process of the undercooled austenite, defined as $t_{8/5} = t_{500} - t_{800}$, the average cooling rate of the corresponding position is $v_c = (800 - 500)/t_{8/5}$.

The driving force for austenitization in the X100 pipeline steel is the difference between the volume-free energies of the original microstructure and the austenite. The austenitization process can be divided into four stages: (1) nucleation, (2) nucleus growth, (3) dissolution of residual carbides, and (4) homogenization of austenite, which is affected by the heating rate, the peak temperature, and the residence time at high temperature. Generally, austenite nucleates inhomogeneously on the subcrystal boundaries of bainite and ferrite grains or on carbonitrides of the alloying elements, which forms two new interfaces after nucleation. It grows through the transformation of ferrite to austenite and the dissolution of the carbonitrides [24–28,43]. The rate of formation of austenite depends on the nucleation rate and the growth rate. The rate of formation of the austenite and the inhomogeneity of the austenite composition increases with a growing heating rate. The initial austenite grains are refined with the heating rate [22,25,30,44,45].

Alloying elements affect the formation rate of austenite, the dissolution of carbides, and the homogenization of austenite. The diffusion coefficient of carbon is affected by alloying elements. In addition, the diffusion coefficient of carbon in austenite is reduced, and the formation rate of austenite is slowed by strong carbides, such as those of Cr, Mo, W, V, Ti, and Nb [43–47].

The cooling rate determines the continuous cooling transformation of undercooled austenite, which determines the final structure and mechanical properties of a laser-welded joint in X100 pipeline steel. With increasing cooling rates, the diffusion of carbon and alloying elements decreases gradually. The microstructure transformation types are diffusive phase transition, semi-diffusive phase transition, block transformation, and shear transformation. The products of the transformations are ferrite, GB, upper bainite, M-A components, Weiss structures, BF, and lath martensite.

4.2. Microstructure and Quantification of Welded Joints

The laser-welded joint of X100 pipeline steel exhibits a clear microstructure gradient from the base metal to the weld, which can be divided into the following sections, as shown in Figure 5. The sections include the base metal (BM), a banded-microstructure heat-affected zone (BMHAZ), a fine-grained heat-affected zone (FGHAZ), a transitional-microstructure heat-affected zone (TMHAZ), a coarse-grained heat-affected zone (CGHAZ), and a columnar zone.

The microstructure of the X100 pipeline steel in the laser-welded joint mainly includes the AF, GB, BF, and M-A constituents. The relative proportions of these four microstructural components have important effects on the properties of the joint. The M-A constituent from carbon-enriched and untransformed austenite surrounded by bainite is known to deteriorate the toughness of bainitic steel [24,25,27,30]. In addition, when the AF and GB contents in X100 pipeline steel are too low, the original austenite grains are not separated, and the growth of the lath beams is not controlled, which results in large effective grain sizes for the lath bundles and diminished elongation and impact properties. When the AF and GB contents are too high, the strength of the steel plate is greatly reduced. AF and GB easily nucleate and grow along grain boundaries, which segments the original austenite grains so that clear substructures appear in the deformed grains near the grain boundaries in addition to the original large-angle boundaries. The original austenite grains are divided into numerous regions with different orientations and uneven sizes by the AF and GB. The orientations in the region are approximately the same, and the orientation differences are small. The growth of the lath beams is limited in the segmented region, and the AF and GB microstructures refine the grains through the phase-transformation processes. The GB and LB contents decrease with increasing cooling rate, whereas the AF content shows the opposite effect [24,27,28,47,48]. Furthermore, the decrease in the amount of brittle and hard phases in steel leads to a decrease in the strength.

The composition and microstructure segregations in X100 under thermomechanically controlled processing are the fundamental reasons for the formation of banded microstructures in laser-welded joints. According to a comparative analysis of the microstructures observed in the SEM micrographs and those observed in the OM images, the banded microstructure is composed of bainite bands and M-A bands. The GB in the bainite bands is smaller than in the base material, and the solid-solution alloying elements are partially precipitated, which indicates that fewer grain defects are present, the overall energy is lower, and the corresponding BM area is more stable. The M-A bands form in the microstructure of the BM, which is characterized by a small amount of alloying elements, the strong aggregation of the alloying elements, and a relatively high energy stability. The bainite band appears as a dark band under an optical microscope, and the granular M-A constituent is clearly observed via high-powered SEM. With a decreasing distance from the center of the weld, the bainite content decreases gradually (Figure 6b).

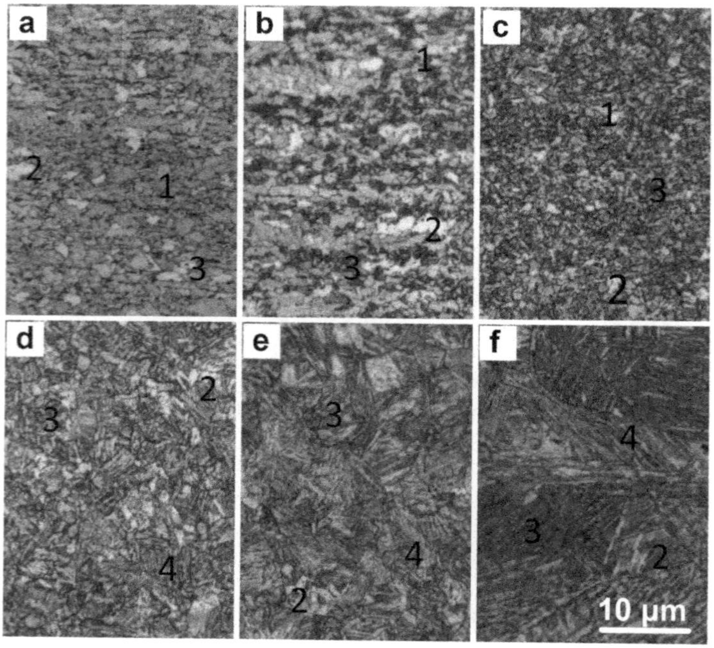

Figure 6. Optical images of the microstructures in the X100 laser-welded joint (1-AF, 2-GB, 3-MA, 4-BF): (**a**) BM, (**b**) BMHAZ, (**c**) FGHAZ, (**d**) TMHAZ, (**e**) CGHAZ, and (**f**) WELD.

OM revealed a large gradient from fine grains to coarse grains in the HAZ. With a decreasing distance from the center of the weld, the grain size of the original austenite gradually increases and the white bainite microstructure and the grain size gradually increase. Combined with the SEM analysis results, these results reveal that the grain boundaries of primary austenite are clear and that a small amount of M-A constituent exists at the grain boundary, which consists mainly of a small amount of GB dispersed within BF (Figure 6c–e). However, the coarse-grained area near the fusion line is relatively coarse. The sizes of the original austenite grains and GB increase. The block bainite appears and the BF lath length increases. In addition, a small amount of lath martensite is present.

The weld of the laser-welded joint of X100 pipeline steel mainly consists of coarse columnar crystals. In addition, large bulk bainitic microstructures are present, the BF lath beams are longer, and a large-angle relationship exists between the laths (Figure 6f).

According to the EDS results, the carbon contents of the carbon-rich bainite and lath martensite in the BF and the lath martensite are approximately 1.18% and 0.72%, respectively. As shown in Figure 7d, the solid-solution alloying elements in the BF and martensite are Cr, Mn, and Si.

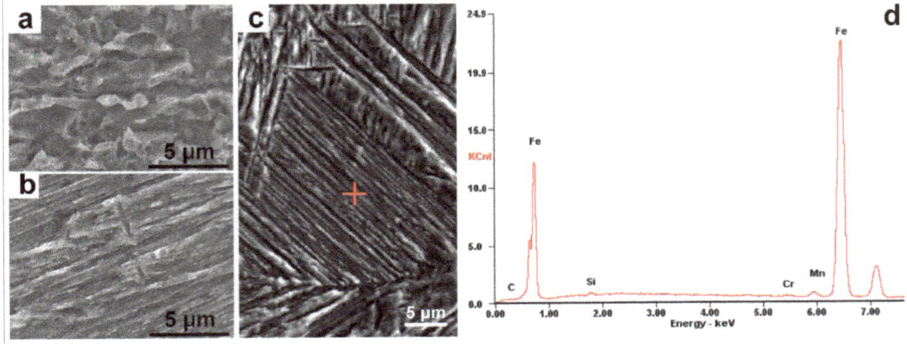

Figure 7. (**a**–**c**) X100 characteristic microstructures (SEM) of the laser-welded joint and (**d**) the energy spectrum of bainitic ferrite (BF).

According to the comparison between the microstructure analyses and the energy-dispersive spectrometer analyses of the X100 pipeline steel, the microstructure in the HAZ is mainly granular bainite and acicular ferrite. As shown in Figure 6b–e, the dark white matrix is acicular ferrite, and the bright white granular microstructure is granular bainite. The microstructure of the weld is relatively complex. For example, the bright white granular microstructure in Figure 7a is the M-A constituent, and the bainite matrix in Figure 7b is bainitic ferrite.

4.3. *Quantitative Relationship between Thermal Cycling and the Microstructure*

The microstructures of welded joints are determined by the microstructure transformations from the welding thermal cycle. The degree of austenitization determines the starting state for the continuous cooling transformation of undercooled austenite, and $t_{8/5}$ determines the type of continuous cooling transformation for the undercooled austenite, which leads to differences in the products of the microstructure transformations.

As shown in Figure 8, when the distance from the center of the weld increases, the degree of austenitization (austenitization index D_A: 433.4–92.8) decreases linearly, and $t_{8/5}$ (1.3–0.6 s) increases exponentially. The corresponding cooling rate decreases sharply. From BM to WELD, the microstructure changes from AF, GB, and MA to GB, MA, and BF, and the transformation occurs in the TMHAZ. From BM to FGHAZ, the AF:GB:MA ratio changes from 9.16:69.03:6.06 in BM to 2.04:75.4:18.58 in BMHAZ and to 2.22:60.35:17.5 in FGHAZ. From TMHAZ to WELD, GB:MA:BF changes from 4.8:11.84:83.36 in TMHAZ to 11.74:7.9:80.36 in CGHAZ and to 2.17:6.7:82.47 in WELD.

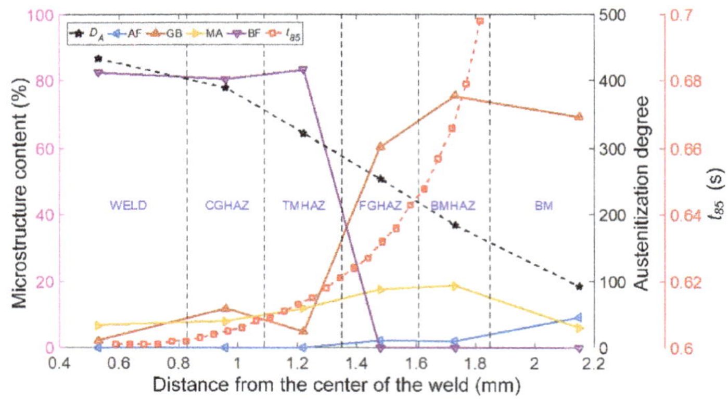

Figure 8. Quantitative relationship between the degree of austenitization and $t_{8/5}$ and the microstructure distribution in the laser-welded joint of X100 pipeline steel.

From the symmetry of the temperature field contours and the $t_{8/5}$ contours, the austenitization degree index and the $t_{8/5}$ distribution of the X100 laser-welded joint can be deduced to be equivalent. Thus, the microstructure of the welded joint from BM to WELD is also consistent with the characteristic region studied.

5. Conclusions

(1) Accurate heat-source model parameters are the core of the accurate simulation of temperature fields resulting from the laser-welding process. The Gaussian-distributed rotating body heat-source model presented in this paper accurately simulates the welding temperature field, and the calculation process is stable and reliable.

(2) The final microstructure, which determines the strength and toughness of a laser-welded joint, is determined by the welding thermal cycle process, which affects the austenitization and continuous cooling transformation of the original microstructure of the BM. The degree of austenitization is more significant than the cooling rate.

(3) The degree of austenitization in the microstructures at different positions in welded joints can be expressed by the parameter $D_A = \int_{t_0}^{t_{800}} T(x, y, z, t) - T_0 dt$, which accurately reflects the effects of the heating rate, the peak temperature, and the residence time at high temperature on the original microstructure of the BM in the heated region and in the high-temperature region.

(4) With a decreasing distance from the center of the weld and the upper surface of the weld, the heating rate, the peak temperature, and the residence time at high temperature of the welding heat cycle increase gradually. The austenitization degree and the cooling rate increase significantly, which results in increasing grain sizes for the original austenite. However, the final microstructure has less austenite overall.

(5) The microstructure of the X100 pipeline steel in the HAZ is mainly composed of AF, GB, and BF, with small amounts of lath martensite, upper bainite, and bulk microstructure found in the columnar zone of the weld.

(6) The microstructure gradient of the HAZ of the welded joint of X100 high-grade pipeline steel is mainly affected by the different degrees of austenitization in the BM and in locations where the change in the cooling rate is relatively small. The contribution of the cooling rate to the microstructure transition from FGHAZ to CGHAZ is limited.

(7) Welding the same material under the same austenitization degree and cooling rate yields the same final microstructure and phase proportions, and the strength and toughness properties of the welded joints are the same.

Author Contributions: Methodology, G.W., L.Y., J.W., and Z.Y. Investigation, J.W., H.H., and Z.Y. Data curation, G.W., J.W., and H.H. All the authors contributed to writing the original draft of the paper. Writing—review & editing L.Y. and J.W. Supervision, G.W. and L.Y. Project administration, L.Y. Funding acquisition, L.Y. All authors have read and agreed to the published version of the manuscript.

Funding: The National Natural Science Foundation of China (No. 51674056), the Frontier and Applied Basic Research Projects of Chongqing (No. cstc2018jcyjAX0108), the Science and Technology Planning Project of Guangdong Province (2018A050506060), and the Science and Technology Program of Guangzhou (201704030068, 201907010006), and the Program for Creative Research Groups in University of Chongqing (Grant No.CXQT19031) supported this research.

Conflicts of Interest: The authors declare no conflict of interest.

References

1. Witek, M. Possibilities of using X80, X100, X120 high-strength steels for onshore gas transmission pipelines. *J. Nat. Gas Sci. Eng.* **2015**, *27*, 374–384. [CrossRef]
2. Pourahkakbar, H.; Jandaghi, M.R. Predictions of toughness and hardness by using chemical composition and tensile properties in microalloyed line pipe steels. *Neural Comput. Appl.* **2014**, *25*, 1993–1999.
3. Shinohara, Y.; Madi, Y.; Besson, J. Anisotropic ductile failure of a high-strength line pipe steel. *Int. J. Fract.* **2016**, *197*, 127–145. [CrossRef]
4. Wang, G.; Yin, L.; Yao, Z.; Wang, J.; Jiang, S.; Zhang, Z.; Zuo, C. The Evolution and Distribution of Microstructures in High-Energy Laser-Welded X100 Pipeline Steel. *Materials* **2019**, *12*, 1762. [CrossRef]
5. Yin, L.; Wang, J.; Chen, X.; Liu, C.; Siddiquee, A.N.; Wang, G.; Yao, Z. Microstructures and their distribution within HAZ of X80 pipeline steel welded using hybrid laser-MIG welding. *Weld. World* **2018**, *62*, 721–727. [CrossRef]
6. Gu, X.; Liu, Y.; Li, W.; Han, Y.; Zheng, K. Physical Characteristics of Coupled Plasma and Its Influence on Weld Formation in Hybrid Laser-Double-Arc Welding. *Materials* **2019**, *12*, 4207. [CrossRef]
7. Pouriamanesh, R.; Dehghani, K.; Vallant, R. Friction stir welding of API X70 steel incorporating Ti-dioxide. *Can. Metall. Q.* **2019**, *58*, 69–81. [CrossRef]
8. Guo, N.; Cheng, Q.; Zhang, X.; Fu, Y.; Huang, L. Microstructure and Mechanical Properties of Underwater Laser Welding of Titanium Alloy. *Materials* **2019**, *12*, 2703. [CrossRef]
9. Xu, G.; Li, P.; Li, L.; Hu, Q.; Zhu, J.; Gu, X.; Du, B. Influence of Arc Power on Keyhole-Induced Porosity in Laser + GMAW Hybrid Welding of Aluminum Alloy: Numerical and Experimental Studies. *Materials* **2019**, *12*, 1328. [CrossRef]
10. Xin, Z.; Yang, Z.; Zhao, H.; Chen, Y. Comparative Study on Welding Characteristics of Laser-CMT and Plasma-CMT Hybrid Welded AA6082-T6 Aluminum Alloy Butt Joints. *Materials* **2019**, *12*, 3300. [CrossRef] [PubMed]
11. Zhou, X.; Duan, J.A.; Zhang, F.; Zhong, S. The Study on Mechanical Strength of Titanium-Aluminum Dissimilar Butt Joints by Laser Welding-Brazing Process. *Materials* **2019**, *12*, 712. [CrossRef] [PubMed]
12. Chen, X.; Lei, Z.; Chen, Y.; Jiang, M.; Tian, Z.; Bi, J.; Lin, S. Microstructure and Mechanical Properties of Laser Welded Al-Si Coated Hot-Press-Forming Steel Joints. *Materials* **2019**, *12*, 3294. [CrossRef] [PubMed]
13. Casalino, G.; D'Ostuni, S.; Guglielmi, P.; Leo, P.; Palumbo, G.; Piccininni, A. Off-set and focus effects on grade 5 titanium to 6061 aluminum alloy fiber laser weld. *Materials* **2018**, *11*, 448. [CrossRef] [PubMed]
14. Silva Leite, C.G.; da Cruz Junior, E.J.; Lago, M.; Zambon, A.; Calliari, I.; Ventrella, V.A. Nd: YAG Pulsed Laser Dissimilar Welding of UNS S32750 Duplex with 316L Austenitic Stainless Steel. *Materials* **2019**, *12*, 2906. [CrossRef] [PubMed]
15. Hipp, D.; Mahrle, A.; Beyer, E.; Jäckel, S.; Hertel, M.; Füssel, U. Thermal Efficiency Analysis for Laser-Assisted Plasma Arc Welding of AISI 304 Stainless Steel. *Materials* **2019**, *12*, 1460. [CrossRef] [PubMed]
16. Yu, Q.B. Effect of cooling rate on microstructures and mechanical properties of X80 pipeline steel. *Adv. Mater. Res.* **2012**, *12*, 535–537. [CrossRef]
17. Gook, S.; Gumenyuk, A.; Rethmeier, M. Hybrid laser arc welding of X80 and X120 steel grade. *Sci. Technol. Weld. Join.* **2014**, *19*, 15–24. [CrossRef]
18. Grünenwald, S.; Seefeld, T.; Vollertsen, F.; Kocak, M. Solutions for joining pipe steels using laser-GMA-hybrid welding processes. *Phys. Proced.* **2010**, *5*, 77–87. [CrossRef]

19. Lei, Z.; Tan, C.; Chen, Y.; Sun, Z. Microstructure and mechanical properties of fiber laser-metal active gas hybrid weld of X80 pipeline steel. *J. Press. Vessel Technol. Trans. ASME* **2013**, *135*, 011403.
20. Inose, K.; Kanbayashi, J.; Abe, D.; Matsumoto, N.; Nakanishi, Y. Design and welding method for high-strength steel structure using laser-arc hybrid welding. *Weld. World* **2013**, *57*, 657–664. [CrossRef]
21. Costa, A.; Quintino, L.; Yapp, D.; Iordachescu, D. Characterization of fiber laser welds in X100 pipeline steel. *Mater. Des.* **2009**, *30*, 2701–2707.
22. Nafisi, S.; Arafin, M.A.; Collins, L.; Szpunar, J. Texture and mechanical properties of API X100 steel manufactured under various thermomechanical cycles. *Mater. Sci. Eng. A* **2012**, *531*, 2–11. [CrossRef]
23. Shanmugam, N.S.; Buvanashekaran, G.; Sankaranarayanasamy, K.; Kumar, S.R. A transient finite element simulation of the temperature field and bead profile of T-joint laser welds. *Int. J. Model. Simul.* **2010**, *30*, 108–122. [CrossRef]
24. Lan, L.; Chang, Z.; Kong, X.; Qiu, C.; Zhao, D. Phase transformation, microstructure, and mechanical properties of X100 pipeline steels based on TMCP and HTP concepts. *J. Mater. Sci.* **2017**, *52*, 1661–1678. [CrossRef]
25. Wu, H.D.; Miyamoto, G.; Yang, Z.G.; Zhang, C.; Chen, H.; Furuhara, T. Incomplete bainite transformation in Fe-Si-C alloys. *Acta Mater.* **2017**, *133*, 1–9. [CrossRef]
26. You, Y.; Shang, C.; Nie, W.; Subramanian, S. Investigation on the microstructure and toughness of coarse grained heat affected zone in X-100 multi-phase pipeline steel with high Nb content. *Mater. Sci. Eng. A* **2012**, *558*, 692–701. [CrossRef]
27. Lan, L.; Qiu, C.; Zhao, D.; Gao, X.; Du, L. Analysis of martensite–austenite constituent and its effect on toughness in submerged arc welded joint of low carbon bainitic steel. *J. Mater. Sci.* **2012**, *47*, 4732–4742. [CrossRef]
28. Furuhara, T.; Chiba, T.; Kaneshita, T.; Wu, H.; Miyamoto, G. Crystallography and interphase boundary of martensite and bainite in steels. *Metall. Mater. Trans. A* **2017**, *48*, 2739–2752. [CrossRef]
29. Moore, P.L.; Howse, D.S.; Wallach, E.R. Microstructures and properties of laser/arc hybrid welds and autogenous laser welds in pipeline steels. *Sci. Technol. Weld. Join.* **2004**, *9*, 314–322. [CrossRef]
30. Silva, R.D.A.; Souza, L.F.G.D.; Morales, E.V.; Rios, P.R.; Bott, I.D.S.; Silva, R.D.A. Formation of microphases and constituents from remaining austenite decomposition in API X80 steel under different processing conditions. *Mat. Res.* **2015**, *18*, 908–917. [CrossRef]
31. Cho, W.I.; Na, S.J.; Thomy, C.; Vollertsen, F. Numerical simulation of molten pool dynamics in high power disk laser welding. *J. Mater. Process. Technol.* **2012**, *212*, 262–275. [CrossRef]
32. Di, X.J.; Cai, L.; Xing, X.X.; Chen, C.X.; Xue, Z.K. Microstructure and mechanical properties of intercritical heat-affected zone of X80 pipeline steel in simulated in-service welding. *Acta Metall. Sin. (Engl. Lett.)* **2015**, *28*, 883–891. [CrossRef]
33. Wang, R.; Lei, Y.; Shi, Y. Numerical simulation of transient temperature field during laser keyhole welding of 304 stainless steel sheet. *Opt. Laser Technol.* **2011**, *43*, 870–873. [CrossRef]
34. Abderrazak, K.; Bannour, S.; Mhiri, H.; Lepalec, G.; Autric, M. Numerical and experimental study of molten pool formation during continuous laser welding of AZ91 magnesium alloy. *Comput. Mater. Sci.* **2009**, *44*, 858–866. [CrossRef]
35. Kubiak, M.; Piekarska, W.; Stano, S. Modelling of laser beam heat source based on experimental research of Yb:YAG laser power distribution. *Int. J. Heat Mass Transf.* **2015**, *83*, 679–689. [CrossRef]
36. Kazemi, K.; Goldak, J.A. Numerical simulation of laser full penetration welding. *Comput. Mater. Sci.* **2009**, *44*, 841–849. [CrossRef]
37. Zain-Ul-Abdein, M.; Nelias, D.; Jullien, J.F.; Deloison, D. Prediction of laser beam welding-induced distortions and residual stresses by numerical simulation for aeronautic application. *J. Mater. Process. Technol.* **2009**, *209*, 2907–2917. [CrossRef]
38. Zain-Ul-Abdein, M.; Nélias, D.; Jullien, J.F.; Boitout, F.; Dischert, L.; Noe, X. Finite element analysis of metallurgical phase transformations in AA 6056-T4 and their effects upon the residual stress and distortion states of a laser welded T-joint. *Int. J. Press. Vessel. Pip.* **2011**, *88*, 45–56. [CrossRef]
39. Luo, Y.; You, G.; Ye, H.; Liu, J. Simulation on welding thermal effect of AZ61 magnesium alloy based on three-dimensional modeling of vacuum electron beam welding heat source. *Vacuum* **2010**, *84*, 890–895. [CrossRef]

40. Shanmugam, N.S.; Buvanashekaran, G.; Sankaranarayanasamy, K. Some studies on weld bead geometries for laser spot welding process using finite element analysis. *Mater. Des.* **2012**, *34*, 412–426. [CrossRef]
41. Zain-Ul-Abdein, M.; Nélias, D.; Jullien, J.F.; Deloison, D. Experimental investigation and finite element simulation of laser beam welding induced residual stresses and distortions in thin sheets of AA 6056-T4. *Mater. Sci. Eng. A* **2010**, *527*, 3025–3039. [CrossRef]
42. Tsirkas, S.A.; Papanikos, P.; Kermanidis, T. Numerical simulation of the laser welding process in butt-joint specimens. *J. Mater. Process. Technol.* **2003**, *134*, 59–69. [CrossRef]
43. Xia, Y.; Miyamoto, G.; Yang, Z.G.; Zhang, C.; Furuhara, T. Direct measurement of carbon enrichment in the incomplete bainite transformation in Mo added low carbon steels. *Acta Mater.* **2015**, *91*, 10–18. [CrossRef]
44. Chen, X.W.; Qiao, G.Y.; Han, X.L.; Wang, X.; Xiao, F.R.; Liao, B. Effects of Mo, Cr and Nb on microstructure and mechanical properties of heat affected zone for Nb-bearing X80 pipeline steels. *Mater. Des.* **2014**, *53*, 888–901. [CrossRef]
45. Ivanov, A.Y.; Sulyagin, R.V.; Orlov, V.V.; Kruglova, A.A. Formation of structure in the heat-affected zone and properties of welded joints of pipe steels of strength classes X80 and X90. *Met. Sci. Heat Treat.* **2012**, *53*, 560–566. [CrossRef]
46. Isasti, N.; Jorge-Badiola, D.; Taheri, M.L.; Uranga, P. Phase transformation study in Nb-Mo microalloyed steels using dilatometry and EBSD quantification. *Metall. Mater. Trans. A Phys. Metall. Mater. Sci.* **2013**, *44*, 3552–3563. [CrossRef]
47. Gianetto, J.A.; Fazeli, F.; Chen, Y.; Shalchi-Amirkhiz, B.; Smith, T. Microstructure and toughness of simulated grain coarsened heat affected zones in X80 pipe steels. In Proceedings of the 10th International Pipeline Conference American Society of Mechanical Engineers, New York, NY, USA, 29 September–3 October 2014.
48. Yang, J.H.; Liu, Q.Y.; Sun, D.B.; Li, X.Y. Microstructure and transformation characteristics of acicular ferrite in high niobium-bearing microalloyed steel. *J. Iron Steel Res. Int.* **2010**, *17*, 53–59. [CrossRef]

materials

Article

Steel Sheets Laser Lap Joint Welding—Process Analysis

Hubert Danielewski * and **Andrzej Skrzypczyk**

Faculty of Mechatronics and Mechanical Engineering, Kielce University of Technology, Al. 1000-lecia P.P. 7, 25-314 Kielce, Poland; tmaask@tu.kielce.pl

* Correspondence: hdanielewski@tu.kielce.pl

Received: 7 April 2020; Accepted: 11 May 2020; Published: 14 May 2020

Abstract: This article presents the results of steel-sheet lap-joint-welding using laser beam radiation. The use of a laser beam and keyhole effect for deep material penetration in lap joint welding was presented. Thermodynamic mechanism of laser welding is related to material properties and process parameters. Estimation of welding parameters and joint properties' analysis was performed through numerical simulation. The article presents a possibility of modeling laser lap-joint welding by using Simufact Welding software based on Marc solver and thermo-mechanical solution. Numerical calculation was performed for surface and conical volumetric heat sources simulating laser absorption and keyhole effect during steel sheet welding. Thermo-mechanical results of fusion zone (FZ), heat-affected zone (HAZ) and phase transformations calculated in numerical simulation were analyzed. The welding parameters for partial sealed joint penetration dedicated for gas piping installations were estimated from the numerical analysis. Low-carbon constructional steel was used for numerical and experimental analyses. A trial joint based on the estimated parameters was prepared by using a CO_2 laser. Numerical and experimental results in the form of hardness distributions and weld geometry were compared. Metallographic analysis of the obtained weld was presented, including crystallographic structures and inclusions in the cross section of the joint.

Keywords: laser welding; steel sheets; numerical simulation; lap joints; mechanical properties; microstructure analysis

1. Introduction

Welding methods are based on the thermal effect of melting and crystallization process. Conventional methods use electric arc as a heat source. Alternative to gas metal arc welding (GMAW) are beam welding methods [1,2], in which the concentrated energy of focused electrons or photons achieves high power density. The energy distribution factor allows high-speed welding, and the quantity of thermal energy absorbed in the materials is low. Electron beam welding (EBW) has high energy distribution, but it has to be performed in a vacuum, which is problematic in some welding applications. Laser beam welding (LBW) is an alternative technology in which a high power density of a focused photon beam shielded by inert gas can be used for numerous types of joining applications [3,4]. The keyhole effect in LBW enables deep penetration of the welded material. Moreover, laser beam penetration is possible through more than one material. Rapid development of laser technology has determined LBW for use in advanced joint configuration. Currently, researchers are focusing on welding dissimilar materials, where low-carbon and austenitic steels are welded, and some works are related to joining advanced aluminum, nickel and titanium alloys [5–9]. Nonconventional joining methods, such as laser welding in butt, lap and T-joint configurations, are also being widely studied [10–14]. Numerical analysis of the laser welding process has been undertaken, including lap joints, for a wide range of materials by many researchers [15–18]. However, laser lap welding of low-carbon steel has been reported in only a few works concentrated mostly on numerical analysis or joints properties [19–24]. The works cited above lack a more comprehensive study of lap joints, laser beams applied to the welding of commonly used constructional steel, or analysis based on mechanical and metallographic study, supported by numerical simulation. The selection of manually programmed laser welding parameters is problematic and requires performing a number of trial joints and an experienced operator. It thus seems reasonable to use some aided methods for supporting welding parameters estimation, such as analytical method, where thermal conduction calculation is based on solving the moving heat-sources equation proposed by Rosenthal [9]. Evolution of analytical computation relies on improving the mathematical description of heat sources. Solving the moving-heat-sources equation enables the use of the thermal distribution to estimate the shape of weld geometry. Analytical solution allows for the estimation of welding parameters in simple cases. In applications such as lap joints of steel sheets, it is more complex and requires using numerical methods [25–27].

Numerical simulation of welding process can be performed by using dedicated software such as SYSWELD or Simufact Welding or advanced multiphase heat-mass flow programs such as ANSYS with FLUENT module. Calculations are based on the Finite Elements Method (FEM) and solver engine. CAD geometry is discretized, and finite elements mesh is generated. During discretization of the area where significant heat effect can occur, FE refinement is performed. Welding simulation requires defining heat-source dimensions and heat-energy volume related to welding parameters [28–33]. If we consider thermo-mechanical simulation, in addition to the results of temperature distribution, a stress–strain analysis can be obtained. Considering lap welding, material properties in the upper plate of joint will differ from those in the lower plate. An analysis of this phenomenon is presented in this article. Properties of stress–strain distribution in the welded material are related to thermo-physical material properties, temperature distribution, heat expansion coefficient and phase transformation, changing in time. Defining these properties is complex; nevertheless, using numerical computation accurate estimation of the results is possible.

When performing a welding simulation, we must remember that, no matter how accurate, the results obtained are just estimations. The quality level depends on the defined boundary conditions and programmed welding parameters; therefore, experimental verification is required. The welding process requires a shielding atmosphere of inert gas. In order to confine ionization effect, helium is recommended as a reference gas. In the case of a sealed lap joint with partial penetration of the lower plate, no shielding atmosphere for the weld root is needed. However, the space between the plates contains some oxygen and may cause inclusions and welding defects; therefore, metallographic analysis was performed in this study. A programming numerical model verification stage needs to be considered, especially for defining some process properties, such as heat-source efficiency and energy distribution (TEM), related to the laser type, which must be included in the simulation [34–38]. The programmed heat-source dimension can be verified by comparing the simulated weld geometry with the trial joint.

In this paper, a numerical-simulation-aided analysis of laser welding of sealed steel sheets' lap joints is presented.

2. Methodology

2.1. Materials

Low-carbon constructional steel S235JR in the form of 2.5 mm thick steel sheets was used as a base material for both the simulation and trial joint welding. The commonly used low-carbon steel sheets were welded in lap joint configuration, using an advanced heat source in the form of a laser beam. The S235JR is unalloyed steel with carbon content up to 0.2% and a trace amount of other alloying elements (Table 1).

Table 1. Chemical composition of low-carbon constructional steel S235JR.

Element	Mn	Si	Cu	Cr and Ni	Nb	Mo	B
Percentage	1.65	0.5	0.4	0.3	0.06	0.08	0.0008

The structure of S235JR steel is typically ferritic–pearlitic. Low-carbon content and the trace amount of alloying elements reduce steel hardening, though some strengthening effect may occur via steel phase transformations. Thermo-physical properties of the material affect heat expansion and welding results (Table 2). The S235JR steel has high thermal conductivity, with the phase transformation temperature of 725 °C for AC_1 and 863 °C for AC_2 [39,40].

Table 2. Thermal properties of low-carbon constructional steel S235JR.

Property	Thermal Conductivity (W/cm·°C)	Specific Heat (J/kg·°C)	Latent Heat (J/g)	Solidus-Liquidus Range (°C)	Austenite Therm. Exp. Coeff. (1/°C)	Ferrite Therm. Exp. Coeff. (1/°C)
Value	0.45	480	256	1466.8–1517.1	2.54×10^{-5}	1.71×10^{-5}

2.2. Numerical Simulation

A numerical simulation of laser welding generally uses two types of heat-source models simulating laser interaction with the material: a surface heat source and a conical heat source. The surface-heat-source model (disc shape) is more accurate and is used for simulating laser energy absorption by steel sheets' surface (in some applications, for conduction welding solutions) [41]. The conical heat source is dedicated to simulating the keyhole effect and is related to energy being directed inside the material through the keyhole walls. Simufact Welding software (Simufact Engineering GmbH, Hamburg, Germany) uses a combination of surface and conical volumetric heat sources with uneven energy intensity distribution (Gaussian parameter), as shown below (Figure 1).

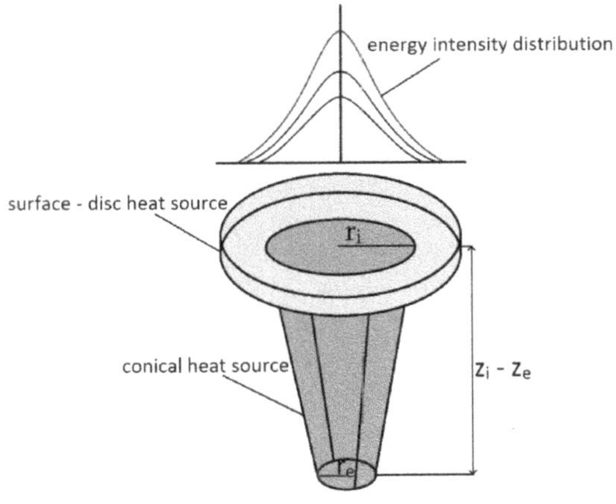

Figure 1. Heat-source model used in laser-welding simulation.

Conical volumetric heat source with the Gaussian distribution can be described by the following equation:

$$Q(x, y, z) = Q_0 \exp\left(-\frac{x^2 + y^2}{r_0^2(z)}\right) \quad (1)$$

Moreover, $r_0(z)$ is defined as follows:

$$r_0(z) = r_e + \frac{r_i - r_e}{z_i - z_e}(z - z_e) \tag{2}$$

where Q_0—maximum volumetric heat flux density; $r_i - r_e$—upper and lower conical radius dimension; $z_i - z_e$—conical heat source depth; and x, y, z—coordinates of heat source.

The numerical simulation was performed by using Simufact software with a Marc solver. The program is dedicated to welding applications; however, some physics phenomena, such as solidification, are simplified. This phenomenon is solved by using the assumption that the latent heat is uniformly released within the solidus and liquidus temperature range, where the solver uses the modified specific heat to model the latent heat effect based on material experimental data calculated by using JMatPro. Thermal conductivity is the dominant heat-transfer method; therefore, the governing equation is based on this phenomenon. Based on Fourier's law, three-dimensional heat conduction is given by the following governing equation:

$$\rho c(T)\frac{\partial T}{\partial t} = \frac{\partial}{\partial x}\left(k(T)\frac{\partial T}{\partial x}\right) + \frac{\partial}{\partial y}\left(k(T)\frac{\partial T}{\partial y}\right) + \frac{\partial}{\partial z}\left(k(T)\frac{\partial T}{\partial z}\right) + q_v \tag{3}$$

where $c(T)$—temperature dependent specific heat capacity; $k(T)$—temperature dependent thermal conductivity; q_v—volumetric internal energy; x, y, z—space coordinates; T—temperature; ρ—density; and t—time.

The simulation includes the convection effect and uses the Petro–Galerkin convection–diffusion model and nodal velocity vectors, as shown below.

$$\frac{\partial T}{\partial t} + v \cdot \nabla T = \nabla \cdot (\kappa \nabla T) + Q \tag{4}$$

where v—nodal velocity vector; T—temperature; κ—diffusion tensor; and Q—source term.

The obtained numerical model accounts for convection but does not account for surface tension or the Marangoni effect; therefore, some differences in heat transfer compared to the experimental process will occur, reflected in the fusion zone shape.

In the boundary conditions, a rigid restraint for welded elements, using fixed geometry, was programmed (Figure 2). Two 2.5 mm thick sheets were meshed by using finite elements hexahedral in shape. A preliminary research of mesh convergence was carried out, where welding simulations were performed with the same process parameters and different mesh sizes, starting with FE size equal to 1.25 mm, and then 0.5, 0.25 and 0.125 mm up to 0.0625 mm. For 0.0625 and 0.125 mm, no significant differences in the weld or HAZ geometry were observed; however, some differences between 0.125 and 0.25 mm were detected. In order to confirm mesh convergence, a study of temperature distribution was carried out. For all mesh sizes, measurement point, placed 2.5 mm from weld axis (approximately center of HAZ) were sets, and temperature changes shown in graph form were compared (Figure 2). The graphs in Figure 2 show similar results as the HAZ geometry analysis.

Therefore, in order to save simulation time, a general FE size was programmed as 0.25 mm. In order to obtain more accurate and realistic results, a refinement procedure with FE size equal to 0.125 mm was performed near the weld zone (at the temperature exceeding 400 °C). Sheets of S235JR low-carbon constructional steel were selected for the simulation (Table 2). The material multiphase library allowed the calculation of the overall thermo-mechanical joint properties.

The performed research assumed keyhole welding and deep material penetration; however, in the first stage before the keyhole effect appears, laser beam reflectivity from the metal surface is high, and when the heat-source efficiency is programmed, this phenomenon must be included. Therefore, for the simulation of CO_2 laser welding, the heat source efficiency coefficient was assumed as 0.77.

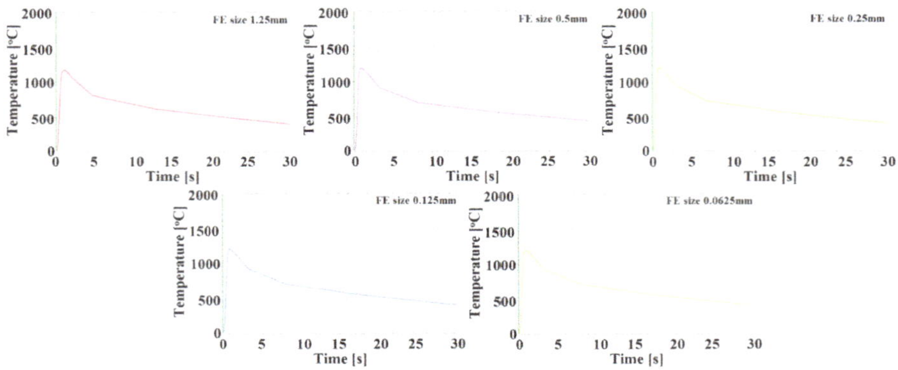

Figure 2. Graphs showing temperature changes for different FE mesh size.

The geometry of the heat source is related to the focal length and focusing power of welding optics, and for the performed research, a single spot spherical mirror with a focal point diameter of 0.3 mm and a focal length equal to 200 mm was used. The disc-shaped heat source with a radius equal to 0.65 mm and a depth of 0.2 mm, and the conical heat source with the upper radius of 0.5 mm and the lower radius of 0.2 mm and the depth of 4 mm were programmed. Geometry of the heat source (HS) is related to welding optics. Nevertheless, some calibration for more accurate results is required, and test welding at the speed of 1 m/min and output power equal to 1 kW was performed. Comparison with the simulation results showed some discrepancy, and the heat-source geometry was adjusted by reducing the HS radius by approximately 10% and Gaussian distribution parameter from 2.9 to 2.8 [42,43].

Numerical calculation of the laser lap-joint welding using a single pass process simulated by the heat source moving through steel sheets was carried out (Figure 3). Welding simulations with constant speed rate of 1 m/min and changing output power from 1 to 5 kW, changing with a step of 0.5 kW, were performed until the assumed sealed lap joint geometry was obtained. Phase transformation requires cooling time, so for welding equal to 1.2 s, the time for the complete process was programmed as 30 s. The simulation process was performed on a Dell PC class station with the i7 processor and 64 GB RAM, and the simulation calculation time was about 26 h.

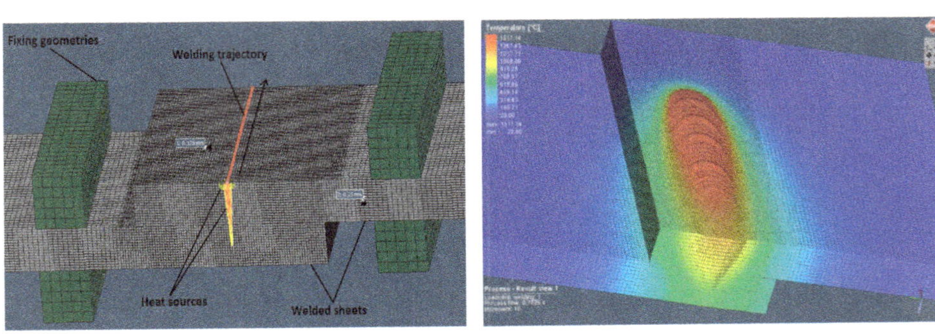

Figure 3. Laser-welding simulation model and the weld forming during the lap joint welding process.

2.3. Experimental Procedure

Verification of the numerical model was performed by welding the trial joint with parameters estimated at the simulation stage (laser power equal to 4 kW, with the welding speed of 1 m/min). For the configuration of the sealed steel sheet lap joint with partial penetration, welding conditions equal to those defined in the numerical simulation were established. In order to reduce the plasma ionization effect, helium as a shielding gas, with a flow rate equal to 20 L/min, was used. The welding process was performed with a CO_2 laser Trumpf TruFlow 6000 integrated with a 6 axis LaserCell 1005 work station (Figure 4).

Figure 4. Laser-welding station with lap joint low-carbon steel configuration.

The welding head with a focal length of 200 mm and coaxial shielding gas delivery system was used. The focal point was placed on the upper steel-sheet surface, and the welding of the trial joint was performed. The upper sheet was welded through, and the lower plate had the fusion zone approximately in the middle of its thickness. Therefore, the assumed lap joint partial penetration of two 2.5 mm thick S235JR steel sheets was obtained [44–46].

Weld strength characteristics were defined by the mechanical properties of the obtained joint. The weld properties were investigated by using destructive tests. The hardness test was carried out according to PN-EN ISO 6507-1 standard, using an Innovatest Nexus 4303 machine [47], and the hardness test point distribution for the steel sheets' lap joint is shown below (Figure 5).

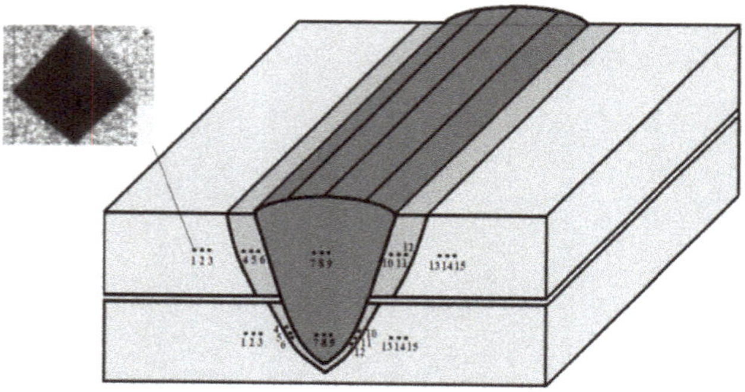

Figure 5. Hardness test point distributions in the lap joint.

The three-point test was performed according to the standards for all characteristic zones: base material (BM), heat-affected zone (HAZ) and fusion zone (FZ) in the upper and lower plates.

The material hardness resulting from the crystallographic structure of the welded material affects joint-strength characteristics. Moreover, the properties are related to the thermal cycles and chemical composition of the material. In order to investigate the joint-strength properties, the tensile-strength test was carried out, using an MTS-100 testing machine (MTS Systems Corporation, Eden Prairie, MN, USA) (Figure 6) [48].

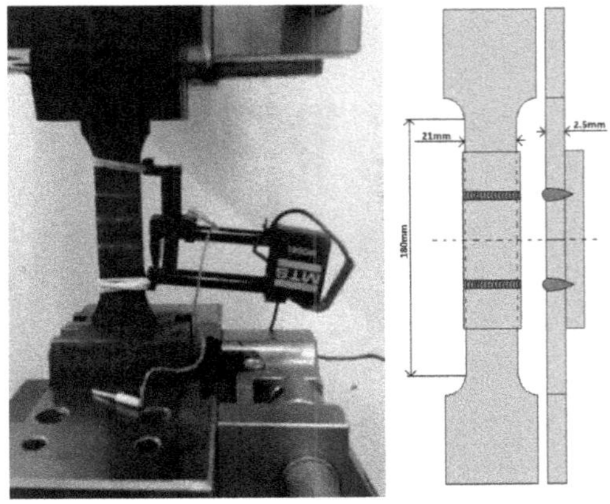

Figure 6. Tensile-strength test stand and specimen scheme.

The specimen was prepared by welding two upper plates to a lower plate, with the same process parameters as presented in the scheme (Figure 6). In this configuration, stretching, as well as shearing phenomena, will occur during the test. To confirm the obtained joint properties, an additional test of a specimen welded with the same parameters was performed. No-uniaxial complex-force distribution will certainly affect the test results. Nevertheless, the bonding force of welded sheets in the sealed lap joint will be related to the weld properties; therefore, metallographic analysis is required.

Metallographic tests were carried out according to PN-EN ISO 17639 [49]. A microscopic and macroscopic test, using a HiroxKH-8700 confocal digital microscope (Hirox Co Ltd., Tokyo, Japan), was performed, in order to investigate the crystallographic structure of the welded material.

The obtained welded lap joint was analyzed. The low-carbon S235JR steel is not typical hardening steel; nevertheless, phase transformation during laser welding affects the material, and the structure changes. A visual microscopic test was carried out to confirm the weld quality and to detect any defects. The crystallographic structure in the upper and lower plate after welding was investigated [50,51].

Laser welding of overlap joints is complex. Thin interspace between welded sheets can affect porosity and oxides' formation. To confirm the uniform structure of the weld, the qualitative and quantitative analysis was carried out, using energy-dispersive X-ray spectroscopy with a scanning electron microscope JSM-7100F (JEOL Ltd., Tokyo, Japan).

3. Results

3.1. Simulation Analysis Results

Laser-welding parameters were estimated by using numerical simulation analysis. Welding-simulation parameters with a speed rate of 1 m/min, output power 4 kW, heat source efficiency of 0.77 and the Gaussian parameter of conical and surface heat source equal to 2.8 were programmed (according to laser TEM01* mod). According to those parameters, the assumed partial penetration in the joint was obtained. Results of the simulation showed that the output power equal to 4 kW provided the partial welding penetration (Figure 7a). Further analysis showed that, by increasing the output power by 0.5 kW, the complete penetration was obtained (Figure 8). For the assumed geometry of heat sources and programmed boundary conditions, partial penetration of the sealed lap joint was achieved and considered in further investigations. The thermo-mechanical simulation, which took into account phase transformation, gave realistic results of the welding process, with a convex face of the weld and material deformation obtained by a solver mechanism. The recalculation of the contact tolerance or remeshing of the distorted region was performed (Figures 7a and 8). Macroscopic examination of the cross-section of the trial joint was performed by using a confocal digital microscope HiroxKH-8700 with a magnification of ×35 (Figure 7b).

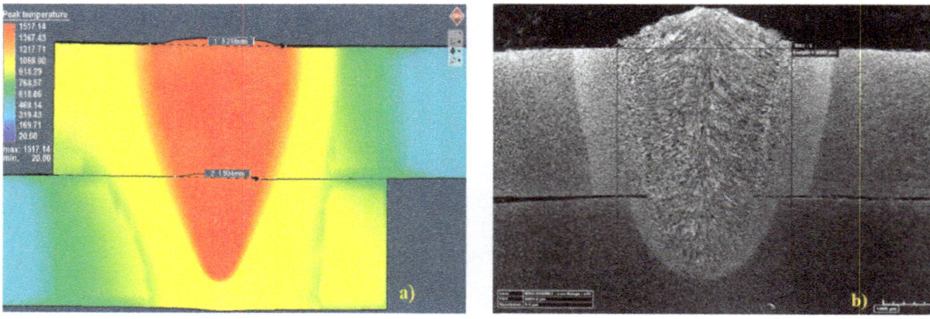

Figure 7. Laser lap-joint welding results from (**a**) simulation, (**b**) experiment.

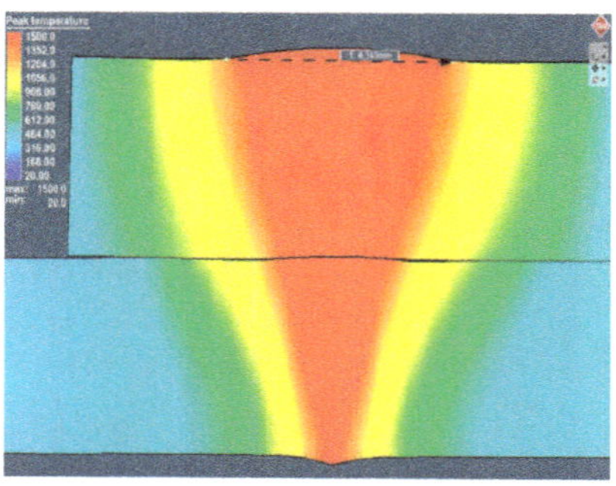

Figure 8. Laser lap-joint welding results for complete penetration.

Parameters estimated in the numerical simulation gave similar results: The face of the weld in the simulation was 3.22 mm in width, and in the experimental welding, it was 3.38 mm (for the complete penetration this value was equal to 4.35 mm). Moreover, the weld width in the overlap area for the simulation was equal to 1.99 mm, and it was 2.05 mm for the trial joint. The depth of the obtained welds was 4.33 mm for the trial joint and 4.38 for the simulation. Macroscopic analysis showed differences in heat expansion; the HAZ in the upper plate was wider. This effect is related to dumping and energy decrement during the penetration of the lower sheet. Although the welded surfaces adjoin each other, the spot size of the laser beam on the lower sheet during surface penetration was bigger, indicating lower power density. This phenomenon was also related to the heat-expansion direction. In the upper plate, heat expanded only in the XY direction, and in the lower sheet, it expanded in the XYZ direction. The macroscopic examination confirmed the accuracy of the simulation results, and it was possible to perform further numerical analyses. A thermo-mechanical simulation and a stress–strain analysis were carried out. The overall calculated displacement (Figure 9a) was 0.32 mm, with maximum principal stress (Figure 10a) of 1120 MPa. The changes in the total displacement (Figure 9b) and maximum principal stress (Figure 10b) were recorded against the defined measurement points. Higher displacement values occurred in the steel sheet plate edge (at points 1 and 3). The higher maximum principal stress concentration was measured on the opposite side of the weld (at points 2 and 4).

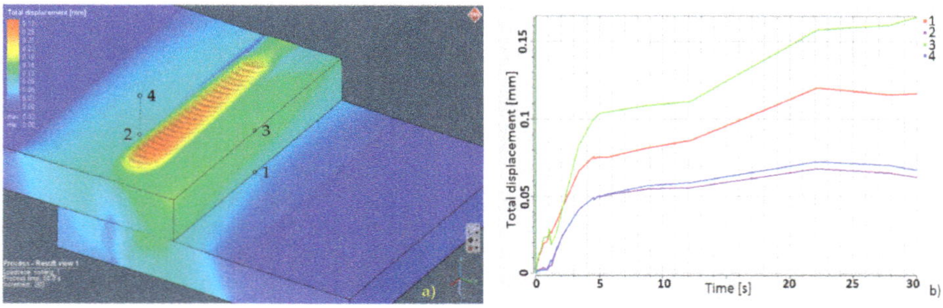

Figure 9. Total displacement: (**a**) distribution map and (**b**) displacement chart according to measurement points.

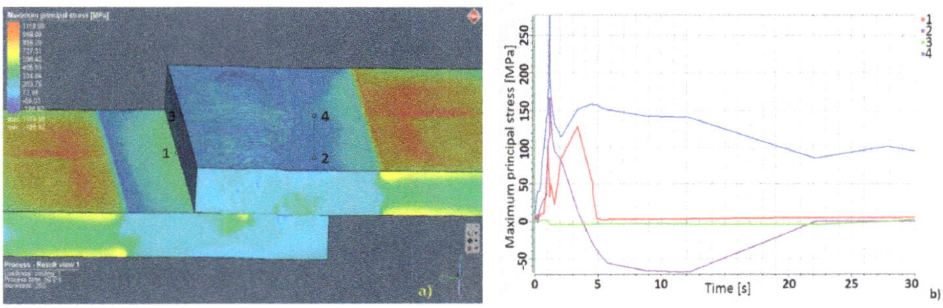

Figure 10. Maximum principal stress: (**a**) distribution map and (**b**) stress chart according to measurement points.

3.2. *Analysis of the Results for Properties*

Numerical simulations with phase transformation allowed for the calculation of material phase change and hardness distribution. As set forth in the standards, the three-point test was performed for all characteristic zones in the upper and down plates, in the cross-section, as defined in measurement point distribution (Figure 5). Results from the simulation are shown in Figure 11a, and the results obtained for the trial joint are shown in Figure 11b.

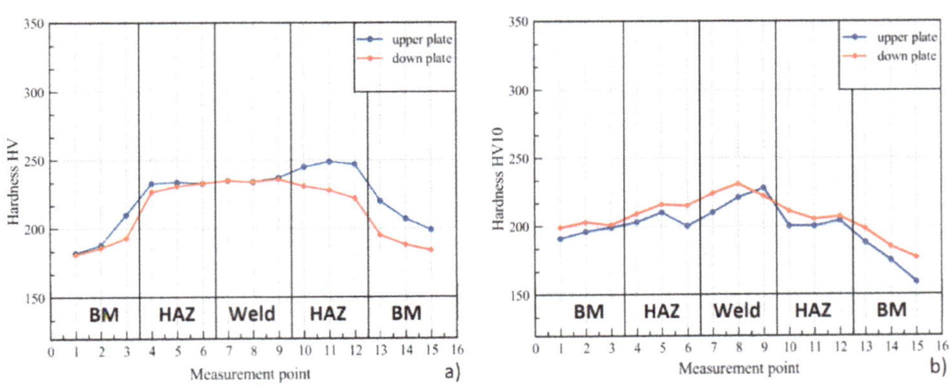

Figure 11. Results of hardness distribution in cross-section from (**a**) simulation and (**b**) experimental welding.

The hardness test results show strengthening in the weld and HAZ. The hardness distribution obtained from the simulation differs from the results measured on the trial joint. The calculated hardness values are greater than those of the welded materials. Hardness simulated for the weld zone does not exceed 240 HV (maximum value is equal to 237 HV). Discrepancies between the upper- and lower-plate values can be observed. In the HAZ for the upper plate, it takes the value from 233 to 249 HV, and for the lower plate, from 222 to 233 HV. The experimental hardness values (trial joint) are lower than the calculated ones. The highest measured value is 231 HV10 and occurs in the weld. In HAZ, the hardness value ranges from 200 to 218 HV10, and in the BM, it is from 160 to 200 HV10. Differences between measured and simulated values amount to 6 HV in the weld and 31 HV in HAZ. Despite the high accuracy of the obtained weld geometry, the results from the simulation and experiment vary. The discrepancies depend on phase-transformation phenomena (temperature gradient and chemical composition of welded material) and affect the calculated and measured hardness values [52,53]. According to PN-EN ISO 15614-11, the maximum allowable limit of Vickers hardness HV10 after welding is 350. Neither calculated nor measured hardness exceeded the allowed value. Therefore, no additional post-weld heat treatment was carried out in the simulation or on the trial joint.

Material-strength characteristics change during the welding process, depending on the phase transformation phenomena, and are different for the fusion zone, heat-affected zone and base material. The hardness test showed differences in the trial joint properties relative to the simulation results. Therefore, in order to prove the high quality of the joint obtained by using the estimated parameters, a static tensile test was performed. The properties of the obtained joint were confirmed by performing an additional verifying test of joint strength. Manufactured specimens were stretched by increasing the loading force until failure at the tensile test rate equal to 2 mm/min. The results of the tensile test were compiled as a force vs. displacement graph (Figure 12).

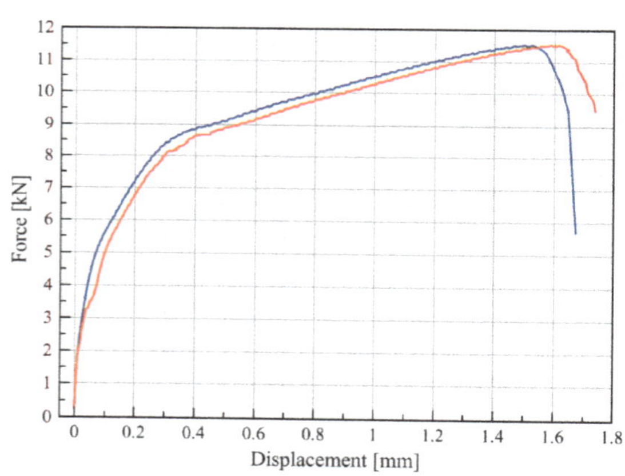

Figure 12. Tensile strength test results.

The static tensile test results show failure of the lap joint at the maximum force equal to 11.5 kN for the first specimen, and at 11.49 kN for the other specimen. The strength of the obtained joint was 110 and 108 MPa, respectively. The tensile strength of BM was 360 MPa. The failure of both samples occurred along the fusion-zone line. The joint configuration (Figure 6) affected the measurement results. Tension and shear occurred, and the results obtained were mostly dependent on the weld strength [54–57].

3.3. Metallographic Analysis

Optical and electron microscopes were used for crystallographic structure analysis. The structure of the base material (Figure 13) was examined by using a HiroxKH-8700 confocal digital microscope at a magnification of ×800. It showed a typical low-carbon structure. The microstructure of BM was characterized as fine-grained ferritic–pearlitic (dark areas: ferrite and bright: pearlite), with no explicitly shown banding, which is characteristic of S235JR steel.

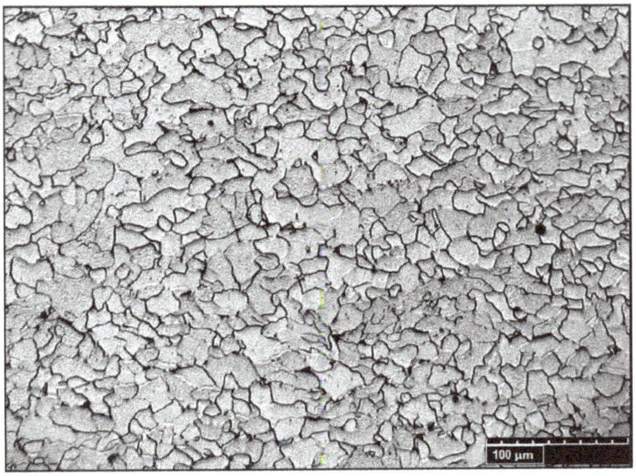

Figure 13. Microstructure of BM.

The low-carbon S235JR steel is not a typical hardening material, but phase transformation during laser welding at the speed rate of 1 m/min affected the crystallographic structure. The microscopic examination showed a good-quality weld with no defects (Figure 7b).

The HAZ structure analysis using an optical microscope with a magnification of ×140 was performed (Figure 14). In the weld interface, assuming a direction from the fusion (I) toward the BM (V), three characteristic areas were identified: II—overheated zone, III—normalization zone and IV—partial recrystallization zone.

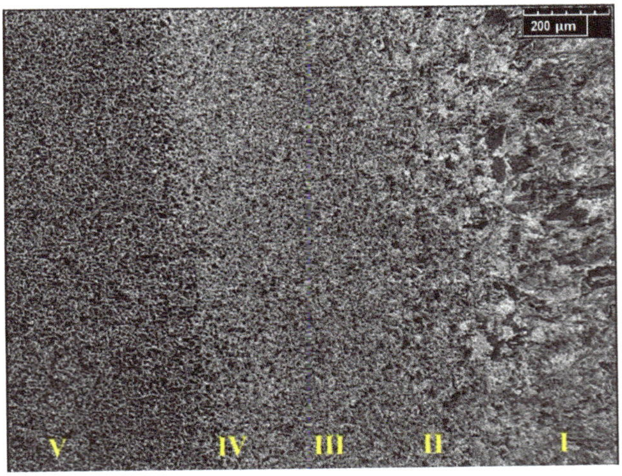

Figure 14. Microstructure of HAZ.

The weld-structure investigation was performed by using an optical microscope at a magnification of ×400. The weld exhibited a coarse-grained dendritic structure (Figure 15), which is typical for the laser-welding process.

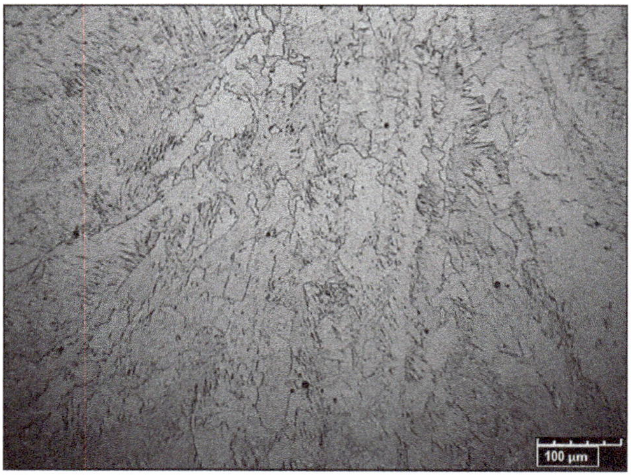

Figure 15. Microstructure of the weld.

The structure, precipitations and oxides in the weld across the interspace line between the welded sheets were determined by using energy-dispersive X-ray spectroscopy performed on a JSM-7100F scanning electron microscope. The chemical composition in the cross-section of HAZ to FZ (Figures 16 and 17) and weld (Figures 18 and 19) was analyzed [58].

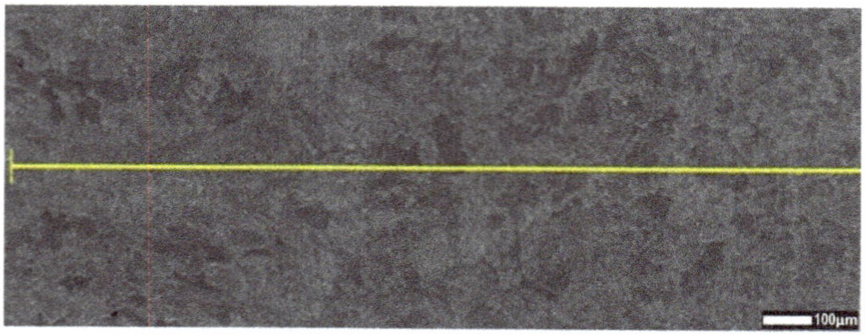

Figure 16. HAZ-to-FZ measurement line for chemical composition analysis.

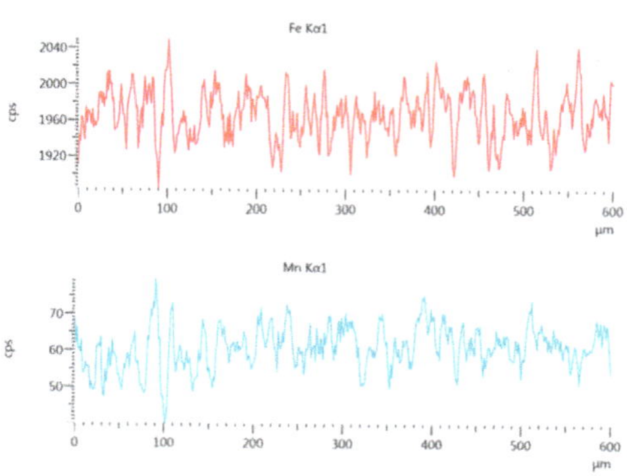

Figure 17. Ferrite and manganese content along the HAZ-to-FZ measurement line.

A chemical composition analysis of HAZ to FZ, based on ferrite and manganese distribution, was performed (Figure 16). A uniform mixture of ferrite and manganese was found across the measured line (Figure 17).

Using the ferrite and manganese distribution to identify fusion-zone uniformity, the weld in the overlap area was analyzed (weld transition) (Figure 18).

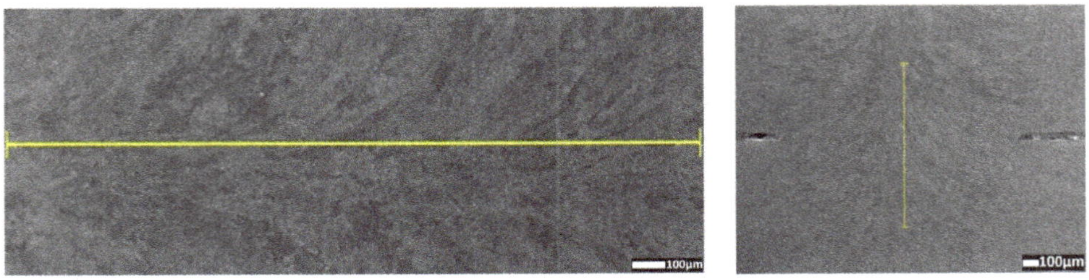

Figure 18. Measure line of fusion zone in overlap transition.

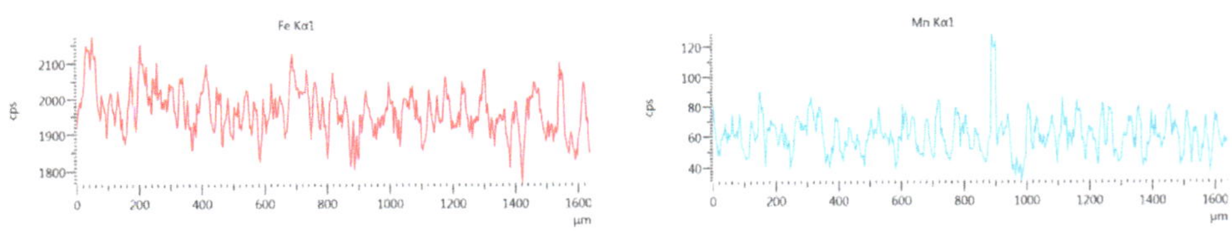

Figure 19. Ferrite and manganese amount along the overlap transition measurement line.

Uniform distribution of the measured elements showed a high mixing factor. Precipitation analysis of fusion zone was performed, and some inclusions were detected (Figures 20 and 21).

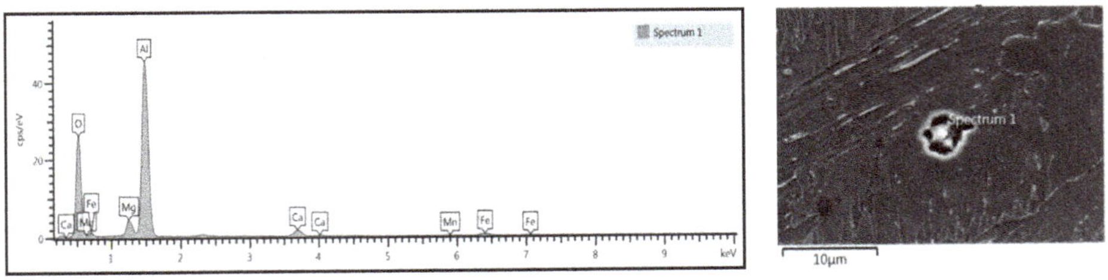

Figure 20. Spectroscopy analysis of identified aluminum oxide in the weld.

Analysis of weld transition zone revealed the presence of aluminum oxide (Figure 20).

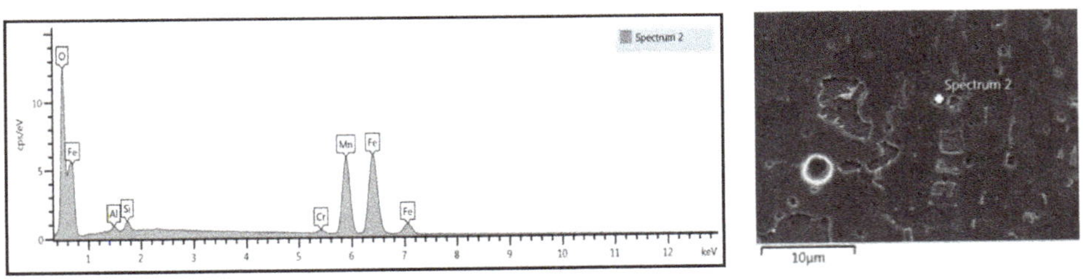

Figure 21. Spectroscopy analysis of identified oxide in the weld.

Further analysis revealed other oxides (Figure 21). In addition to aluminum oxide (Figure 20), manganese oxides in significant quantities were detected in the overlap fusion zone. No oxides or inclusions were found in the face and root of the weld. No impurities affecting the mechanical properties of the weld, such as phosphorus or sulfur, were detected.

4. Discussion

The numerical simulation of laser welding in lap joint specimens was performed, and welding parameters for obtaining partial penetration for the sealed joint were estimated. According to simulation results, 4 kW of output power with a speed ratio equal to 1 m/min gave enough linear power density to obtain a partial joint penetration weld of 4.38 mm. The programmed simulation provided realistically accurate results. The difference in weld face width between the simulation and experimental results was 0.16 mm, approximately 0.06 mm in the overlapping zone, and 0.05 mm was the depth difference. Therefore, the programmed heat-source geometry and boundary conditions can be assumed to be accurate [59]. Results of the stress–strain analysis showed that the maximum value of principal stress was equal to 1120 MPa and the total displacement was equal to 0.32 mm. The calculated total displacement maximum value was related to the face of weld geometry; however, for the determined measurement points, the displacement was more than 0.15 mm (Figure 9b, point 3). The maximum principal stress at the measurement points exceeded 250 MPa (Figure 10b, point 4) and was related to sheet restraint, energy dumping factor, thermal gradient resulting from the heat absorbed by welded materials, and the material thermo-mechanical properties [60,61]. The highest displacement value occurred in the fusion zone. The maximum principal stress was related to the fixed geometry.

The simulation and trial joint analysis indicated differences in the cross-sectional hardness distribution, with higher values in the simulation results. The hardness values in BM from the simulation ranged from 181 to 220 HV and varied from the measured values. Moreover, the weld and HAZ achieved higher hardness values in the simulation. The highest value in the weld from the simulation was 237 HV, with the measured value of 231 HV10. In the HAZ, hardness was between 222 and 249 HV, and the measured values ranged from 200 to 218 HV10. The differences may result from the thermal gradient and phase-transformation velocity factor [62]. Moreover, in the numerical simulation, the load applied during the hardness test was not defined. The measured hardness of the trial joint was lower, and the weld zones showed smaller differences in hardness compared to the values calculated in the numerical simulation. The maximum measured value for the trial joint did not exceed 350 HV10, and no additional post-weld heat treatment was carried out.

The tensile strength of the tested joint was 110 MPa, and compared to the BM, it is lower by about 250 MPa. Both tested specimens failed along the fusion zone line at the maximum load of 11.5 kN. During the performed tensile-strength test, according to specimen configuration, both tensile and shear phenomena occurred (Figure 6). Not-uniaxial complex-force distribution affected the test results and the joint strength obtained was related to the tensile–shear strength of the weld. The stress–strain curve did not have the serrated flow region that is characteristic of low-carbon steels, and the plastic–elastic joint character was observed [63,64].

The crystallographic structure of the base material was identified as fine-grained ferritic–pearlitic. The material structure in the fusion zone changed during melting and solidification processes. The metallographic analysis showed a coarse-grained dendritic structure of the weld. Separate dendrite groups formed pillar crystals, with the growth direction related to the fusion line. No impurities or welding defects were detected in the pillar crystals' contact area. HAZ consist of three areas: the overheated area with a characteristic coarse-grained structure, the normalization area with a uniform fine-grained structure and the partial recrystallization area (incomplete annealing) heated to $Ac_1 \div Ac_2$ transformation point during the welding process. The partial recrystallization area consisted of non-transformed ferrite grains and a fine-grained ferritic–pearlitic structure established from the austenite range [65].

The uniform weld structure was analyzed by using ferrite and manganese distributions. The quantity analysis of the distribution of alloying elements in the joint showed a uniform weld structure. Lack of differences along the measurement line of overlap transition (Figures 18 and 19) confirmed the obtainment of a weld of high quality. Ferrite and manganese distribution from the BM to the weld line confirmed a uniform chemical composition of the laser-welded trial lap joint (Figures 16 and 17) [66].

The energy-dispersive X-ray spectroscopy analysis showed some precipitation. No porosity defects were detected in the obtained weld; nevertheless, some oxides in the transition zone were observed. Precipitation in the form of aluminum oxide was detected (Figure 20), which is typical of low-carbon steels and probably related to steel deoxidizing in the metallurgical process, not to the welding process. Further investigations showed other inclusions, in the form of manganese oxide precipitations (Figure 21). The presence of oxides in the weld is related to the absence of shielding gas between the welded steel plates, and the types of oxides are related to the composition of the welded material.

5. Conclusions

A numerical simulation of laser welding allowed us to estimate the parameters for the lap joint with partial penetration. By programming properly calibrated heat-source geometry and boundary conditions, accurate results can be obtained. Welding simulation based on thermo-mechanical solution with phase transformation gave realistic results with a convex face of the weld. A single-pass laser-welded lap joint based on calculated parameters was produced, and the properties of the obtained joint were investigated. Hardness in the measured trial joint was lower than that from the calculated results and did not exceed 227 HV10. Therefore, according to restrict B quality level, no additional heat treatment was applied. The tensile test results showed the joint strength was 110 MPa, and it is under nominal strength of BM. The not-uniaxial position resulted in the occurrence of tensile-shearing phenomena. Crystallographic analysis confirmed the typical ferritic–pearlitic structure of the BM, grain refining in the heat affected zone and the weld having a characteristic dendritic structure. No welding defects were detected. The energy dispersive X-ray spectroscopy analysis showed good mixture factor of alloying elements. The fusion zone had a uniform structure. Spectroscopy showed that oxides precipitates in the weld. Manganese oxides were detected in the overlap transition line. No additional shielding gas between welded sheets was used, and atmospheric oxygen affected inclusions. The vacuum atmosphere or complete shielding of the fusion zone by using inert gas could reduce the oxidation process. High-quality welding of the sealed lap joint for a gas pipe system was performed by using a laser beam.

Further research of lap joint welding will consider using twin spot-welding optics for widening the fusion zone, improving strength characteristic and producing a lap joint by using a zigzag-shaped welding trajectory. Further research will apply heat-source enhancement to decrease the differences in hardness values between numerical simulation and measured results. Development of the shielding system for overlap region is planned. Further work on laser lap-joint welding of low-carbon steel based on these assumptions can provide a comprehensive analysis of the investigated problem.

Author Contributions: Conceptualization, H.D. and A.S.; methodology, H.D. and A.S.; software, H.D.; validation, H.D. and A.S.; formal analysis, H.D. and A.S.; investigation, H.D.; resources, H.D.; writing—original draft preparation, H.D.; writing—review and editing, H.D.; visualization, H.D.; supervision, A.S.; project administration, H.D.; funding acquisition, H.D. All authors have read and agree to the published version of the manuscript.

Funding: This research was funded by NCBiR, grant number LIDER/31/0173/L-8/16/NCBR/2017: Technology of manufacturing sealed weld joints for gas installation by using concentrated energy source.

Conflicts of Interest: The authors declare no conflict of interest.

References

1. Steen, W.M.; Mazumder, J. *Laser Material Processing*; Springer: Berlin/Heidelberg, Germany, 2010.
2. Pilarczyk, J. *Engineer's Guide—Welding Engineering*; WNT: Warszawa, Poland, 2003.
3. Unt, A.; Poutiainen, I.; Grünenwald, S.; Sokolov, M.; Salminen, A. High Power Fiber Laser Welding of Single Sided T-Joint on Shipbuilding Steel with Different Processing Setups. *Appl. Sci.* **2017**, *7*, 1276. [CrossRef]

4. Piekarska, W.; Kubiak, M. Modeling of thermal phenomena in single laser beam and laser-arc hybrid welding processes using projection method. *Appl. Math. Model.* **2013**, *37*, 2051–2062. [CrossRef]
5. Quiroz, V.; Gumenyuk, A.; Rethmeier, M. Laser Beam Weldability of High-Manganese Austenitic and Duplex Stainless Steel Sheets. *Weld. World.* **2012**, *56*, 9–20. [CrossRef]
6. Yuce, C.; Karpat, F.; Yavuz, N. Investigations on the microstructure and mechanical properties of laser welded dissimilar galvanized steel–aluminum joints. *Int. J. Adv. Manuf. Technol.* **2019**, *104*, 2693–2704. [CrossRef]
7. Gu, J.; Yang, S.; Duan, C.; Xiong, Q.; Wang, Y. Microstructure and Mechanical Properties of Laser Welded Al-Mg-Si Alloys Joints. *Mater. Trans.* **2019**, *60*, 230–236. [CrossRef]
8. Shiegel, A.N.; Evtikheev, N.N.; Gusev, D.S.; Ivanchenko, A.B. Modeling of butt and lap joint welding of aluminum alloys and construction steel sheets. *Non-Ferrous Met.* **2016**, *1*, 27–32.
9. Shanmugam, N.S.; Buvanashekaran, G.; Sankaranarayanasamy, K. Experimental Investigation and Finite Element Simulation of laser lap welding of SS304 sheets. *Int. J. Mech.* **2013**, *7*, 120–127.
10. Tasalloti, H.; Kah, P.; Martikainen, J. Laser Overlap Welding of Zn-Coated Steel on Aluminium Alloy for Patchwork Blank Applications in the Automotive Industry. *Rev. Adv. Mater. Sci.* **2015**, *40*, 295–302.
11. Oussaid, K.; El Ouafi, A.; Chebak, A. Experimental Investigation of Laser Welding Process in Overlap Joint Configuration. *Chem. Mater. Sci.* **2019**, *7*, 16–31. [CrossRef]
12. Masoumi, M.; Marashi, S.P.H.; Pouranvari, M. Metallurgical and Mechanical Characterization of Laser Spot Welded Low Carbon Steel Sheets. *Mater. Technol.* **2010**, *81*, 1144–1150. [CrossRef]
13. Meco, S.; Ganguly, S.; Williams, S.; McPherson, N. Design of laser welding applied to T joints between steel and aluminium. *J. Mater. Process. Technol.* **2019**, *268*, 132–139. [CrossRef]
14. Enz, J.; Khomenko, V.; Riekehr, S.; Ventzke, V.; Huber, N.; Kashaev, N. Single-sided laser beam welding of adissimilar AA2024–AA7050 T-joint. *Mater. Des.* **2015**, *76*, 110–116. [CrossRef]
15. Kumar, S.K. Numerical Modeling and Simulation of a Butt Joint Welding of AISI316L Stainless Steels Using a Pulsed Laser Beam. *Mater. Today Proc.* **2015**, *2*, 2256–2266. [CrossRef]
16. Meco, S.; Cozzolino, L.; Ganguly, S.; Williams, S.; McPherson, N. Laser welding of steel to aluminium: Thermal modelling and joint strength analysis. *J. Mater. Process. Technol.* **2017**, *247*, 121–133. [CrossRef]
17. Lee, C.; Lee, J.B.; Park, D.H.; Na, S.J. Finite Element Modeling of Laser Spot Welded Lap Joint. *Mater. Sci. Forum.* **2008**, *580–582*, 291–294.
18. Zhanga, Y.; Chena, Y.; Zhoua, J.; Sunb, D.; Lib, H. Experimental and numerical study on microstructure and mechanical properties for laser welding-brazing of TC4 Titanium alloy and 304 stainless steel with Cu-base filler metal. *J. Mater. Res. Technol.* **2020**, *9*, 465–477. [CrossRef]
19. Ma, J.; Kong, F.; Kovacevic, R. Finite-element thermal analysis of laser welding of galvanized high-strength steel in a zero-gap lap joint configuration and its experimental verification. *Mater. Des.* **2012**, *36*, 348–358. [CrossRef]
20. Tawfi, T.A. Parametric Optimization of Pulsed Nd:YAG Laser Lap Welding of Stainless Steel ASTM A240/316L with Carbon Steel ASTM A570/Gr30. *Al-Nahrain J. Eng. Sci.* **2017**, *20*, 27–35.
21. Sabbaghzadeh, J.; Azizi, M.; Torkamany, M.J. Numerical and experimental investigation of seam welding with a pulsed laser. *Opt. Laser Technol.* **2008**, *40*, 289–296. [CrossRef]
22. He, Q.; Wei, H.; Chen, J.S.; Wang, H.P.; Carlson, B.E. Analysis of hot cracking during lap joint laser welding processes using the melting state-based thermomechanical modeling approach. *Int. J. Adv. Manuf. Technol.* **2018**, *94*, 4373–4386. [CrossRef]
23. Vicentin, L.C.; Ierardi, M.C.F.; Garcia, A.; Vilar, R. Laser welding of low carbon steel blanks. *Lasers Eng.* **2000**, *10*, 183–192.
24. Keskitalo, M.; Hietala, M.; Mäntyjärvi, K. The normal and shear strength properties of laser lap weld. *Procedia Manuf.* **2019**, *36*, 224–231. [CrossRef]
25. Sindhu, R.A.; Park, M.K.; Lee, S.J.; Lee, K.D. Effects of residual stresses on the static and fatigue strength of laser-welded lap joints with different welding speeds. *Int. J. Automot. Technol.* **2010**, *11*, 857–863. [CrossRef]
26. Rizz, D.; Sibillano, T.; Calabrese, P.P.; Ancona, A.; Lugarà, P.M. Spectroscopic, energetic and metallographic investigations of the laser lap welding of AISI 304 using the response surface methodology. *Opt. Lasers Eng.* **2011**, *49*, 892–898. [CrossRef]
27. Meco, S.; Pardal, G.; Ganguly, S.; Williams, S.; McPherson, N. Application of laser in seam welding of dissimilar steel to aluminium joints for thick structural components. *Opt. Lasers Eng.* **2015**, *67*, 22–30. [CrossRef]

28. Evdokimov, A.; Springer, K.; Doynov, N.; Ossenbrink, R.; Michailov, V. Heat source model for laser beam welding of steel-aluminum lap joints. *Int. J. Adv. Manuf. Technol.* **2017**, *93*, 154–196. [CrossRef]
29. Ki, H.; Mazumder, J.; Mohanty, P.S. Modeling of laser keyhole welding: Part, I. mathematical modeling, numerical methodology, role of recoil pressure, multiple reflections, and free surface evolution. *Metall. Mat. Trans. A* **2002**, *33*, 1–14. [CrossRef]
30. Fey, A.; Ulrich, S.; Jahn, S.; Schaaf, P. Numerical analysis of temperature distribution during laser deep welding of duplex stainless steel using a two-beam method. *Weld. World* **2020**, *64*, 623–632. [CrossRef]
31. Bag, S.; Kiran, D.V.; Syed, A.A.; De, A. Efficient Estimation of Volumetric Heat Source In Fusion Welding Process Simulation. *Weld. World* **2012**, *56*, 88–97. [CrossRef]
32. Wang, R.; Lei, Y.; Shi, Y. Numerical simulation of transient temperature field during laser keyhole welding of 304 stainless steel sheet. *Opt. Laser Technol.* **2011**, *43*, 870–873. [CrossRef]
33. Ribic, B.; Rai, R.; DebRoy, T. Numerical simulation of heat transfer and fluid flow in GTA/Laser hybrid welding. *Sci. Technol. Weld. Join.* **2008**, *13*, 683–693. [CrossRef]
34. Danielewski, H. Laser welding of pipe stubs made from super 304 steel. Numerical simulation and weld properties. *Tech. Trans.* **2019**, *1*, 167–176. [CrossRef]
35. Kik, T. Computational Techniques in Numerical Simulations of Arc and Laser Welding Processes. *Materials* **2020**, *13*, 608. [CrossRef] [PubMed]
36. Kumar, U.; Gope, D.K.; Srivastava, J.P.; Chattopadhyaya, S.; Das, A.K.; Krolczyk, G. Experimental and Numerical Assessment of Temperature Field and Analysis of Microstructure and Mechanical Properties of Low Power Laser Annealed Welded Joints. *Materials* **2018**, *11*, 1514. [CrossRef] [PubMed]
37. Chunming, W.; Bin, L.; Ping, J.; Xiang, X.; Gaoyang, M. Numerical and experimental investigation of vacuum-assisted laser welding for DP590 galvanized steel lap joint without prescribed gap. *Int. J. Adv. Manuf. Technol.* **2018**, *94*, 4177–4185.
38. Kubiak, M.; Piekarska, W.; Stano, S.; Saternus, Z. Numerical modelling of thermal and structural phenomena in Yb:Yag laser butt-welded steel elements. *Arch. Metall. Mater.* **2015**, *60*, 821–828. [CrossRef]
39. Yilbas, B.S.; Arif, A.F.M.; Abdul Aleem, B.J. Laser welding of low carbon steel and thermal stress analysis. *Opt. Laser Technol.* **2010**, *42*, 760–768. [CrossRef]
40. Podany, P.; Reardon, C.; Koukolikova, M.; Prochazka, R.; Franc, A. Microstructure, Mechanical Properties and Welding of Low Carbon, Medium Manganese TWIP/TRIP Steel. *Metals* **2018**, *8*, 263. [CrossRef]
41. Nagel, F.; Simon, F.; Kümmel, B.; Bergmann, J.P.; Hildebrand, J. Optimization Strategies for Laser Welding High Alloy Steel Sheets. *Phys. Procedia* **2014**, *56*, 1242–1251. [CrossRef]
42. Andersson, O.; Budak, N.; Melander, A.; Palmquist, N. Experimental measurements and numerical simulations of distortions of overlap laser-welded thin sheet steel beam structures. *Weld. World* **2017**, *61*, 927–934. [CrossRef]
43. Hatami, N.; Babaei, R.; Dadashzadeh, M.; Davami, P. Modeling of hot tearing formation during solidification. *J. Mater. Process. Technol.* **2008**, *205*, 506–513. [CrossRef]
44. Hafez, K.M.; Ramadan, M.; Fathy, N.; Ismail, M. Microstructure and Mechanical Properties of Laser Welded Dual Phase and Mild Steel Joints for Automotive Applications. *Appl. Mech. Mater.* **2017**, *865*, 81–87. [CrossRef]
45. Górka, J.; Stano, S. Microstructure and Properties of Hybrid Laser Arc Welded Joints (Laser Beam-MAG) in Thermo-Mechanical Control Processed S700MC Steel. *Metals* **2018**, *8*, 132. [CrossRef]
46. Zdravecka, E.; Slota, J. Mechanical and Microstructural Investigations of the Laser Welding of Different Zinc-Coated Steels. *Metals* **2019**, *9*, 91. [CrossRef]
47. PN-EN ISO 6507-1:2018. *Metallic Materials—Vickers Hardness Test—Part 1: Test Method*; International Organization for Standardization: Geneva, Switzerland, 2018.
48. PN-EN ISO 9018:2003. *Destructive Tests on Welds in Metallic Materials—Tensile Test on Cruciform and Lapped Joints*; International Organization for Standardization: Geneva, Switzerland, 2003.
49. PN-EN ISO 17639:2003. *Destructive Tests on Welds in Metallic Materials—Macroscopic and Microscopic Examination of Welds*; International Organization for Standardization: Geneva, Switzerland, 2003.
50. Yan, F.; Fang, X.; Chen, L.; Wang, C.; Zhao, J.; Chai, F.; Wang, W. Microstructure evolution and phase transition at the interface of steel/Al dissimilar alloys during Nd: YAG laser welding. *Opt. Laser Technol.* **2018**, *108*, 193–201. [CrossRef]

51. David, S.A.; Vitek, J.M. Correlation between solidification parameters and weld microstructures. *Int. Mater. Rev.* **1989**, *34*, 213–245. [CrossRef]
52. Partes, K.; Schmidt, M.; Gorny, S. Prediction of Preheating Temperatures for S690QL High Strength Steel Using FEM-Simulation for HighPower Laser Welding. *Lasers Manuf. Mater. Process.* **2020**. [CrossRef]
53. Sun, J.; Liu, X.; Tong, Y.; Deng, D. A comparative study on welding temperature fields, residual stress distributions and deformations induced by laser beam welding and CO2 gas arc welding. *Mater. Des.* **2014**, *63*, 519–530. [CrossRef]
54. Kang, M.; Jeon, I.-H.; Han, H.N.; Kim, C. Tensile–Shear Fracture Behavior Prediction of High-Strength Steel Laser Overlap Welds. *Metals* **2018**, *8*, 365. [CrossRef]
55. Sinha, A.K.; Kim, D.Y.; Ceglare, D. Correlation analysis of the variation of weld seam and tensile strength in laser welding of galvanized steel. *Opt. Lasers Eng.* **2013**, *51*, 1143–1152. [CrossRef]
56. Hietala, M.; Järvenpää, A.; Keskitalo, M.; Jaskari, M.; Mäntyjärvi, K. Tensile and fatigue properties of laser-welded ultra-high-strength stainless spring steel lap joints. *Procedia Manuf.* **2019**, *36*, 131–137. [CrossRef]
57. Evin, E.; Tomas, M. The Influence of Laser Welding on the Mechanical Properties of Dual Phase and Trip Steels. *Metals* **2017**, *7*, 239. [CrossRef]
58. Gorka, J. Properties of thermomechanically treated welds of high yield point steel. *Weld. Technol. Rev.* **2011**, *83*, 31–35.
59. Lee, K.D.; Ho, K.I.; Park, K.Y. Analysis of the Local Stresses at Laser Welded Lap Joints. *Weld. J.* **2014**, *93*, 351–361.
60. Deng, D. FEM prediction of welding residual stress and distortion in carbon steel considering phase transformation effects. *Mater. Des.* **2009**, *30*, 359–366. [CrossRef]
61. Hsu, C.; Albright, C.E. Fatigue analysis of laser welded lap joints. *Eng. Fract. Mech.* **1991**, *39*, 575–580. [CrossRef]
62. Salminen, A.; Baskutis, S.; Petronis, E. Influence of welding modes on weldability of structural steel lap joints in laser welding. *J. Laser Appl.* **2017**, *29*, 022419. [CrossRef]
63. Goyal, R.; El-zein, M. Influence of laser weld shape on mechanical and fatigue behaviour of single lap laser welded joints. *J. Adv. Join. Process.* **2020**, *1*, 100018. [CrossRef]
64. Furusako, S.; Miyazaki, Y.; Akiniwa, Y. Tensile shear strength of laser lap joints. *Weld. Int.* **2015**, *29*, 838–846. [CrossRef]
65. Farabi, N.; Chen, D.L.; Li, J.; Zhou, Y.; Dong, S.J. Microstructure and mechanical properties of laser welded DP600 steel joints. *Mater. Sci. Eng. C* **2010**, *527*, 1215–1222. [CrossRef]
66. Zhang, C.; Li, G.; Gao, M.; Zeng, X.Y. Microstructure and Mechanical Properties of Narrow Gap Laser-Arc Hybrid Welded 40 mm Thick Mild Steel. *Materials* **2017**, *10*, 106. [CrossRef] [PubMed]

materials

MDPI

Article

Investigation of Welds and Heat Affected Zones in Weld Surfacing Steel Plates Taking into Account the Bead Sequence

Miloš Mičian [1,*], Jerzy Winczek [2,*], Marek Gucwa [2], Radoslav Koňár [1], Miloslav Málek [1] and Przemysław Postawa [2]

1 Department of Technological Engineering, Faculty of Mechanical Engineering, University of Žilina, 010 26 Žilina, Slovakia; radoslav.konar@fstroj.uniza.sk (R.K.); miloslav.malek@fstroj.uniza.sk (M.M.)

2 Department of Technology and Automation, Faculty of Mechanical Engineering and Computer Sciences, Czestochowa University of Technology, 42-201 Czestochowa, Poland; mgucwa@spaw.pcz.pl (M.G.); postawa@ipp.pcz.pl (P.P.)

* Correspondence: milos.mician@fstroj.uniza.sk (M.M.); winczek@imipkm.pcz.czest.pl (J.W.)

Received: 4 November 2020; Accepted: 8 December 2020; Published: 11 December 2020

Abstract: In this paper, the experimental investigation results of the bead sequence input on geometry, structure, and hardness of surfaced layers after multi-pass weld surfacing are analyzed. Three S355 steel plates surfaced by GMAW (Gas Metal Arc Welding) were tested with three different combinations of six beads. The geometric, structural, and hardness analysis was carried out in the cross-section of the plates in the middle of the welded layers. The dimensions of padded layers, fusion and heat-affected zone, as well as the individual padded weld were evaluated. On the basis of metallographic samples, qualitative and quantitative structure analysis was performed. Hardness measurements in surfacing welds and heat-affected zones in the tested cross-sections of the surfacing layers were carried out. A comparative analysis of structure and hardness, taking into account the thermal implications of the bead sequence, allowed for the formulation of conclusions. Comparative studies have shown differences in properties between heat-affected zones (HAZ) for individual surfacing sequences. These differences were mainly in the dimensions of the surfacing layers, the share of structural components, as well as the uniformity of hardness distributions. Finally, the most favorable sequence in terms of structure and hardness distribution, maximum hardness, and range of hardness has been indicated.

Keywords: weld surfacing; welding sequence; weld geometry; structure analysis; phase shares; hardness distribution

1. Introduction

Multi-pass welding (joining, rebuilding, surfacing) is widely used in industrial practice and is of interest to many researchers. The thermo-mechanical and metallurgical phenomena occurring during the multi-pass welding process are complicated and coupled, and their effects are still the subject of research and analysis for scientists.

Numerical methods, mainly FEM (Finite Element Method), are most often used in multi-pass welding analysis. Relatively few works concern the results of experimental research. McDonald et al. [1] measured residual stresses in repaired low ferritic CrMoV steel plates using X-ray diffraction techniques and the deep hole drilling method. Vemanaboina et al. [2] evaluated residual stresses in multi-pass dissimilar butt joints using X-ray diffraction.

An experimental analysis of surfaced layers, mainly in the structural aspect as well as corrosion and tribological resistance, was performed in [3–11]. The results of metallographic analysis were presented in [3–8]. In work [9] displacements (strains) were measured, and in [10] non-destructive tests of welded joints were performed. The authors [11] proposed optimization of multi-pass welding parameters using the Taguchi method. Corrosion resistance was analyzed in [8]. The influence of bead sequences on the properties of welds and heat-affected zones (HAZ) was not discussed in these works. Romek et al. [12] investigated the impact of padding weld shape on their abrasive wear.

The bead sequence has a great influence on the temperature distribution, resulting in welding distortions and residual stresses, as well as changes in the structural and properties of the material after welding. In their analysis of the influence of welding sequences on thermomechanical states, the researchers' interests focus mainly on predicting the displacements (deformations) and the residual stresses after welding. This scope is dominated by research and optimization of the bead sequence when making welded joints, dividing the weld path into multiple seams and taking into account different directions of their execution [13–30]. In the works [13–18] only the results of numerical analysis were presented for various sequences of beads in simulated states of strain (deformation) and stress in welded joints. In some works, the results of numerical simulations were compared with the experimental results published by other researchers [19,20]. In other studies, the verification of the calculation results was carried out based on experimental investigations. In the works [21–23], the verification of the simulated thermal cycles was carried out with thermocouple measurements. Deng [24] used the electric strain gauge method to measure strains in longitudinal and transverse directions. The values of calculated residual stresses using the X-Ray method were verified in [22,25,26]. Displacements and distortions were measured in [27–30].

Jiang et al. [31] investigated the effect of the number of beads on the quality of a joint in stainless steel clad plates. The results of numerical simulations were compared with experimental data presented in the literature. They showed that with increasing the number of welding layers, the residual stresses are reduced.

There are relatively few works containing the numerical or experimental analysis of the influence of the surfacing bead sequence on the dimensions of heat-affected zones, structure, mechanical, or other properties [32,33].

Kohandehghan and Serajzadeh [32] conducted an analysis of the imposing subsequent welds influence on the development (change in value) of longitudinal and transverse stresses in AA5251 aluminum alloy plates with a thickness of 2 mm. During the tests, only one (serial) surfacing sequence was used. In addition, aluminum alloys do not undergo phase changes (structural changes), so they intensely affect the final structure of the material, mechanical properties, as well as the state of internal stress after welding.

Tomków et al. [33] analyzed the impact of the surfacing sequence in underwater welding on cracks tendency. The hardness structure tests of two hardfacing layers made with different sequence of five stitches were carried out. The first sample was made with serial welds, the second with alternating beads to the left and right of the center of the layer (first bead). As a result of the research, it was found that the susceptibility of steel to cold cracking can be reduced by changing the sequence of the beads during the process.

According to the literature review, the current state of knowledge on the input of the surfacing bead sequence on the geometric parameters of the surfacing layers, their structure and hardness are insufficient. The influence of the weld surfacing sequence on material properties (mainly in weld and HAZ) requires further study considering a larger number of surfacing sequence combinations as well as a larger number of beads. The outermost beads of the surfacing layer, due to the cooling conditions other than for the inner beads of the layer, can differ both structural and mechanical (functional) properties.

In this paper, the influence of three different combinations of six-bead sequences on the geometrical, structural, and hardness analysis of heat-affected zones in three S355 steel plates surfaced by GMAW (Gas Metal Arc Welding) is analyzed. The materials used and research methods were described. Comparative tests were carried out on metallographic specimens (macro and micro) made in half the length of the surfaced layers. The macro samples made it possible to measure the geometric parameters of the surfacing layers (width, surface area, reinforcement height), as well as HAZ. Then, a structural analysis was carried out determining the type of structure. Geometric and structural analysis was performed using a microscope equipped with specialized software. Based on the size and number of grains, the shares of structural components were estimated. The last stage of the experimental tests was the hardness measurements. The article ends with conclusions from the analysis of the results of the experiments performed and the directions of further research.

In welding practice, the choice of surfacing bead sequences is often made on the basis of experience and intuition. Although tests of multi-pass welded joints are conducted [1–30], detailed information on the effects of using certain sequences of surfacing beads is very scarce [31–33]. In welding handbooks (e.g. [34]) there are multi-pass surfacing schemes for selected machine parts without justifying such a solution. The article deals with this difficult subject by analyzing the influence of the sequence of surfacing beads on the geometrical dimensions of the surfacing layer, structure, and hardness distributions. Based on the analysis of the considered order and place of weld application, the most favorable sequence in terms of structure and hardness distribution, maximum hardness, and range of hardness has been indicated. The results of the research allowed for the designation of further investigation directions.

2. Materials and Methods

The investigation was carried out on three S355 steel plates welded with different sequences of six beads. Square plates with a side length of 30 cm and a thickness of 30 mm were surfaced using the Robot CLOOS QRC 350 (CLOOS-Polska, Świdnica, Poland) and the Qineo PULSE 450 (CLOOS, Haiger, Germany) welding source.

The six beads imply 720 combinations of their execution (without taking into account the change of direction—the electrode velocity vector). Therefore, the authors had to select the number of samples, their size, and the technological parameters. The selection of the sequence was made taking into account the increase in the inter-pass times (temperatures) in the B and C plates, based on our own experience and the surfacing sequences adopted, among others in [35], and not to repeat the alternative sequence adopted at work [33].

The size of the welded plates (as well as the number and size of beads) was determined by the dimensions of the samples for further tests, depending on the capabilities of the test equipment.

Before proceeding to a substantial part of the experiment, tests were performed to determine the parameters for a single weld dimensions: a width of 2 cm and a height of approximately 3 mm. Such dimensions of a single padding weld allowed to make a layer of padding welds (taking into account the technological overlap) up to 10 cm wide. The boundary dimensions of the welded zone (in the central part of the plate) resulted from the assumptions of the condition of no heat exchange with the environment through the bottom and side walls of the welded plates.

The additional material was selected so that its chemical composition differed slightly from the chemical composition of the deposited material. The technological parameters were established on the basis of single-pass welding tests, also using the software of the Qineo PULSE 450 welding source.

Each surfacing sequence was performed once. No attempts were made repeatedly because the purpose of the experience was the comparative analysis of qualitative data, not quantitative. Statistical processing of the results was not planned.

All three GMAW welding tests were carried out while maintaining the same welding parameters: voltage 28.3 V, current 290 A, welding speed 0.4 m/min, torch tilt angle 90°, and CTDW (Contact Tip to Work Distance) 15 mm. The welding was made using a Lincoln Ultra MAG G3Si1 (Lincoln Electric Europe, Nijmegen, The Netherlands) wire with a diameter of 1.2 mm, a wire feeding speed of 9.5 m/min, and in MIX 21 (82% Argon/18% CO_2) shielding gas with a flow rate of 12 L/min. Twenty centimeter-long welds were applied in the central parts of the plates, and the order of individual beads for individual sequences A, B, and C is shown in Figure 1. The inter-operative time associated with the movement of the robot arm to the beginning of the next bead was 3 s.

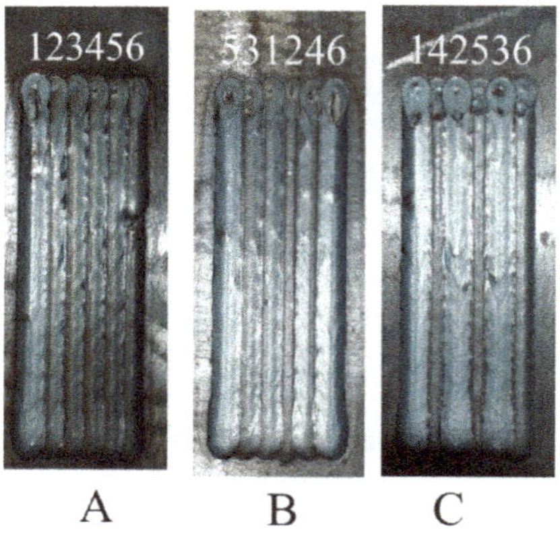

Figure 1. Photographs of plates with padded welds marked by a sequence of beads.

From the weld-surfaced plates, for metallographic examination and hardness measurements, material strips about 25 mm thick were cut perpendicular to the surfacing direction. Samples were taken at mid-length.

Macroscopic examination of cross-sections of the deposited layers were performed on a Zeiss LSM 700 confocal optical microscope. Metallographic (structural) analysis in the areas of welds and heat-affected zones as well as native material was carried out using Olympus GX51 (Olympus Europe, Hamburg, Germany) metallographic microscope with Olympus Stream Essentials software.

Hardness measurements of individual heat-affected zones in the cross-sections of surfacing welds were made with an automatic hardness tester ZHVμ Micro Vickers hardness device from the company Zwick Roell (Ulm, Germany) in accordance with EN ISO standards [36,37].

3. Results and Discussion

3.1. Macroscopic Examination

In Figure 2, the macro-panoramas of the specimens in the central cross-section of the plates have been presented.

Figure 2. Macro-panoramas of the metallographic specimens in the middle cross-section of the plates.

The measured geometrical values of the welds for individual bead sequences are summarized in Tables 1–3. The numbers without asterisks indicate the order of the welds in the layers presented in Figure 2, whereas the numbers with asterisks (in half-round brackets) indicate the order of the bead sequences.

Table 1. The geometrical characteristics of the deposition zone in plate A.

Width of padded area, mm	69.11					
Number of the weld	1(1*)	2(2*)	3(3*)	4(4*)	5(5*)	6(6*)
Depth of penetration (fusion), mm	5	5.27	5.13	5.13	5.8	6.21
eight of reinforcement, mm	2.72	2.86	2.99	2.72	2.86	2.99
Depth of HAZ, mm	0.33	0.844	0.756	1.07	1.2	1.56
Reinforcement field, mm^2	157.36					
Penetration area, mm^2	192.03					
Total weld padding area, mm^2	349.39					
Area of HAZ, mm^2	185.81					

Table 2. The geometrical characteristics of the deposition zone in plate B.

Width of padded area, mm	72.13					
Number of the weld	1(5*)	2(3*)	3(1*)	4(2*)	5(4*)	6(6*)
Depth of penetration (fusion), mm	5.61	5.34	5.2	4.8	5.2	6.02
Height of reinforcement, mm	2.81	3.08	2.71	2.81	2.9	3.08
Depth of HAZ, mm	1.43	1.16	1.12	1.29	1.21	1.29
Reinforcement field, mm^2	186					
Penetration area, mm^2	204.88					
Total weld padding area, mm^2	390.88					
Area of HAZ, mm^2	211.36					

Table 3. The geometrical characteristics of the deposition zone in plate C.

Width of padded area, mm	71.03					
Number of the weld	1(1*)	2(4*)	3(2*)	4(5*)	5(3*)	6(6*)
Depth of penetration (fusion), mm	4.81	6.18	5.15	5.72	5.38	5.65
Height of reinforcement, mm	2.99	3.11	2.65	2.99	2.54	2.77
Depth of HAZ, mm	0.852	0.963	1.04	1.33	0.852	0.889
Reinforcement field, mm^2	172.74					
Penetration area, mm^2	202.99					
Total weld padding area, mm^2	375.73					
Area of HAZ, mm^2	190.64					

Figures 3–5 compare reinforcement height, penetration, and HAZ depths of individual beads in samples A, B, and C numbered in their order in the surfaced layer shown in Figure 2.

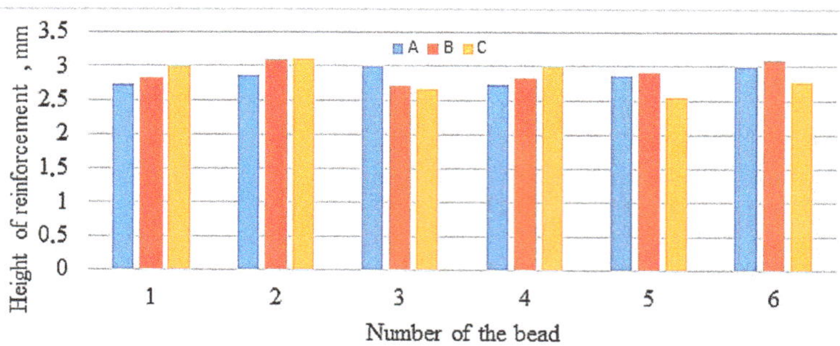

Figure 3. Reinforcement height of particular beads.

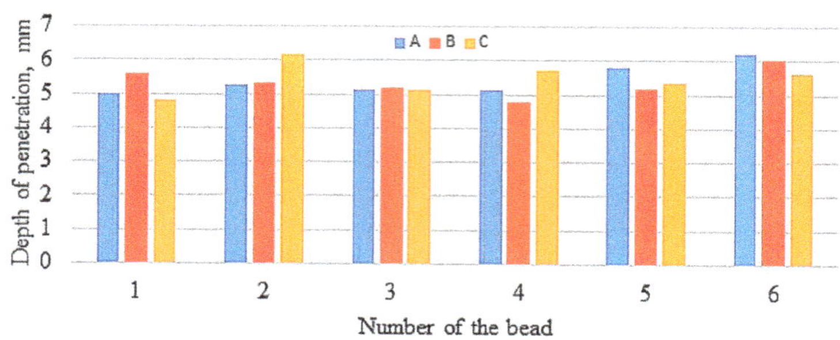

Figure 4. Penetration depth of particular beads.

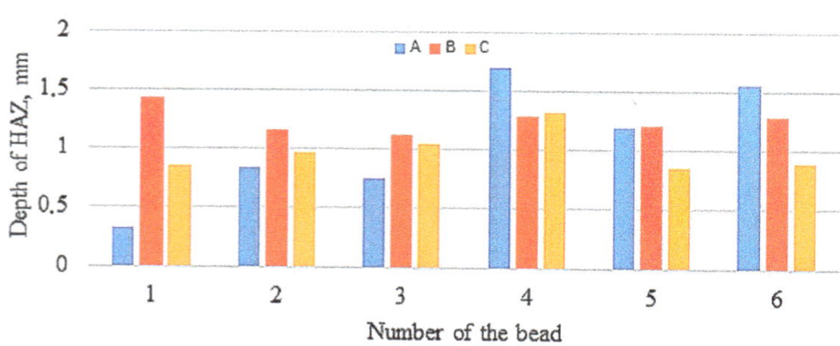

Figure 5. Heat-Affected Zone (HAZ) depth of particular beads.

The values of reinforcement height and depth of HAZ of beads both between samples and in samples show great variation, with the largest differences in depth of HAZ observed in sample A. Relatively small differences occur in depth of penetrations. Figures 6–8 compare reinforcement height, penetration depth, and HAZ depth for samples A, B, and C in the order of beads in accordance with welding sequences.

Although the measured values are varied (also within the samples), an increase in the geometrical parameters can be observed along with the subsequent beads, regardless of their sequence. Especially in sample A, an increase in the depth of penetration is noticeable with subsequent weld beads. The welding thermal cycle of the surfacing caused a successive heating of subsequent areas before laying the next beads.

The geometrical analysis of single beads is of general informative nature about the height of the overlay of the surfacing layer and its fusion depth (HAZ depth). It is obvious that the height of the reinforcement, as well as the fusion depth of the welds depends on the presence of adjacent beads and the technological weld overlay, which is approximately 50% of the bead width. An important piece of information is the smallest reinforcement height among the seams of the applied layer, which may determine the conditions of possible machining (e.g., in the case of rebuilding). In the case of hardfacing, these dimensions are of secondary importance. More useful information is the geometric dimensions of the padding layer, such as its width (Tables 1–3), reinforcement, penetration, HAZ, and total weld padding areas.

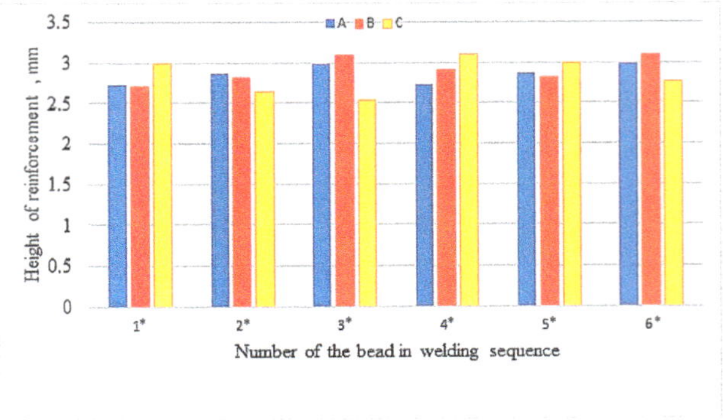

Figure 6. Height of the reinforcement in welding sequences.

Figure 7. Depth of the penetration in welding sequences.

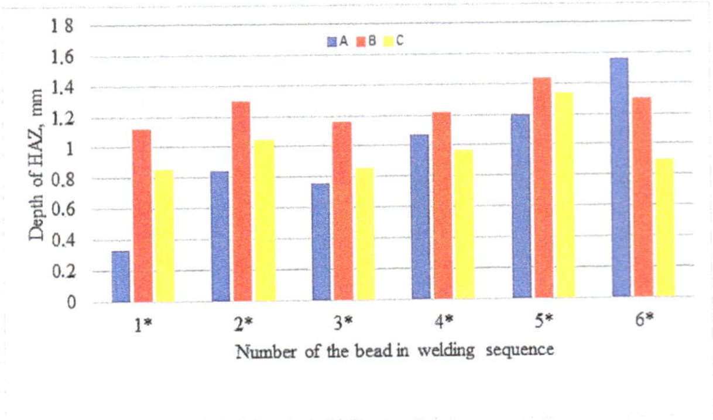

Figure 8. Depth of the HAZ in welding sequences.

In Figure 9, the areas of individual zones in the weld layers: reinforcement, penetration, HAZ, and total weld padding areas are presented. The sizes of these areas are the largest for sample B (sequence 5-3-1-2-4-6) and the lowest for sample A (sequence 1-2-3-4-5-6. However, it is difficult to say what the reason is. In the case of sample C (sequence 1-4-2-5-3-6), intermediate values (between samples A and B) were recorded for all measured areas.

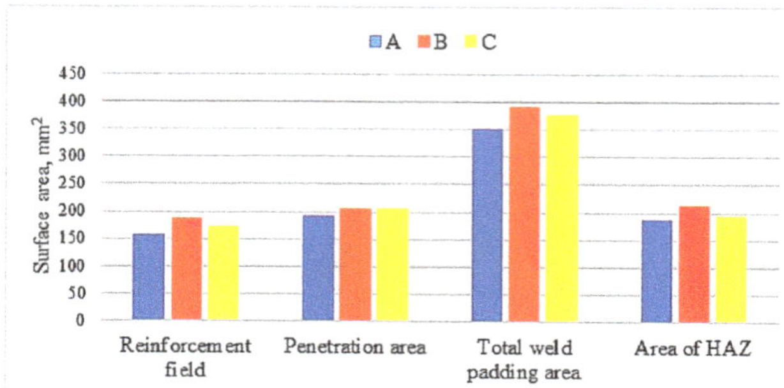

Figure 9. Surface areas of individual zones in the weld layer cross-sections.

3.2. *Microscopic Investigation*

The parent material was S355 steel plates with a ferritic–pearlitic structure in a banded system after rolling (Figure 10). Structural analysis in the areas of welds and heat-affected zones as well as native material was performed using Olympus GX51 metallographic microscope with Stream software. The structure of padding welds was tested in their central part; Figure 11 indicates places in HAZ where structural components are assessed. Shares of structures in indicated places of the surfaced welds for individual plates are listed in Table 4 (sample A), Table 5 (sample B), and Table 6 (sample C).

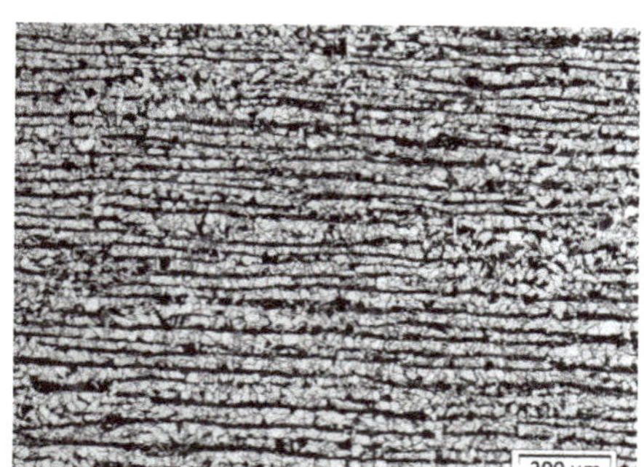

Figure 10. Native material: ferritic–pearlitic in a banded system after rolling.

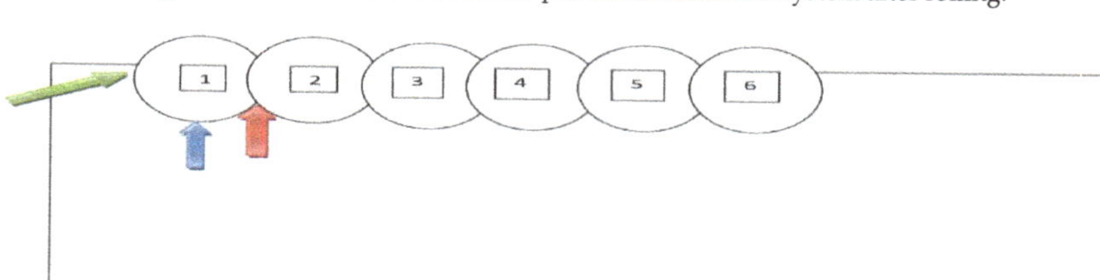

Figure 11. Places in HAZ where structural components were assessed.

Table 4. Shares of structures in the surfaced welds on plate A.

Type of Structure	Structure Share%								
	M	B	F + P	M	B	F + P	M	B	F + P
Number of bead		1			2			3	
In front of bead	20	40	40						
In the bead axis	5	50	45	0	60	40	0	60	40
Between current and next bead	30	40	30	20	50	30	20	40	40
Number of bead		4			5			6	
In the bead axis	10	50	40	0	40	60	10	50	40
Between current and next bead	10	40	50	20	40	40			
Behind bead							10	40	50

Table 5. Shares of structures in the surfaced welds on plate B.

Type of Structure	Structure Share%								
	M	B	F + P	M	B	F + P	M	B	F + P
Number of bead		5			3			1	
In front of bead	20	60	20						
In the bead axis	0	40	60	0	40	60	0	70	30
Between current and next bead	30	40	30	5	60	35	10	60	30
Number of bead		2			4			6	
In the bead axis	20	40	40	0	60	40	0	50	50
Between current and next bead	10	40	50	10	50	40			
Behind bead							5	45	50

Table 6. Shares of structures in the surfaced welds on plate C.

Type of Structure	Structure Share%								
	M	B	F + P	M	B	F + P	M	B	F + P
Number of bead		1			4			2	
In front of bead	30	50	20						
In the bead axis	10	40	50	0	40	60	0	70	30
Between current and next bead	10	40	50	0	70	30	10	60	30
Number of bead		5			3			6	
In the bead axis	0	40	60	0	40	50	0	50	50
Between current and next bead	10	50	40	5	55	40			
Behind bead							10	40	50

Figure 12 shows the typical changes in structure after welding. There is a small but visible amount of the martensite close to the fusion line in front of the first bead and in the axis of the bead. The structure in HAZ is not homogenous due to the effect of the welding cycle and some chemical segregation of the base material, which is visible in Figure 10 (lamellar structure after plastic deformation and Mn segregation). Farther from the fusion line structure is a mix of bainite and perlite.

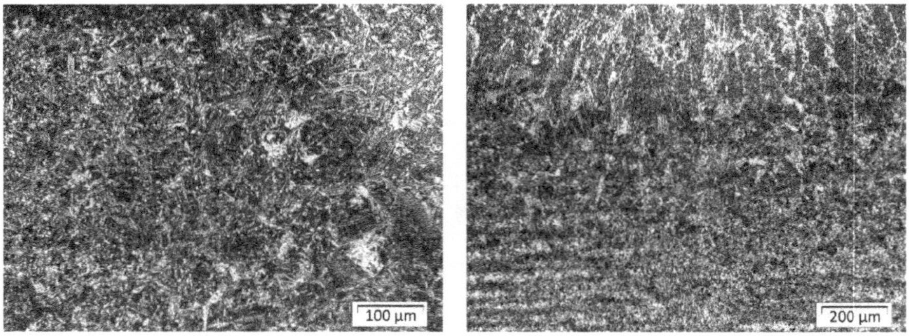

(a) (b)

Figure 12. Sample A: (**a**) HAZ in front of bead number one; (**b**) fusion line and HAZ in axis of bead number one.

In sample A, in addition to the effect of increased penetration depth, you can see changes in the HAZ width and changes in its structure. As a result, near the fusion line for the fifth and sixth bead you can observe elements of the microstructure similar in appearance to the Widmanstätten ferrite, which testify to a large overheating of the structure (Figure 13a).

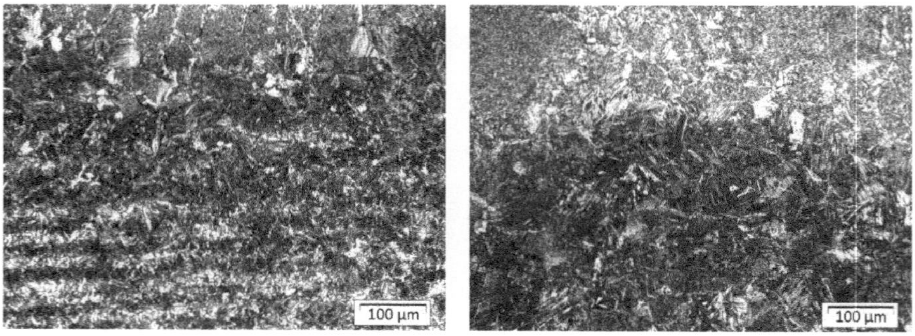

(a) (b)

Figure 13. Sample A: (**a**) the structure of the HAZ in the axis of bead number six; (**b**) the structure of the HAZ between the fifth and sixth bead.

Additionally, the structure between the last beads is gradually degraded, because in addition to the mixture of martensite and bainite, Widmanstätten ferrite grains appear (Figure 13b), which can potentially lead to weakening of the structure in this area.

In the HAZ structure, the bainitic structure dominates, with a small amount of martensite between successive beads. Martensite can be mainly seen near the fusion line and its quantity decreases significantly with increasing distance from the fusion line. Depending on the observation site, a change in bainite morphology can be seen.

In the central areas of sample B, which were first surfaced, a large amount of lower bainite can be seen (Figure 14a), whereas in the HAZ between the outer stitches made at the end, lower bainite occurs in a smaller amount (Figure 14b).

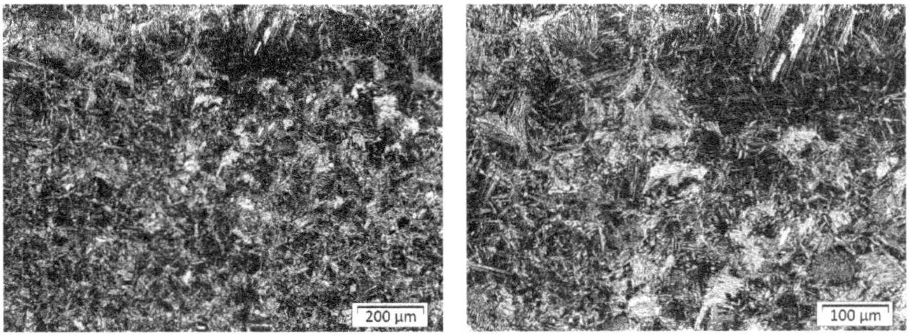

(a) (b)

Figure 14. Sample B: (**a**) the structure of the HAZ between the first and second bead; (**b**) the structure of the HAZ between fifth and third bead.

In Figure 15a, the structure of the HAZ near the fusion line in front of stitch 1 for sample C is shown. The structure in this area consists mainly of martensite, which accounts for about 70% of the structure in the analyzed area. This is also confirmed by the hardness test results (please compare with Figure 19), where the highest hardness value was measured for this area.

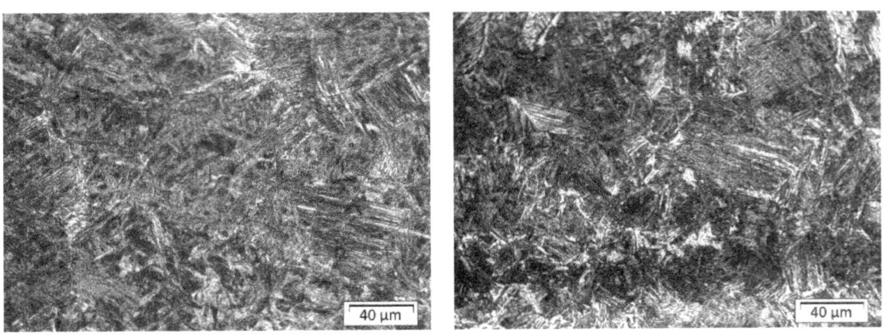

(a) (b)

Figure 15. Sample C: (**a**) the structure of the HAZ in front of bead number one; (**b**) the structure of the HAZ close to the fusion line in the axis of bead number one.

The structure in Figure 15b is different in the quantity of fraction (Table 6) compared with the structure shown in Figure 15a. Because this area (in the axis of the bead number one) has the longest cooling time, the structure consists mainly of bainite with a small amount of the martensite. Farther from this area, the structure consists of mixed bainite, perlite, and ferrite. In the axis of bead number five (Figure 16), there is mainly fine bainitic structure close to the fusion line. This bead was done in the middle of the plate and was the penultimate bead, so the heat accumulated in the plate increased the time of cooling and prevented formation of martensite.

The changes in structure of investigated plates are mainly in the quantity of the phases in the characteristic areas of the HAZ. Martensite structure is obtained almost in whole researched regions after the cladding process. The differences in the quantity of the martensite structure is because of the level of the heat input, heat dissipation, and sequence of the cladding. For the first external beads in each case (plates A, B, and C), there are noticeable amounts of the martensite in front of the beads. In the middle of the plates, especially for plate C, the amount of the martensite decreases because of the accumulated heat and extended cooling time. The martensite was observed close to the fusion line and there were some areas very rich in this structure. This can be explained by the segregation of the C and Mn in the structure of the base material (the texture after plastic deformation). The short heating time and fast cooling time limited the diffusion of the elements and created conditions for the formation of the martensite in certain areas of the HAZ.

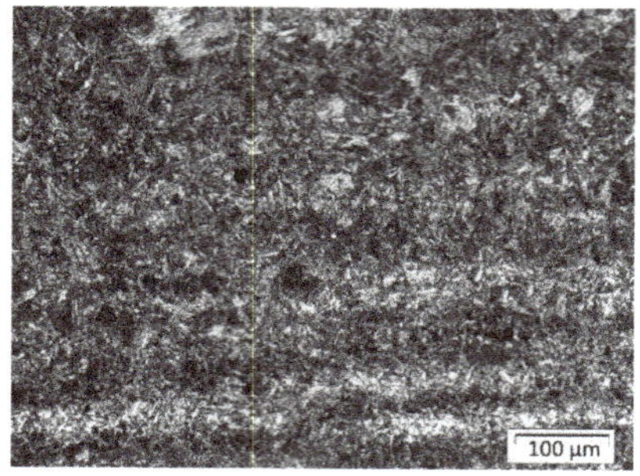

Figure 16. Sample C: the structure of the HAZ in the axis of bead number five.

3.3. *Hardness Measurement Results*

The hardness HV0.5 was measured in a line 2 mm away from the surface of the beads. It passes from the basic material through the individual welds and their HAZ. The measurement was performed on an automatic hardness tester ZHVµ Micro Vickers hardness device from the company Zwick Roell with automatic reading of the size of the indentation. The distance between indentations was 250 µm.

The sequence of welding beads has a significant effect on the hardness distribution in the measured line. From the course of hardness, the mutual influence of individual beads is visible. The hardness of HAZ is reduced by the effect of annealing with another weld bead. Figures 17–19 show macroscopic images of the individual sequences with the measurement line and a graphical representation of the hardness distribution.

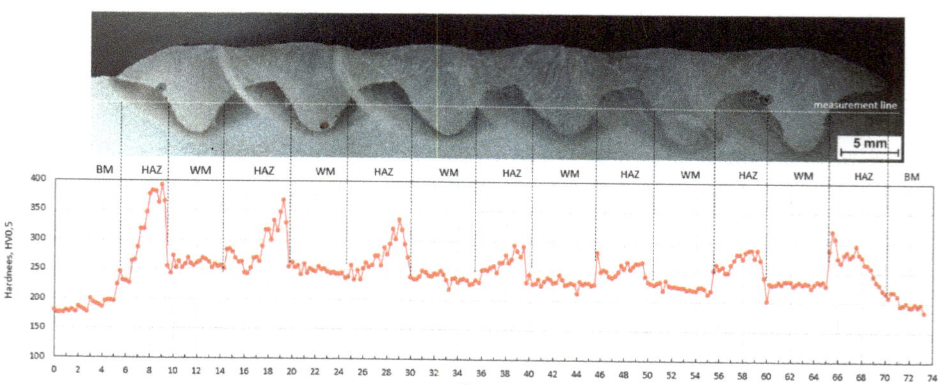

Figure 17. Hardness $HV_{0.5}$ distribution in plate A; BM—base material, W—weld metal, HAZ—heat-affected zone.

External weld beads reach higher values of hardness in HAZ than internal ones. This effect was observed for all sequences. In each of them a high content of martensite is observed near the fusion line, which results from the rapid heat dissipation during the laying of the first weld beads.

In the case of sequence A, the hardness of the weld bead and its HAZ decrease with the consecutive application of further beads. Heat accumulates in the base material and increases inter-pass temperature.

This increases the time $t_{8/5}$, leading to the transformation of austenite into softer microstructures. The maximum individual hardness value is 391 HV (in the area of first HAZ), the minimum is 201 HV (in the area of HAZ between the fifth and sixth welding beads).

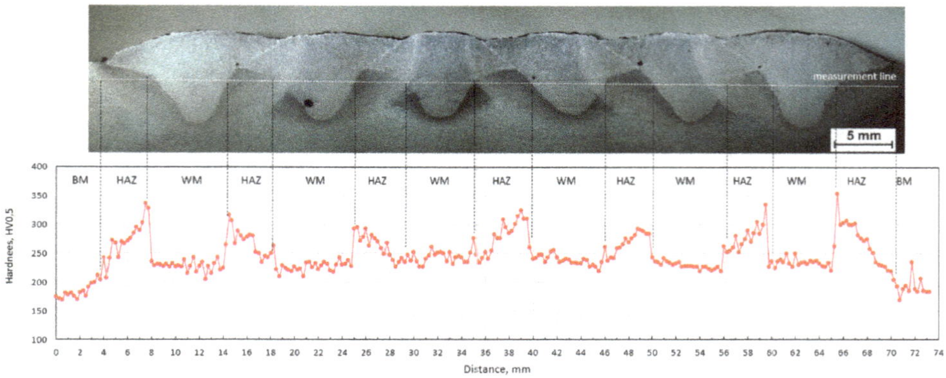

Figure 18. Hardness $HV_{0.5}$ distribution in plate B; BM—base material, WM—weld metal, HAZ—heat-affected zone.

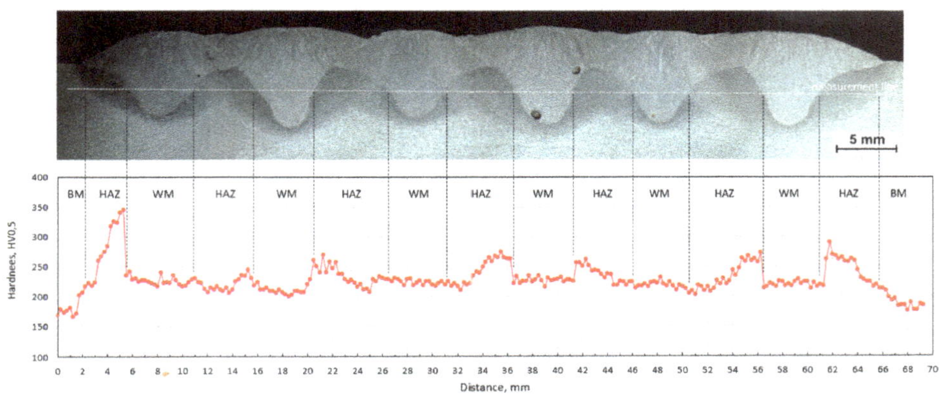

Figure 19. Hardness $HV_{0.5}$ distribution in plate C; BM—base material, WM—weld metal, HAZ—heat-affected zone.

In the case of sequence B, the welds are deposited from the center alternately on one and the other side. The inner beads are affected by others, which leads to the annealing of microstructure. The hardness of the welding beads is approximately the same (about 240 HV), and the HAZ reach even higher hardness values. External welds reach significantly higher values in HAZ (with maximum individual value of 350 HV) because they were made last. The effect of increased hardness in these places was not affected by the rising inter-pass temperature.

In the case of sequence C, an alternating application of beads is utilized. Beads four and five have a significant annealing effect and the hardness in this section is relatively balanced. The maximum hardness value is 345 HV (in HAZ of first bead). In terms of hardness distribution in the measured line, maximum hardness, and range of hardness, sequence C has the most suitable properties. The alternating (1-4-2-5-3-6) sequence of depositing weld seams had a positive effect on the hardness distribution in the joint. The hardness values had a low variance, the cooling time $t_{8/5}$ was extended. In the bead axis, the dominant structure is bainite. This applies not only to this case, but for almost every bead in the tested samples.

The hardness values from the measured lines were used to determine the average hardness of individual HAZ and welding beads. Figure 20a shows the average hardness values for individual welding beads for all welding sequences. In addition, Figure 20b shows the average hardness values for the individual heat-affected zones and all welding sequences. These values also support the microscopic determination of the volume percentage of structural phases in the individual zones evaluated in Section 3.2.

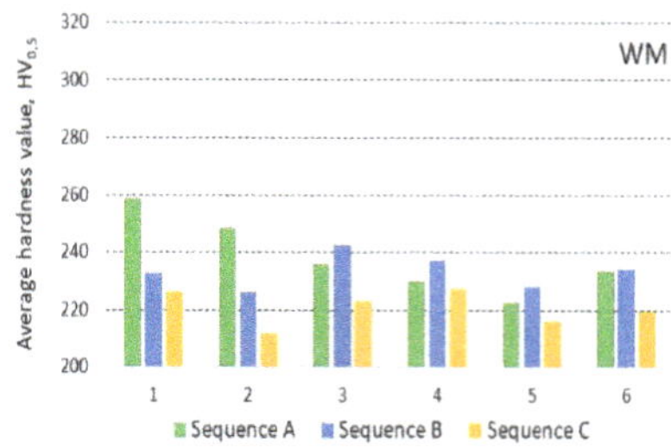

(**a**)

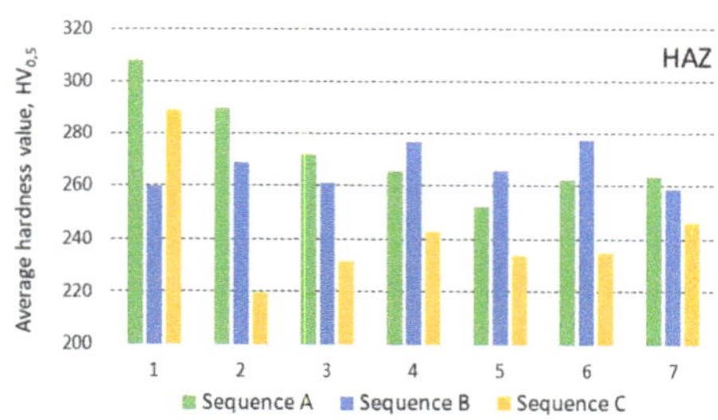

(**b**)

Figure 20. Average hardness value $HV_{0.5}$ of (**a**) weld metal (WM) and (**b**) of heat-affected zone (HAZ), measured on cross-sections of plates from left to right.

4. Conclusions

The experimental investigation results of the bead sequence on characteristic geometric dimensions, structure, and hardness of welds and heat-affected zones allow one to formulate the following conclusions:

- The serial sequence of the beads results in an increase in the depth of fusion along with the laying of subsequent surfacing beads. This is probably due to the length of the welds (time of making a single run), leading to a very short inter-pass time (increasing inter-pass temperature in the consecutive beads) or not obtaining a quasi-stationary temperature distribution in the padding welds. Determining the inter-pass temperature is important not only for the structure and hardness of the material after surfacing but also for the depth of penetration (HAZ).
- The calculated reinforcement, penetration, HAZ, and total weld padding areas are the largest for sample B (sequence 5-3-1-2-4-6) and the lowest for sample A (sequence 1-2-3-4-5-6). For sample C (sequence 1-4-2-5-3-6), intermediate values (between samples A and B) were recorded for all measured fields.

- Analysis of the geometrical dimensions of the weld and HAZ does not allow to formulate unambiguous conclusions regarding the influence of the welding sequence on the depth of penetration and depth of HAZ for individual beads.
- The greatest width, reinforcement, penetration, and total weld area in the cross-sections of the padded layer were obtained from sample B, and the lowest from sample A.
- The dominant structure in HAZ is bainite and does not depend on the bead sequence in contrast to the martensitic structure, which is most visible in the HAZ before the first bead of the weld and in the HAZ areas between successive overlapping beads.
- The most favorable structure was observed in sample C, where the alternating (1-4-2-5-3-6) sequence of depositing bead welds had a positive effect on the hardness distribution (maximum values and range) in the joint by extending the cooling time, creating a uniform temperature distribution and limiting the steel's tendency to hardening.

The achieved research results, their analysis, and their conclusions do not give an unequivocal answer to the question of which sequence of beads may be the most advantageous. Therefore, experimental analysis of the influence of the bead sequence on the properties of surfacing welds and HAZ should be the subject of further research and analysis, as it may be particularly important in the case of rebuilding and hardfacing layers. In the case of hardfacing with the use of appropriate electrode wires, the aim should be to obtain a regular distribution of high hardness (avoiding tempered zones). The use of additional research methods to determine the effect of the surfacing sequence on the type and magnitude of stresses can be very helpful in determining areas with a potentially higher probability of crack formation (rebuilding) and areas with reduced wear resistance (hardfacing). An important aspect would also be to determine the impact of the surfacing sequence on the type and size of welding distortions.

Author Contributions: Conceptualization, methodology, and formal analysis, J.W. and M.M. (Miloš Mičian); investigation, M.M. (Miloš Mičian), M.G., R.K., and M.M. (Miloslav Málek); resources, M.G., R.K., and P.P.; original draft preparation, M.G., M.M. (Miloslav Málek), and R.K.; review and editing, M.G. and P.P.; visualization, M.M. (Miloslav Málek) and P.P.; supervision, J.W., M.M. (Miloš Mičian), and M.G.; project administration, J.W.; funding acquisition, M.M. (Miloš Mičian). All authors have read and agreed to the published version of the manuscript.

Funding: This research was funded by The Slovak Research and Development Agency (APVV), grant number APVV-16-0276; Cultural and educational grant agency (KEGA), grant number KEGA, 009ŽU-4/2019; and Scientific Grant Agency (VEGA), grant number VEGA 1/0951/17.

Conflicts of Interest: The authors declare no conflict of interest.

References

1. McDonald, E.J.; Hallam, K.R.; Bell, W.; Flewitt, P.E.J. Residual stresses in multi-pass CrMoV lov alloy ferritic steel repair weld. *Mater. Sci. Eng.* **2002**, *A325*, 454–464. [CrossRef]
2. Vemanaboina, H.; Naidu, G.G.; Abhinav, D.S.; Krishna, K.; Reddy, D.R. Evaluation of residual stress in multipass dissimilar butt joints using X-ray diffraction. *Mater. Today.* **2019**, *19*, 283–288. [CrossRef]
3. Kyriakongonas, A.P.; Papazoglou, V.J.; Pantelis, D.I. Complete investigation of austenitic stainless steel multipass welding. *Ships Offshore Struc.* **2011**, *6*, 127–144. [CrossRef]
4. Sun, Y.; Wu, X.; Wu, X.; Li, J.; Jiang, Y. Influence of multi-pass welding on the microstructure evolution and corrosion resistance of a super duplex stainless steel. *Int. J. Electrochem. Sci.* **2016**, *11*, 9666–9675. [CrossRef]
5. Ivanow, Y.; Gromov, V.; Kormyshev, V.; Konovalov, S.; Kapralov, E.; Semin, A. Electric arc surfacing on low carbon steel: Structure and properties. *AIP Conf Proc.* **2016**, *1783*, 020077. [CrossRef]
6. Winczek, J.; Gucwa, M.; Mičian, M.; Koňár, R.; Parzych, S. The evaluation of the wear mechanism of high-carbon hardfacing layers. *Arch. Metall. Mater.* **2019**, *64*, 1111–1115.
7. Bermejo, M.A.V.; Hurtig, K.; Eyzop, D.; Karlsson, L. A new approach to the study of multi-pass welds–microstructure and properties of welded 20-mm-tick superduplex stainless steel. *Appl. Sci.* **2019**, *9*, 1050. [CrossRef]

8. Górka, J. Assessments of the corrosion resistance, properties and the structure of TIG braze welded galvanized steel sheets. *Arch. Metall. Mater.* **2020**, *65*, 47–54.
9. Golański, D.; Dymny, G.; Kujawińska, M.; Chmielewski, T. Experimental investigation of displacement/strain fields in metal coatings deposited on ceramic substrates by thermal spraying. *Solid State Phenom.* **2016**, *240*, 174–182. [CrossRef]
10. Tomków, J.; Janeczek, A. Underwater in situ local heat treatment by additional stitches for improving the weldability of steel. *Appl. Sci.* **2020**, *10*, 1823. [CrossRef]
11. Bhatti, S.; Chhabra, N.; Singh, P. Experimental investigation of multi-pass, welding current & arc travel speed on AISI 1020 on weld joint mechanical properties during GMAW. *Int. J. Res. Eng. Technol.* **2017**, *4*, 2024–2029.
12. Romek, D.; Selech, J.; Ulbrich, D.; Felusiak, A.; Kieruj, P.; Janeba-Bartoszewicz, E.; Pieniak, D. The impact of padding weld shape of agricultural machinery tools on their abrasive wear. *Tribologia* **2020**, *14*, 55–62. [CrossRef]
13. Kadivar, M.H.; Jafarpur, K.; Baradaran, G.H. Optimizing welding sequence with genetic algorithm. *Comput. Mech.* **2000**, *26*, 514–519. [CrossRef]
14. Voutchkov, I.; Keane, A.J.; Bhaskar, A.; Olsen, T.M. Weld sequence optimization: The use of surrogate models for solving sequential combinatorial problems. *Comput. Method. Appl. M.* **2005**, *194*, 3535–3551. [CrossRef]
15. Mohammed, M.B.; Sun, W.; Hyde, T.H. Welding sequence optimization of plasma arc for welded thin structures. *WIT Trans. Built. Env.* **2012**, *125*, 231–242.
16. Syahroni, N.; Hidayat, M.I.P. 3D finite element simulation of T-joint fillet weld: Effect of various welding sequences on the residual stresses and distortions. In *Numerical Simulation. From Theory to Industry*; Andriychyk, M., Ed.; InTechOpen: Rijeka, Croatia, 2012; pp. 583–606.
17. Jiang, W.; Yahiaoui, K. Effect of welding sequence on residual stress distribution in a multipass welded piping branch junction. *Int. J. Pres. Ves. Pip.* **2012**, *95*, 39–47. [CrossRef]
18. Viswanath, M.; Vijayanand, P.; Suresh, V.; Kumar, S.N.; Seshappa, A. Gas Metal Arc Welding sequence effect on residual stress and distortion by using Finite Element analysis. *SSRG Int. J. Mech. Eng. Spec. Issue ICETEST* **2018**, *5*, 15–19.
19. Gannon, L.; Liu, Y.; Pegg, N.; Smith, M. Effect of welding sequence on residual stress and distortion in flat-bar stiffened plates. *Mar. Struct.* **2010**, *23*, 385–404. [CrossRef]
20. Teng, T.L.; Chang, P.H.; Tseng, W.C. Effect of welding sequences on residual stresses. *Comput. Struct.* **2003**, *81*, 273–286. [CrossRef]
21. Chen, B.Q.; Soares, C.G. Effect of welding sequence on temperature distribution, distortions, and residual stress on stiffened plates. *Int. J. Adv. Manuf. Tech.* **2016**, *86*, 3145–3156. [CrossRef]
22. Fu, G.; Lourenço, M.I.; Duan, M.; Estefen, S.F. Influence of the welding sequence on residual stress and distortion of fillet welded structures. *Mar. Struct.* **2016**, *46*, 30–55. [CrossRef]
23. Moradi, M.; Pasternak, H. A study on the influence of various welding sequence schemes on the gain in strength of square hollow section steel T-Joint. *J. Weld. Join.* **2017**, *35*, 41–50. [CrossRef]
24. Deng, D. Influence of deposition sequence on welding residual stress and deformation in an austenitic stainless steel J-groove welded joint. *Mater. Des.* **2013**, *49*, 1023–1033. [CrossRef]
25. Fu, D.; Zhou, C.; Li, C.; Wang, G.; Li, L. Effect of welding sequence on residual stress in thin-walled octagonal pipe−plate structure. *T. Nonferr. Metall. Soc.* **2014**, *24*, 657–664. [CrossRef]
26. Shao, Q.; Xu, T.; Yoshino, T.; Song, N.; Yu, Z. Optimization of the welding sequence and direction for the side beam of a bogie frame based on the discrete particle swarm algorithm. *I. Mech Eng. B-J. Eng.* **2018**, *232*, 1423–1435. [CrossRef]
27. Sattari-Far, I.; Javadi, Y. Influence of welding sequence on welding distortions in pipes. *Int. J. Pres. Ves. Pip.* **2008**, *85*, 265–274. [CrossRef]
28. Islam, M.; Buijk, A.; Rais-Rohani, M.; Motoyama, K. Simulation-based numerical optimization of arc welding process for reduced distortion in welded structures. *Finite Elem. Anal. Des.* **2014**, *84*, 54–64. [CrossRef]
29. Rupani, R.; Barai, A. Influence of welding sequence on distortion of circumferential pipe joint—a review. *J. Prod. Res. Manag.* **2016**, *6*, 1–5.
30. Isaac, H.A.; Eduardo, G.E.; Joaquín, P.M. Welding sequence analysis in three dimensional weldments with experimental verification. In Proceedings of the XXII Annual Congress of the Association of Mexican Mechanical Engineers, Merida, Mexico, 28–30 September 2016; pp. 376–387.

31. Jiang, W.C.; Wang, B.Y.; Gong, J.M.; Tu, S.T. Finite element analysis of the effect of welding heat input and layer number on residual stress in repair welds for a stainless steel clad plate. *Mater. Des.* **2011**, *32*, 2851–2857. [CrossRef]
32. Kohandehghan, A.R.; Serajzadeh, S. Experimental investigation into the effects of weld sequence and fixture on residual stresses in arc welding process. *J. Mater. Eng. Perform.* **2012**, *21*, 892–899. [CrossRef]
33. Tomków, J.; Fydrych, D.; Rogalski, G. Role of bead sequence in underwater welding. *Materials* **2019**, *12*, 3372. [CrossRef] [PubMed]
34. Klimpel, A. *Cladding and Thermal Spraying*, 1st ed.; WN-T: Warsaw, Poland, 2000; pp. 10, 107–111, 140, 258, 297. (In Polish)
35. Winczek, J. Modeling of temperature field during multi-pass GMAW surfacing or rebuilding of steel elements taking into account the heat of the deposit metal. *Appl. Sci.* **2017**, *7*, 6. [CrossRef]
36. EN ISO 9015-1:2018. *Destructive Tests on Welds in Metallic Materials-Hardness Testing-Part 1: Hardness Test on arc Welded Joints*; European Committee for Standardization: Brussels, Belgium, 2018; p. 24.
37. EN ISO 6507-1:2018. *Metallic Materials-Vickers Hardness Test-Part 1: Test Method*; European Committee for Standardization: Brussels, Belgium, 2018; p. 14.

Publisher's Note: MDPI stays neutral with regard to jurisdictional claims in published maps and institutional affiliations.

materials

MDPI

Article

Heat Source Models in Numerical Simulations of Laser Welding

Tomasz Kik

Department of Welding Engineering, Silesian University of Technology, Konarskiego 18A, 44-100 Gliwice, Poland; tomasz.kik@polsl.pl; Tel.: +48-32-237-1681

Received: 25 May 2020; Accepted: 5 June 2020; Published: 10 June 2020

Abstract: The article presents new possibilities for modifying heat source models in numerical simulations of laser welding processes conducted using VisualWeld (SYSWELD) software. Due to the different power distributions and shapes of a laser beams, it was necessary to propose a modification of heat source models and methods of defining the heat introduced into a welded material in the case of simulations of welding processes using solid-state and high-power diode lasers. A solution was proposed in the form of modification of predefined heat source models in the case of simulations of welding processes using solid-state disc lasers and high-power diode lasers (HPDL). Based on the results of metallographic tests and the acquisition of thermal cycles of real laser welding processes, the process of calibration and validation of the proposed models of heat sources depending on the type of device used as well as the obtained shapes of fusion beads was carried out. The purpose and assumptions of this approach towards creating heat sources were also reported, comparing exemplary stresses and cumulative plastic strain distributions for the calculation variant using a standard and modified heat source model.

Keywords: finite element method (FEM); welding; laser; heat source model; keyhole welding; thermal analysis

1. Introduction

The basics of lasers were presented in 1917 by Albert Einstein in his theory of stimulated emission, where it was stated that stimulated radiation and stimulating radiation have the same properties, i.e., the same direction of propagation and the same phase of vibration and polarization. These properties therefore indicate the fact that with a beam of laser light, a concentrated portion of energy can be transported. However, a long time passed before this could be applied at an industrial scale. This was mainly due to the problem of building devices that would be able to provide adequate power so that the laser beam focused on the surface of the material would be able to heat it to a temperature suitable for the selected technological processes. At present, as this possibility now exists, the undoubted advantages of concentrated, monochromatic light combine the precision of the process with product quality and efficiency inaccessible to traditional arc welding methods. Joining material technologies using a laser beam are characterized by smaller deformations and the effect of heat on the joined elements. The use of laser devices has caused many technologies to change and has expanded the areas of application, offering an up-to-now unattainable level of product quality. With their advantages, laser technologies somehow meet ever-increasing quality requirements and are also able to address the increasing complexity of the welding technologies used [1,2].

Depending on the internal design of the device and the principle of operation, it is an option to use devices with different radiation intensity distributions across the laser beam cross-section [2]. The power density of the device depends mainly on the energy distribution in the focus of the laser beam. The shape of the electromagnetic field generated on the beam cross-section is referred to as transverse electromagnetic mode (TEM). The multitude of energy distributions that it is possible to obtain on the surface of the laser beam and in devices used in the industry, on the one hand, introduces incredible flexibility in adapting these devices to a specific application. On the other hand, it allows for many devices with significantly different properties despite, e.g., the similar maximum laser beam power offered. The existing variety of types of devices as well as beam focus shapes, radiation wavelengths, and power distributions on the laser beam focus surface means that the modern engineer has considerable possibilities for controlling the dimensions of the beads and the penetration shape obtained in processes using laser devices. However, this also causes a problem in choosing the right device for a particular application, especially taking into account the constantly shortening response time required by the market. The degree of complexity of current, modern welding technologies based on this type of equipment, along with the mentioned advantages and requirements, forces engineers and constructors to take a different approach to the design stage and search for modern tools supporting the preparation of welding and heat treatment technologies themselves [2–7].

One modern tool supporting their work is (among other programs for numerical simulation of manufacturing processes) based on the finite element method (FEM). The idea of supporting calculations was created quite a long time ago. The first attempts at prediction of residual stresses and deformations distributions were presented in 1938 by Rodger and Fletcher [8]. However, the insufficient computing power at that time blocked the use and development of these techniques for years. In 1970, Brust et al. presented the first welding numerical simulations using the finite element method [9–11]. Over the past several years, mainly due to the development of computers and the significant increase in available computing power, there has also been a huge improvement in computational techniques in the field of welding and heat treatment simulation. New computational techniques and a method for the mathematical description of the heat introduced into the welded material were developed. This resulted in a significant extension of the possibility of using modern computational programs in the everyday engineering practice of designers and welding engineers in industry [12–17].

Currently, numerical analyses of welding processes are usually used to optimize and validate these processes both from a technological and economic point of view, but in some cases they are used for the modification of the welded structure itself. Thanks to the continuous development of this field of knowledge, it is now possible to conduct local welding analyses, and then based on the received heterogeneous distributions of structures and material properties determine stresses and plastic deformations as well as the loads of the analyzed structure with external forces, pressures, and deformations. Based on the results obtained, it is possible to determine potential places of fatigue damage in the analyzed welded constructions. Modern software packages for numerical analyses create completely new opportunities for the modern engineer. The engineer may receive help that allows for a significant reduction in the time of production preparation or the introduction of changes in the technological process, thanks to the possibility of simulating their impact on the process result. In addition, this operation is completely safe because it takes place in the computer's memory and does not affect the actual technological process or security. These possibilities are however also associated with a certain degree of inconvenience, which is possible to resolve. The abovementioned possibilities for conducting analyses and the amount of input data and related information make numerical simulations of welding processes difficult and complicated to carry out using FEM simulations [17].

2. Description of the Problem

2.1. Basics of Formulating the Problem of Temperature Distribution Calculations in Numerical Simulations

Before FEM computer simulations of welding processes came into general use, the only way to present temperature fields during welding was through analytical methods. However, the emergence of modern technologies, new materials, and the increasing complexity of the designed structures have made the application of analytical methods very difficult or even impossible in some cases. The development of packages for computer simulation of welding processes based on the finite element method has opened up new possibilities in the field of design support for these processes. Since numerical simulations of temperature distributions are calculated on the basis of Fourier's differential formula, to perform calculations it is necessary to obtain the values of the heat conduction coefficient, specific heat, and density depending on the temperature [18–20]:

$$\frac{\partial T}{\partial t} = \frac{\lambda}{c\rho}\left(\frac{\partial^2 T}{\partial x^2} + \frac{\partial^2 T}{\partial y^2} + \frac{\partial^2 T}{\partial z^2}\right) = a\nabla^2 T \tag{1}$$

where T is the temperature, t is time, x, y, and z refer to the point coordinates, a is the thermal diffusivity coefficient, λ is the heat conductivity coefficient, c is specific heat, and ρ is mass density.

In the analyses presented later in the article, the VisualWeld 15.5 (SYSWELD) package of the ESI Group (Paris, France), was used, where the coupled thermo-metallurgical analysis is described by a modified heat convection equation [19,20]:

$$\left(\sum_i P_i(\rho C)_i\right)\frac{\partial T}{\partial t} - \nabla\left(\left(\sum_i P_i\lambda_i\right)\nabla T\right) + \sum_{i<j} L_{ij}(T)\cdot A_{ij} = Q \tag{2}$$

where $P_{i,j}$ refers to the proportions of phases with indexes i and j, respectively, $L_{ij}(T)$ is the latent heat of i to j transformation, A_{ij} is the proportion of phase i transformed to j in a time unit, and Q is the total amount of introduced heat.

The consequence of using Fourier's differential equation is the non-linearity of the calculated temperature field distributions. It is also necessary to consider that the temperature field distributions are calculated based on the mathematical description of a heat source model represented by thermal flow density into material. In many cases it is a moving heat source and the coordinates describing its location are related not only to the space coordinates but also to the duration of the process. The formation and development of welding strains and stresses is associated with several very closely related factors such as clamping conditions, thermo-mechanical material properties, type of used welding technology, process parameters, ambient temperature, and cooling conditions, among others [14–19,21,22].

2.2. History of Heat Source Models in Numerical Simulations

Models of heat sources used in numerical simulations of welding processes can be generally divided into two groups:

— Models with a homogeneous distribution;
— Models with heterogeneous distribution [23]

In 1941, Daniel Rosenthal was the first person to apply Fourier's law to a moving heat source [24]. This resulted in the first model of a heat source with a homogeneous distribution. Unfortunately, the biggest disadvantage of this solution was the fact that all the energy of the process was concentrated in one point, which in effect meant that the use of this model did not in any way allow for the determination of the shape of the molten pool and the depth of penetration. This solution was then modified, but an acceptable compliance of temperature distribution prediction was obtained only at a distance from the heat source itself. A certain extension was the linear model of the heat source proposed in 1969 by Pavelic [25]. Thanks to this solution, an attempt was made to determine the temperature distribution in a two-dimensional element. Still, a problem was the inability to precisely map the temperature gradient over the thickness of the model. A few years later, Friedmann proposed a heat source model that had a volume distribution. Equations describing cylindrical and square models of the heat source were successively developed. Thanks to these activities, the results obtained became closer and closer to those obtained from real tests. It should be noted that at that time processes with high power density were already being described using rectangular (electron beam) and cylindrical heat source models (laser beam). Nevertheless, the degree of simplification of these models did not ensure a complete mapping of the nature of these processes [23,26]. In 1983, Eagar and

Tsai introduced the first model with the heterogeneous distribution. The proposed model to some extent allowed for the weld pool geometry to be shown by using the Gaussian heat distribution on the surface [27]. On this basis, a disk model was later created. Its use was still somewhat limited due to the inability to determine the depth of operation of the model [23].

Over the years, in association with the development of numerical analyses of welding processes, many mathematical descriptions have been created regarding models of heat sources. These range from the simpler point models that limit all energy of the welding process to one point (mesh node), through to linear and circular models, up to more advanced forms. However, it was in the early 1980s that John Goldak proposed a widely used model in the shape of a double ellipsoid. Although it is currently mainly used for simulation of arc welding processes, i.e., a technique with a liquid metal pool, after appropriate modifications it can also be used to simulate selected laser welding processes, as shown later in the article [16,21–23,28,29].

2.3. Typical Models of Heat Sources Used in Numerical Simulations

In commercially available software dedicated to solving welding process problems, three types of predefined heat source models are usually used:

— The Gaussian surface heat source model;
— The Double ellipsoid heat source model (Goldak's model);
— The conical heat source model

The Gaussian distribution model, in which energy is distributed according to the course of the Gaussian curve, is usually used in simulations of welding processes with high power densities, i.e., laser, plasma, and microplasma or electron beam welding (Figure 1). With the help of this heat source model it is possible to map deep penetration while maintaining a small width, characteristic of "keyhole" welding techniques. In the VisualWeld (SYSWELD) environment, a normal distribution heat source is a predefined and recommended model used in simulations of a surface heat treatment processes, including laser beam hardening [20]. The Gaussian surface heat source model is described as in [20]:

$$Q(x,y,z) = Q_0 exp\left(-\frac{x^2+y^2}{r_e^2}\right) \quad (3)$$

where Q_0 is the amount of heat introduced by the source, and x, y, and r_e are the geometrical parameters describing the heat source position.

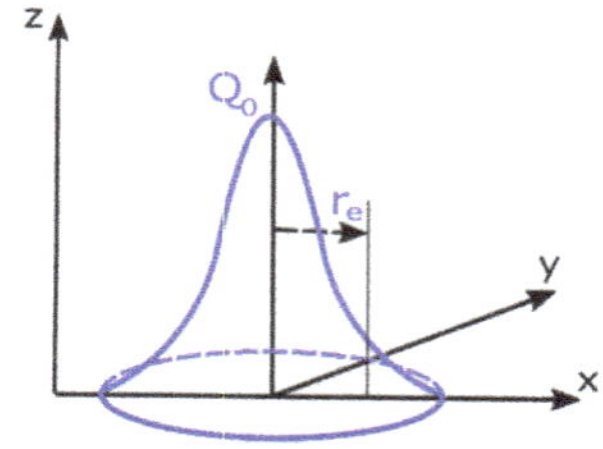

Figure 1. Gaussian surface heat source model [20].

The most popular heat source model currently used in commercial programs for numerical analysis of welding processes is the volumetric double-ellipsoid heat source model, also known as Goldak's model. This model is built of two ellipsoids placed perpendicular to each other, allowing the size of the source to be determined in the plane resulting from the position of the source perpendicular to the direction of welding (Figure 2).

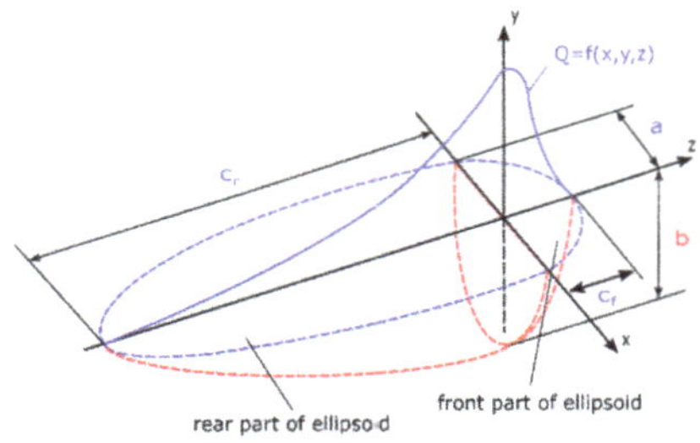

Figure 2. Double ellipsoid heat source models [17,18,20,21].

Unlike other volumetric models used for arc welding simulations, Goldak's model is described by two equations individually for each part of the ellipsoid separately [17,18,20,21]. For the front part of the ellipsoid, the equation is:

$$Q_f(x,y,z) = \frac{6\sqrt{3}f_f Q}{abC_f\pi\sqrt{\pi}} exp\left(\frac{-kx^2}{a^2}\right) exp\left(\frac{-ly^2}{b^2}\right) exp\left(\frac{-mz^2}{c^2}\right) \tag{4}$$

and for the rear part of the ellipsoid, the equation is:

$$Q_r(x,y,z) = \frac{6\sqrt{3}f_r Q}{abC_r\pi\sqrt{\pi}} exp\left(\frac{-kx^2}{a^2}\right) exp\left(\frac{-ly^2}{b^2}\right) exp\left(\frac{-mz^2}{c^2}\right) \tag{5}$$

where Q_f, Q_r are the volumetric heat flux density in front and rear part of the model, respectively, (W/m^3), Q is the total introduced power, and a b, c_f, and c_r are respectively the width, depth, and length of the front and the rear part of the estimated molten pool.

The coefficient of heat transfer efficiency to the welded material is determined depending on the welding method being analyzed [17,20].

How much of the total heat is transferred to the front and back of the ellipsoid is determined by the parameters f_f and f_r. These are constants that affect the distribution of energy flow to the material, which can somewhat compensate, e.g., the tilting effect of the welding torch. As practice has shown, the relation of f_f to f_r is usually accepted as a ratio of 60:40. It is important to remember that these parameters must satisfy the dependence $f_f + f_r = 2$ [17,30]. Determining the values of coefficients k, l, and m is a much more complex issue. When using an unmodified heat source model, all three factors assume a value of 3. An example of such an application can be e.g., simulation of the Metal Manual Arc (MMA) welding process. However, in the case of simulation, e.g., of the Gas Metal Arc (GMA) welding process, the values of these coefficients often must be modified experimentally to obtain the correct shape of the molten pool. Currently, this model so far best describes the actual state of arc welding methods [16–18,20–22].

In analyses of welding processes characterized by high energy densities, i.e., laser, electron beam welding or plasma welding, a conical source model with a normal distribution is usually used (Figure 3).

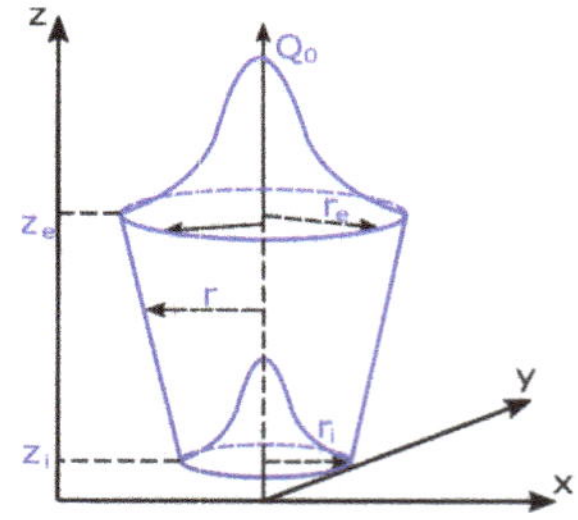

Figure 3. Three-dimensional conical heat source models [17,20].

The mathematical description of volumetric heat flow density into the material is described as follows [17,20]:

$$Q(x, y, z) = Q_0 exp\left(-\frac{x^2 + y^2}{r_0^2(z)}\right) \quad (6)$$

$$r_0(z) = r_e + \frac{r_i - r_e}{z_i - z_e}(z - z_e) \quad (7)$$

where Q_0 is the maximum value of volumetric heat flux density, r_e, and r_i are respectively the upper and lower cone radius dimensions parameters, z_e, and z_i are the cone length parameters, and x, y, and z are point coordinates (Figure 3).

The use of a moving heat source model in numerical simulations of welding processes requires the use of an appropriate calibration procedure. Only in this case is it possible to correctly determine the stresses and deformation levels and distributions, calculated based on previously obtained results of temperature fields and individual metallurgical phase distributions. Incorrectly performed calibration of the heat source model usually leads to significant differences in the results obtained as compared to the reality. The use of predefined models does not mean, however, that the user cannot create a combination of these sources to obtain e.g., a set suitable for the simulation of hybrid welding consisting of a conical model for the laser beam and the double-ellipsoid model for mapping the effect of arc welding [31]. It is possible to use self-defined heat source models as well as other computational techniques that do not use moving heat source models. Unfortunately, they often require initial calibration based on the analyses performed using them [17].

2.4. The Shapes of Beads Geometry Obtained in the Case of Laser Welding Processes—Formulation of the Research Problem

As mentioned before, the conical heat source model is the most commonly used heat source model in the case of the "keyhole" welding technique. However, the model does not always allow for correct results as compared to the reality. While it works relatively well when calculating a process partial penetration, when there is full penetration obtained it is not possible to model the bead expansion characteristic of the laser welding process in the root area when welding thicker parts. In the laser welding process, essentially three different bead geometry shapes can be obtained. In the case of welding without full penetration, the shape is usually narrow, cylindrical, or slightly expanded upwards—the so-called bell shape (Figure 4a).

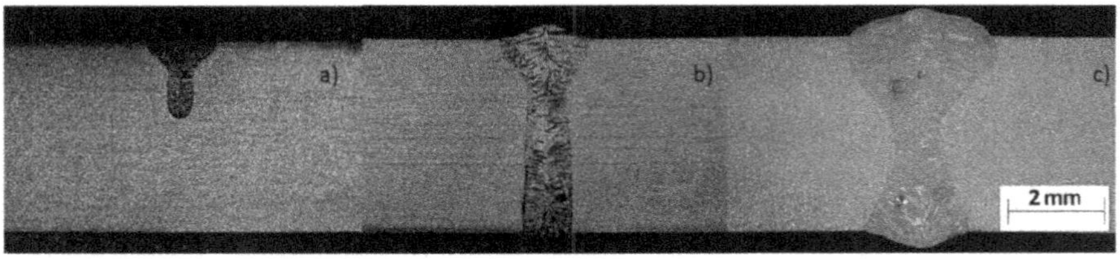

Figure 4. Examples of macrographs performed with disc laser welding (**a**) with partial penetration (bell shape), and (**b**) and (**c**) with full penetration (with a narrow and wide root known as the hourglass shape).

The situation changes after achieving full penetration of the bead. In this case, basically two shapes of the melted metal area are obtainable. One has a wide face area and slight widening in the area of the root (Figure 4b). The situation changes after achieving full remelting. In this case, two shapes of the melted metal area are obtainable. The first is due to wide melting in the face area and slight expansion in the ridge area. The second, in turn, shows wide remelting in both the face and root areas, with a narrow area melted in the middle of the bead (hourglass shape) (Figure 4c). The bead shape obtained when welding with full penetration depends on the type of device used, the process parameters, and sheet thickness as well as the type of material being welded [29,31–35]. The information found in the literature shows that the nature of the bead geometry changes at full penetration results from several complex phenomena that determine the flow of liquid metal in the molten pool and the convective heat that affects the shape of the bead [29,34–36]. A complete solution to this type of problem requires numerical simulations using complex fluid mechanics models. Unfortunately, the aforementioned simple models used in commercial software for simulation of welding processes are models based on thermal conductivity, where this flow is not taken into account. Several studies have appeared in the literature on attempts to solve this problem. In 2017, Flint proposed a model that is an extension of the double-ellipsoid model that can be used for narrow-groove and keyhole weld configurations [37]. Wu et al. also proposed some modifications of the conical model where a modified three-dimensional cone model (TDC) was used to simulate the plasma welding process with the mesh technique [38]. Similarly proposed by Farrokhi et al. in the case of hybrid welding, combined models based on a double three-dimensional conical heat source model for the reconstruction of "keyhole" welding conditions with partial and full penetration indicate the fact that in order to obtain the correct shape of the melted area, it is necessary to combine two or more volumetric models of heat sources [29]. In 2018, for the investigation of a dual-beam laser welding of aluminum sheets, He et al. proposed modification of the cone-shaped heat source model for the lap joint configuration. Based on the proposed model and the experimental validation procedure, investigations of hot cracking were conducted [39].

From a scientific point of view, this is the most correct reasoning aimed at precisely and accurately reproducing the obtained bead shape in the heat source model. For a typical commercial user of mathematical notation of this type, proposed models of heat sources can be a real obstacle to everyday use. Not every user has the programming knowledge needed to create their own heat source model description. Therefore, research should be continued to propose the simplest and most effective way to reproduce the actual geometry of the bead during numerical simulations of welding processes.

3. Modification of Predefined Heat Source Models in the Case of Laser Bead-On-Plate Welding

3.1. Test Stand and Preliminary Attempts to Use Predefined Heat Source Models

To propose a solution that would be simple to use for a typical commercial user, welding tests were carried out on two diametrically opposed laser devices. The first test stand was equipped with a Trumpf TruDisk 3302 (Ditzingen, Germany) disk laser with a maximum beam power of 3300 W (Figure 5a). This laser allowed us to obtain a high quality beam with a wavelength of 1030 nm while maintaining a beam spot diameter of 200 μm. The laser head was equipped with a 200-mm collimator lens and a 200-mm focusing lens. The beam parameter product (BPP), as claimed by the manufacturer of this laser, was <8.0 mm × mrad.

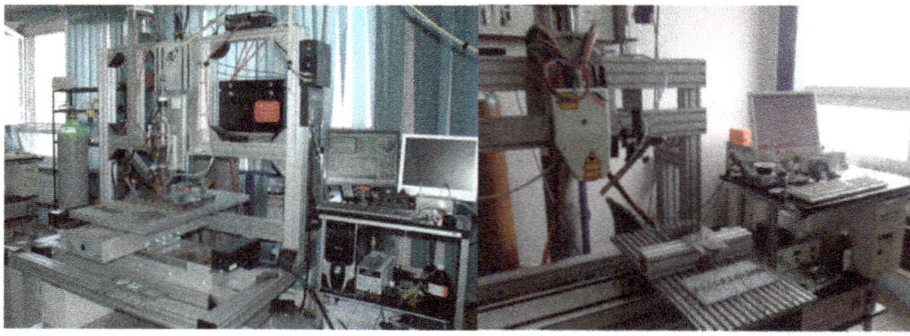

(**a**) (**b**)

Figure 5. View of test stands equipped with (**a**) TruDisk 3302 disk laser with 3300 W laser beam power and (**b**) a Rofin DL020 high-power diode laser with 2200 W beam power.

The second type of laser used in the research was a high-power diode laser (HPDL) (Figure 5b). The ROFIN DL 020 direct diode laser from ROFIN Sinar (Hamburg, Germany) allows for the following rectangular beam spot dimensions to be obtained: 1.8 mm × 6.8 mm at 82 mm focal length or 1.8 mm × 3.8 mm at 32 mm focal length (with an additional focusing lens). The maximum output power of the HPDL ROFIN DL 020 is 2200 W with the wavelength at 808 nm, with very even energy distribution across the beam spot (multimode TEM).

For preliminary tests using a disk laser, bead-on-plate welding tests were performed on a 6.0-mm-thick sample of S355 steel at 1500 W with a welding speed of 8.3 mm/s, and on a 4.0-mm sample of AISI304 steel (X5CrNi18-10) at 1000 and 1500 W and 8.3 mm/s welding speed (Tables 1 and 2).

Table 1. Chemical composition of S355 (EN10025-2) and X5CrNi18-10 (EN10088-3) steel [40,41].

	Cr (%)	Ni (%)	Mn (%)	N (%)	Si (%)	C (%)	S (%)	P (%)
S355	-	-	1.5	-	0.55	0.20	0.035	0.035
AISI304	17.5–19.5	8.0–10.5	max. 2.0	max. 0.11	max. 0.75	max. 0.08	max. 0.03	max. 0.045

To simulate bead-on-plate welding tests, a three-dimensional numerical model of a 6.0-mm-thick flat sheet was prepared. To ensure the correct mapping of heat dissipation conditions, a boundary condition was adopted describing heat exchange with the environment on all external surfaces of the model. Boundary conditions for clamping were adopted as standard conditions corresponding to welding of the sample placed without additional fastening on the laser station table. The calculations were performed in the "heat source input fitting" module as a "steady state"-type analysis (analysis carried out at only one moment of time after the process was established), using a normal distribution heat source model (Gaussian surface heat source model) with the following parameters: laser beam power, 1500 W; welding speed, 8.33 mm/s; r_0 = 2.0 mm (Figure 1). As a result of tests carried out at the laser stand, a bead width of 2.08 mm and a maximum penetration depth of 5.20 mm were obtained. For comparison, the results of numerical simulation allowed the determination of these dimensions at the level: bead width 2.02 mm and maximum penetration depth 5.31 mm (Figure 6). In the case of the parameters used, a similar molten pool shape was obtained in the numerical simulation. The tests carried out showed that as the laser beam power (and penetration) increased, the compatibility of the numerical simulation with the reality for this heat source model decreased significantly.

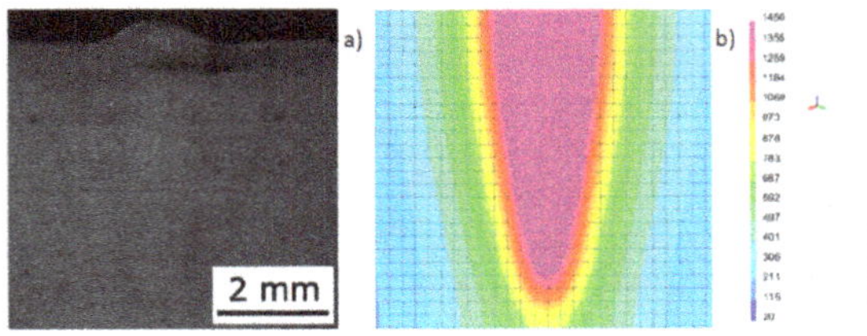

Figure 6. View of the macrostructure of the bead (**a**) in bead-on-plate welding tests performed on a 6.0-mm-thick sample of S355 steel using the Trumpf TruDisk 3302 disk laser at laser beam power of 1500 W and a welding speed of 8.3 mm/s; (**b**) cross-sectional view of the molten pool obtained as a result of numerical simulation; and (**c**) the 1450 °C iso-surface.

In the case of AISI304 steel samples, a three-dimensional numerical model of a flat sheet with a thickness of 4.0 mm was made analogously. The boundary conditions were also set to simulate no sample fixing. The calculations were performed using a conical source heat source model using the following parameters of the heat source model (Figure 3):

DL1.0_8.33—Laser beam power: 1000 W; welding speed: 8.33 mm/s; $r_e = 2.44$ mm; $r_i = 1.12$ mm; $z_e = 0$; $z_i = 3.16$ mm

DL1.5_8.33—Laser beam power: 1500 W; welding speed; 8.33 mm/s; $r_e = 3.1$ mm; $r_i = 2.4$ mm; $z_e = 0$; $z_i = 4.0$ mm.

As a result of the tests, beads with the aforementioned shapes were obtained, with bell and hourglass shapes with the geometrical dimensions presented in Table 2 and Figure 7. While there is considerable agreement between the dimensions measured on real samples and the same dimensions obtained from the results of numerical analyses carried out using a conical model of the heat source, the shape of the real and calculated beads varied significantly (Figure 7).

Table 2. Comparison of geometric dimensions of beads in bead-on-plate welding tests performed on a 4.0-mm-thick sample of AISI304 steel using the Trumpf TruDisk 3302 disk laser with values obtained from numerical analysis (values in brackets) (Figure 7).

Specimen Designation	Bead Widthin the Face Area (mm)	Bead Width in the Root Area (mm)	Penetration Depth (mm)
DL1.0_8.33	2.44 (2.36)	1.12 (1.01)	3.16 (3.15)
DL1.5_8.33	3.04 (3.09)	2.3 (2.07)	4.0 (4.0)

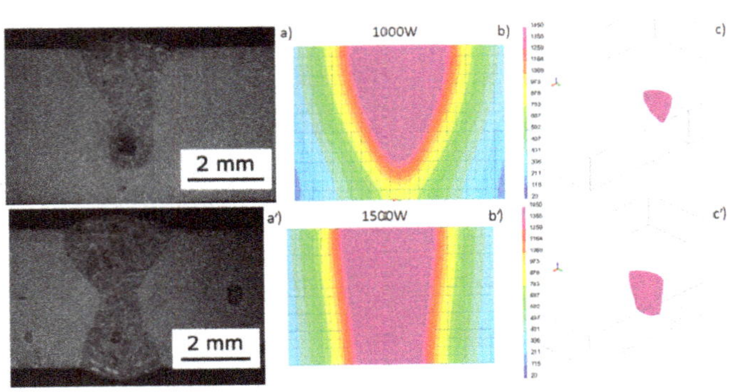

Figure 7. View of the macrostructure of the bead (**a, a′**) in bead-on-plate welding tests performed on a 4.0-mm-thick sample of AISI304 steel using the Trumpf TruDisk 3302 disk laser at laser beam power accordingly 1000 and 1500 W and welding speed of 8.3 mm/s; (**b, b′**) a cross-sectional view of the molten pool obtained as a result of numerical simulation; (**c, c′**) the 1450 °C iso-surface.

In summary, the modeling tests carried out using implemented models of heat sources, i.e., the Gaussian surface and conical heat source model, showed that there is a need to develop modifications to obtain the actual shape of calculated beads.

3.2. Assumptions and Construction of a Modified Numerical Model in the Case of a Disk Laser

It is possible to create a combined model consisting for example of two conical sources or one conical and other cylindrical sources (Figure 8). However, in this case it is problematic to divide the amount of energy for each of these parts (as shown by the author's experience, not from the volume ratio of both models). This is due to different heat distribution conditions in individual batches of materials as well as heat dissipation conditions. Additionally, this requires defining two trajectories along which the heat source model moves (Figure 8). Nonetheless, it should be emphasized that this simulation method allows for a significant improvement in the results obtained and a better mapping of the fusion line shape.

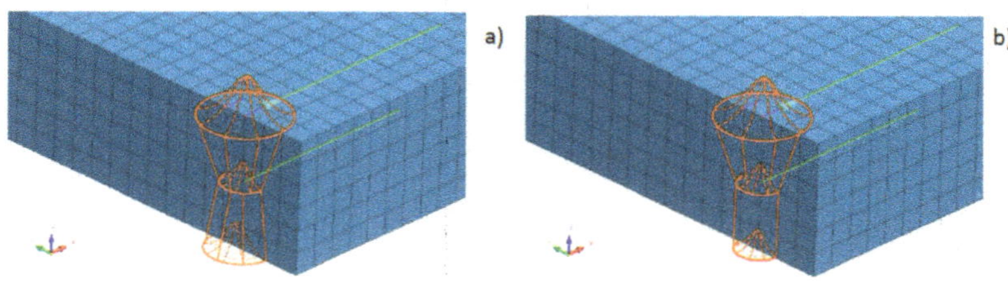

Figure 8. A view of heat sources models consisting of (**a**) two conical sources and (**b**) a conical and cylindrical source (green lines correspond to trajectories along which the sources move).

Considering the above, the simplest methods for creating a model with a high correlation coefficient of results with reality are still being sought as simply as possible. To solve this problem, it is possible to use the functionality offered by the VisualWeld 15.5 package from the ESI Group, France. In this software, it is possible to specify elements in the model mesh that will be only affected by a standard heat source model. Simply put, only the selected mesh elements covered by the standard heat source model will be heated. Using this function, it is possible to modify the shape of the fusion line in a relatively convenient and simple way (Figure 9).

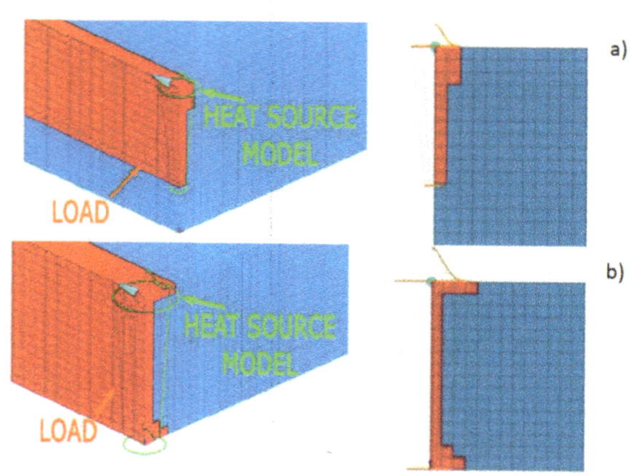

Figure 9. Examples defining the loaded areas of the model (network) and location of the heat source model in the case of laser welding with the mesh technique with (**a**) incomplete and (**b**) full penetration.

To use this method, however, it must be assumed that in principle, it applies to all numerical analyses of welding processes. The shape of the bead fusion line (based on the macroscopic metallographic picture) should be treated as input data, not the result of the analysis. The task of numerical analysis performed based on the heat conduction equation, not the problem of fluid mechanics, is to obtain additional information on the distribution of metallurgical phases, hardness, stresses, or joint deformations during and after the process. The lack of consideration of complicated mathematical models describing the convective movements of liquid metal in the weld pool itself means that the shape of the fusion line, which is highly influenced, must be set by the user. This is justified by the fact that in simulations with available programs for numerical analyses of welding and heat treatment processes many factors are ignored that in welding processes have a significant impact on the shape of the fusion line. This even includes the impact of the shielding gas used or the method of transferring metal in the arc during GMA welding, which are factors that have a huge impact on the nature and intensity of the abovementioned movements of liquid metal in the pool. Thus, the shape of the fusion line is a very important input parameter at the calibration stage of the calculation model. Following the proposed assumption, the calibration of the heat source model is not only about entering the source power and dimensions of the heat source model corresponding to actual or expected values, but also manipulation of the heat source model shape and the area it affects (red-colored area in Figure 9 marked as "LOAD") so that the obtained result is as close as possible to the result of metallographic tests of the bead used in model calibration. The user, by appropriately selecting the parameters of the heat source model and the mesh elements that he exposes, has much greater possibilities of mapping the actual shape of the melted area.

3.3. *Calibration and Validation of the Model with a Modified Method of Heat Input in the Disk Laser Bead-On-Plate Welding Process*

To confirm the assumptions of the proposed simulation method, a TruDisk 3302 disk laser bead-on-plate welding process on 6.0-mm-thick AISI304 steel samples was carried out (Table 3). Data from metallographic tests of the obtained fusion beads were used to calibrate the heat source model. To validate the calibrated model, 4 samples with a thickness of 4.0 mm from AISI304 steel were also analyzed, to which the bottom surface was welded with three K-type thermocouples (Ni–NiCr wire with diameter of 0.2 mm) to record the heat cycles of the welding process. Thermocouples with accuracy ±0.0075 × T in a temperature range from −40 to +1200 °C (according to EN 60584-1 standard) were connected via compensation cables with an Agilent 34970A recorder by Agilent Technologies Inc., Santa Clara, CA, USA [42]. The arrangement scheme was as follows: the TC3 thermocouple was placed on the bead axis in the middle of its length, the TC2 thermocouple was placed 5.0 mm from the bead axis in the middle of its length, and the TC1 thermocouple was also placed in the bead axis at 5.0 mm from TC3 (Figure 10).

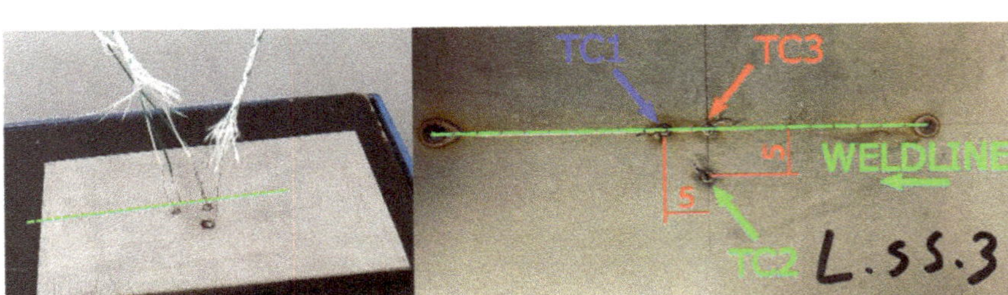

Figure 10. Diagram of the location of thermocouples relative to the bead axis at the bottom of the test sample.

Table 3. Parameters of bead-on-plate laser welding of the 4.0- and 6.0-mm-thick AISI304 plates with the TRUMPF TruDisk 3302 laser.

Specimen Designation	Laser Beam Power (w)	Welding Speed (mm/s)	Energy Per Unit Length (J/mm)	Remarks
DL_T1	1000	8.33	120	Partial penetration, bell shape
DL_T2		16.6	60	
DL_T3	2000	8.33	240	
DL_T4		16.6	120	
DL_T5	3000	8.33	360	Full penetration, narrow root
DL_T6		16.6	180	
DL_TC_1	1400	16.6	84	Partial penetration, bell shape
DL_TC_3	1800	33.3	54	
DL_TC_2	1800	16.6	108	Full penetration, narrow root
DL_TC_4	2400	33.3	72	

Remarks—Argon flow rate via cylindrical nozzles: 15 L/min, laser spot diameter: 200 μm, location of the laser beam focus: on the upper surface of the welded sample. TC1_4 samples: 4.0-mm-thick samples equipped with sets of 3 thermocouples welded to the bottom of the sample according to Figure 10.

Then the symmetrical three-dimensional discrete model, containing 28,080 3D elements and 30,622 nodes for 6.0-mm-thick samples and 21,518 3D elements (6 node linear wedges and 8 node linear bricks) and 23,706 nodes for 4.0-mm-thick samples was prepared. Mesh was refined in the weld and heat-affected zone (HAZ). The applied boundary conditions simulating sample attachment corresponded to its free placement on the laser table throughout the whole process of welding and then its cooling. In addition, a boundary condition of symmetry was added to the surface of the model division (Figure 11). The boundary condition describing heat dissipation to the environment was implemented by defining convective heat dissipation to the environment at 20 °C and radiation on all external surfaces of the model. As a heat source model, proposed as volumetric, a conical model with a modified "LOAD" area was used (Figure 11). The calculations to define the material database contained the thermal, mechanical, and metallurgical material properties, which were dependent on temperature and metallurgical phase proportions.

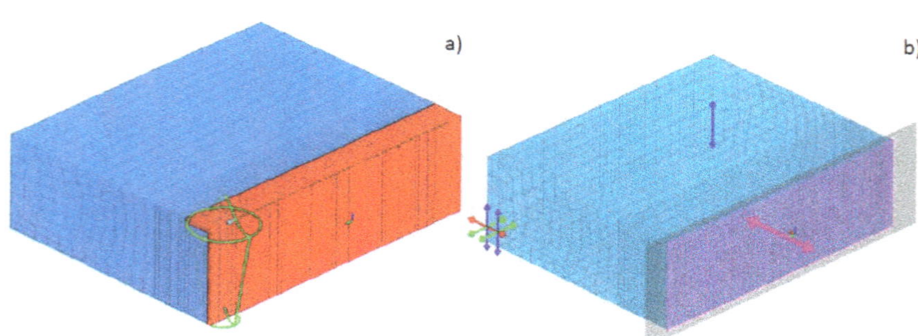

Figure 11. View (**a**) of the symmetrical three-dimensional model of the laser bead-on-plate welding process together with a conical model of the heat source and the "LOAD" area, and (**b**) boundary clamping conditions.

Based on the results of metallographic tests, the calculation models were calibrated, and the final results of numerical analyses were compared with the corresponding shapes of the actual beads obtained in the welding tests (Figures 12–14). On this basis, a comparison was also made of the measured and calculated basic geometric dimensions of the beads obtained (Table 4).

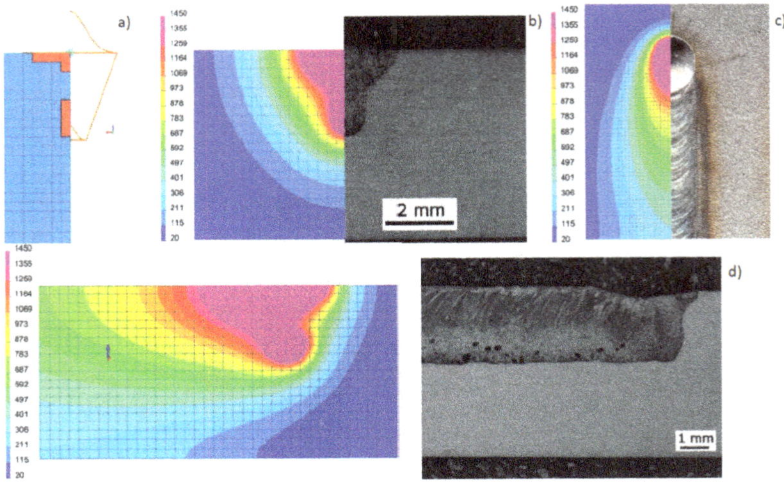

Figure 12. View (**a**) of the method of loading the mesh elements and placement of the conical model of the heat source, comparing (**b**) the shape of the bead on the cross-section, (**c**) the shape and size of the weld pool, and (**d**) the longitudinal section of DL_T1 test bead with the results of numerical analyses (Tables 3 and 4).

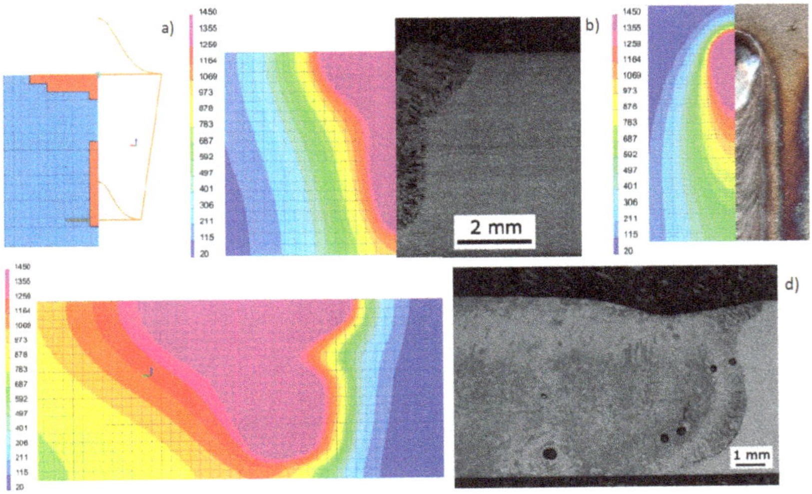

Figure 13. View (**a**) of the method of loading the mesh elements and placement of the conical model of the heat source, comparing (**b**) the shape of the bead on the cross-section, (**c**) the shape and size of the weld pool, and (**d**) the longitudinal section of DL_T3 test bead with the results of numerical analyses (Tables 3 and 4).

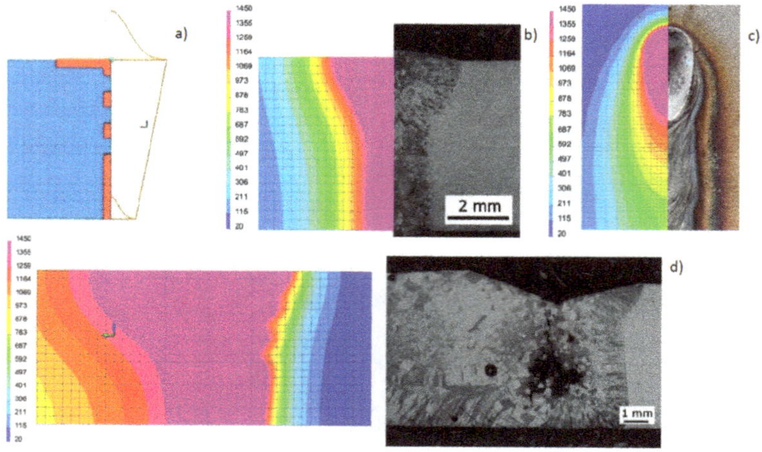

Figure 14. View (**a**) of the method of loading the mesh elements and placement of the conical model of the heat source, comparing **(b)** the shape of the bead on the cross-section, (**c**) the shape and size of the weld pool, and (**d**) the longitudinal section of DL_T5 test bead with the results of numerical analyses (Tables 3 and 4).

Table 4. Comparison of geometric dimensions of selected beads on the bead-on-plate laser welding process performed on 6.0-mm AISI 304 steel samples using the Trumpf TruDisk 3302 disk laser with values obtained from numerical analysis (values in brackets) (Table 3, Figures 12–14).

Specimen Designation	Bead Widthin the Face Area (mm)	Bead Width in the Root Area (mm)	Maximum Penetration Depth (mm)	Molten Pool Length (mm)
DL_T1	3.07 (3.18)	1.04 (0.96)	2.71 (2.71)	4.3 (4.6)
DL_T3	4.44 (4.46)	1.56 (1.44)	5.11 (5.33)	7.0 (7.08)
DL_T5	4.38 (4.37)	1.88 (1.79)	6.0 (6.0)	7.5 (7.54)

Following the assumptions of the proposed model preparation method, re-calculations were also performed for the DL1.5_8.33 sample analyzed in sub-chapter 3.1 (Figure 7). The results of the repeated numerical analysis showed that there was a significant improvement in the quality of the fusion line mapping using the same model of a conical heat source but with a modified area of the loaded elements of the mesh (previously all mesh elements covered by the cone were loaded), (Figures 7 and 15).

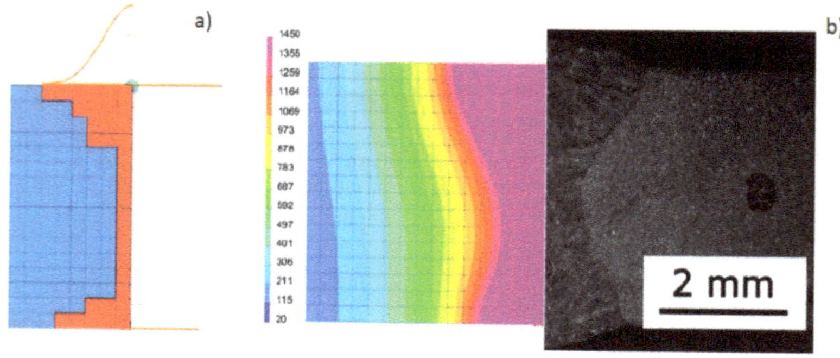

Figure 15. View (**a**) of the method of loading the mesh elements and placement of the conical model of the heat source, comparing (**b**) the shape of the DL1.5_8.33 bead on the cross-section with the results of numerical analysis (Table 1, Figure 7).

TC_1-TC_4 samples were prepared to collect information for model validation using thermal cycle runs. After calibrating the calculation model (Figure 16), the registered and calculated points corresponding to the location of thermocouples were also compared, as was the course of the process thermal cycles (Figure 17, Table 3).

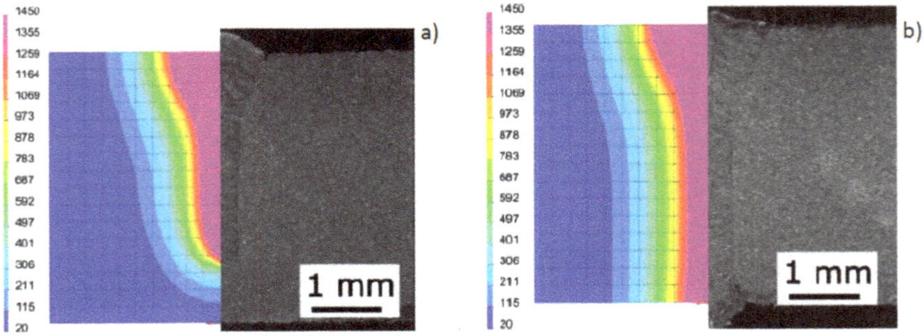

Figure 16. Comparison of the bead shape in the cross-section of sample (**a**) DL_TC3 and (**b**) DL_TC4 with the results of numerical analyses (Table 3).

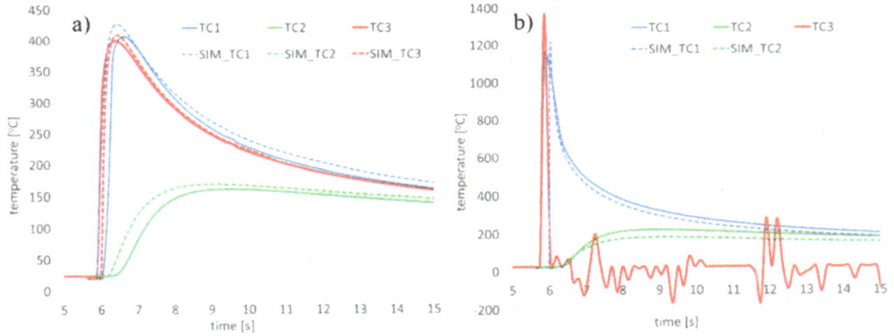

Figure 17. Comparison of recorded and calculated thermal cycles with the bead-on-plate laser welding process performed on 4.0-mm AISI 304 steel samples using the Trumpf TruDisk 3302 disk laser for sample (**a**) DL_TC3 and (**b**) DL_TC4. TCx: cycles recorded during the welding process; SIM_TCx: cycles obtained from numerical analyses.

To show the purpose of such precise calibration of the model and reproduction of the actual fusion line shape, for selected DL_T3 beads the von Mises reduced stresses and cumulative plastic strain distribution were also calculated for the simulation case with a standard heat source model and unmodified and modified load element area (Figures 18 and 19).

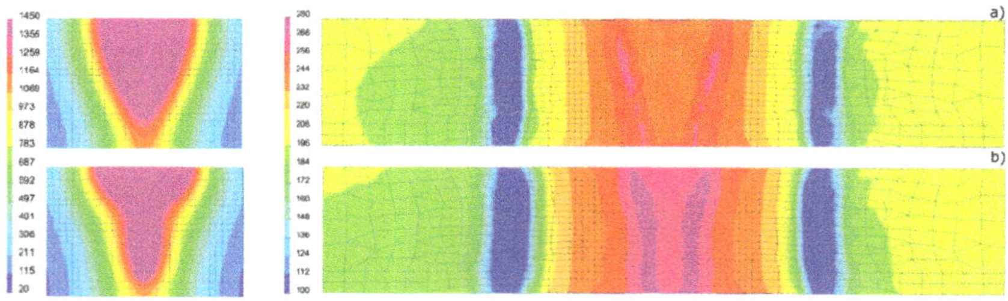

Figure 18. View of the molten pool and von Mises stresses distribution in the case of calculations for (**a**) the model without modification of the area of mesh elements loaded with the heat source model and (**b**) the model based on the proposed calculation modification for the DL_T3 sample (Table 3).

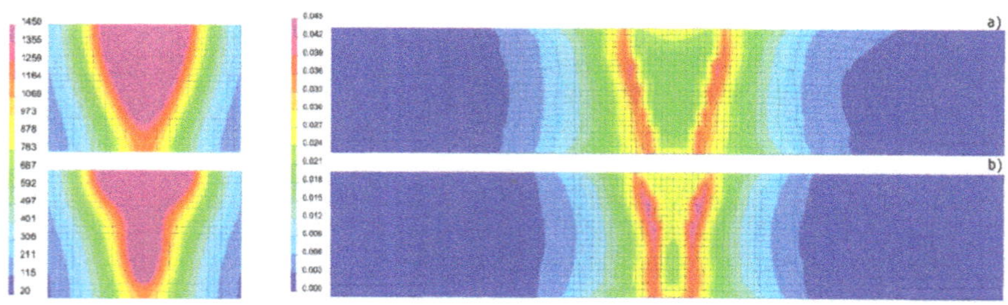

Figure 19. View of the molten pool and the distribution of cumulative plastic strains in the case of calculations for (**a**) the model without modification of the area of mesh elements loaded with the heat source and (**b**) the model developed based on the proposed modification for the DL_T3 sample (Table 3).

3.4. Assumptions and Construction of a Modified Numerical Model in the Case of a High-Power Diode Laser

The second laser device analysed for numerical simulations in the presented research was a high-power diode laser. This source is an interesting device that is noteworthy from a numerical simulations point of view as due to its internal structure (laser diode packages in the form of laser rods) it emits a rectangular laser beam, with beam spot dimensions of 1.8 mm × 6.8 mm at a focal length of 82 mm or 1.8 mm × 3.8 mm using an additional focusing lens and focal length of 32 mm. Due to the multimode energy distribution on the surface of the focus of the laser beam, the value of this energy is almost constant on the entire surface (Figure 20).

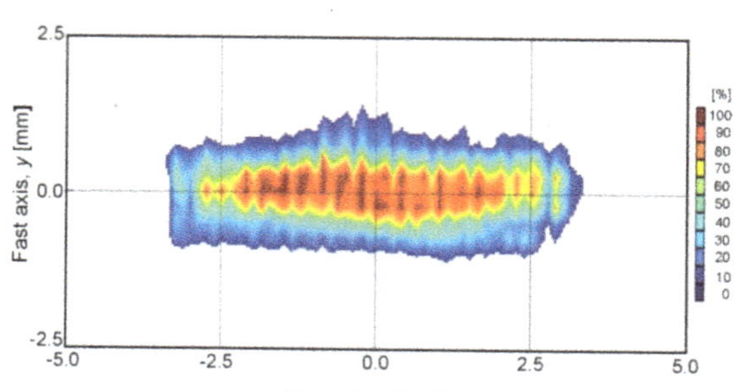

Figure 20. The 2D beam profile of the used HPDL laser in the focal plane [43]. (Reprinted with permission, copyright 2017, EBSCO Industries).

This type of distribution caused the power density on the surface of a relatively large focus of the laser beam to be much lower than on the previously used disk laser (maximum 3.2 × 10^4 W/cm^2 as compared to 1.05 × 10^7 W/cm^2). The inability tc conduct the welding process with the "keyhole" technique in the case of these lasers is compensated by their advantages when used in surface treatment and surfacing processes (wide beam with uniform energy distribution) [43].

The shape of the laser beam focus and the even distribution of energy of the laser radiation on the surface of this focus also require a special approach in the case of numerical analyses due to the fact that models that are usually not well represented are not available. While the historical rectangular heat source model described here corresponds to the beam shape, as already mentioned the user of modern programs dedicated to welding and heat treatment processes usually has three predefined models of heat sources: the Gaussian surface model, the double-ellipsoid model, and the conical model. As before, in this case it is possible to load the model's heat source with selected elements of the model's mesh and utilize its dimensions to control the width and depth of bead fusion.

Taking into account the internal structure of this type of laser (single diode emitters with single-mode distribution), the first approximation of this heat source can be a model in the form of two rows (due to the longitudinal dimension of the laser beam focus of 1.8 mm) of individual normal (Gaussian) distributions moving along the trajectory placed in the load area, specifying the bead obtained with some approximation (Figure 21a). This brings with it the aforementioned disadvantage associated with the need to define a large number of trajectories and heat source models as well as to divide the total amount of heat introduced by the appropriate number of sources used. Such an action does not reflect the construction of the laser rod in such a way where there are emitters with a power of 1–2 W. Therefore, with a laser power of 2000 W the grid of the discussed model and the number of heat sources necessary to define the model would cause the model to cease to be real due to the quantity of grid elements used in the numerical model (hundreds or even thousands of small heat source models corresponding to individual diode emitters). The tests carried out showed that even with a much rarer mesh of this type (and a reduced number of used heat sources), the solution ensured high compatibility of the shapes of the calculated beads with those obtained as a result of real tests (Figure 22, Table 5). The aforementioned much lower power density on the surface of the laser beam focus and the use of processes with the "weld pool" technique (conduction mode) in a similar way to arc welding inspired an attempt to adapt the most popular model, which is the double-ellipsoid model. Due to the ease of use in modeling the "weld pool" technique, the need to define only one trajectory of heat source model movement, and the ability to determine the depth of model interaction, attempts were made to confirm the usefulness of this solution. For this purpose, a model was built in which, as in the previous case, the "LOAD" area roughly corresponded to the dimensions of the obtained bead, while the double-ellipsoid heat source model was much wider than this area to ensure uniform energy distribution on the surface of the loaded area (Figure 21b). Other dimensions of the heat source model are related to the depth of penetration of the bead and the longitudinal dimension of the laser beam focus.

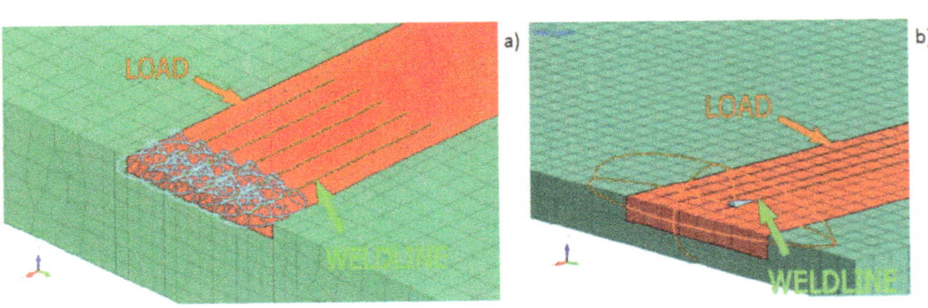

Figure 21. View of loading of the elements of the model mesh and placement of (**a**) heat source models with a normal distribution (Gaussian surface heat source model) and (**b**) a double-ellipsoid heat source model.

To verify the assumptions of the presented model, a three-dimensional discrete model in VisualWeld (SYSWELD) package was created, containing 11,120 3D elements and 13,243 nodes. The boundary conditions used, simulating the clamping of the sample, corresponded to its free placement on the table of the laser stand throughout the welding process and then its cooling. The boundary condition regarding heat dissipation was met by convective dissipation to the environment at 20 °C and radiation was defined on all external surfaces of the model. Numerical analyses were carried out for a model consisting of 12 pieces of the Gaussian surface heat source models (in two rows placed behind each, 6 models in each) and 6 trajectories determining the movement of these sets, located at a distance of 1.0 mm from each other. The second model was a model consisting of 22 Gaussian surface heat source models (in two rows placed behind each with 11 models in each) and 11 trajectories determining their movement at a distance of 0.5 mm from each other (Figure 21a). In the case of the described model, built using a single double-ellipsoid heat source model, analyses of 6 variants differing in the dimensions of the heat source model itself were performed (Figure 22, Table 5). In each of the analyzed cases the values of laser beam power and welding velocity were the same and were respectively 1400 W and 2.5 mm/s (Table 5).

Table 5. Comparison of geometric dimensions of selected beads on the bead-on-plate laser welding process performed on 6.0-mm AISI 304 steel samples using the ROFIN DL020 high-power diode laser with values obtained from numerical analyses (values in brackets) (Figures 21 and 22).

Specimen Designation	Heat Source Model	Heat Source Dimension * (mm)	Energy Per Unit Length ** (J/mm)	Bead Width (mm)	Penetration Depth (mm)	Molten Pool Length (mm)
HPDL_1	Gaussian surface heat source model	1.0/1.0	45 for each of 12 models (total 560)	4.50	1.05	2.72
HPDL_2		1.0/0.5	25 for each of 22 models (total 560)	4.35	1.13	2.84
HPDL_3	Double ellipsoid model	15/1.8/0.1	560	5.01	1.01	2.74
HPDL_4		25/1.8/0.1		5.07	0.99	2.68
HPDL_5		25/1.8/0.5		4.98	1.00	2.66
HPDL_6		25/1.0/0.1		5.03	1.02	2.46
HPDL_7		50/1.8/0.1		5.09	0.97	2.66
HPDL_8		25/1.8/1.0		4.95	1.02	2.64

* Dimensions for the Gaussian surface heat source model: diameter/distance between trajectories (mm); for the double-ellipsoid model: width/length/height of model. ** Heat transfer efficiency coefficient: 0.5 (determined on the basis of model calibration based on actual bead dimensions).

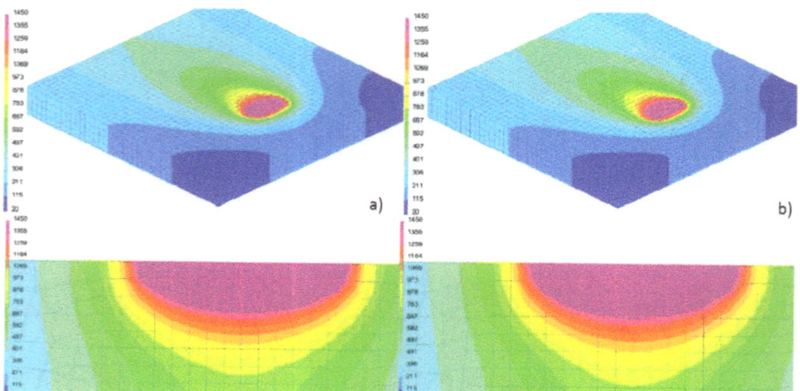

Figure 22. Comparison of the view of the molten pool surface and cross-sections (**a**) when using a model consisting of normal distributions and (**b**) a double-ellipsoid heat source model.

3.5. Calibration and Validation of the Model with a Modified Method of Heat Input in the High-Power Diode Laser Bead-On-Plate Welding Process

To verify the proposed model, numerical analyses were performed using the model of assumptions proposed earlier, and a three-dimensional discrete model was created containing 38,320 3D elements and 41,148 nodes. The applied boundary conditions simulating sample clamping corresponded to its free placement on the laser table throughout the whole process of welding and then its cooling. The boundary condition regarding heat dissipation was met by convective dissipation to the environment at 20 °C and radiation defined for all external surfaces of the model. As a heat source model, a double-ellipsoid model was used which acted on selected elements of the model's mesh (Figure 23). In this case, the calculations using the defined material database containing the thermal, mechanical, and metallurgical material properties depended on temperature and metallurgical phase proportions.

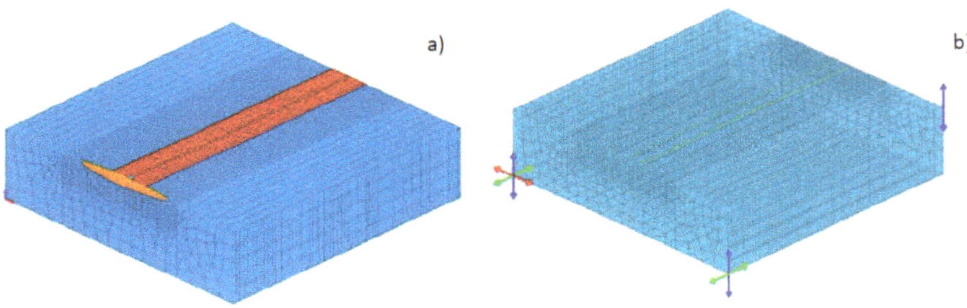

Figure 23. View (**a**) of a three-dimensional model of the bead-on-plate laser welding process performed on 6.0-mm steel samples using a high-power diode laser together with a double-ellipsoid heat source model and the loaded area (red color), and (**b**) clamping conditions.

For the calibration of the proposed model, the results of the bead-on-plate high-power diode laser welding process on 10-mm X37CrMoV5-1 tool steel samples were used (Tables 6 and 7). Based on the results of metallographic tests, the calculation models were calibrated and the final results of numerical analyses were compared with the corresponding shapes of actual beads obtained in the real tests (Figure 24). On this basis, a comparison was also made of the measured and calculated basic geometric dimensions of the beads obtained (Table 7).

Table 6. Chemical composition of X37CrMoV5-1 steel according to ISO 4957:2018 standard [44].

	C	Mn	Si	P	S	Cr	Mo	V
WCL (X37CrMoV5-1)	0.32–0.42	0.2–0.5	0.8–1.2	max. 0.030	max. 0.030	4.5–5.5	1.2–1.5	0.3–0.5

Table 7. Process parameters and comparison of geometric dimensions of beads made on 10 -mm WCL (X37CrMoV5-1) steel samples using a ROFIN DL020 high-power diode laser with values obtained from numerical analyses (values in brackets) (Figure 24).

Specimen Designation	Laser Beam Power (W)	Welding Speed (mm/s)	Energy Per Unit Length (J/mm)	Bead Width (mm)	Penetration Depth (mm)
HPDL_R1	1000	3.33	300	4.15 (4.10)	0.30 (0.35)
HPDL_R2	1250	2.5	500	5.18 (5.13)	0.50 (0.52)

Argon flow rate via cylindrical nozzle: 15 L/min; laser beam spot dimensions: 1.8 mm × 6.8 mm at 82-mm focal length; location of the laser beam focus: on the upper surface of the samples.

To validate the model, similarly to the case of disk laser welding simulations the recorded thermal cycles were used. K-type thermocouples (Ni–NiCr wires with a diameter of 0.2 mm) were welded in the holes made at the bottom of the samples so that the measuring points were located at 0.5, 1.0, 2.0, and 3.0 mm from the welded sample surface (Figure 25). Thermocouples with accuracy ±0.0075 × t in a temperature range from −40 to +1200 °C (according to PN-EN 60584-2) were connected via compensation cables with Agilent 34970A recorder. After calibrating the calculation model (Figure 24), the registered and calculated thermal cycles in the points corresponding to the location of thermocouples were also compared (Figure 26).

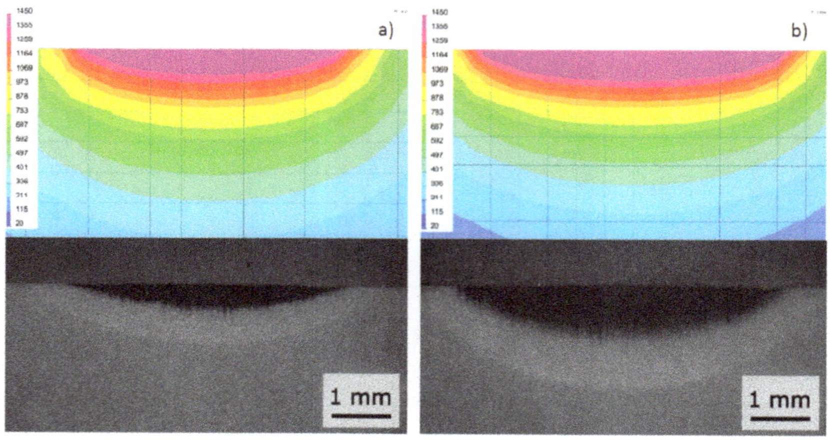

Figure 24. Comparison of the bead shape on the cross-section of sample (**a**) HPDL_R1 and (**b**) HPDL_R2 with the results of numerical analyses.

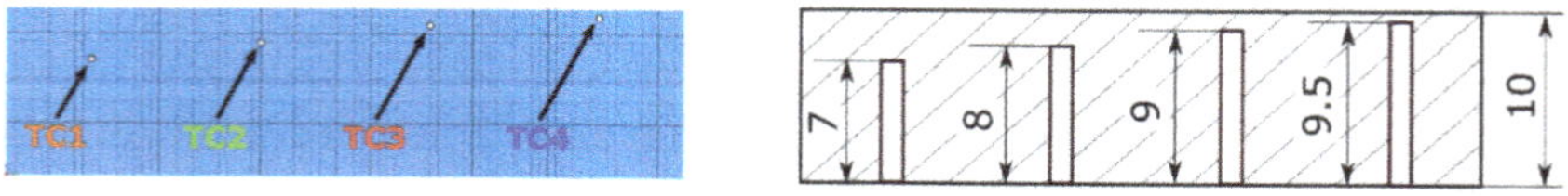

Figure 25. Scheme of thermocouple placement on the cross-section of samples in the axis of welded beads.

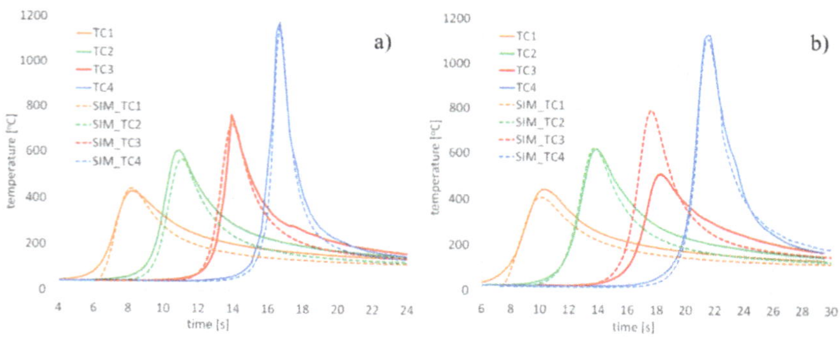

Figure 26. Comparison of recorded and calculated thermal cycles on the bead-on-plate laser welding process performed on 10-mm AISI304 steel samples using the ROFIN DL020 high-power diode laser with (**a**) TC3 and (**b**) TC4 samples. TCx: thermal cycles recorded during the welding process; SIM_TCx: thermal cycles obtained from numerical analyses.

4. Conclusions

A properly prepared and conducted numerical analysis consists of three important stages: precise preparation of the material base, calibration, and model validation. A properly developed material base, created based on tests on the behavior of the material under the influence of various temperatures and thermal cycles, allows for a faithful reproduction of the behavior of materials considering its thermometallurgical and mechanical nature. The stage of building and calibrating the model is important because the variety of methods used and the effects obtained as a result do not allow for the creation of one universal model. Each process must be treated individually by the user, considering the special features of the welding heat source used and the most accurate description of the conditions of the process. Model validation allows for obtaining an appropriate degree of model reliability in further applications.

The presented results of preliminary tests using a disk laser proved the need to supplement the portfolio of implemented models of heat sources with those allowing the correct geometry of the beads in laser welding to be obtained (Figures 6 and 7). Bearing in mind the mentioned desire for the greatest convenience and ease of use of models, it was proposed to extend the method of laser beam modeling or another method of welding with the "keyhole" technique (characterized by high power density). The combination of the already-present heat source model with the possibility of defining selected mesh elements which are affected by it allowed for easy and effective control of the shape of the fusion line of the obtained bead (Figures 9 and 12, Figures 13–15). On knowing the shape of the fusion line from real tests it was possible to calibrate the heat source model in such a way that the calculated shape of the molten pool corresponded to that from real tests. This was well-visible from the comparison of the bead cross-sections and calculated results. Interestingly, on the longitudinal cross-sections and in the corresponding results of numerical analyses, the temperature distributions corresponding to the molten pool showed considerable convergence with the obtained fusion shapes, especially when conducting the process using the "keyhole" technique. The liquid metal moved in a specific way in the gas channel formed by the beam, which was not considered in the numerical analyses based on the heat conduction equation.

The results of the validation seem to confirm that a properly calibrated model corresponds thermally to the situation recorded during actual welding tests (Figure 17). It should also be noted that in the case of thermocouple measurements, there may be disturbances in the registration of thermal cycles. In Figure 17b for the cycle recorded by TC3 it can be seen that, after recording the signal increase as a result of melting through the sample, the thermocouple located in the bead axis was damaged by liquid metal. For this reason, the corresponding course of the calculated thermal cycle was not provided. Incorrect registration of a thermal cycle with a thermocouple could also occur with incorrect positioning of the thermocouple, i.e., plane positioning or installation in drilled holes at the wrong depth, as with thermocouple TC3 in Figure 26b. These types of errors can lead to incorrect indications and in extreme situations it is impossible to register the correct thermal cycles. Therefore, in such cases it is also important that the validation is carried out based on more than one signal recorded in the actual process.

When comparing the recorded and calculated thermal cycles, there are sometimes also visible differences in the value of both the maximum value of the registered/calculated temperature as well as the course of the curves themselves (cooling speed). These differences may of course result from a simplified definition of heat exchange with the environment, convection, and emissivity coefficient values (even when they are dependent on temperature changes) [45]. This may also be due to the fact that the thermocouple tip has a certain diameter which is not taken into account when the thermal cycle is collected at a single mesh node. This is especially visible when fine mesh is used where the thermocouple tip can cover few nodes, and in case of laser welding the temperature gradients are much higher in comparison to arc welding and therefore this phenomenon is noticeable. In this situation, several nodes of the grid, differing in the maximum temperatures of the calculated thermal cycle, may be in the area covered by the thermocouple tip in the case of a real test. This should be kept in mind, especially in the case of process validation with numerical simulations, where the thermal cycles are very short and as already mentioned earlier, the temperature gradients are large so the values in the adjacent mesh nodes change very quickly.

Therefore, it is necessary to pay attention to the fact that the correct calibration of the heat source model, ensuring the convergence of the model with reality, is necessary because the heat distribution as well as the cooling speed in individual areas of the welded elements later affect metallurgical phases distribution, , grain size, and plastic deformation, and thus also stresses and strains distributions. This was confirmed by the calculated distribution of reduced stresses in the sample for the variant using the standard heat source model and the proposed extension. There are visible differences in the calculated maximum stress values as well as their distribution (Figure 18). The situation is similar in terms of cumulative plastic strain case distribution in both analyzed cases (Figure 19). It is particularly important that in the second case higher stress values were obtained, and thus a more rigorous assessment of the level of stresses and the possible danger associated with it was found.

The analyzed case of the high-power diode laser with a characteristic rectangular shape of the laser beam focus shows that more and more often the users of this type of program have to resort to the use of more complex calculation techniques to obtain correct results. The combination of knowledge on the construction of the device itself and actual welding tests carried out at the laser station allowed us to propose a simple and effective method for the simulation of an even distribution of laser radiation in this type of device (Figures 21 and 22). Similar values of geometric dimensions obtained for the tested variants of the model testify to the considerable flexibility of this method and thus also guarantee its considerable versatility, as confirmed in this case by tests and calculations carried out to validate the proposed model (Figures 24 and 26, Table 5). The aforementioned lower temperature recorded for TC3 in Figure 26b was caused by the incorrect position of the thermocouple in the measuring hole. As later tests showed, the thermoelement tip was welded on the wall of the hole and not on its bottom as assumed.

As a summary it should be stated, as the presented considerations have shown, that the implemented models of heat sources in commercially available programs for the numerical analyses of welding processes are not the only way to solve the analyzed problems. The possibility of introducing often simple changes or modifications to the built models allows them to be adapted in a much better way to the expected results. However, it is very important in this case that each of these modifications requires accurate calibration and validation of the proposed model based on the real experimental results. Otherwise, the results of numerical analyses may lead to erroneous conclusions.

Funding: This research was financed from the own resources of the Silesian University of Technology.

Conflicts of Interest: The author declares no conflict of interest.

References

1. Bagger, C.; Olsen, F. Comparison of plasma, metal inactive gas (MIG) and tungsten inactive gas (TIG) processes for laser hybrid welding. In Proceedings of the ICALEO 2003, Jacksonville, FL, USA, 13–17 October 2003.
2. Lisiecki, A. Welding of titanium alloy by different types of lasers. *Arch. Mater. Sci. Eng.* **2012**, *58*, 209–218.
3. Górka, J. Assessment of the Effect of Laser Welding on the Properties and Structure of TMCP Steel Butt Joints. *Materials* **2020**, *13*, 1312. [CrossRef] [PubMed]
4. Orozco, N.J. Fully integrated hybrid-laser welding control process, Processes for Laser Hybrid Welding. In Proceedings of the ICALEO 2003, Jacksonville, FL, USA, 13–17 October 2003.
5. Lisiecki, A.; Piwnik, J. Tribological characteristic of titanium alloy surface layers produced by diode laser gas nitriding. *Arch. Metall. Mater.* **2016**, *61*, 543–552. [CrossRef]
6. Murakami, T.; Shin, M.H.; Nakata, K. Effect of welding direction on weld bead formation in high power fiber laser and MAG arc hybrid welding. *Trans. JWRI* **2010**, *39*, 175–177.
7. Tomkow, J.; Fydrych, D.; Rogalski, G. Role of bead sequence in underwater welding. *Materials* **2019**, *12*, 3372. [CrossRef] [PubMed]
8. Rodger, E.D.; Fletcher, R.P. The Determination of Internal Stresses from the Temperature History of a Butt Welded Pipe. *Weld. J. Res. Suppl.* **1938**, *17*, 4–7.
9. Masubuchi, K. Control of Distortion and Shrinkage in Welding. *Weld. Res. Counc. Bull.* **1970**, *149*, 1–30.

10. Brust, F.W.; Rybicki, F.E. Computational Model of Backlay Welding for Controlling Residual Stresses in Welded Pipes. *J. Press. Vessel Technol.* **1981**, *103*, 294–299.
11. Barber, T.E.; Brust, F.W.; Mishler, H.W.; Kanninen, M.F. *Controlling Residual Stresses by Heat Sink Welding*; EPRI Report NP-2159-LD; Electric Power Research Institute: Palo Alto, CA, USA, December 1981.
12. Danielewski, H.; Skrzypczyk, A. Steel Sheets Laser Lap Joint Welding—Process Analysis. *Materials* **2020**, *13*, 2258. [CrossRef] [PubMed]
13. Winczek, J.; Skrzypczak, T. Thermomechanical states in arc weld surfaced steel elements with heat of melted electrode material taken into account. *Arch. Metall. Mater.* **2016**, *61*, 1277–1288. [CrossRef]
14. Kong, F.; Ma, J.; Kovacevic, R. Numerical and experimental study of thermally induced residual stress in the hybrid laser—GMA welding process. *J. Mater. Process. Technol.* **2011**, *211*, 1102–1111. [CrossRef]
15. Zimmerman, J.; Lindemann, Z.; Golański, D.; Chmielewski, T.; Włosiński, W. Modeling residual stresses generated in Ti coatings thermally sprayed on Al_2O_3 substrates. *Bull. Pol. Acad. Sci. Tech. Sci.* **2013**, *61*, 515–525. [CrossRef]
16. Sajek, A. Application of FEM simulation method in area of the dynamics of cooling AHSS steel with a complex hybrid welding process. *Weld. World* **2019**, *63*, 1065–1073. [CrossRef]
17. Kik, T. Computational Techniques in Numerical Simulations of Arc and Laser Welding Processes. *Materials* **2020**, *13*, 608. [CrossRef] [PubMed]
18. Kik, T.; Moravec, J.; Novakova, I. New method of processing heat treatment experiments with numerical simulation support. Modern technologies in industrial engineering V. In Proceedings of the ModTech 2017 International Conference, Sibiu, Romania, 14–17 June 2017.
19. Devaux, J.; Leblond, J.B.; Bergheau, J.M. Numerical study of the plastic behaviour of a low alloy steel during phase transformation. *J. Shanghai Jiaotong Univ.* **2000**, *5*, 206–212.
20. Welding Simulation User Guide, Sysweld Manual ESI Group. Available online: https://www.esigmbh.de/downloads/ESI/Dokumente/Welding/The_Welding_Simulation_Solution_210408 (accessed on 10 June 2020).
21. Kik, T.; Moravec, J.; Novakova, I. Numerical simulations of X22CrMoV12-1 steel multilayer welding. *Arch. Metall. Mater.* **2019**, *64*, 1441–1448.
22. Kik, T.; Moravec, J.; Novakova, I. Application of numerical simulations on 10GN2MFA steel multilayer welding, dynamical systems in applications. In *Dynamical Systems Theory and Applications*; Springer: Paris, France, 2018; pp. 193–204.
23. Rochalski, D.; Golański, D.; Chmielewski, T. Welding heat source models in the analysis of temperature field. *Weld. Technol. Rev.* **2017**, *89*, 5.
24. Rosenthal, D. The Theory of Moving Sources of Heat and Its Application to Metal Treatments. *Trans. ASME* **1946**, *68*, 849–866.
25. Pavelic, V.; Tanbakuchi, R.; Uyehara, O.A.; Myers, F.S. Experimental and computed temperature histories in Gas Tungsten Arc Welding of thin plates. *Weld. J.* **1969**, *48*, 295–305.
26. Komanduri, R.; Hou, Z.B. Thermal Analysis of the Arc Welding Process: Part II. Effect of Variation of Thermophysical Properties with Temperature. *Metall. Mater. Trans.* **2001**, *32*, 483–499. [CrossRef]
27. Eagar, T.W.; Tsai, N.S. Temperature Fields Produced by Traveling Distributed Heat Sources. *Weld. J.* **1983**, *62*, 346–355.
28. Goldak, J.; Chakravarti, A.; Bibby, A. A New Finite Element Model for Welding Heat Sources. *Metallurgical Transaction B* **1984**, *15*, 299–305. [CrossRef]
29. Farrokhi, F.; Endelt, B.; Kristiansen, M. A numerical model for full and partial penetration hybrid laser welding of thick-section steels. *Opt. Laser Technol.* **2019**, *111*, 671–686. [CrossRef]
30. Denga, D.; Liang, W.; Murakawa, H. Determination of welding deformation in fillet-welded joint by means of numerical simulation and comparison with experimental measurements. *J. Mater. Process. Technol.* **2007**, *183*, 219–225. [CrossRef]
31. Kik, T.; Górka, J. Numerical simulations of laser and hybrid S700MC T-joint welding. *Materials* **2019**, *12*, 516. [CrossRef] [PubMed]
32. Zhang, Z.; Wu, C. Effect of fluid flow in the weld pool on the numerical simulation accuracy of the thermal field in hybrid welding. *J. Manuf. Process.* **2015**, *20*, 215–223. [CrossRef]
33. Zhang, L.; Zhang, J.; Gumenyuk, A.; Rethmeier, M.; Na, S. Numerical simulation of full penetration laser welding of thick steel plate with high power high brightness laser. *J. Mater. Process. Technol.* **2014**, *214*, 1710–1720. [CrossRef]

34. Haug, P.; Rominger, V.; Speker, N.; Weber, R.; Graf, T.; Weigl, M.; Schmidt, M. Influence of laser wavelength on melt bath dynamics and resulting seam quality at welding of thick plates. *Phys. Procedia* **2013**, *41*, 49–58. [CrossRef]
35. Powell, J.; Ilar, T.; Frostevarg, J.; Torkamany, M.J.; Na, S.J.; Petring, D.; Zhang, L.; Kaplan, A.F.H. Weld root instabilities in fiber laser welding. *J. Laser Appl.* **2015**, *27*, S29008. [CrossRef]
36. Naito, Y.; Katayama, S.; Matsunawa, A. Keyhole behaviour and liquid flow in molten pool during laser-arc hybrid welding. In Proceedings of the International Congress on Laser Advanced Materials Processing, Osaka, Japan, 3 March 2003.
37. Flint, T.F.; Francis, J.A.; Smith, M.C.; Balakrishnan, J. Extension of the double-ellipsoidal heat source model to narrow-groove and keyhole weld configurations. *J. Mater. Process. Technol.* **2017**, *246*, 123–135. [CrossRef]
38. Wu, C.S.; Wang, H.G.; Zhang, Y.M. A new heat source model for keyhole plasma arc welding in FEM analysis of the temperature profile. *Weld. J.* **2006**, *85*, 284–291.
39. He, Q.; Wei, H.; Chen, J.S.; Wang, H.P.; Carlson, B.E. Analysis of hot cracking during lap joint laser welding processes using the melting state-based thermomechanical modeling approach. *Int. J. Adv. Manuf. Technol.* **2018**, *94*, 4373–4386. [CrossRef]
40. CEN. *Hot Rolled Products of Structural Steels—Part 2: Technical Delivery Conditions for Non-alloy Structural Steels*; EN 10025-2; CEN: Brussels, Belgium, 2019.
41. CEN. *Stainless Steels—Part 3: Technical Delivery Conditions of Semi-finished Products, Bars, Rods and Sections for General Purposes*; EN 10088-3; CEN: Brussels, Belgium, 2014.
42. CEN. *Thermocouples - Part 1: EMF Specifications and Tolerances.*; EN 60584-1; CEN: Brussels, Belgium, 2013; Available online: https://thermal-detection.com/storage/app/media/Tech%20Data/Thermocouple%20tolerance%20chart.pdf (accessed on 1 June 2020).
43. Janicki, D. Fabrication of High Chromium White Iron Surface Layers on Ductile Cast Iron Substrate by Laser Surface Alloying. *Strojniški vestnik. J. Mech. Eng.* **2017**, *63*, 705–714.
44. ISO 4957:2018 Tool Steels. Available online: https://www.iso.org/obp/ui/#iso:std:iso:4957:ed-3:v1:en (accessed on 1 June 2020).
45. Perić, M.; Garašić, I.; Tonković, Z.; Vuherer, T.; Nižetić, S.; Dedić-Jandrek, H. Numerical prediction and experimental validation of temperature and residual stress distributions in buried-arc welded thick plates. *Int. J. Energy Res.* **2019**, *43*, 3590–3600. [CrossRef]

materials

MDPI

Article

Computational Techniques in Numerical Simulations of Arc and Laser Welding Processes

Tomasz Kik

Department of Welding Engineering, Silesian University of Technology, Konarskiego 18A, 44-100 Gliwice, Poland; tomasz.kik@polsl.pl; Tel.: +48-32-1681

Received: 2 December 2019; Accepted: 20 January 2020; Published: 29 January 2020

Abstract: The article presents a comparison of modern computational techniques used in numerical analyses of welding processes. The principles of the "transient" technique calculations with a moving heat source, the "macro-bead" (MBD) technique, with an imposed thermal cycle on a selected weld bead section and the "local–global" approach with shrinkage calculation technique were described. They can be used, depending on the variant chosen, both for individual, simple weld joints and those made of many beads or constructions containing dozens of welds and welded elements. Differences in the obtained results and time needed to perform calculations with four different calculation examples of single and multipass arc and laser beam welding processes were presented. The results of calculations of displacements and stresses distributions in the welded joints using various computational techniques were compared, as well as the calculation times with the described techniques. The numerical analyses in the SYSWELD software package have shown the differences between the described computational techniques, as well as an understanding of the benefits and disadvantages of using each of them. This knowledge allows preparing an efficient and fast optimization of the welding processes, often aimed at minimizing deformations in the first place, as well as detection of potential defects of both simple and complex welded structures. In general, the possibilities and flexibility of modern numerical calculation software have been presented.

Keywords: FEM; numerical analyses; simulations; computational techniques; SYSWELD; displacements; stresses

1. Introduction

Terms such as simulation, calculations or numerical solutions have been known to engineers for a long time, not only in technical fields but also in medicine or natural sciences. Of course, with the development of electronics, at this time, great progress was made both in terms of the hardware itself and the possibilities that the modern software offers us. Many branches of technology can now show almost comprehensive numerical solutions, from the design stage, production preparation, through the stage of determining technological and production conditions, to the overall optimization of the production process. It is also possible to simulate the actual or assumed conditions of use of the product together with determining its durability. The use of modern manufacturing techniques such as laser or electron beam welding places requirements that are increasingly difficult to meet. Increasing requirements regarding the accuracy of element preparation and maintaining minimum dimensional deviations during and after the welding process mean that we readily use modern computational tools [1–7].

The development of numerical analysis software has recently focused mainly on advanced second-generation simulation calculations. These types of simulations support a unique way of creating the final product using various industrial technologies. In this way, it is possible to simulate the entire production process and its relationships with the limit states of structures and materials.

This is a typical path, e.g., in the automotive industry, where the processes of sheet metal forming, body welding and in the last stage of crash tests are analyzed. However, in the case of numerical analyses of welding processes, despite the long and intensive development of computational techniques, it is still difficult to achieve all of the mentioned operations in one calculation program [1–5,8].

Since welding is one of the most widespread technological processes, this technology has also received a lot of attention when developing modern simulation tools. The problem with the simulation of this process is the fact that the most frequently moving heat source is the reason for the uneven distribution of displacements and stresses in welded elements [5–7]. However, they are not only dependent on the moving heat source itself. Their values and distributions are also strongly dependent on many other technological factors such as clamping condition, mechanical and thermal properties, type of the welding technology and welding parameters, preheating temperatures, weld joints design, temperature of surroundings, etc. The residual stresses in the structure after welding have a negative influence on the durability of the structure and its reliability under various operating conditions. The degree of complexity of the process itself and the non-linearity of phenomena that occur in the welded element during its heating and cooling, and their dependence on the type of material and technology used, significantly increase the degree of complexity of numerical analyses [5–7,9,10].

Currently, there are two views on the issue of conducting and solving tasks in the field of numerical analysis of welding processes. The first of them is associated with a detailed, often partial solution of selected effects of the welding process, consisting of determining the shape and dimensions of the weld molten pool, heat affected zone (HAZ) and related metallurgical and mechanical changes in the area of the welded joint [1–5]. The result of this type of analysis is always a detailed description of the solution, unfortunately only in relation to a specific, individually considered data and process parameters. The second approach is to apply numerical analyses to large, complete parts and production units. Due to the complexity of the structure, this method must contain a number of simplifications, which to some extent reduce the resulting accuracy of calculations. Despite this disadvantage, this allows conducting analyses that in the first approach, even despite the currently available computing power, are practically impossible to perform, precisely because of the number of variables that should be taken into account. Not without significance, in this case, is the constant focus on reducing production costs and reducing to the absolute minimum the time needed to prepare a new product [11–14].

The use of numerical analyses today allows for a better understanding of the production process, finding relationships that connect individual process parameters to its results and increases product quality. It also leads to a significant cost reduction in the production preparation stage by limiting and often excluding costly prototypes. This is particularly important when making samples or prototypes involving a lot of material, time and are energy-consuming. Based on the results of numerical simulations, it is possible to determine the optimal solution, even if their results are not 100% consistent (in terms of value) with the results of real tests. Thanks to the constant development of modern simulation software, it offers us today various possibilities of approach to the issue of numerical analyses, allowing for the mentioned analyses within both one joint and entire constructions. In the following, based on the SYSWELD software from ESI Group, typical computational techniques are presented. Each of them are successfully used in today's modern computational packages [1–5,8,11–18].

2. Description of the Problem

As it was already mentioned, depending on the selected calculation method, numerical analyses of welding processes carried out in the SYSWELD environment can be divided into the following groups:

Local analysis: Used to determine the distribution of temperature fields, metallurgical phases as well as stresses and deformations within one welded joint;

Global analysis: This applies to the entire structures and usually involves changing the size of the structures (displacements) and stresses distributions around the welds [1–4].

In the case of welding, essentially the only load on the workpiece is the non-stationary distribution of the temperature field both on its surface and in many cases in thickness. Thus, the temperature in the model is determined by means of a variable function in time and dimensional space. Accurate determination of the temperature field distribution is, therefore, the basic and most important step in determining its influence on the metallurgical properties of the material and the stresses and deformation distributions of the welded structure. In order to refine the record of changes in temperature function in time and space, it is necessary to define a mathematical model of a heat source [1]. This description, however, brings with it a lot of problems due to the complexity of the issue. There is a need to take into consideration many input parameters and apply some simplifications in the description of this model. When creating it, the influence of convection in the liquid metal pool, active elements, type of protective gas, wire diameter or method of transferring metal in the arc are not taken into account. Models of heat sources can be divided into three basic groups while defining the degree of complexity of the analyses carried out [1]:

One-dimensional;

Two-dimensional;

Three-dimensional.

Therefore, depending on the dimension of the heat source model used, point, linear or flat models of heat sources were used in the analyses. Their advantage was low "resource consumption" in the calculations. The limitations of these heat source models resulted primarily from the available computing power. Nowadays, when we have sufficient computing power, in most cases we almost exclusively use three-dimensional models [5,11–18].

2.1. Transient Technique: Moving Heat Source

In the case of local analyses, when the moving heat source model is used, the calculations are carried out by the so-called "transient" technique, also called the "step-by-step" technique. Calculations are made in this case for each subsequent time increased by a given time step, which is automatically adjusted depending on the mesh density of the model or set by the user. The mathematical model of the heat source in this type of analysis moves along the line that determines the welding trajectory. The parameters used to describe the heat source model, in this case, are the actual welding process parameters, i.e., welding current, arc voltage, welding speed or thermal efficiency of the method, etc. So, this is mostly data that can be found, among others, at the Welding Procedure Specification Form (WPS) but also are part of the knowledge of every welding engineer. One of the most popular models used in current numerical analyses is the volumetric double ellipsoid model, also known as Goldak's model [1,5,15–20]. It consists of two ellipsoids which, placed in two perpendicular planes, create an image of the weld pool in both horizontal and vertical space (Figure 1). Transferred heat into a volume is described by Equations (1) and (2) [1,5,19,21]:

For the front part of the heat source model, the Equation is:

$$Q_f(x,y,z) = \frac{6\sqrt{3}f_fQ}{abC_f\pi\sqrt{\pi}}exp\left(\frac{-kx^2}{a^2}\right)exp\left(\frac{-ly^2}{b^2}\right)exp\left(\frac{-mz^2}{c^2}\right) \tag{1}$$

and for the rear part of the heat source model, the Equation is:

$$Q_r(x,y,z) = \frac{6\sqrt{3}f_rQ}{abC_r\pi\sqrt{\pi}}exp\left(\frac{-kx^2}{a^2}\right)exp\left(\frac{-ly^2}{b^2}\right)exp\left(\frac{-mz^2}{c^2}\right) \tag{2}$$

where Q_f, Q_r are volumetric heat flux density in front and rear part of the model [W/m^3]; Q is total power source; a, b, c_f, c_r are width, depth and length of the front and the rear part of the molten pool; f_f, f_r are constants which influence energy flow intensity into the material ($f_f + f_r = 2$); and k, l, m are coefficients enabling modification of the liquid metal pool shape.

The efficiency of the heat transfer into parent material is given by the applied welding method [20]. For the proper heat source calibration, it is necessary to compare the results with experimentally determined and measured values. It is mainly the overall heat, the geometrical parameters of the molten pool and the efficiency of the heat transfer from the source into the material. After specifying the input parameters, it remains to add the constants f_f, f_r affecting the distribution of energy flow to the material and choose the coefficients k, l, m. Practice shows that we usually accept intercourse of f_f to f_r as 60:40. The choice of coefficients k, l, m is more complicated. If the unmodified heat source model is used, all three factors will be equal to three. The unmodified source is particularly suitable for simulated welding with coated electrodes. In the case of welding simulation, e.g., with the GMAW method, the model must be modified to obtain the correct shape of the liquid metal pool and the values of the coefficients chosen experimentally [1,5,15–22].

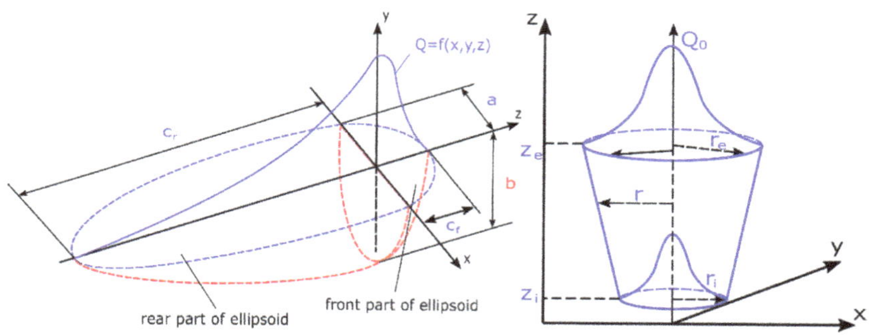

Figure 1. Double ellipsoid and 3D conical heat source models [1,5,19].

While Goldak's source successfully covers the scope of numerical analyses of arc welding processes, in the case of laser or electron beam welding, the conical model with a Gaussian distribution is currently most commonly used. Its conical shape allows good reproduction of welding processes carried out using welding heat sources with a high concentration of energy (for example laser and electron beam welding). The mathematical description of this model can be represented by two Equations [1,5,22]:

$$Q(x,y,z) = Q_0 exp\left(-\frac{x^2+y^2}{r_0^2(z)}\right) \tag{3}$$

$$r_0(z) = r_e + \frac{r_i - r_e}{z_i - z_e}(z - z_e) \tag{4}$$

where Q_0 is the maximum value of volumetric heat flux density; r_e, r_i are upper and lower 3D cone radius dimensions parameters; z_e, z_i are the 3D cone length parameters; and x, y, z are point coordinates (Figure 1).

Equation (3) describes the volumetric heat flow density into the material in the dependence of coordinate data. Equation (4) supplements Equation (3) by the definition of the radius change in the direction of the depth.

This way of defining a heat source model and conducting numerical analyses based on a moving heat source requires its appropriate calibration procedure (Figure 2). Only in this case will it be possible to recreate with it the actual distribution of temperature fields that affect the metallurgical transformations, stress and deformation distributions that occur. It is a long-term process but can be carried out on partially simplified models (including 2D cross-section models). Incorrect calibration of the heat source model leads to duplication of errors during this stage and simulation, which in turn leads to incorrect results.

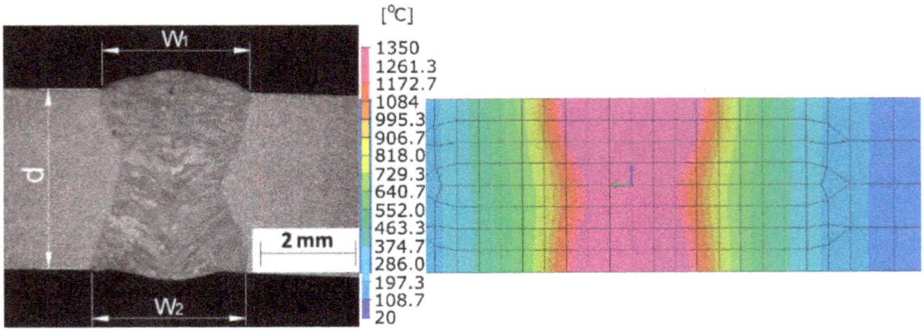

Figure 2. An example of heat source model parameters calibration. Comparison of the calculated shape and size of the molten zone in the laser welding process with a real macrograph of laser weld bead (laser beam power 2000 W, welding speed 0.5 m/min).

The analysis carried out using the "transient" technique is divided into three parts: preparation of the material base, calculation of thermal phenomena and related metallurgical changes as well as mechanical phenomena (i.e., stresses and displacement distributions, etc.) [1–4]. Taking into account the fact of performing calculations at each subsequent time, the user after completing the analysis receives a powerful set of thermo-metallurgical and mechanical data related to the simulated process. The price of such a complex solution is the extended duration of calculations. This is the main reason why the calculation of large welded structures with many welds uses other calculation techniques. However, such a large amount of data, including distribution of temperature fields, hardness, metallurgical phases, stresses and deformations makes it an ideal method by which the already mentioned local effects of the welding process can be determined. Based on the results of these analyses, it is also possible to conduct further simulations regarding the impact of external loads, i.e., forces, moments and pressures, as well as e.g., assessment of fatigue life of a structure based on the Dang Van criterion [1–4,16,18].

2.2. Macro-Bead Technique: Imposed Thermal Cycle

As it was already mentioned, calculations using the "transient" technique are practically impossible to use, from the point of view of the amount of data generated during them, in the case of numerical analyses of large or complex welded structures. In such cases, the "macro-bead" (MBD) technique is used [1,3,4,23,24], which is in a way an extension of the "transient" technique and involves the use of a properly prepared thermal cycle immediately on one or several areas (elements) of the model simultaneously (Figure 3).

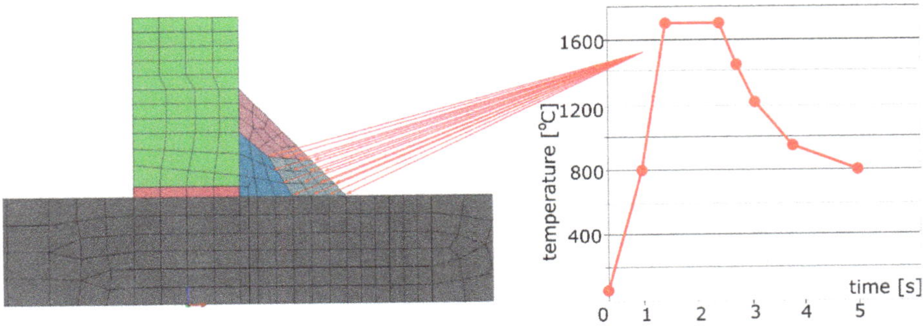

Figure 3. Imposed thermal cycle set on selected nodes of the model in calculations using the macro-bead (MBD) technique.

In this technique, the actual welding trajectory is divided into sub-areas so that the welding order and directions are maintained. The number of these sub-areas and the time step are defined on the basis of technological parameters of the welding process as well as the experience of the person working with this method [1–4]. The standard procedure is to prepare the thermal cycle based on a simple model "transient" analysis result (often using also 2D cross-section models) or as a result of thermocouple measurements. This cycle is then applied successively to the mentioned sub-areas of welds.

The difference is that in the case of the "transient" technique with a moving heat source, it is needed as many time steps (calculation cards) as there are elements in the grid on the welding trajectory. When applying the imposed thermal cycle, required are as many steps as the points describing the given thermal cycle in each of the sub-areas (Figure 4). In addition, the mesh can be modified and divided into a significantly smaller number of finite elements because there are no temperature gradients in the middle of the weld section at the time of the applied thermal cycle. They only occur at the beginning and end of the bead (sub-section). Therefore, in the MBD technique, the mesh is only refined at the edges of the section of elements on which the thermal cycle is applied. In their center, due to the lack of thermal gradient, the mesh is much thinner. Thus, for example, 100-time steps in the calculation with the "transient" technique can be replaced with a maximum of 20–40 time steps in the MBD technique, which significantly reduces the calculation time (Figure 4).

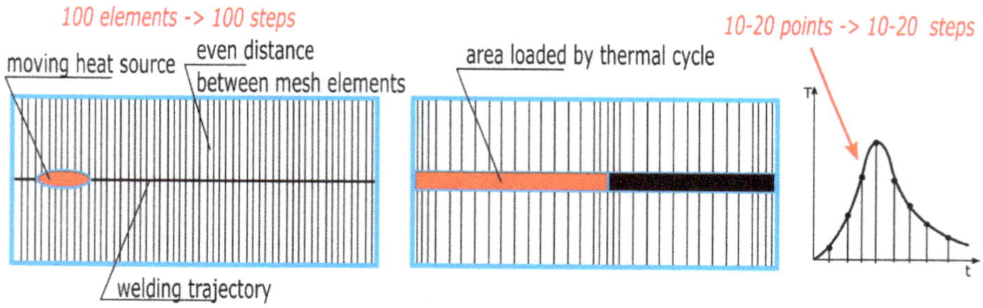

Figure 4. Differences between the number of simulation time steps for the "transient" and MBD analysis.

Based on the MBD technique, it is also possible to perform calculations using the "local–global" method [1,3,4]. Its main idea is the assumption that the welding process leads to local changes in stresses distribution and plastic deformations, while the effect of this on a global scale is a specific state of deformations. The local effects of the welding process are determined in this case by means of precise calculation models of single welded joints using the "transient" or MBD techniques. When the joints are repetitive (welded with the same parameters), the calculations are significantly accelerated. The results of local analyses are then transferred to the global model to determine the total deformations of the structure. The limitation of this method, however, is that the results of the analyses carried out on the global model are only structural deformations as well as internal forces and moments under specific mounting conditions. The stresses level and distribution of individual metallurgical phases are determined by local models obtained as a result of analyses carried out using the "transient" or MBD technique [1,3,4,23,24].

2.3. Shrinkage Technique: Distortion Engineering

The application of the "local–global" method in the case of very large and complicated welded constructions may have been problematic even despite the acceleration of calculations, which is provided by the use of the MBD technique [1,3,4]. In this case, it is also possible to use the "shrinkage" technique, which, despite many simplifications, allows to determine the deformation of the structure in situations where the standard techniques and even MBD is not possible to carry out due to too long calculations time. In accordance with the assumptions of the "local–global" method, in the case of calculations carried out using the "shrinkage" technique, calculations of thermal phenomena and related metallurgical changes are not carried out. Only the results of the analysis of mechanical phenomena occurring during the welding process are used. In a simplified way, it can be described that the thermal cycle first causes an expansion of a certain area of the material as a result of its heating and then shrinkage, which is caused by specific cooling conditions and heat dissipation from this zone. If during the cooling process some areas of the joint are subject to shrinkage, it is the user's task to find a representative value of the force that causes this shrinkage. In order to find this representative force causing the contraction, it is necessary to carry out the calibration process. For this reason, a special local model is created (for individual, selected joints from the whole structure), on which a classic "transient" analysis is performed and deformations are determined in specific places of the model. Then, the same model is calculated using the "shrinkage" technique. The deformations obtained as a result of the analysis are compared to the model from the classical analysis and the input data are calibrated to obtain similar values. The parameter that changes the value of the representative force for this calculation method is the radius of the pipe/sphere encompassing the model grid nodes at a certain distance from the assumed trajectory of the weld (weld axis) (Figure 5). All mesh nodes inside the area enclosed by its radius will be subjected to contraction. By increasing the value of this radius, it is possible to set the conditions of the shrinkage impact in such a way that the results of the local model of this technique come close to the results obtained in the local analysis of the "transient" type. After the calibration, it can be transferred to the model of the entire structure. In the case where the structure has a large number of the same joints (joint type and lateral dimensions of the weld), due to its repeatability, the calculation process is significantly accelerated in time as it was the case with the "local–global" method using MBD techniques [1–4,23,24].

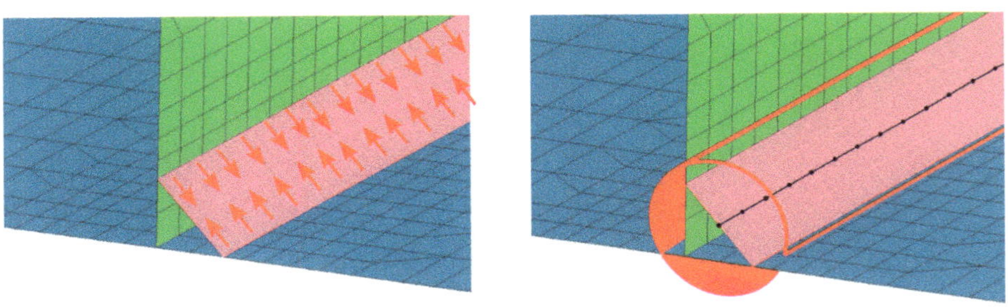

(**a**) (**b**)

Figure 5. View of a fragment of the T-joint model with (**a**) a representative shrinkage force and (**b**) marked trajectory (weld axis) with shrinkage area (tube) [4].

Analyses carried out using the "shrinkage" technique allows for quick analysis of structure deformations with a very large number of welded and welded joints, e.g., with a laser beam, as well as optimization of the order of their execution in terms of minimizing deformations. The method allows for an efficient search for a compromise between the clamping conditions (the degree of stiffening of the structure) and the level of stresses and deformations as well as determining the optimal welding sequence, which can additionally be a complicated, complex sequence of variable fastening of elements and increasing rigidity of the structure as the assembly progresses [1,3,4,23,24].

3. Calculation Examples

To show the differences between the results and the duration of calculations obtained using individual techniques, as well as to introduce the described issues, several examples of analyses were carried out using the techniques described above. All calculations were made on a computer with an Intel i5-8750H processor and 16 GB of operational memory in VisualWeld 14.5 (SYSWELD 2019.0 solver produced by ESI Group, Paris, France).

3.1. Numerical Analysis of a T-Joint with One Fillet Weld Arc Welding Process

In order to compare the results obtained with the described techniques, a calculation model of the T-joint of sheets 4.0 mm thick and 200 mm long made of S355 steel GMAW welded with one fillet weld was prepared. As a material database was used standard S355 database included in the SYSWELD material database. The finite element (FE) model consisted of 26,172 elements and 20,923 nodes in the case of calculations using the "transient" method, and the heat source model used was a double ellipsoid model. Heat source parameters were set as follows: energy per unit length, 380 J/mm; welding speed, 5.0 mm/s; arc heat efficiency, 0.7; length/width/depth of Goldak's model, 8.0/4.0/2.0 mm. In the case of calculations with the imposed thermal cycle, the FE model consisted of 13,982 elements and 11,193 nodes, and in the case of calculations using the "shrinkage" method, respectively, 4550 2D elements and 4464 nodes (Figure 6a–c). The thermal cycle used in the calculation of the MBD technique and the calibration of the shrinkage method model was carried out for auxiliary analysis using the 2D cross-section model (Figure 7). Mesh was fined in the weld area to increase calculation accuracy. Clamping conditions during welding were set to simulate welding without any additional mounting. It means that were used three nodes in which the possibilities of displacement in three, two and one of the directions were blocked (Figure 6d). This type of boundary condition simulates a situation where the elements lie freely on the welding table without additional clamping during the welding and cooling process. Boundary conditions correspond to heat dissipation to the environment on each external surface and the heat transfer coefficient correspond to the cooling conditions in the free air with an ambient temperature of 20 °C.

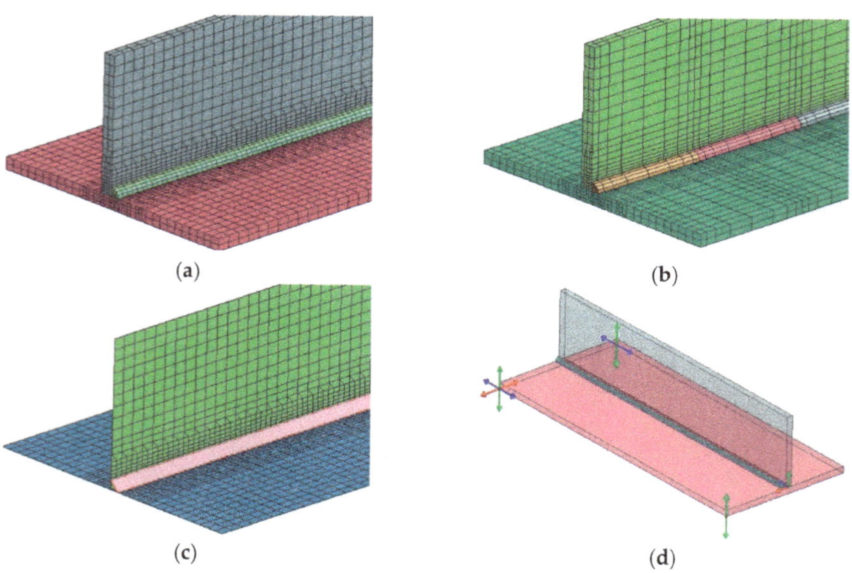

Figure 6. Finite element (FE) models prepared for the comparison of computational techniques for (**a**) "transient", (**b**) MBD, (**c**) "shrinkage" techniques and (**d**) clamping conditions of welded elements.

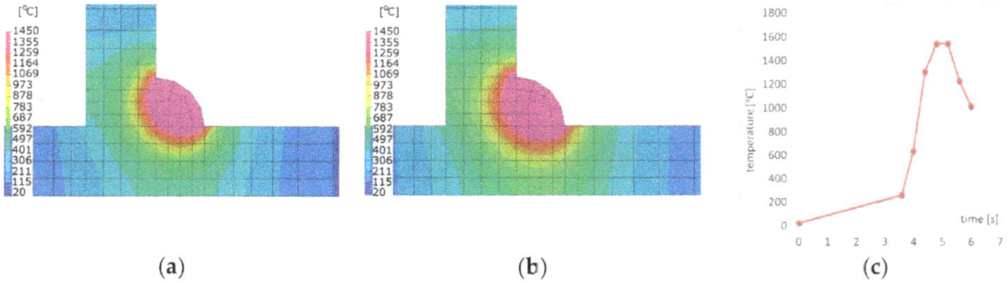

Figure 7. Temperature fields distributions comparison on the cross-sections of arc welded T-joint: (**a**) calculated using the "transient", (**b**) MBD technique and (**c**) an example of imposed thermal cycle used in the MBD technique calculations.

The calculated distortion distributions in all three techniques differed to some extent in the size distribution, but the maximum values and the deformation tendency of the joint were similar (Figures 8–10). As was observed, the stresses values in the case of the "transient" technique were higher; however, the maximum value is usually influenced in this case by the beginning and end of the weld, while in the remaining areas they are much lower (Figure 8). It also results in a sense from the definition of the start and end of the heat source model movement (certain inaccuracies related to the definition of the amount of heat in these areas can be improved by using a power ramp). This difference cannot be seen in the other two other techniques. However, when the "macro-bead" technique is used, there are characteristic peaks of the stresses values at the points where the subsequent sections are joined at which the heat cycle is given (Figures 9 and 10). The precision of stresses state calculations for these last two methods is much lower due to the simplifications used and should rather be treated as approximate values.

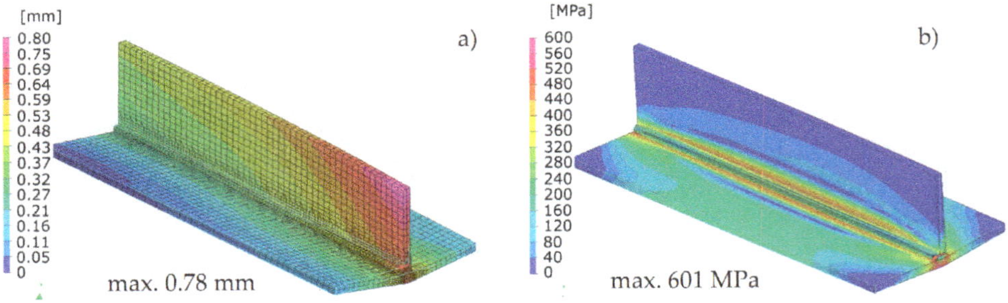

Figure 8. Distributions of (**a**) normal displacements and (**b**) von Mises stresses in arc welded T-joint calculated using the "transient" technique.

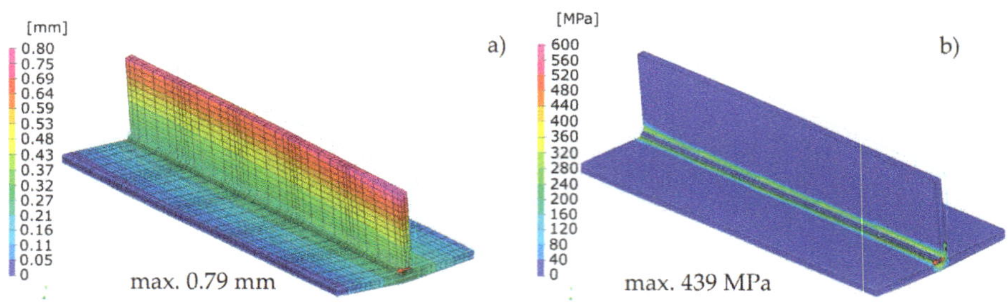

Figure 9. Distributions of (**a**) normal displacements and (**b**) von Mises stresses in arc welded T-joint calculated using the Macro-Bead Distortion Engineering (MBD DE) technique.

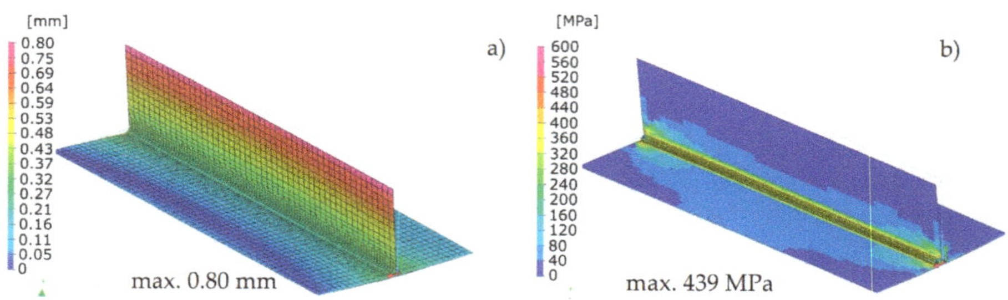

Figure 10. Distributions of (**a**) normal displacements and (**b**) von Mises stresses in arc welded T-joint calculated using the "shrinkage" technique.

Depending on the calculation method used, obtaining the result requires a certain time. A comparison of calculation times is presented for all variants in Table 1. If the MBD technique is used, it is also possible to carry out the "local–global" method (without the calculation of the temperature and metallurgical phases distributions), therefore the time for this case is also given as Macro-Bead Distortion Engineering (MBD DE). It should also be added that the times needed to calibrate the computational models in MBD and the "shrinkage" technique were not included. However, these operations can be carried out on simplified models, e.g., 2D cross-section (same as for the "transient" heat source model calibration), which significantly reduces the computation time compared to the variant using the classical "transient" technique. In the case for the complex structure wherein there are many of the same joints/welds, a significant acceleration of calculations is visible because the calibration on a separate model is made only for one, especially when the changes do not apply to technological parameters (welding current, welding speed, etc.).

Table 1. Summary of the duration times of numerical analyses for tested calculation techniques of arc welding of a T-joint model with one fillet weld.

Transient	MBD (MBD DE) [1]	Shrinkage
2h 44 min 2 s	12 min 10 s (119 s)	55 s

[1] MBD DE—Macro-Bead Distortion Engineering for the "local–global" approach using the MBD calculation technique.

3.2. *Numerical Analysis of the T-Joint Multipass Arc Welding Process with Three Fillet Welds*

In the case of multipass welded joints, it is not possible to use the models made in the shell technique (made of 2D elements). However, it is still possible to use the MBD technique as a tool that significantly reduces computation time allowing the analyses of more complex structures. Therefore, a T-joint model was prepared for S355 steel sheets 10 mm thick and 200 mm long, GMAW welded on one side, with three fillet welds. In both computational techniques, the same model (without changing the density of the mesh inside the sections, characteristic of the MBD technique) was used as a computational model consisting of 49,780 elements and 41,814 nodes. For the material database, the standard S355 database included in the SYSWELD material database was used. Boundary conditions replies to clamping conditions during welding were set to simulate welding without any additional mounting as it was described in the previous example and shown in Figure 11. Boundary conditions corresponding to heat dissipation to the environment also were on each external surface as free air cooling and the model was cooled to ambient temperature at a level of 20 °C. The calibration of the heat source model was carried out on the 2D cross-section model and on this basis, the heat cycle for MBD calculation was also prepared (Figures 11 and 12). As the heat source model in the "transient" technique was used Goldak's heat source model. Heat source parameters were set as (similar for each bead): energy per unit length, 450/350/390 J/mm; welding speed, 8.0 mm/s; arc heat efficiency, 0.7; length/width/depth of Goldak's model, 12.0/8.0/4.0 mm. The area of elements affected by the heat source model (in the SYSWELD nomenclature called LOAD) was additionally determined separately to get the right expected shape of the molten area of each bead.

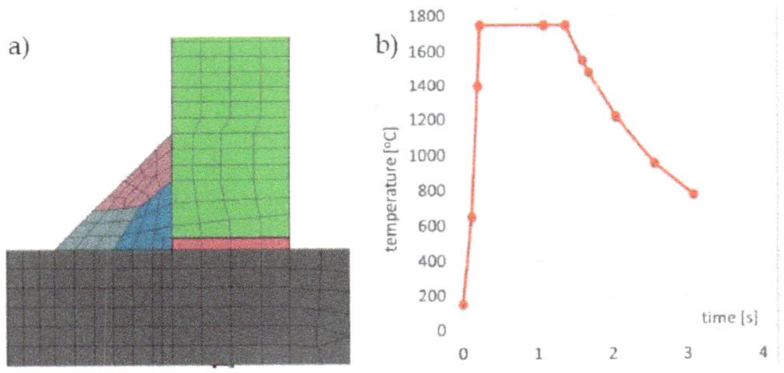

Figure 11. A view of (**a**) the 2D cross-section of the FE model, (**b**) imposed thermal cycle prepared for the calculation by the MBD technique and (**c**) clamping conditions of welded elements.

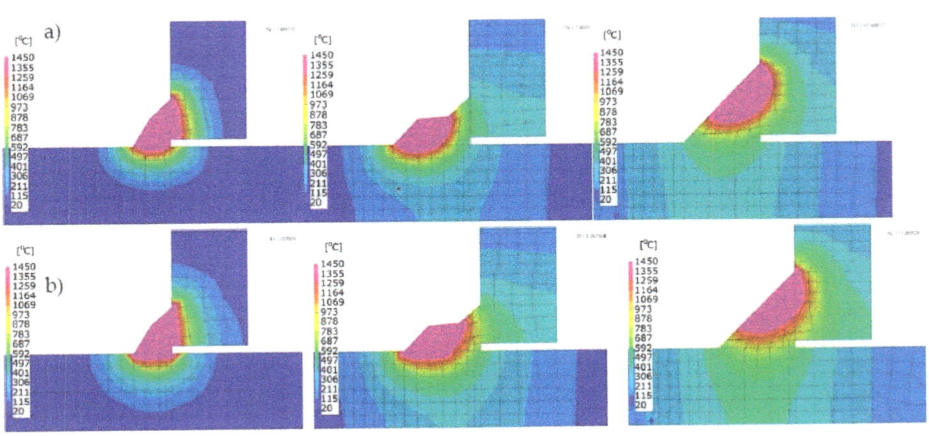

Figure 12. Comparison of temperature field distributions on the cross-sections of the multipass T-joint calculated using the (**a**) "transient" and (**b**) MBD techniques.

The calculated displacements, similarly to the previous case, slightly differed in the size and value distribution; however, they are similar to the place and direction of occurrence (Figures 13 and 14).

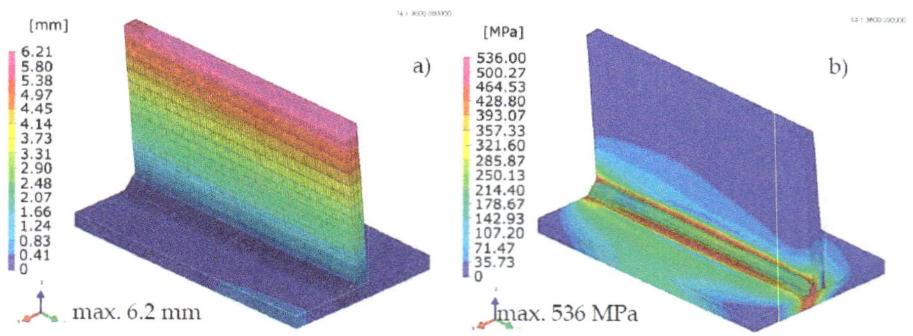

Figure 13. Distributions of (**a**) normal displacements and (**b**) von Mises stresses in arc welded multipass T-joint calculated using the "transient" technique.

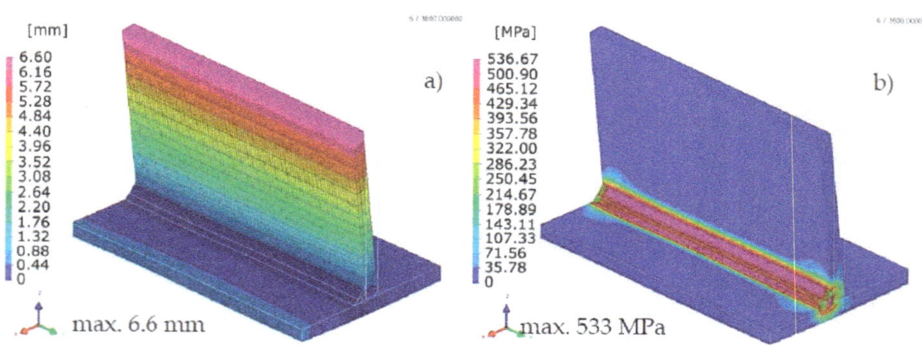

Figure 14. Distributions of (**a**) normal displacements and (**b**) von Mises stresses in arc welded multipass T-joint calculated using the MBD technique.

The calculated stresses values were also similar. As for the distribution of these stresses, also in this example, the "transient" technique offers more precise results, while in the case of the "macro-bead" technique, the user also receives more general information about the order of height and location of these stresses in the joint (Figures 13 and 14). In addition, in the case shown, no value peaks are visible along the length of the joint because the heat cycle has been applied along the entire length of each simulated beads (Figure 14).

The summary of calculation times by the compared techniques/methods are presented in Table 2. A comparison of thermal cycles of the welding process calculated by the "transient" and MBA techniques showed considerable similarities in their course as well as values (Figure 15).

Table 2. Summary of the duration times of numerical analyses for tested calculation techniques of T-joint with three arc welded beads.

Transient	MBD	MBD DE [1]
1h 8 min 2 s	20 min 26 s	11 min 28 s

[1] MBD DE—Macro-Bead Distortion Engineering for the "local–global" approach using the MBD calculation technique.

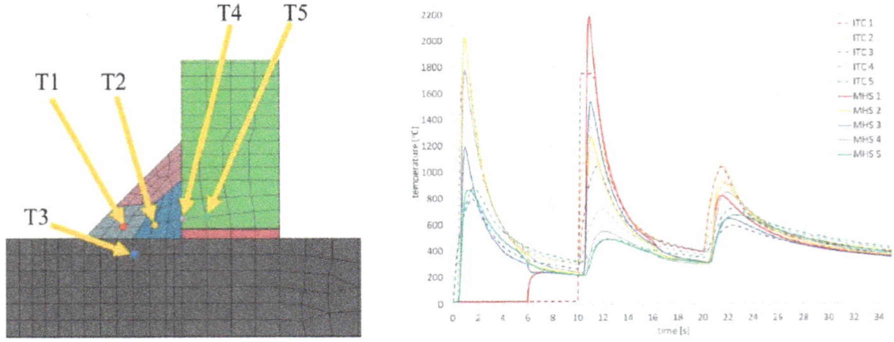

Figure 15. Thermal cycles in selected nodes of the model, calculated using the "transient" (moving heat source) and MBD (imposed thermal cycle (ITC)) techniques.

3.3. Numerical Analysis of the Laser Welding Process of Pipes Butt Joints

When using laser techniques, it is also possible to use the described computational techniques. Especially in this case, when the number of welds made using a laser beam is large and they are often repetitive welds, optimization of the order of the execution of individual beads is an ideal task to perform using the technique of "shrinkage" in the "local–global" approach. The use of the "transient" technique due to the need to use a very dense mesh (due to the size of the laser beam focus point) causes a drastic increase in the calculation time due to a large number of analysis time steps. However, typical issues can also be solved using the MBD technique, which efficiently reduces this time.

For numerical simulations of 70 × 5.0 mm pipes laser butt welding, a calculation model consisting of 16,576 elements and 14,232 nodes was prepared. Simulations were done for two different materials: AISI304 stainless steel and Inconel 625 nickel superalloy, which are also implemented in the standard SYSWELD material database. The calculation model was made as a symmetrical model (only half of the joint separated by a plane of symmetry was calculated to reduce the calculation time). Calibration of the heat source model was carried out on the 2D cross-section model, and on this basis, the imposed heat cycle used further in the MBD calculation was also determined (Figure 16a–c). Boundary conditions replies to clamping conditions were set as a complete stiffening of the two nodes in every four places (quarters) of the tube as shown in Figure 16d. After cooling the element in the final phase of calculations, the element was released from its mounting. For the clamping boundary conditions, in this case, a symmetry condition was also defined at the contact surface of the axisymmetric model. Heat dissipation to the environment was set on all external surfaces of the model as the heat transfer coefficient corresponding to the cooling conditions in the free air with an ambient temperature of 20 °C. In this case, described earlier, the conical source with a Gaussian distribution was used as a moving heat source model in the "transient" calculation technique. Heat source parameters were set as follows: laser beam power–3.0 kW, welding speed–39 mm/s, laser beam heat efficiency–1.0, z_e–5.0 mm, z_i–0 mm, r_e–1.0 mm, r_i–0.4 mm. For MBD calculation was set the thermal cycle obtained as a result of heat source calibration on a simple 2D model.

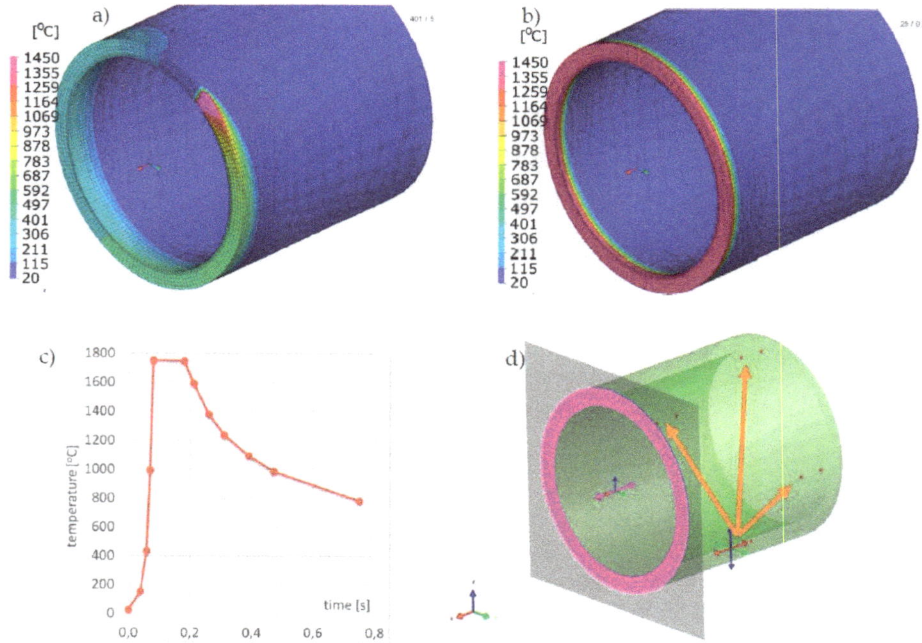

Figure 16. Temperature fields distributions during the laser welding process of pipes butt joints: (**a**) calculated with "transient", (**b**) MBD technique, (**c**) imposed thermal cycle used in MBD technique calculations and (**d**) clamping conditions during welding.

Comparison of the results calculated with both techniques showed similarities in the values and the distributions of displacements and stresses in the analyzed welded joints (Figures 17–20). The thermal cycles of the welding process calculated by the "transient" and MBD techniques showed a high convergence of their course as well as values (Figure 21). The summary of calculation times achieved by different calculation techniques/methods are presented in Table 3.

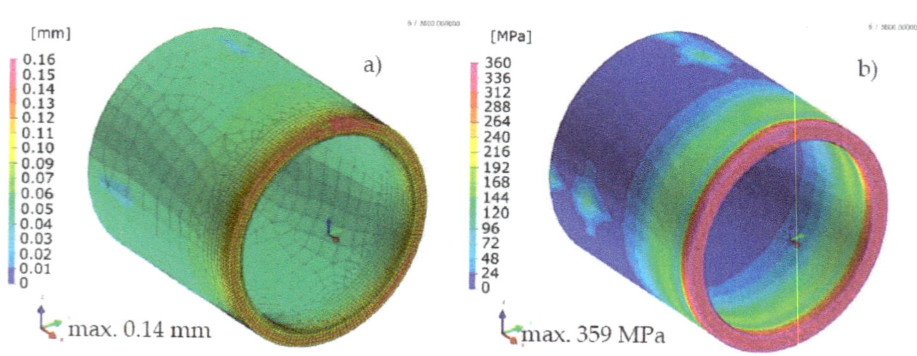

Figure 17. Distributions of (**a**) normal displacements and (**b**) von Mises stresses in stainless steel laser-welded pipes butt joints calculated using the "transient" technique.

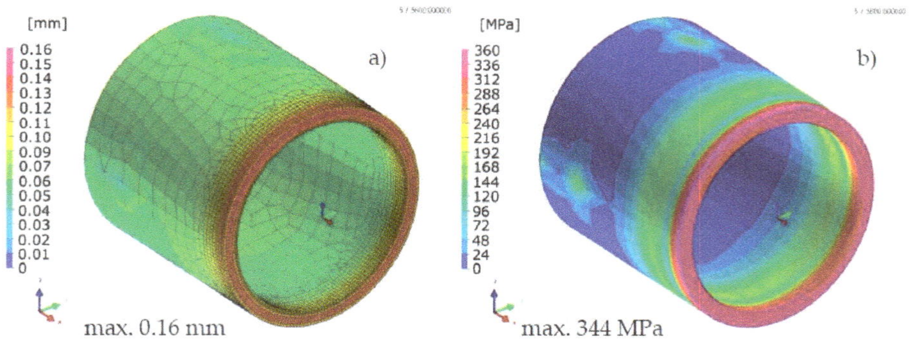

Figure 18. Distributions of (**a**) normal displacements and (**b**) von Mises stresses in stainless steel laser-welded pipes butt joints calculated using the MBD technique.

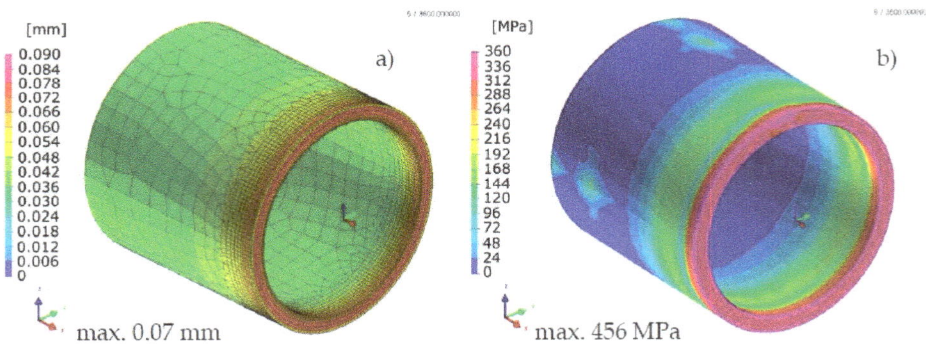

Figure 19. Distributions of (**a**) normal displacements and (**b**) von Mises stresses in Inconel 625 nickel superalloy laser-welded pipes butt joints calculated using the "transient" technique.

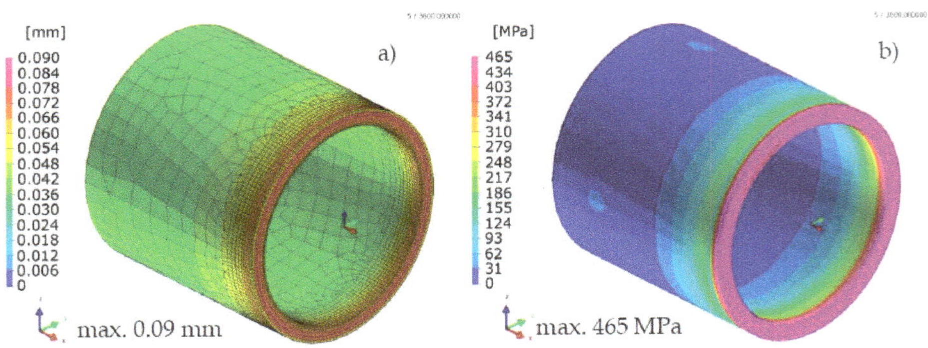

Figure 20. Distributions of (**a**) normal displacements and (**b**) von Mises stresses in Inconel 625 nickel superalloy laser-welded pipes butt joints calculated using the MBD technique.

Table 3. Summary of the duration times of numerical analyses for tested calculation techniques of laser-welded stainless steel and Inconel 625 nickel superalloy pipes.

	Transient	MBD	(MBD DE) [1]
AISI 304	45 min 8 s	5 min 1s	50 s
Inconel 625	33 min 12 s	3 min 48 s	55 s

[1] MBD DE—Macro-Bead Distortion Engineering for the "local–global" approach using the MBD calculation.

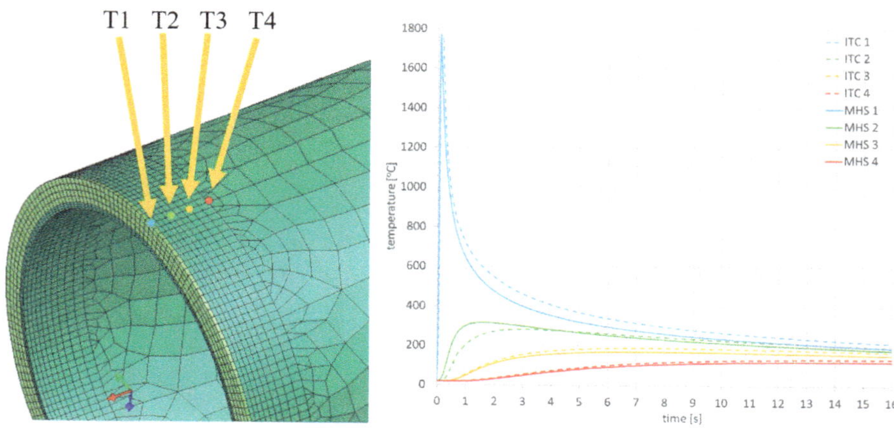

Figure 21. Thermal cycles in selected nodes of the model, calculated using the "transient" (moving heat source) and MBD (imposed thermal cycle (ITC)) techniques for AISI 304 pipes laser welding simulation example.

4. Conclusions

This article describes three calculation techniques for welding processes numerical simulations using the finite element method. The described methods are both the "transient" method that gives the largest amount of precise information and details. Due to their specificity, they require conducting step-by-step calculations, which increases the calculation time, as well as the MDB or "shrinkage" technique, which in turn allow for significant reducing the calculation time and allows conducting analyses of complex welded constructions at the expense of quality and precision of obtained results. Significantly longer calculation times in the "transient" technique result from the aforementioned calculation method, where the moving heat source travels over the grid elements generating subsequent time moments (calculation cards) as well as the fact that all quantities describing the welding process are calculated in these steps (temperature fields, phase distributions, heat fluxes, etc.). In the case of complicated structures with a large number of welds/joints, this makes it impossible to carry out analyses due to the real-time needed for their implementation and the amount of data generated in the resulting output file. However, analyses of this type result in a number of important, precise data on the so-called local welding process effects for the selected welded joint. They also support the creation of simple (and therefore requiring shorter calculation times) models for calibrating the input data used in the MBD or "shrinkage" technique. They can be both 2D cross-section models as well as fully three-dimensional models, however, with less complexity or smaller size (primarily length but sometimes also transverse dimensions).

As the presented examples have shown, the use of the MBD technique significantly reduces the time of calculations due to the fact that in this technique the nodes of the mesh are affected by a given thermal cycle consisting of several to several points (steps) instead of the amount resulting from the density of the mesh and the size of the liquid metal pool (and hence also in some sense the dimensions of the heat source model). The cost incurred by the user in the event of its application results in the lower precision of the results obtained, caused by the specifics of thermal load (e.g., increases in stresses values at the end of the section into which the weld bead is divided (Figure 9b). Some differences also occur as a distribution of calculated quantities, although the maximum and minimum values and their distribution are usually similar to "transient" calculations results (Figures 8–10, 13 and 14).

In addition, in the case of resignation from calculations of the temperature fields and metallurgical phase distributions and the imposed heat cycle that was previously calibrated using a simple model calculated using the "transient" technique, it is observed a significant reduction of the calculation times. For the presented simulation examples, the calculation time reduction (without taking into account the calibration stage of the heat source model, which must also be carried out for the "transient" method at heat source calibration stage) is from a few to several times (Tables 1–3). In the case of very complex structures, where the use of the "transient" technique is impossible due to the aforementioned long calculation time, the use of the MBD DE or "shrinkage" technique supports the very fast prediction of the deformations of the weld structure. In this case, the abandonment of calculations of temperature field and metallurgical phase distribution and the use of shell elements instead of 3D solid elements, on the one hand, limits the result to the approximate deformation distribution, but on the other hand, it allows to quickly check different clamping variants and the order and direction welding individual beads. It can, therefore, be said that in a situation when, after proper calibration, the displacements distributions are similar, and the analysis mainly concerns structure deformations, the use of this computational technique brings measurable time benefits. This will be particularly evident when a large element or entire structure containing a large number of welds/beads is being analyzed. The mentioned repeatability of the welding parameters, allows for calibration on one model using the "transient" technique, and then transferring them to the structure and finally calculating the total deformation using the MBD or "shrinkage" technique. In such cases, the results of the analyses are significantly accelerated and this is often the only possible solution in terms of hardware and time limitations.

However, not only the mesh density or the complexity of the structure affects the calculation time. The type of material of the welded parts also affects the calculation time. It is because the whole simulation process is divided into time cards. Their amount depends on the density of the mesh, i.e., the number of steps that the moving heat source and cards related to the next cooling of the welded element to the set temperature must follow. Individual iterations in the calculation process of the selected time card must meet the assumed convergence criterion. Only then does the solver move to the next card. The number of iterations strictly depends on the non-linearity of the calculations. In turn, the non-linearity of the calculations is greater when the plastic deformations in the material are higher. Thus, the lower the yield strength of a material, the faster it gets into plastic deformations but also convergence problems. This is very well visible on the presented example of with laser welding of AISI 304 steel and Inconel 625 nickel superalloy pipes. The almost twice lower yield strength for AISI 304 steel makes it so that the calculation time is longer compared to the calculation time for Inconel 625 nickel superalloy. The differences in the calculation time for both materials in the case of the "transient" method are approximately 26% and 24% for the MBD technique (Table 3). The slight difference in the calculation times when using the MBD DE technique is confirmed by the fact that, in this case, they are not taken take into account with the distribution of metallurgical phases and changes in material properties, and hence stresses and strains values.

In summary, in a situation where currently the main challenge for the industry is to reduce the time needed for design work and research and development activities, the described modern calculations methods create possibilities to increase efficiency and reduce the waiting time for the final product. In addition, it is possible to detect positioning problems and incorrect design, placement or use of fasteners. If we add the possibility of detecting the risk of defects during assembly, there is a real advantage of using numerical analyses in the form of improving production efficiency as well as significant economic benefits.

Funding: This publication was co-funded by the statutory subsidy of the Faculty of Mechanical Engineering at the Silesian University of Technology in 2019.

Conflicts of Interest: The author declares no conflict of interest.

References

1. Welding Simulation User Guide, Sysweld Manual ESI Group. Available online: https://www.esigmbh.de/downloads/ESI/Dokumente/Welding/The_Welding_Simulation_Solution_210408 (accessed on 15 February 2016).
2. Slováček, M.; Kik, T. Use of Welding Process Numerical Analyses as Technical Support in Industry. Part 1: Introduction to Welding Process Numerical Simulations. *Biul. Inst. Spaw.* **2015**, *4*, 25–31.
3. Slováček, M.; Vaněk, M.; Kik, T. Use of Welding Process Numerical Analyses as Technical Support in Industry. Part 2: Methodology and Validation. *Biul. Inst. Spaw.* **2015**, *59*, 38–43.
4. Kik, T. Metody analiz numerycznych procesów spawania złożonych konstrukcji. *Technol. Rev.* **2017**, *89*, 41–45. [CrossRef]
5. Kik, T.; Górka, J. Numerical simulations of laser and hybrid S700MC T-joint welding. *Materials* **2019**, *12*, 516. [CrossRef] [PubMed]
6. Janicki, D. Fiber laser welding of nickel based superalloy Inconel 625. In *Laser Technology 2012: Applications of Lasers*; International Society for Optics and Photonics: Bellingham, WA, USA, 2013; Volume 8703, p. 87030R. [CrossRef]
7. Lisiecki, A. Titanium Matrix Composite Ti/TiN Produced by Diode Laser Gas Nitriding. *Metals* **2015**, *5*, 54–69. [CrossRef]
8. Stadnicki, J.; Wróbel, I. Numerical effectiveness of the simulation of an automotive body part stamping. *Adv. Mech. Eng.* **2015**, *7*, 708434. [CrossRef]
9. Chmielewski, T.; Golański, D. The role of welding in the remanufacturing process. *Weld. Int.* **2015**, *29*, 861–864. [CrossRef]
10. Tomkow, J.; Fydrych, D.; Rogalski, G. Role of bead sequence in underwater welding. *Materials* **2019**, *12*, 3372. [CrossRef] [PubMed]
11. Ferro, P.; Porzner, H.; Tiziani, A.; Bonollo, F. The influence of phase transformations on residual stresses induced by the welding process—3D and 2D numerical models. *Model. Simul. Mater. Sci. Eng.* **2006**, *14*, 117–136. [CrossRef]
12. Wu, W.; Liang, N.; Gan, C.; Yu, G. Numerical investigation on laser transformation hardening with different temporal pulse shapes. *Surf. Coat. Technol.* **2006**, *200*, 2686–2694. [CrossRef]
13. Shanmugam, N.S.; Buvanashekaran, G.; Sankaranarayanasamy, K.; Ramesh Kumar, C. A Transient Finite Element Simulation of the Temperature and Bead Profiles of T-joint Laser Welds. *Mater. Des.* **2010**, *31*, 4528–4542. [CrossRef]
14. Winczek, J.; Skrzypczak, T. Thermomechanical states in arc weld surfaced steel elements with heat of melted electrode material taken into account. *Arch. Metall. Mater.* **2016**, *61*, 1277–1288. [CrossRef]
15. Chiocca, A.; Frendo, F.; Bertini, L. Evaluation of Heat Sources for the Simulation of the Temperature Distribution in Gas Metal Arc Welded Joints. *Metals* **2019**, *9*, 1142. [CrossRef]
16. Moravec, J.; Bradác, J.; Neumann, H.; Nováková, I. Grain size prediction of steel P92 by help of numerical simulations. In Proceedings of the METAL 2013 22nd International Conference on Metallurgy and Materials, Brno, Czech Republic, 15–17 May 2013; pp. 508–513.
17. Kong, F.; Ma, J.; Kovacevic, R. Numerical and experimental study of thermally induced residual stress in the hybrid laser-GMA welding process. *J. Mater. Process. Technol.* **2011**, *211*, 1102–1111. [CrossRef]
18. Devaux, J.; Leblond, J.B.; Bergheau, J.M. Numerical study of the plastic behaviour of a low alloy steel during phase transformation. In Proceedings of the 1st International Conference on Thermal Process Modelling and Computer Simulation, Shanghai, China, 28–30 March 2000; pp. 206–212.
19. Goldak, J.; Chakravarti, A.; Bibby, A. A New Finite Element Model for Welding Heat Sources. *Metall. Trans.* **1984**, *15*, 299–305. [CrossRef]
20. DuPont, J.N.; Marder, A.R. Thermal efficiency of arc welding processes. *Weld. J. Incl. Weld. Res. Suppl.* **1995**, *74*, 406s.

21. Bradáč, J. Calibration of Heat Source Model in Numerical Simulations of Fusion Welding. *Mach. Technol. Mater.* **2013**, *11*, 9–12.
22. Wu, C.S.; Wang, H.G.; Zhang, Y.M. A new heat source model for keyhole plasma arc welding in FEM analysis of the temperature profile. *Weld. J.* **2006**, *85*, 284–291.
23. Slováček, M. Numerical Simulation of Welding, Calculation and Evaluation of Distortions and Residual Stresses. Ph.D. Thesis, Brno Military Academy, Brno, Czech Republic, September 2005.
24. Slováček, M.; Diviš, V.; Ochodek, V. Numerical simulation of the welding process, distortion and residual stresses prediction, heat source model determination. In Proceedings of the 58th Annual Assembly Meeting, Prague, Czech Republic, 10–15 July 2005; pp. 1586–2005.

materials

MDPI

Article

Behavior of Weld to S960MC High Strength Steel from Joining Process at Micro-Jet Cooling with Critical Parameters under Static and Fatigue Loading

Tadeusz Szymczak [1], Bożena Szczucka-Lasota [2], Tomasz Węgrzyn [2,*], Bogusław Łazarz [2] and Adam Jurek [3]

[1] Department of Vehicle Type-Approval & Testing, Motor Transport Institute, Jagiellońska 80, 03-301 Warszawa, Poland; tadeusz.szymczak@its.waw.pl

[2] Faculty of Transport and Aviation Engineering, Silesian University of Technology, 40-119 Katowice, Poland; bozena.szczucka-lasota@polsl.pl (B.S.-L.); boguslaw.lazarz@polsl.pl (B.Ł.)

[3] Novar Sp. z o. o., Towarowa 2, 44-100 Gliwice, Poland; adam.jurek@novar.pl

* Correspondence: tomasz.wegrzyn@polsl.pl; Tel.: +48-32-603-4389

Abstract: The paper is focused on testing the weld of the S960MC steel produced at the micro-jet cooling under static and fatigue loading at critical parameters. This kind of material was in the form of a sheet with a thickness equal to 2 mm. The joint was obtained using three different types of welding wires: EDFK 1000, Union NiMoCr and Union X96 at the same parameters of the process. The joints were examined using non-destructive and destructive tests. The results from non-destructive experiments enable us to assess the quality of the welds directly before the joining process. In contrast, the destructive one allows following welds behavior under different loading conditions with their critical parameters. The bending experiments confirmed the good plastic properties of the weld, expressed by no cracks in the region tested in many variants of the joint manufactured. The results from static tests indicated a significant reduction of mechanical parameters of the weld in comparison to the base metal, expressed by 50% differences. Fatigue data have enabled us to follow the welding behavior at the increasing amplitude of axial stress up to fracture at constant amplitude value covering the following values of stress 650 MPa–100 MPa. Variations of total energy are presented at different values of several cycles up to fracture. Fracture regions are collected for analysis of the joint region features under cyclic loading. They have indicated differences in weld cracking depended on the stress level. Finally, the Wöhler S-N curve of the weld was determined, indicating the value of the fatigue limit of the weld tested, i.e., 100 MPa. The weld at the Union NiMoCr welding wire was indicated as the joint having the highest resistance on static and fatigue loadings.

Keywords: high-strength steel; components; joining; transport; micro-jet; welding; mechanical resistance; parameters; microstructure; mini-specimen; fracture; fatigue limit

Citation: Szymczak, T.; Szczucka-Lasota, B.; Węgrzyn, T.; Łazarz, B.; Jurek, A. Behavior of Weld to S960MC High Strength Steel from Joining Process at Micro-Jet Cooling with Critical Parameters under Static and Fatigue Loading. *Materials* **2021**, *14*, 2707. https://doi.org/10.3390/ma14112707

Academic Editor: Bolv Xiao

Received: 16 April 2021
Accepted: 18 May 2021
Published: 21 May 2021

1. Introduction

High-Strength Steels (HSS) belong to modern metals used in a lot of branches of the industry concerning their mechanical properties, such as ultimate tensile strength and yield stress, which are higher compared to typical structural materials [1,2]. Among them, the following material types can be indicated: Strenx [1,3], Docol [4,5], Amstrong Ultra [6] and Optim QC [7]. Their application is very wide, and it is selected based on comparing the value of stress due to operational conditions and mechanical parameters at microstructure features [1,2]. Therefore, they are employed for manufacturing different components called: car bodies, stringers [8], frames [9], bumper reinforcement and door beams [5], Rear Underrun Protective Device [1,10], booms [6,7], platform [8], tippers and aerial platforms [6].

Until now, unalloyed steel with a minimum yield strength of 355 MPa and minimum impact strength KV-20 = 27 J was used for the construction of the container part of the container-type vehicle [11]. Due to low strength, the walls of the containers were massive, with a thickness of up to 5 mm. Recently, attempts have been made to use high-strength steel (e.g., S960 grades) for the construction of containers, which has a strength of 2.5 times greater, which makes it possible to significantly reduce the thickness of the container wall, and thus the entire weight of the container [12]. This is very important in transport, as reducing the weight of the container will allow you to carry more loads. It is a very economical solution [13]. Worth noticing, the production process for elements made of HSS materials is very different, and it can be conducted using welding, screwing, or bolting techniques. Moreover, in the case of the last connection methods, the sub-components used for the joining have certificates, while the quality of the first one depends on the qualifications of the welding group. This problem is considered by a lot of research teams [9,14], which examined parameters of the joining technique or modified ones applying additional sub-elements such as micro-jet cooling device [15,16] or hybrid welding process [17,18]. These efforts are transferred for various kinds of high-strength steel focusing on the S960MC [15,17,18], which is an attractive kind of metal because it is produced in a lot of forms such as sheets, tubes and plates [1,6]. The Strenx 960MC is very good weldable, cold-formed and machined by cutting. Therefore, typical applications are represented by advanced lifting devices such as mobile cranes and lighter transport solutions and components [3,6]. They indicate the steel and their joints can be subjected to various types of loading, including: static and cyclic, which influence mechanical parameters important for engineering practice and inspection stages for a correct designing and service life, respectively. Therefore, different tests are used for determining the steel reaction [19]. They follow the behavior of the welded zones with different filler materials up to fatigue limits indicating its value close to 100 MPa at a very narrow stress range represented by the following values 100–300 MPa, [18]. Moreover, taking these data, differences between the role of metal used for filling can be easily indicated. Besides fatigue tests, hardness experiments are still used as the fundamental probe for assessing the quality of a welded joint made of the S960MC steel [18]. This method is very helpful at the calibration of the welding process as well as the hybrid joining technique. Moreover, the technique is very typical; nevertheless, some differences between welded regions can be easily indicated [18]. This type of results enables to follow concludes on the quality of joint if sufficient numbers of measurement regions are collected [20]. This approach can also be evidence in [17,21], which presents data from micro-hardness tests [17] on the joint manufactured at the Gas Metal Arc Weld (GMAW). The same sentence, as it has been formulated in the case of the hardness probe, can be addressed to impact test, which has enabled indicating advantages of the laser and laser welding process, expressed by the following values of accumulated energy reaching 57 J and 49 J, respectively. The same kind of test was also used for differencing features of the S960MC welds produced by various types of the process [22]. In this case, the experiment type was very effective and has enabled us to distinguish variations of impact toughness. Some authors [23] have used a combined approach such as fracture toughness results to predict fatigue endurance of a weld of the S960MC. They have presented a fatigue curve at stress value close to data shown in [18], indicating good agreement between the method and experimental efforts.

Welding S690 steel grades is not easy. For making joints with a thickness of about 6 mm, it is absolutely necessary to use preheating and control the inter-stitch temperature. Joints made of these steels have low fatigue strength and low relative elongation. The softening of the HAZ (Heat Affected Zone) is one of the major issues of HSS welding [24,25]. There is no information in the technical literature on the fatigue strength of welded joints made of S960 steel grade. Thin-walled joints are a major welding problem, as they can be prone to welding cracks.

Pipe joints made of HSS steel with variable wall thickness were tested. It has been shown that a thick-walled joint has better fatigue properties than a thin-walled joint due to the reduction of the stress concentration factor. In fact, the presence of fatigue cracks or otherwise induced defects in such connections poses a potential safety risk to the structure.

The authors found the possibility of cold cracks, especially in joints of greater thickness, where they recommend using preheating. It is also related to the higher hydrogen content in the joint obtained without preheating [26,27].

In welding AHSS (Advanced High Strength Steel) steels, there is a need to modify the technology to improve the strength and plastic properties of the joint. This is due to the slightly different metallographic structure of the weld and the base material (expanded ferrite and martensite in the weld). For this purpose, for the first time, it was decided to test the micro-jet cooling method during S960MC welding (research gap), which can lead to the fragmentation of the joint structure, which is related to the improvement of the mechanical properties of the joint. For the first time, fatigue tests of such a joint were performed, which are the most important information leading to the verification and validation of the new design and technology.

The aim of the article was to develop a material and technological solution enabling the production of a thin-walled structure made of difficult-to-weld steel, characterized by the best mechanical properties. Therefore, in the article, it was decided to very carefully check the possibility of MAG (Metal Active Gas) welding S960MC steel using various parameters, materials and welding methods. It was also decided to check the possibility of welding S960MC steel to obtain the best mechanical properties. It was decided to test three electrode wires with a variable content of carbon and other alloying elements, and it was decided to test the use of two different shielding mixtures in the MAG welding process. Independently, it was decided to check the possibility of making a thin-walled structure of the S960MC steel joint with the use of micro-jet cooling, believing that this type of cooling would allow for the fragmentation of the dominant martensitic and ferritic structures. It was decided to perform numerous non-destructive and destructive tests to determine the most favorable welding and cooling parameters of the micro-jet, which allows the best mechanical properties of the joint to be obtained. Taking into account the results discussed above, the two research paths for examining the S960MC weld are evidenced. The first one is represented by typical experiments such as hardness and fatigue tests but at a narrow range of experimental procedures, while the second manner is expressed by complex approach, i.e., theoretic-experimental. From the practical and scientific point of view, these manners are suited for selected loading regimes, without details on the energetic aspects of the fatigue process as well as cases at high value of stress, including an increase of amplitude related to the fractured moment. The most important point of the research was to check the fatigue strength after welding S960MC steel. For this purpose, the size of the samples was specially designed, and fatigue strength tests were performed on very sensitive equipment.

In addition to the use of micro-jet cooling, the type of shielding gas plays an important role. Only the use of the Ar–CO_2 mixture allows obtaining the correct joints. Further research focused on the evaluation of joints made with three different welding wires. M21 (Ar-18% CO_2) cover and micro-jet cooling were always used. Tensile and fatigue tests were performed. The influence of the selection of the electrode wire on the mechanical properties of the joint was analyzed. The wires had a comparable chemical composition but differed slightly in carbon content. It was believed that the carbon content of the wire could affect the mechanical properties of the joint: both tensile strength and fatigue strength.

2. Materials and Methods

2.1. Specimen for Welding and Parameters of the Joining Process

Welded butt joints of S960MC steel with a thickness of 2 mm, length 200 mm and width 300 mm were made. The MAG (Metal Active Gas) welding process was used according to the standard requirements (EN 15614-1). The preparation of the material for single-stitch welding is shown in Figure 1.

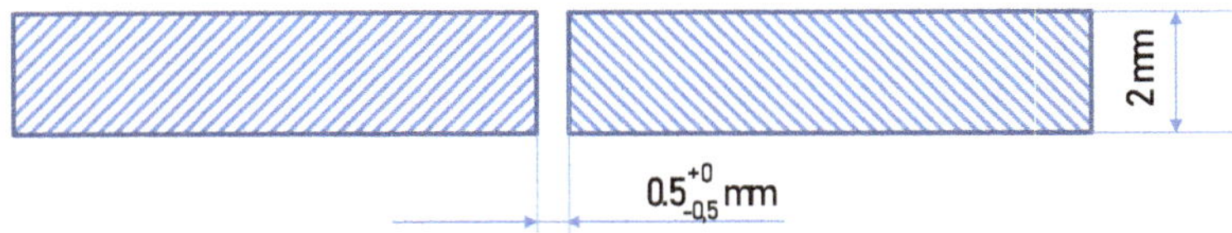

Figure 1. Preparation of the element for metal active gas (MAG) welding with micro-jet cooling.

It was decided to make test joints using two shielding gases: CO_2 and gas mixture M21, i.e., Ar + 18% CO_2 (according to the PN-EN 14175 standard). All of the sheet metal samples were welded with three different electrode wires (ED-FK 1000, Union NiMoCr and UNION X96) with a similar chemical composition but with a slightly increasing carbon content in each of them (0.08% C; 0.09% C; 0.11% C):

- EN ISO 12534-A G 69 5 M Mn4Ni1.5Cr-Union (Voestalpine Böhler Welding, Hamm, Germany) NiMoCr (C 0.08, Si 0.6, Mn 1.7, Cr 0.2, Mo 0.5 and Ni 1.5);
- EN ISO 16834-A G 89 6 M21 Mn4Ni2CrMo-ED-FK 1000 (FLIESS, Duisburg, Germany) (C 0.09, Si 0.8, Mn 1.8, Cr 0.31, Mo 0.55 and Ni 2.2);
- EN ISO 16834-A G 89 5 M21 Mn4Ni2.5CrMo-UNION X96 (Voestalpine Böhler Welding, Hamm, Germany) (C 0.11, Si 0.78, Mn 1.9, P 0.01, S 0.009, Cr 0.35, Mo 0.57, Ni 2.23, V 0.004, Cu 0.02, Ti 0.057, Zr 0.001 and Al 0.002).

The chemical composition of the three selected electrode wires was comparable. A slight difference in the carbon amount was noted. The carbon content influences the steel weldability and tensile strength. There were also slight differences in the chemical composition between the base material and the electrode wires. Molybdenum and nickel, which are not present in steel, were additionally introduced in the electrode wires. This is conducted in order to increase the plastic properties of the joint. Vanadium and copper in both the base material and the filler wires are low. However, there are noticeable differences in the content of aluminum, especially titanium, which is responsible for the precipitation strengthening of the joint [28].

In the initial stage, the MAG welding parameters were established: current 95–105 A, arc voltage (17–22 V) and welding speed (250–350 mm/min). Observations were made 48 h after welding to ensure that no cold cracks would form (by the recommendations of the standard PN-EN 1090-2). Welding tests were performed with the assumed gas flow of 14 L/min. The input energy during the welding of thicker sheets was below the recommended value of 4 kJ/mm. All welding tests were carried out without preheating. The best results (correct joint form) were put in Table 1.

Table 1. The welding current of S960MC steel.

Diameter of the Electrode, mm	Current Intensity, A	Voltage, V	Polarization + or −	Welding Speed, mm/min
1.0	101	19	DC "+"	300

The chemical composition and mechanical properties of the S960MC steel are presented in [3]. The gap between sheets was varied in the range 0–1 mm. The correct results were obtained for the gap equal to 0.5 mm, Figure 1. This case was only taken for further investigation. The most important other parameters of the welding process are given in Table 1.

S960MC steel was welded using micro-jet cooling. Micro-jet cooling parameters were varied in the following way:

- micro-jet gas: argon;
- stream diameter: 60 and 70 μm;
- gas pressure: 0.6 and 0.7 MPa.

Table 3. Details in the form of mechanical parameters of the welds to experiment under increasing stress amplitude.

No.	Specimen	Yield Stress (MPa)	Ultimate Tensile Strength (MPa)	Time for the Envelop (s)
1.	Weld at EDFK 1000	530	660	600
2.	Weld at Union NiMoCr	550	700	
3.	Weld at Union X96	530	660	

The selected value of time to fracture under the stress cycles was represented by 600 s. This was directly used in the Envelop section for cyclic signals in the Wave Matrix software of the 8874 Instron servo-hydraulic testing machine. In all cases, the frequency was equal to 10 Hz. In all experimental efforts, the value of yield stress was reached at the same time; next, the stress cycles were conducted at an increasing value of stress amplitude at the minimum and maximum values in a range of time 0–600 s (Figure 3). The tests were conducted up to specimen fracture starting from the value of yield stress.

Fatigue tests on the Wöhler's curve were conducted at stress signal in the form of a sinusoidal function at frequency 10 Hz; stress ratio R = 0.1 and values of amplitude as follows: 650 MPa, 600 MPa, 550 MPa, 500 MPa, 400 MPa, 300 MPa, 200 MPa and 100 MPa (Figure 4).

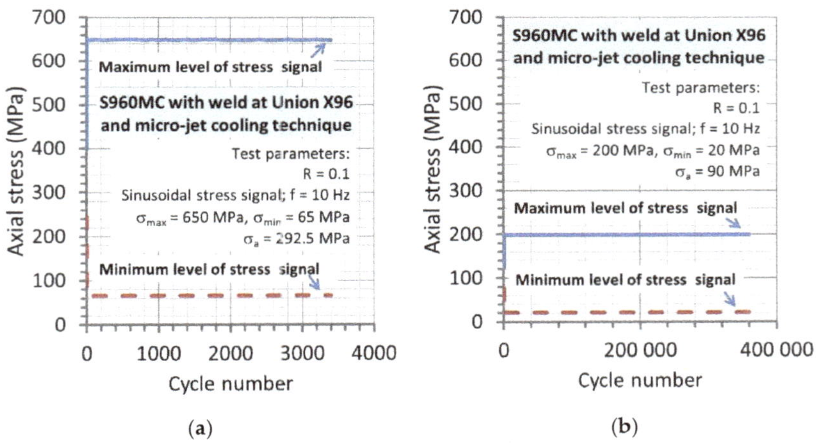

Figure 4. Maximum and minimum values of stress signals used for determining the Wöhler curve of the joint from the welding process supported by the micro-jet technique at the Union X96 welding wire: (**a**) the highest and (**b**) the lowest stress range applied in the test.

3. Results and Discussion

3.1. The Results of Non-Destructive Tests

Non-destructive test (NDT) results are presented in Table 4. The results of NDT tests allow for a preliminary assessment of the quality of the weld. They expressed differences in the weld quality and indicate the NDT is very important for determining the quality of the zone examined because defects such as cracks can be evidenced.

Table 4. The assessment of non-destructive testing on the movable platform joint.

Designation	Shielding Gas	Wire, C Amount (%)	Micro-Jet Stream Pressure (MPa)	Micro-Jet Stream Diameter (μm)	Micro-Jet Gas	Observation Acceptability: (× or √, B) *
UNm1	Ar + 18% CO_2	Union NiMoCr, (0.08)	without	without	without	×
UNm2	Ar + 18% CO_2	Union NiMoCr, (0.08)	0.6	60	Ar	×
UNm3	Ar + 18% CO_2	Union NiMoCr, (0.08)	0.7	60	Ar	√, B
UNm4	Ar + 18% CO_2	Union NiMoCr, (0.08)	0.6	70	Ar	√, B
UNm5	Ar + 18% CO_2	Union NiMoCr, (0.08)	0.7	70	Ar	×
EDm1	Ar + 18% CO_2	ED-FK 1000, (0.09)	without	without	without	×
EDm2	Ar + 18% CO_2	ED-FK 1000, (0.09)	0.6	60	Ar	×
EDm3	Ar + 18% CO_2	ED-FK 1000, (0.09)	0.7	60	Ar	√, B
EDm4	Ar + 18% CO_2	ED-FK 1000, (0.09)	0.6	70	Ar	√, B
EDm5	Ar + 18% CO_2	ED-FK 1000, (0.09)	0.7	70	Ar	×
UXm1	Ar + 18% CO_2	Union X96, (0.11)	without	without	without	×
UXm2	Ar + 18% CO_2	Union X96, (0.11)	0.6	60	Ar	×
UXm3	Ar + 18% CO_2	Union X96, (0.11)	0.7	60	Ar	√, B
UXm4	Ar + 18% CO_2	Union X96, (0.11)	0.6	70	Ar	√, B
UXm5	Ar + 18% CO_2	Union X96, (0.11)	0.7	70	Ar	×
UNc1	CO_2	Union NiMoCr, (0.08)	without	without	without	×
UNc2	CO_2	Union NiMoCr, (0.08)	0.6	60	Ar	×
UNc3	CO_2	Union NiMoCr, (0.08)	0.7	60	Ar	×
UNc4	CO_2	Union NiMoCr, (0.08)	0.6	70	Ar	√, B
UNc5	CO_2	Union NiMoCr, (0.08)	0.7	70	Ar	×
EDc1	CO_2	ED-FK 1000, (0.09)	without	without	without	×
EDc2	CO_2	ED-FK 1000, (0.09)	0.6	60	Ar	×
EDc3	CO_2	ED-FK 1000, (0.09)	0.7	60	Ar	√, B
EDc4	CO_2	ED-FK 1000, (0.09)	0.6	70	Ar	×
EDc5	CO_2	ED-FK 1000, (0.09)	0.7	70	Ar	×
UXc1	CO_2	Union X96, (0.11)	without	without	without	×
UXc2	CO_2	Union X96, (0.11)	0.6	60	Ar	×
UXc3	CO_2	Union X96, (0.11)	0.7	60	Ar	×
UXc4	CO_2	Union X96, (0.11)	0.6	70	Ar	×
UXc5	CO_2	Union X96, (0.11)	0.7	70	Ar	×

* Board description: √, B—No cracks in the weld, level of quality-B without welding defects and incompatibilities; ×—Cracks in the weld.

Based on the results of non-destructive testing, the samples in Table 4 were marked as "√"—positive and "×"—negative. Examples of non-destructive testing results are shown in Figures 5 and 6. Figure 5 shows a positive result recorded for sample UXm3. The quality of the sample was on level B. No cracks and welding incompatibilities were found in the analyzed joint. Figure 6 shows a negative result. The result was obtained for sample UXm1, which is characterized by welding incompatibilities.

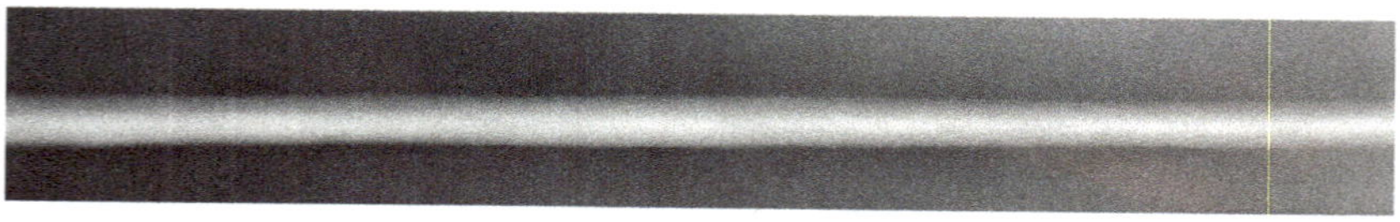

Figure 5. Correctly made joint after welding with micro-jet cooling.

Figure 6. Welding defects (cracks) and incompatibilities after classic MAG welding without micro-jet cooling.

MAG welding in CO_2 shield is more oxidizing than welding in an Ar-18% CO_2 shield. Welding in a CO_2 shield corresponds to the oxygen content in the weld at the level of 500 ppm, which translates into the larger size of non-metallic inclusions that may initiate cracks. Welding in the Ar–CO_2 mixture shield is less oxidizing; it leads to the formation of smaller non-metallic inclusions that may, to a lesser extent, cause cracks in the joint. The classification of welding processes in terms of the oxygen content in the weld and its justification was presented in 1999 [29,30].

Based on the analysis of the table data, it can be noticed that the M21 mixture (Ar + 18% CO_2) is much better than CO_2. From the analysis of the results presented in Table 4, it follows that the test was carried out correctly; the evaluation of the tests is positive only in some cases when micro-jet cooling was not too intense and not too weak. For further tests (destructive test—DT), only those joints that met these requirements were taken into account:

- made with the use of micro-jet cooling;
- no cracks were observed.

Therefore, samples with the following determinations were taken into account (UNm3, UNm4, EDm3, Edm4, UXm3, UXm4 and UNc4, EDc3).

3.2. *Data from the Bending and Hardness Tests*

The test results of bending tests are summarized in Table 5. The process was correct only with medium-power micro-jet cooling because no cracks and other disconformities were found in the samples tested. It reflected the high-quality level of the joint, and it was confirmed in comparison to the regimes of the PN-EN ISO 5817 standard [31] on a crack length $\geq$ 0.5 mm at the highest requirement. Sample section dimension: 2 mm $\times$ 20 mm.

Table 5. Bending tests results.

Sample	Deformed Side	Observation *
UNm3	root	√
UNm3	face	√
UNm4	root	√
UNm4	face	√
EDm3	root	√
EDm3	face	√
EDm4	root	√
EDm4	face	√
UXm3	root	√
UXm3	face	√
UXm4	root	√
UXm4	face	√
UNc4	root	×
UNc4	face	×
EDc3	root	×
EDc3	face	×

* Board description; √: No cracks in the weld, without welding defects and incompatibilities; ×: Cracks in the weld.

The bending tests confirmed that after welding with micro-jet cooling, it is possible to achieve joints with good plastic properties, as measured by the absence of cracks. In addition to the use of micro-jet cooling, the type of shielding gas plays an important role. Only the use of the Ar–CO_2 mixture allows obtaining the correct joints.

3.3. Metallographic Examination and Hardness

Microstructure observations were carried out on the Neophot 32 light microscope (Carl Zeiss Jena, Jena, Germany). The structure is dominated by martensite and ferrite. Figure 7 shows the structure observed in the UXm3 sample. Generally, martensite and coarse-grained ferrite were observed.

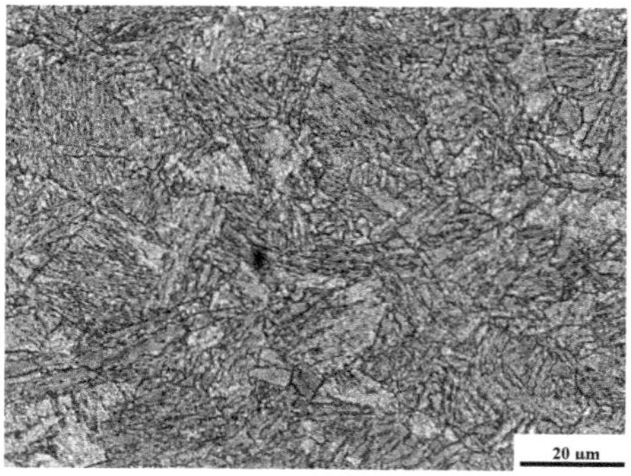

Figure 7. The microstructure of the joint (UXm3) with visible martensite and course ferrite (LM—Light Microscopy).

Only the application of micro-jet cooling after MAG welding made it possible to obtain more fragmented phases: martensite and ferrite (shown in Figure 8). This is due to micro-jet cooling during welding, which limits phase growth during the austenitic transformation. Other authors also draw attention to the possibility of grain fragmentation because of the use of micro-jet cooling [28,32].

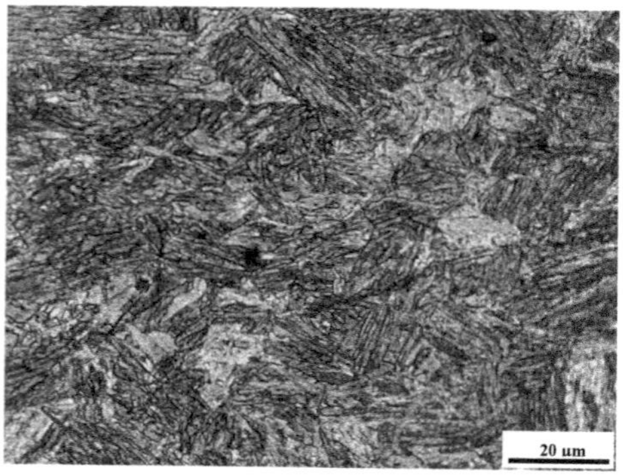

Figure 8. The microstructure of the joint (UXm3) after welding with micro-jet cooling with visible martensite and fine-grained ferrite (LM).

After microscopic observation, it was decided to perform hardness tests. It was made for the same samples. The results are presented in Table 6.

Table 6. The hardness of the joint (UXm3).

MAG Process	Base Material			HAZ			Weld		
With m-j cooling	334	336	337	354	349	351	327	329	332
Without m-j cooling	338	335	339	371	369	373	341	342	344

The table data shows that a more favorable and even distribution of hardness in all tested zones is for a joint made with micro-jet cooling.

3.4. The Steel and Its Micro-Jet Cooled Joint at Different Welding Wire in a Static Test

S960MC steel appears as the material having elastic-plastic with hardening and unstable regions (Figure 9). The proportion between ultimate tensile strength (UTS) and yield stress (YS) reached the value of 1.11, expressing the small distance between the mechanical parameters considered. From the practical point of view, this feature indicates an inspection of the element made of this kind of steel if the stress has equaled yield stress during operation. Another important sign in the S960MC behavior under tensile stress is evidenced in a region of the unstable section, which dominates in 70% comparing to the elastic and elastic-plastic with hardening zones. This informs the steel can carry loading at a long moment after a value of stress exceeds ultimate tensile strength. Taking the long unstable section, the steel can be easy to diagnose at high values of stress without any difficulties and risks for researchers, engineers and diagnostics teams. This sentence is also confirmed by a value of stress at the steel rupture, i.e., 600 MPa, indicating two times difference comparing to the value of ultimate tensile strength and distancing the fracture up to a value of a relative elongation of 9.5%. The behavior of the steel follows advantages concerning the application of this kind of material in a tri-axial stress state because the necking effect is significant, and this prevents brittle cracking.

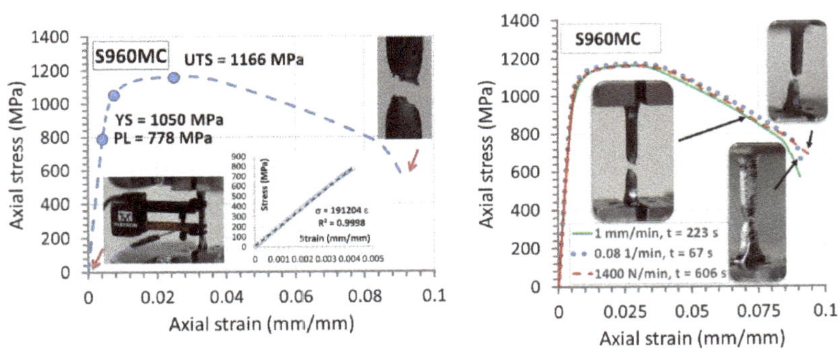

(a) (b)

Figure 9. The tensile characteristic of S960MC steel (**a**) and its variations (**b**) due to three different signals used in the tensile test: displacement at velocity of 1 mm/min, strain at 0.08 1/min and stress reaching 117 MPa/min, proportional limit (PL) = 778 MPa, yield stress (YS) = 1050 MPa, ultimate tensile strength (UTS) = 1166 MPa and elongation 8.75%.

The advantages of the S960MC steel are captured at tensile stress conducted at various types of control signals represented by the following velocity values of displacement, strain and stress signals such as 1 mm/min, 0.08 L/min and 117 MPa/min, respectively (Figure 9). As it can be noticed, the stress–strain curves follow the same course, which expresses insensitivity of the steel tested on the type of tensile signal and the time up to fracture related to the control kind used. This enables to sentence the mechanical parameters of the steel are stable for the wide range of loading signals and their parameters, extending the application of the material examined to operational conditions with different velocity diverse by even 10 times.

The steel behavior as a parent material was also dominant compared to welded joints manufactured by the micro-jet cooling technique (Figure 10). The most differences were visible for all stress parameters, placing the steel as a very attractive material compared to its weld. Nevertheless, the data collected for the micro-jet cooled weld enable to indicate advantages of the joint at the loading type considered. As a first, an attractive value of ultimate tensile strength of 650 MPa at all welding wires used can be evidenced. A second feature of the weld is related to a value of relative elongation because, at the fracture, the parameter reached 12.5% for the EDFK wire. This result and a range of the unstable regions enable to formulate sentences related to the application in a tri-axial stress state, i.e., the micro-jet cooled weld behavior after the necking effect is predictable and will be the same for all joints considered. No influence of the welding wire type on the joint's response was also visible in the elastic and elastic-plastic regions, following almost the same values of stress–strain.

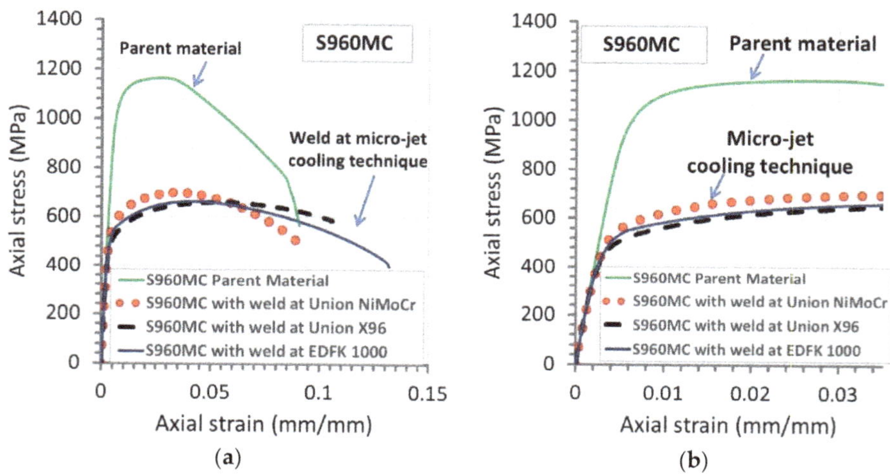

Figure 10. The tensile characteristic of S960MC steel and its weld manufactured using three welding wires at supporting the micro-jet cooling technique: (**a**) up to fracture, (**b**) in a view on the elastic and elastic-plastic relationships.

The results from tensile tests collected using mini-specimens have enabled us to follow the stress–strain relationship and mechanical parameters of the S960MC steel. This kind of specimen was also very sufficient at determining the influence of a control signal as displacement, stress, and strain. Literature [17,19,20] indicates mechanical properties of weld for S960MC without any details on a specimen shape and dimensions [17,20] or presents a bigger specimen, i.e., having 1000 mm length [19]. As it was obtained, the steel tested was not sensitive to the signal type used. The mini-specimen has supported examining the small weld, represented by the width of 0.5 mm under static loading, giving a tensile curve. Comparing data from the experiment on the base metal and welds has allowed us to distinguish differences in responses of the regions tested on static loading. Nevertheless, it is worth noting that the applications of small specimens require more attention during the tuning process because the measurement zone with a mounted extensometer is very close to the grips, which at inducing a testing machine can cause total damage to a sensor.

3.5. *Fatigue Response of the Steel and the Micro-Jet Cooling Weld under Increasing and Constant Stress Amplitude*

The behavior of the base material was expressed by hysteresis loops related to the energy dissipation as a dominant response of the material tested on the cycles used (Figure 11a). The first section of the relationship followed the elastic response of the material tested as an effect of hardening due to cyclic loading, while the second one was connected with permanent deformation. As it can be noticed, this feature occurred up to the cycle before the fracture. Moreover, the welds were loaded within the value of stress from the range between yield stress and ultimate tensile strength, and any hysteresis loops were not visible (Figure 11b–d). In this case, the stress–strain relationship has followed ratcheting up to fracture. This was very significant because it increased very rapidly before the fracture. As it can be noticed, this effect was the same for all cases of the welds considered, indicating the behavior of welds manufactured at the micro-jet cooling technique is insensitive to the type of welding wires used.

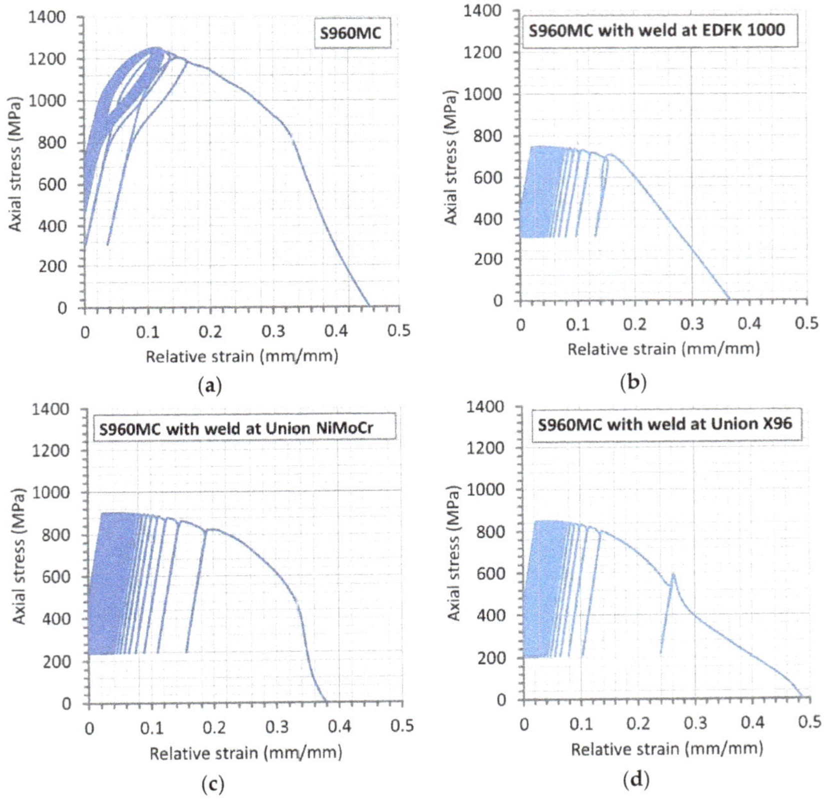

Figure 11. The stress–strain relationship at the final stages of the S960MC steel (**a**) and the joint welded by the micro-jet cooling and the following types of welding wires: (**b**) EDFK 1000, (**c**) Union NiMoCr and (**d**) Union X96.

The fatigue response of the base material under the increasing value of the stress amplitude was analyzed concerning the courses of total energy (Figures 12 and 13) because it demonstrates data representing loading signal (stress, force) and response signal (strain, elongation, displacement). The values of the physical magnitude were followed from the tests' beginning up to fracture, focusing on the last 50-th cycles directly related to the degradation of the parent material and the welds. This quantity was calculated basing on features of the Wave Matrix Instron software. It follows the total work performed on the specimen tested [33]. The values of force and displacement were used for calculations of the energy value. This parameter was followed since the start of the test. The software calculates total energy by continuously integrating force with displacement, using the trapezoidal rule [33].

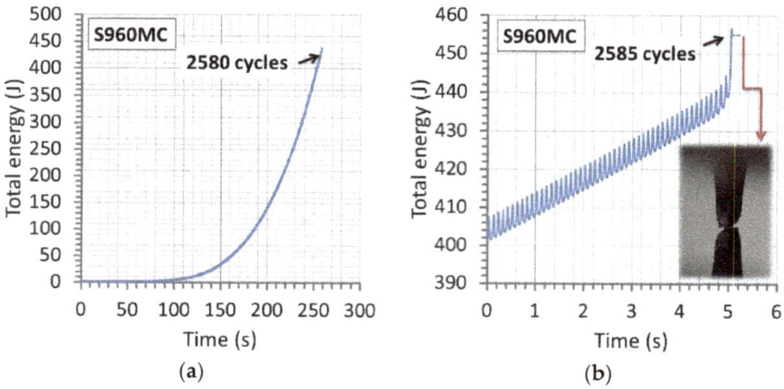

Figure 12. The total energy versus time from tests under increasing amplitude of axial stress before (**a**) and at the fracture (**b**) of S960MC steel, respectively.

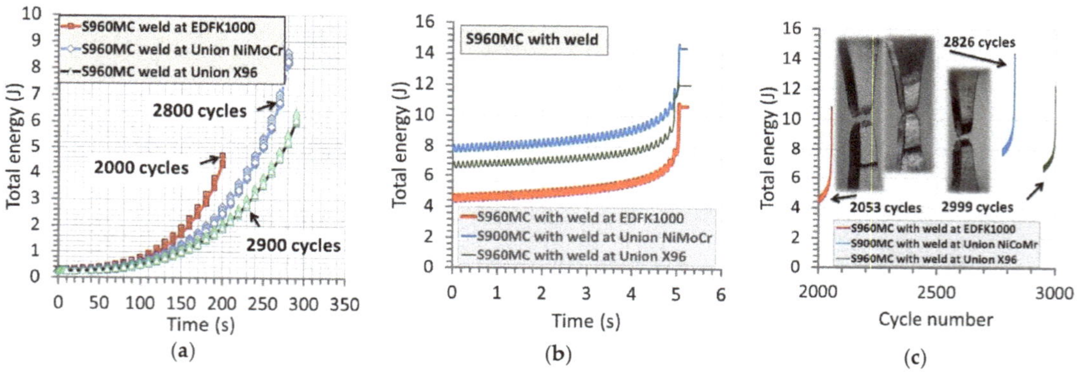

Figure 13. The total energy of the welds versus time (**a**), (**b**) before and at the fracture, respectively, and (**c**) the total energy versus cycle number at fracture for the three welding wires: EDFK 1000, Union NiMoCr and Union X96 for S960MC steel.

The results collected in the test considered have expressed the same tendency of the values of total energy independently on the region tested, represented by an exponential relationship (Figures 12a and 13a). Differences were visible in courses of the physical quantity, expressing not the same mechanical resistance of the welds on the type of loading applied, Figure 13a. They indicated the weld manufactured at micro-jet cooling technique and EDFK 1000 welding wires had obtained the smallest energy values while the biggest one was represented by results of the micro-jet weld at the NiMoCr wire. The same arrangement of the results was observed on the data directly related to the fracture stages of the regions tested (Figure 13b,c), values of the number of cycles, as well as the time to fracture. In the case of the base material, the value of cycles before the fracture reached 2580 at 258 s (Figure 12b), while for the welds, the following values were obtained: 2000 at 200 s (EDFK 1000 welding wire), 2800 at 280 s (Union NiMoCr welding wire) and 2900 at 290 s (Union X96 welding wire) (Figure 13a).

The results from the fracture stage in the form of the relationship of total energy versus the number of cycles have finally expressed the mechanical resistance of the welds examined (Figure 13c). The smallest one was noticed in the case of the micro-jet cooled weld with EDFK 1000 wire, and the biggest one was observed at the weld produced by the same technique but with the Union X96 wire, reaching the following values, respectively: 2053 and 2999 cycles (Table 7). This result strongly corresponds with the data presented in Figure 13a, which arranges the courses of the total energy in the same order. Some details on the behavior of the region tested can be captured based on the fracture zones. As it can be noticed in the photos presented in Figure 13c, the brittle cracking was the dominant feature of the weld at EDFK 1000; the brittle plastic was related to the weld at Union X96; the plastic represented the behavior of the NiMoCr joint. Nevertheless, any relationships were not observed at the comparison of the values of yield stress, ultimate tensile strength and the number of cycles connected with the fracturing of the tested joints (Figure 14a,b). Some information can be collected based on values of total energy and stress at fracture (Figure 14c,d). This is presented by the proportion of the physical quantity, which arrangements the welds with the EDFK 1000 and Union X96 welding wires as the joints with the same and the lowest mechanical resistance of the loading applied, while the weld of the Union NiMoCr wire become the strongest one from the joint examined.

Table 7. The results from tests under increasing stress amplitude.

No.	Specimen	Axial Stress at Fracture (MPa)		Time to Fracture (s)	Number of Cycle to Fracture
		Maximum	Minimum		
1.	Parent material	1253	306	269	2585
2.	Weld at EDFK 1000	745	314	216	2053
3.	Weld at Union NiMoCr	870	229	293	2826
4.	Weld at Union X96	846	214	310	2999

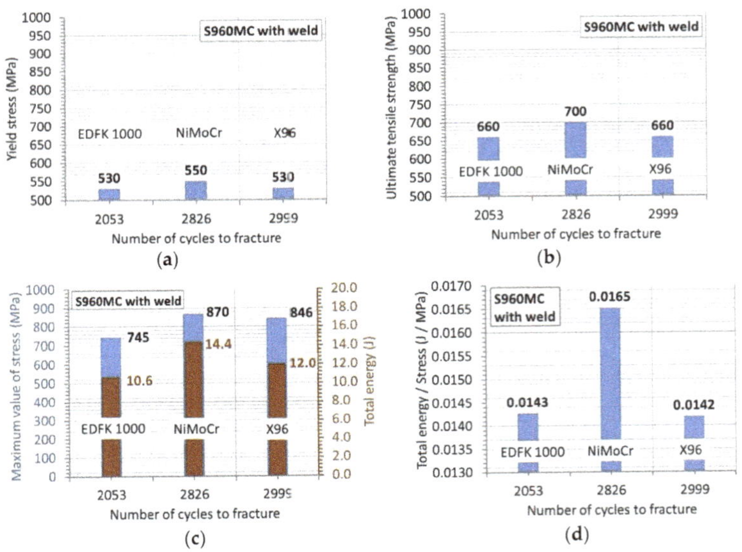

Figure 14. The yield stress (**a**), ultimate tensile strength (**b**), maximum value of stress (**c**) and total energy/the maximum value of stress at a number of cycles (**d**) up to the fracture for welds manufactured by the micro-jet cooling technique and the following welding wires: NiMoCr (Union), EDFK 1000 and X96 (Union).

The results from the tests under increasing stress amplitude were used for selecting the weld for determining the Wöhler curve. For this case, the joint with the Union X96 welding wire as the weakest region welded was tested under cyclic stress at the range of 650 MPa–100 MPa. Data collected from the fatigue experiment are presented in the next sub-chapter.

Fatigue tests at increasing stress amplitude defined by the yield stress and ultimate tensile strength as well as a time have enabled us to distinguish differences in behavior of the welds having very similar tensile characteristics. For this case, the values of the total energy were also significant because they divided into two sections from the beginning of the test up to cycles before and directly at the fracture, giving more information on the regions' decohesion. This kind of test has allowed us to reach differences in fracture regions, giving more details to assess the mechanical resistance of the weld tested. Values of total energy have supported the analysis on the weld quality expressing the joints response on the cyclic loading. Their comparison with the maximum value of stress of the control signal has enabled to indicate the joints for both welding wires (EDFK 1000 and Union X96) expressed the same proportion, even though the stress levels applied as well as the number of cycles to fracture were significantly different by of 99 MPa and 946 cycles, respectively. Concerning the assessment of the weld quality, the following sentence can be formulated: the joints were manufactured at different welding wires, but nevertheless, they have a very similar reaction to the fatigue process. Taking these results into account, the strongest and weakest joints were possible to be indicated, arranging as of NiMoCr weld, EDFK 1000 and X96 welds.

3.6. *Results from Tests of the Micro-Cooled Weld under Constant Stress Amplitude*

The results from the experiments under stress cycles are represented by variations of total energy (Figure 15a), the proportion of total energy to stress (Figure 15b), fracture regions as both parts of specimen directly after cracking (Figure 16) and as the zones on a perpendicular cross-section of the specimens (Figure 17) as well as the Wöhler curve (Figure 18).

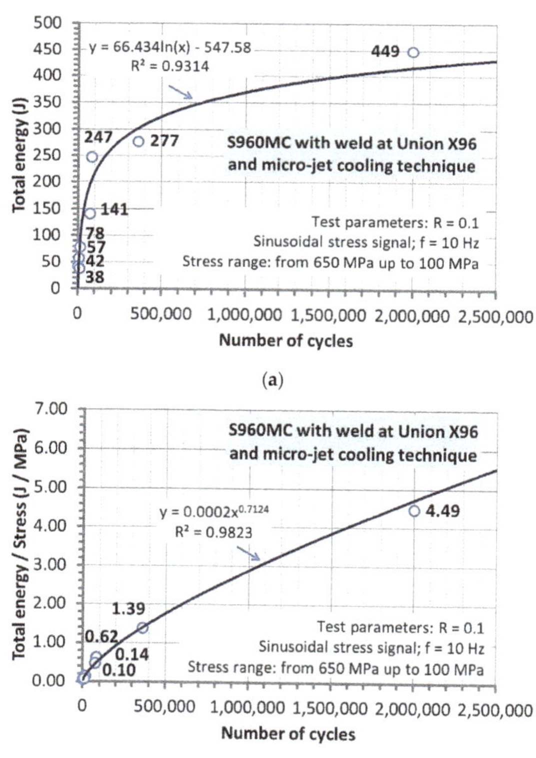

Figure 15. Total energy (**a**) and its value per stress (**b**) versus the number of cycles at the Wöhler curve of S960MC steel welded by the MAG method with the micro-jet cooling technique.

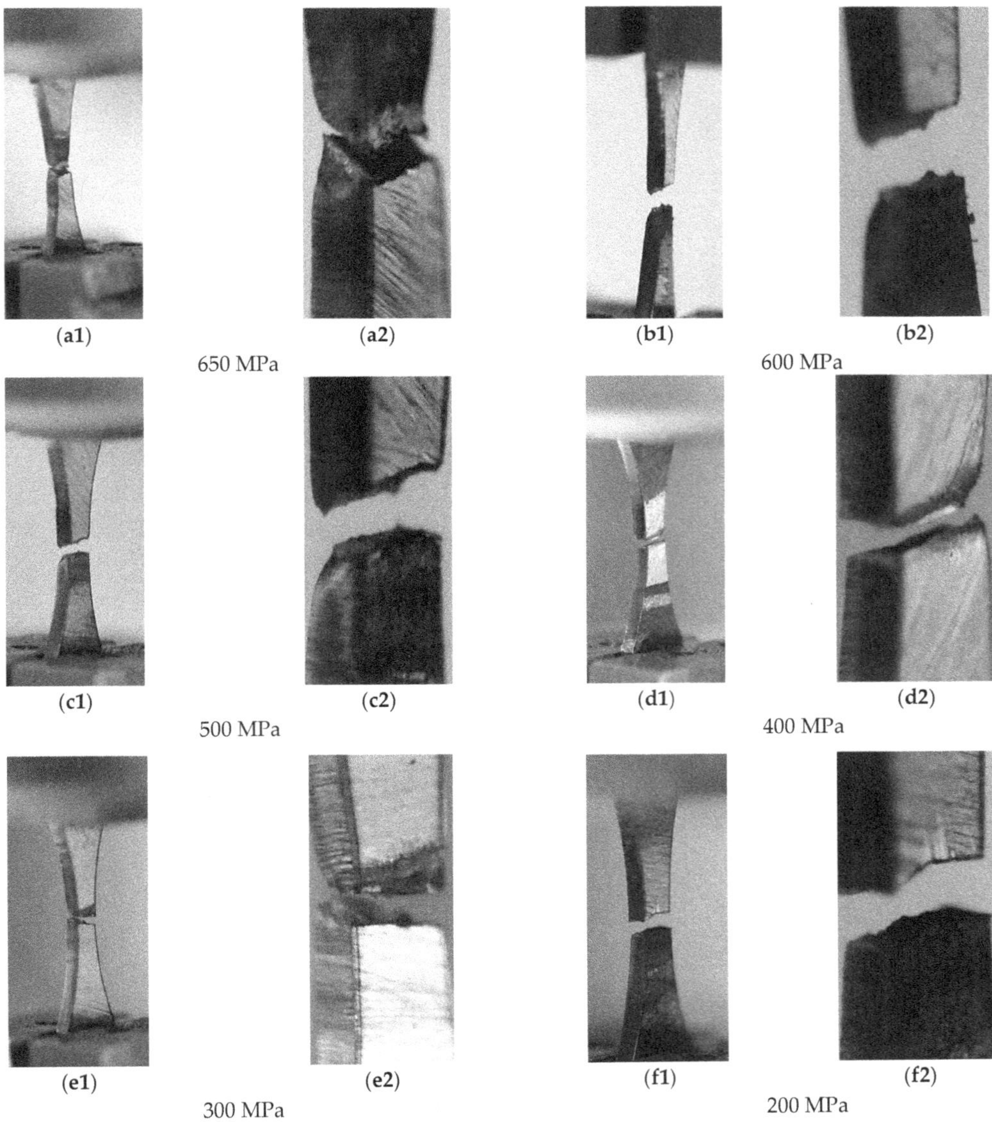

Figure 16. S960MC mini-specimens with the micro-jet cooling weld manufactured with the Union X96 welding wire after the fatigue test for the maximum values of the axial stress amplitude of 650 MPa–100 MPa, (**a1**–**f2**) fracture regions.

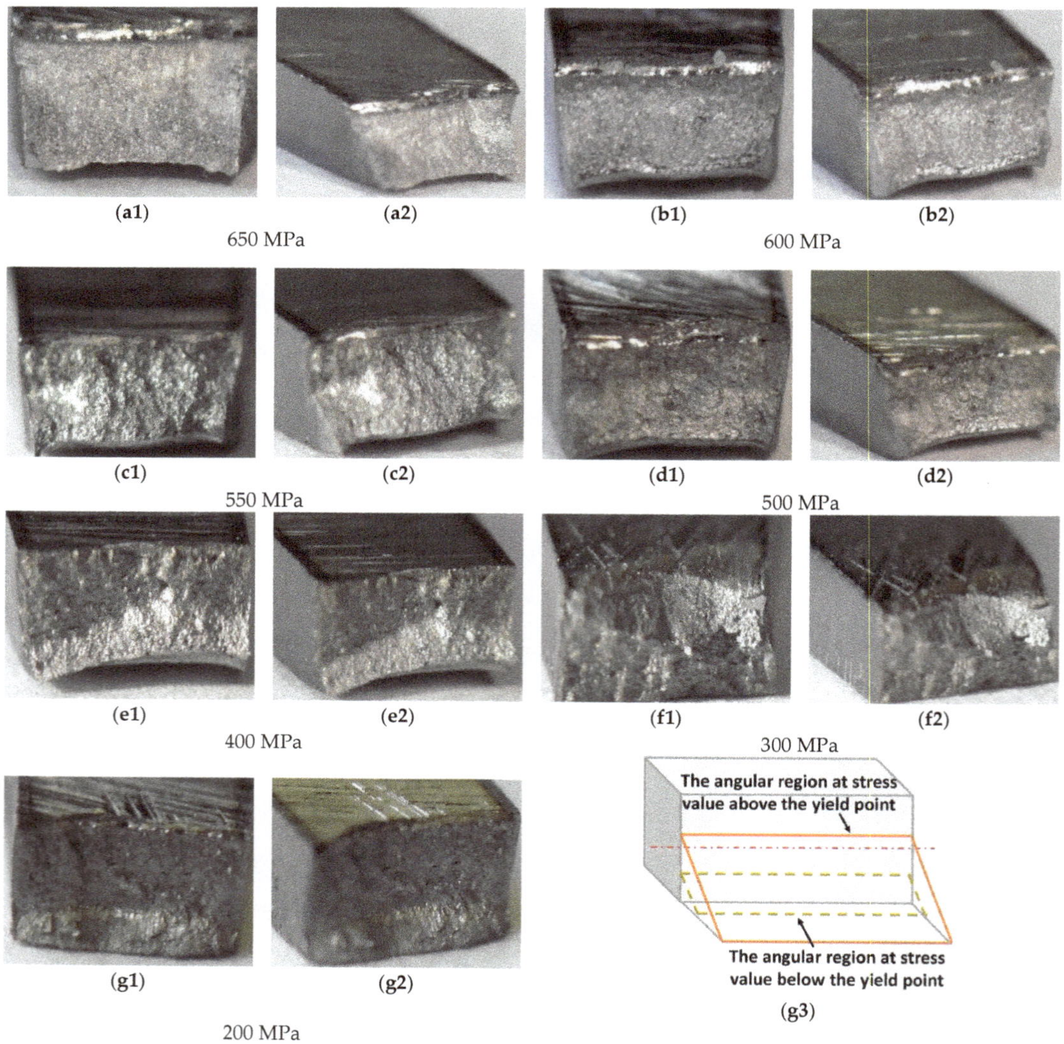

Figure 17. Fracture regions (**a1–g3**) of S960MC weld manufactured at micro-jet cooling technique after fatigue tests for the maximum values of axial stress amplitude of 650 MPa–100 MPa and the fracture scheme (**g3**), number 1 represents a perpendicular view while number 2 follows an axonometric one.

They show a course of total energy up to fatigue limit at values of the number of cycles (Figure 15a). As it can be noticed at the earlier stage of the course up to 5×10^5, the maximum value of total energy is represented by 325 J. For the further section reaching 1×10^6 cycles, the value of 375 J limits, while at the fatigue limit (determined by 2×10^6), an increase of the total energy was expressed by 95 J giving 420 J. If values of total energy and stress are divided then the quantity of Jouls per MegaPascal can be calculated (Figure 15b). This proportion enables to follow the relationship between the physical quantities. In this case, a non-linear tendency was evidenced, indicating the stress values do not have the same influence on the weld fracturing. Moreover, the values of energy can be easily connected with the values of stress (Table 8), covering more practical information for inspection groups. They can be used in comparison with the values of stress as well as total energy from the experiment and captured at operational conditions for predicting the lifetime of components made of the steel grade weld with similar mechanical parameters. The total energy to stress can be used at the current inspection without details on the earlier stage.

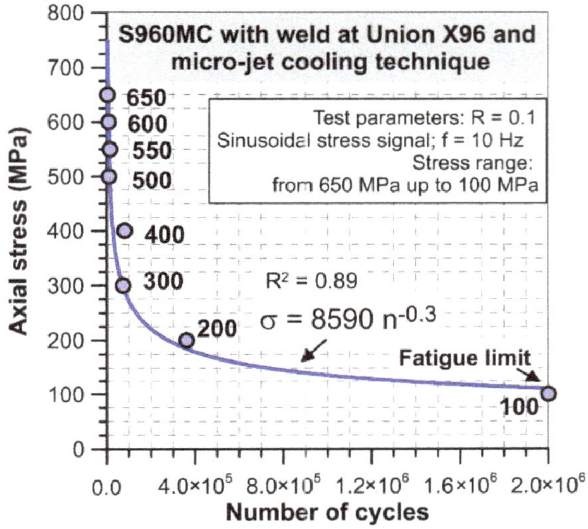

Figure 18. The Wöhler curve of S960MC steel welded by MAG method with the micro-jet cooling technique.

Table 8. Values of stress, total energy and their proportion.

No.	Maximum Value of Stress (MPa)	Total Energy (J)	Total Energy/Maximum Value of Stress (J/MPa)
1.	650	42	0.06
2.	600	57	0.10
3.	550	78	0.14
4.	500	38	0.08
5.	400	247	0.62
6.	300	141	0.47
7.	200	277	1.39
8.	100	449	4.49

The role of the stress level in the weld degradation was determined on fracture zones represented by the general view on the specimens after the test (Figure 16) and exhibited by the view focused on the whole region of degradation (Figure 17). The photos were selected for presenting changes in the fracture regions and for better analysis of the zone degradation at various values of stress. Using them, we can follow the material degradation in a 3D coordinate system and compare it with the orientation of the measured zone. Some differences in the weld degradation due to cyclic loading are visible on the views on both parts of specimens directly after tests (Figure 16). They are expressed by the angular orientation of fracture zones at the stress level of 400–600 MPa, which indicates shear and axial components of stress for the weld decohesion. Comparing these data, a reduction of the angular orientation of the damage region is noticed, indicating the role of shear stress in the fracturing lowers with decreasing the stress value. This is better visible from the cross-section view (Figure 17).

The effect of the booth types of stress in the tested region degradation was also visible on the photos taken with the macro-photography technique (Figure 17). In this case, the same presentation concept for the following decohesion characteristics as was taken for the immediately fracture specimen was used. Using this approach, zone details in relation to fracture mechanics may be easier to collect. The multi-planar cracking was noticed on the perpendicular cross-sections as a major feature of all regions inspected independently of the stress level applied. Nevertheless, the proportion between the angular and horizontal sections of the damage zones was dependent on the values of stress, i.e., the horizontal part becomes more significant with decreasing the stress value (Figure 17g3).

The analysis of the fracture region represented by both zones, i.e., horizontal and angular, enables the selection of the area to be the first for damage occurrence. In this case, the horizontal section of the region considered plays the role of the initial damage. The second area that is damaged further and directly related to the final stage of specimen lifetime is represented by the angular region. This kind of data follows the micro-cooled welded region expresses a mixed (brittle-plastic) cracking at cyclic tensile stress arraigning the axial stress at the beginning of the fracture and shear stress at the final stage of fatigue. In comparison to the good plastic properties of the micro-cooled joint, these results reflect as follows: the weld degradation at plastic cracking is dominant under a stress value exceeding yield stress (530 MPa), while at a smaller one, the brittle behavior appeared.

An analysis of the Wöhler curve has enabled us to select stages with the number of cycles for applications or inspections of the weld tested at a wide range of stress values (Figure 18). For this case, the diagram can be divided at the following regions determined by the stress value and loading cycles: from 650 MPa to 400 MPa at 3×10^4–9×10^4 cycles; from 400 MPa to 100 MPa at 9×10^4–2×10^6 cycles. This enables to formulate as follows: the weld expresses hardening due to cyclic tensile stress close to the ultimate tensile strength reaching the lifetime limited by the 3×10^4. Moreover, the value of fatigue limit can be indicated as 100 MPa, which unfortunately does not express an attractive level for application under cyclic loading at very restricted operational regimes. This kind of result was very similar to data collected by the authors of [19]. Nevertheless, their data did not follow the function proposed as it was obtained in the paper. A comparison of these test results shows that they are complementary due to the different characteristics of the zones containing the welds and can be used for extending the knowledge in the S960MC weld behavior.

Concluding results from tests on the Wöhler curve worth noticing are that this kind of data can be analyzed not only at variations stress versus a number of cycles but also at changes of total energy. This approach collects the relationship between control and response signal, i.e., stress and strain, respectively, supporting results represented by one of the mentioned physical quantities. This kind of data can be used for modifying models for predicting lifetimes as well as elaborating on new ones. Fracture analysis is the stage of the fatigue test for analysis of fracture mechanics due to the type of loading applied. This follows the orientation of a fracture plane compared to a loading direction as well as details related to brittle or brittle-plastic cracking. The first mentioned feature of the region subjected to observation should be captured directly after a specimen fracturing without unmounting the tested object from the grips of the testing machine, while the second one's details are obtained from observations of a fracture plane from various directions. As it was evidenced in the photos of the fracture regions of the weld, the degradation mechanism was related to stress state components, i.e., the shear stress becomes more significant in the weld degradation with an increasing stress level becoming from a value close to the yielding point.

4. Summary

The article deals with MAG welding of thin-walled structures made of S960MC steel. It is a material used in automotive and offshore engineering [34]. Various welding parameters were tested, including micro-jet cooling parameters. To assess the quality of the joints, a series of non-destructive and destructive tests were performed. Welds were made with three different electrode wires and two different shielding gas mixtures. Main welding parameters and parameters of micro-jet cooling were varied. After the results of non-destructive tests, the main information was obtained on the proper welding parameters, where no cracks appeared. The results of non-destructive tests showed that the use of micro-jet cooling is very beneficial, as there were no welding cracks. The condition for a properly made joint was the correct selection of micro-jet cooling parameters. Then bending tests were performed. The results of these tests showed that in addition to cooling the micro-jet, the selection of the shielding gas is important. The use of the Ar gas mixture is beneficial, while the use of CO_2 does not guarantee the possibility of obtaining joints with good plastic properties. Further research focused on the evaluation of joints made with three different welding wires. M21 cover and micro-jet cooling were always used. The U-notched specimen supports experiments for assessment of the quality of the welding process. Moreover, the specimen size was small, and all mechanical parameters of the static and fatigue behavior of the material and joints tested were determined. The results from the static and fatigue tests have extended the knowledge on the weld testing, the role of the three types of welding wires for the joint's quality manufactured by the micro-jet cooling technique, as well as applications and inspections. This was represented by variations of the stress-strain relationship, total energy directly at fracture and versus the number of cycles up to fatigue limit.

5. Conclusions

The results from the non-destructive and destructive experiments have enabled the following conclusions:

- The S960MC steel is very difficult for welding processes even when high technology is used, such as the micro-jet cooling method, and when different welding wires are applied because the weld's mechanical properties can be lowered by 50%;
- The following welding parameters are recommended for MAG-welding sheet metal with a thickness of 2 mm: U = 19 V, I = 101 A, V = 300 mm/min and a mixture Ar-18% CO_2 as shielding gas;
- Elongation (10.50%) of the weld produced at controlling micro-jet cooling with the three welding wires exceeded the value of this parameter (8.75%) captured at the test of the base metal;
- At the same type of mechanical parameters, such as yield stress and ultimate tensile strength, their values were almost the same for all cases of the welds tested, i.e., 537 MPa and 673 MPa, respectively. An increasing stress amplitude enabled following the differences in the behavior of the joints even they represent very similar response in static tests;
- The weld manufactured with the Union NiMoCr welding wire was the joint with the highest resistance on static and fatigue loadings. The response of the weld under cyclic loading should be analyzed not only in the general form represented by the Wöhler curve but also variations of total energy;
- As it was determined in the test, the value of energy at the fatigue limit reached 449 J. In the case of the micro-jet cooled weld with the Union X96 welding wire, the difference between the value of stress from limited and unlimited sections was six times, following the value of fatigue limit of 100 MPa at the final stage of the last-mentioned section;
- The use of micro-jet cooling when welding S960MC steel allows for obtaining joints with better mechanical properties (UTS = 700 MPa, YS = 550 MPa).

Author Contributions: Data curation, A.J.; Formal analysis, B.S.-L., T.W. and B.Ł.; Funding acquisition, T.S., T.W. and B.Ł.; Investigation, T.S., B.S.-L. and A.J.; Methodology, T.S., B.S.-L., T.W. and A.J.; Supervision, B.S.-L., T.W. and B.Ł.. Validation, T.S.; Visualization, T.S. and B.S.-L.; Writing—original draft, T.S., B.S.-L. and T.W.; Writing—review & editing, B.S.-L. and T.W. All authors have read and agreed to the published version of the manuscript.

Funding: This research was funded by Silesian University of Technology, grant number BK-277/RT1 2021.

Institutional Review Board Statement: Not applicable.

Informed Consent Statement: Not applicable.

Data Availability Statement: Data sharing is not applicable to this article.

Acknowledgments: The paper is part of the COST project, CA 18223 and the research project No. 06/21/ZBH/012 in Motor Transport Institute, entitled "Examination of the mechanical resistance of materials and structural components in the conditions of laboratory strength simulation for applications in special vehicles".

Conflicts of Interest: The authors declare no conflict of interest.

References

1. Keeler, S.; Kimchi, M.; Mconey, P.J. *Advanced High-Strength Steels Application Guidelines, Version 6.0*; World Auto Steel: Brussels, Belgium, 2017; 314p.
2. Horvath, C.D. The Future Revolution in Automotive High Strength Steel Usage. In Proceedings of the Great Design in Steel Seminar, Livonia, MI, USA, 18 February 2004.
3. SSAB. *Strenx 960MC*; Data sheet 2015 Strenx 960MC 2017-04-20; SSAB: Borlänge, Sweden, 2017.
4. SSAB. *Docol Press Hardening Steel, Technical Brochure*; 345-en-Docol® Press hardening steel-v1-2020. Confetti; SSAB: Borlänge, Sweden, 2020.
5. Olsson, K.; Gladh, M.; Hedin, J.E.; Larsson, J. Microalloyed high-strength. *Adv. Mater. Proces.* **2006**, *8*, 44–46.
6. ArcelorMittal. *Amstrong Ultra. The Smartest Choice*; PR-BR-AMU-EN-10/2019; ArcelorMittal Europe Communications: Luxembourg, 2019.
7. Ruukki. *Optim QC Structural Steels*; Ruukki: Helsinki, Finland, 2014; 6p.
8. TATA Steel. *High Strength Low Alloy*; 20170314-FVC PVM; TATA Steel: Prague, Czech Republic, 2017.
9. Szymczak, T.; Brodecki, A.; Kowalewski, Z.L.; Makowska, K. Tow truck frame made of high strength steel under cyclic loading. *Mater. Today Proc.* **2019**, *12*, 207–212. [CrossRef]
10. SSAB. *Strenx 960 for Stronger and Safer Rear Underrun Protection Device*; 445-en-Strenx®960 for Stronger and Lighter RUPD V2-2019-A plus M-Österberg; SSAB: Österberg, Sweden, 2019; 2p.
11. Polish Committee of Standardization. *PN-ISO 668:2018-05 Series 1 freight containers—Classification, dimensions and ratings*; Polish Committee of Standardization: Warsaw, Poland, 2018.
12. Wiśnicki, B. *Vademecum of Containerization. Forming the Container Load Unit*; LINK: Szczecin, Poland, 2006.
13. Nordstroem, R.; Andersson, P.; Soekjer-Petersen, S. *Verification of Level of Basic Parameters Important for the Dimensioning Of Cargo Securing Arrangements*; Transport Research Institute: Stockholm, Sweden, 2004.
14. Szymczak, T.; Makowska, K.; Kowalewski, Z.L. Influence of the welding process on the mechanical characteristics and fracture of the S700MC high strength steel under various types of loading. *Materials* **2020**, *13*, 5249. [CrossRef] [PubMed]
15. Węgrzyn, T.; Szymczak, T.; Szczucka-Lasota, B.; Łazarz, B. MAG Welding process with micro-jet cooling as the effective method for manufacturing joints for S700MC steel. *Metals* **2021**, *11*, 276. [CrossRef]
16. Szczucka-Lasota, B.; Węgrzyn, T.; Szymczak, T.; Jurek, A. High martensitic steel after welding with micro-jet cooling in microstructural and mechanical investigations. *Materials* **2021**, *14*, 936. [CrossRef]
17. Jambor, M.; Novy, F.; Mician, M.; Trsko, L.; Bokuvka, O.; Pastorek, F.; Harmaniak, D. Gas metal arc welding of thermo-mechanically controlled processed S960MC steel thin sheets with different welding parameters. *Communication* **2018**, *4*, 29–34. [CrossRef]
18. Siltanen, J.; Tihinen, S.; Kömi, J. Laser and laser gas-metal-arc hybrid welding of 960 MPa direct-quenched structural steel in a butt joint configuration. *J. Laser Appl.* **2015**, *27*, S29007. [CrossRef]
19. Bhatti, A.; Barsoum, Z.; Mee, V.V.D.; Kromm, A.; Kannengiesser, T. Fatigue strength improvement of welded structures using new low transformation temperature filler materials. *Procedia Eng.* **2013**, *66*, 192–201. [CrossRef]
20. Mičian, M.; Maronek, M.; Konar, R.; Harmaniak, D.; Jambor, M.; Trško, L.; Wincek, J. Changes of microstructure and mechanical properties of the HAZ of the S960MC steel sheet weld joint. *Zavar. Zavarene Konstr.* **2020**, *3*, 113–123. [CrossRef]
21. Jambor, M.; Ulewicz, R.; Nový, F.; Bokůvka, O.; Trško, L.; Mičian, M.; Harmaniak, D. Evolution of microstructure in the heat affected zone of S960MC GMAW weld. In Proceedings of the 10th Conference on Terotechnology, Kielce, Poland, 18–19 October 2017; pp. 78–83. [CrossRef]
22. Schneider, C.; Ernst, W.; Schnitzer, R.; Staufer, H.; Vallant, R.; Enzinger, N. Influence of different welding processes on the mechanical properties of structural steel S960MC. *Thermec Graz* **2016**. Available online: https://pure.tugraz.at/ws/portalfiles/portal/3629530/poster.pdf (accessed on 19 May 2021).
23. Steimbreger, C.; Gubeljak, N.; Enzinger, N.; Ernst, W.; Chapetti, M. Influence of static strength on the fatigue resistance of welds. *MATEC Web Conf.* **2018**, *165*, 13010. [CrossRef]
24. Górka, J.; Kotarska, A. MAG welding of 960QL quenched and tempered steel. *IOP Conf. Ser. Mater. Sci. Eng.* **2019**, *591*, 1–8. [CrossRef]

25. Tomków, J.; Fydrych, D.; Wilk, K. Effect of electrode waterproof coating on quality of underwater wet welded joints. *Materials* **2020**, *13*, 2947. [CrossRef] [PubMed]
26. Fydrych, D.; Łabanowski, J.; Tomków, J.; Rogalski, G. Cold cracking of underwater wet welded S355G10+N high strength steel. *Adv. Mater. Sci.* **2015**, *15*, 48–56. [CrossRef]
27. Wang, X.; Wu, J.-F.; Qian, X. Crack driving force and ultimate strength of thick-walled CHS X-joints with near-toe cracks, paper number: ISOPE-I-11-203. In Proceedings of the Twenty-first International Offshore and Polar Engineering Conference, Maui, HI, USA, 19–24 June 2011.
28. Węgrzyn, T.; Szczucka-Lasota, B.; Uściłowska, A.; Stanik, Z.; Piwnik, J. Validation of parameters selection of welding with micro-jet cooling by using method of fundamental solutions. *Eng. Anal. Bound. Elements* **2019**, *98*, 17–26. [CrossRef]
29. Węgrzyn, T. The Classification of Metal Weld Deposits in Terms of the Amount of Oxygen. In *Proceedings of the Conference of International Society of Offshore and Polar Engineers ISOPE'99, Brest, France, 30 May–4 June 1999*; International Society of Offshore and Polar Engineers: Cupertino, CA, USA, 1999; Volume IV, pp. 212–216. ISBN 1-880653-43-5.
30. Górka, J. Assessment of the weldability of T-welded joints in 10 mm thick TMCP steel using laser beam. *Materials* **2018**, *11*, 1192. [CrossRef] [PubMed]
31. ISO. *Welding—Fusion-Welded Joints in Steel, Nickel, Titanium and their Alloys (Beam Welding Excluded)—Quality Levels for Imperfections*; PN-EN ISO 5817:2014; ISO: Warsaw, Poland, 2014.
32. Hadryś, D. Impact load of welds after micro-jet cooling/dynamiczne obciążenie spoin chłodzonych mikrojetowo. *Arch. Met. Mater.* **2015**, *60*, 2525–2528. [CrossRef]
33. WaveMatrixTM. *Software—V1.4 Onwards*; Software Reference Manual M22-16102-EN Revision B; Instron: High Wycombe, UK, 2009; 368p.
34. Aderinola, O.O.; Kah, P.; Martikainen, J. Efficient welding technologies applicable to HSS arctic offshore structures. In Proceedings of the Twenty-third International Offshore and Polar Engineering Conference, Paper Number: ISOPE-I-13-639, Anchorage, AL, USA, 30 June–5 July 2013.

materials MDPI

Article

High Martensitic Steel after Welding with Micro-Jet Cooling in Microstructural and Mechanical Investigations

Bożena Szczucka-Lasota [1], Tomasz Węgrzyn [1,*], Tadeusz Szymczak [2] and Adam Jurek [3]

[1] Faculty of Transport and Aviation Engineering, Silesian University of Technology, Krasińskiego 8, 40-119 Katowice, Poland; bozena.szczucka-lasota@polsl.pl
[2] Department of Vehicle Type-Approval & Testing, Motor Transport Institute, ITS, Jagiellońska 80, 03-301 Warsaw, Poland; tadeusz.szymczak@its.waw.pl
[3] Novar Sp. z o. o., Towarowa 2, 44-100 Gliwice, Poland; adam.jurek@novar.pl
* Correspondence: tomasz.wegrzyn@polsl.pl; Tel.: +48-504-816-362

Abstract: Modern means of transport will play a significant role in the smart city. In the automotive industry, high-strength steels such as Docol are employed more often. This kind of material is relatively not very well weldable. The main reason is related to the Heat Affect Zone, the region in which cracks occur. Another disadvantage is connected with differences in values of ultimate strength of parent and weld material. The differences can be diminished using the correct welding process, which employs nickel and molybdenum electrode wires at much lower sulfur content. The weld metal deposit contains mainly martensite and bainite with coarse ferrite, while the parent material contains mainly martensite and rather fine ferrite. New technology, micro-jet cooling after the joining process enables to obtain the microstructure of weld metal deposit at acceptable parameters. Welding with micro-jet cooling could be treated as a very promising welding Docol steels process with high industrial application. Results of non-destructive inspections on macro samples corresponded with further destructive test results (tensile strength, hardness, fatigue, metallographic structure analyses). This article aims to verify fatigue behavior of Docol 1200 M steel after welding supported by the cooling using the micro-jet technique. For the first time, micro-jet cooling was used to weld this kind of steel to check the mechanical properties of the joint, especially to determine the fatigue limit. This study is formulated as follows: investigating fatigue resistance of the Docol 1200 M weld manufactured at the cooling process with micro-jets. The joints were produced in the MAG (Metal Active Gas) technology modified by micro-jet cooling. The results collected in the fatigue test were processed in the form of the Wöhler's S–N diagram following the fatigue limit of the weld examined. All data have indicated the possibility of obtaining a new method of welded joints with high fatigue limit minimum of 480 MPa. It could be important to achieve a tensile strength of 700 MPa while maintaining the best relative elongation at the level of the base material.

Keywords: smart city; transport; vehicles; mechanical engineering; Docol steel; micro-jet welding; mechanical tests; microstructure; mini-specimen; fracture; fatigue

Citation: Szczucka-Lasota, B.; Węgrzyn, T.; Szymczak, T.; Jurek, A. High Martensitic Steel after Welding with Micro-Jet Cooling in Microstructural and Mechanical Investigations. *Materials* **2021**, *14*, 936. https://doi.org/10.3390/ma14040936

Academic Editor: Andrey Belyakov

Received: 20 January 2021
Accepted: 13 February 2021
Published: 16 February 2021

1. Introduction

The progress in the automotive industry is created by the application of modern materials and types of joining as well as new components and constructions [1–4]. If all science-engineering fields are reached and collated, then a final product is manufactured in the right quality without a waste of time for its prototyping, obtaining beneficial features such as a 90 kg reduction of the nominal mass of the vehicle and 5 kg lowering of Body in White (BIW) in comparison to the previous generation [1]. It is possible by employing beneficial materials such as Advanced High Strength Steel (AHSS) and Ultra High Strength Steel (UHSS), which possess very attractive mechanical parameters [3,5–7]. Therefore, this kind of steel is used not only in typical cars with an internal combustion engine as well as in Hybrid Electric Vehicles (HEV) increasing stiffness by 12% and 10% for bending and torsion loading, respectively [8]. Moreover, it is taken for producing an electric car battery case, being a component of BIW for battery replacement and safety regimes. In this case, the martensitic steel series of Docol 1700 M plays an important role because of high mechanical properties as a result of 3D manufacturing technology. This has been employed since the year 2020 in the Ford Explorer, indicating its actual application [1]. Moreover, this type of steel is employed for producing sill and roof reinforcements as well as door/side intrusion beams, bumpers, and seat structures [9]. With respect to different components and joining processes used for manufacturing of the elements made of this kind of material, and automotive industry regimes for quality and mechanical resistance of this type of components, the steel requires more attention, taking chemical composition, mechanical properties, behavior under fatigue, as well as features due to welding. Joining Docol steels with classic welding methods allows to obtain correct welded joints, but with a much lower fatigue strength than that of the parent material. The aim of the article is to discuss a newly developed welding with micro-jet cooling stage [10,11], ensuring the production of repeatable welded joints with high fatigue properties. The newly developed process made it possible to change the microstructure of the joint, which resulted in an increase in its strength. The aim of this article is to extend the knowledge on this kind of steel and the welding process connecting with micro-jet cooling (Figure 1), and the following sections cover all mentioned material and process aspects.

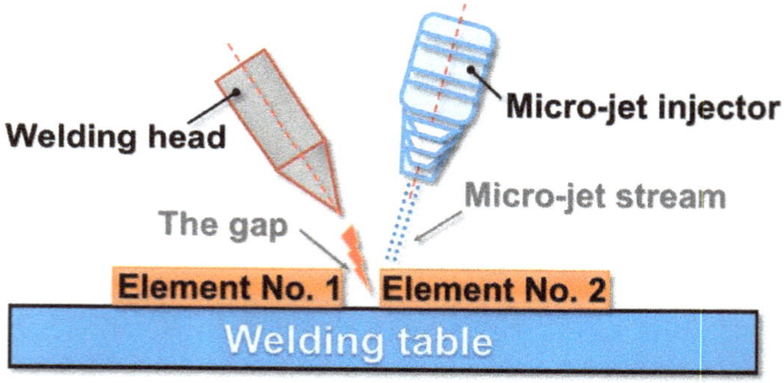

Figure 1. A scheme of Metal Active Gas (MAG) welding with micro-jet technique.

Docol Advanced High-Strength Steel (AHSS) are obtained in the technological process such as: hot and cold rolled as well as hot-dip and electro galvanized. That material range includes thin sheet steel with a thickness range from 0.4 mm to 16 mm. Docol steels have become particularly attractive in the automotive industry for three important reasons [1,4,12]:

- high ultimate tensile strength, up to 1700 MPa,
- high yield stress, up to 1350 MPa,
- elongation, up to 6%.

The application of Docol steels in the automotive branch of industry results from reducing the thickness of the car body at also acceptable or more attractive mechanical parameters in comparison to typical structural materials. For the manufacturing of vehicles, attempts are being made to use high-strength martensitic steels from the Docol group [13–16]. They make it possible to reduce the weight of the vehicle and reduce energy consumption for its production [1,9].

For instance, Reference [17] presents the results of Docol 1200 M martensitic steel welding. In this study, the very thick sheets were joined in a lap joint configuration. The influence of various welding parameters (especially linear heat input) during welding on structural changes was examined by the occurrence mainly of dominant phases: martensite and fine ferrite. Reference [18] presents the weldability of three different modern high-strength steel plates, with a thickness of 6 and 8 mm. Two-passes Metal Active Gas

(MAG) welding process was used with various heat inputs. In this case as well, the test results revealed that increasing the heat input resulted in lower tensile strength of the joint. However, the authors of Reference [19] are of a different opinion. The authors prepared a mixed joint of high-strength steels: S460N and S460ML. The joints were manufactured using various values of heat input for each welding bead. The results of prepared welds showed that using higher heat input has a significant influence on the mechanical properties of the mixed steel joints, as presented in [20].

In Reference [18] the authors have shown that high-strength low alloy (HSLA) steels are quite crucial in weight reduction in the manufacturing of vehicles and constructions with high steel thickness. Quenched and tempered S690QL HSLA steel with 700 MPa yield stress was multi-pass welded with seven passes using MAG (Metal Active Gas) and FCAW (Flux Cored Arc Welding) methods. The effects of the process parameters for both methods on weld metals and heat affected zones (HAZ) were investigated. The authors suggested that welding with higher heat input influenced the mechanical properties.

In summary, it can be concluded that the modern high-strength steel plates have an excellent combination of strength, fracture toughness, and resistance on impact based on micro-alloying and complex microstructure [21]. Retaining this combination of properties in the weld zone is a major challenge for applications in high-demanding structural construction. The problem that has not been solved so far is the development of a repeatable technology of joining these steels, ensuring obtaining joints with sufficiently ultimate tensile strength at the level of 600 MPa and higher without welding defects. With these steels, the plastic properties of the joint were always worse than that of the base material, as evidenced by different values of the relative elongation.

This research aimed to investigate the mechanical properties of the high-strength steel Docol 1200 M after MAG welding with micro-jet cooling. The utilitarian goal of the study is to develop a welding method for these steels. The new technology will be based on classic MIG and TIG (Metal Inert Gas and Tungsten Inert Gas) processes, modernized through the use of cooling attachments, affecting the structure of the connections obtained. It was assumed that the selection of welding and micro-second cooling parameters would allow for crack-free joints with a significant amount of fine-grained martensite, which translates into high-strength properties of the welds.

The newly developed technology for the analyzed steels will allow its use in the automotive industry and should contribute to a wider use of the discussed steels in practice.

These steels are relatively not very well weldable because of the cracks can occur in the weld and Heat Affected Zone [2,22–24]. This is related to carbon content and its equivalent [21]. The CE (Carbon Equivalent) coefficient is very similar to that of low-alloy steels [25,26]. An important disadvantage is that the mechanical parameter of the welded Docol joint is much lower than that of the parent material. For example, the tensile strength of Docol 1200 M steel is on the level of 1200 MPa, whereas the joint strength is nearly half times lower and equals not more than 700 MPa. In this paper, we analyze for the first time the possibilities of AHSS steel welding with micro-jet cooling [7,8,27,28]. Micro-jet cooling enables to control the microstructure of the welds manufactured. The considered method has been applied efficiently in joining low-alloy steels as well as aluminum alloys [2].

Selecting the steel type to the application is conducted based on a range of the material type from Docol 900 M up to Docol 1500 M at details of chemical composition [29] and mechanical properties (Figure 1). As can be noticed in Table 1, the relationship between the C and Mn elements follows the steel type, reaching the value of 0.03 at Docol 900 M and 0.19 at Docol 1500 M. From Figure 2a,b it is easy to note that the yield stress and ultimate tensile strength of Docol steels is represented by the wide range of values of 700 MPa to 1350 MPa and respectively 900 MPa to 1700 MPa at ductility min. 3%. The similar results were presented in [29]. With respect to an engineering point of view, a relationship between the mechanical parameters (Figure 2b) is important to assess the quality of the steel and its welding because the proportion enables to indicate a size of hardening region at the stress that reaches the yield point up to ultimate tensile strength. In this approach, an elongation plays a significant role because the parameter is directly related to ductility, enabling to avoid brittle cracking.

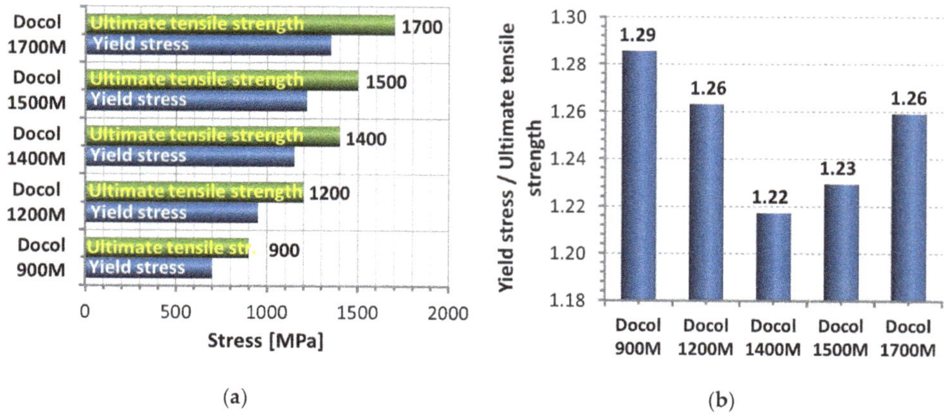

Figure 2. Yield stress and ultimate tensile strength of different types of Docol steel (**a**) on the basis of selected data from the Product Program SSAB, (**b**) the results of own analysis.

During AHSS steels welding, the regime determined by thermodynamic analysis of the process should be reached. The linear energy up to 4 kJ/cm is required to refine the ferrite and reduce stress due to welding process [23–25]. Preheating and control the temperature of the interpass layer is needed for lowering the hydrogen content in the joint. It can be explained as follows: single hydrogen atoms H are created in the weld, which can freely permeate between iron atoms, where they combine into the H_2 molecule (recombination). Hydrogen accumulation in the metal leads to internal pressure, causing an increase of stress, which in turn causes Hydrogen Induced Cracking (HIC) [26,30,31]. HIC in Docol steels appears mainly at the martensite-ferrite grain boundaries as well as at interaction with non-metallic inclusions [2,7]. Hydrogen cracking usually spreads parallel to the surface of the sheet [1,6,8]. The results of Docol steels welding, presented in literature, show that the welding joint does not crack due to stress corrosion cracking (SSCC—Sulfide Stress Corrosion Cracking), because these steels contain traces of sulfur [27]. When welding thin-walled structures, attempts have been made to weld it with varying degrees of success without preheating [2]. The investigation results presented in [2] confirm that the obtained data are not reproducible. In order to manufacture the weld at the attractive technical quality, it was decided to weld Docol 1200 M steel without preheating and micro-jet cooling. Joints with different MAG welding parameters and different parameters of cooling micro-jet were made. When welding various types of steel (ferritic, austenitic, martensitic), it is very important to select the right shielding gas and wire [32].

As can be noticed in [20,22,23], fatigue and tensile experiments are crucial destructive experiments in the assessment of AHSS welds quality because they follow important mechanical parameters used in engineering practice and research efforts. Moreover, this kind of data is significantly expected by vehicle manufacturers. Steel Docol 1200 M joints were not manufactured applying micro-jet cooling process. It is important to capture the fatigue strength because this parameter is the important data for assessment on the quality of the weld [16,31,33]. Therefore, the paper follows the weld behavior under static and fatigue testing.

2. Materials and Methods

A Docol 1200 M welded (BW) butt joints at a thickness of 2 mm was manufactured. The requirements of EN 15614-1 standard were used for the MAG welding method at the low position (PA). The single-stitch welding and a finished 3 mm thick weld are illustrated in Figure 3. The sample size for testing the weldability of Docol 1200 M (HR-Hot Rolled) steel was accomplished according to Figure 3. Sample total dimensions: 3 mm × 200 mm × 400 mm.

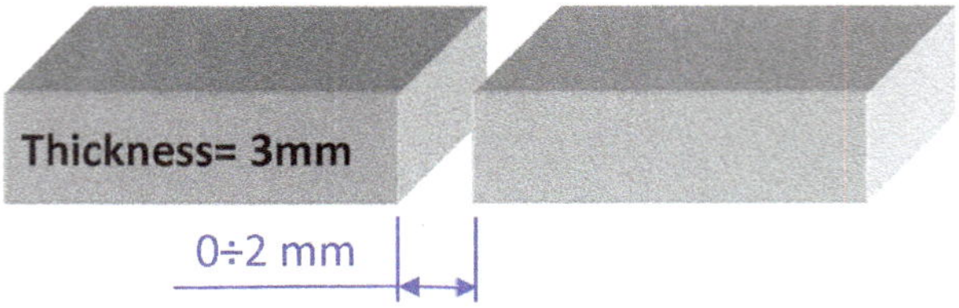

Figure 3. A scheme for preparing elements for MAG welding supported by the micro-jet cooling method, thickness t = 2 mm.

It was decided to produce welds at MAG (Metal Active Gas) process using the gas mixture as of: Ar + 18% CO_2 (the PN-EN 14,175 standard). All of the samples were welded with an electrode wire EN ISO 16834-A G 89 6 M21 Mn4Ni2CrMo−UNION X90 (Böhler Düsseldor, Germany) at the chemical composition presented in [4]. It is easy to note that UNION X90 wire contained Cr and Mo with the amount of providing good plastic properties [4]. The detailed welding process parameters for sheets with a thickness of 3 mm are provided in Table 1.

Table 1. Sample welding details.

Order of Layers	Method of Welding	Diameter of Wire (mm)	Current (A)	Voltage (V)	Polarization	Welding Speed (mm/min)	Energy (kJ/cm)
1	MAG	1.0	109	19	DC "+"	350	3.3

Methods, Scope of Research

The scope of research included non-destructive testing (NDT):

- Visual tests (VT) of welded joints were conducted with the eye fitted with a magnifying glass at 3× magnification—the test was performed according to the requirements of PN-EN ISO 17,638 standard and the assessment criteria according to EN ISO 5817 standard,
- Magnetic-powder tests (MT) were carried out in accordance with the PN-EN ISO 17,638 standard and the assessment criteria according to EN ISO 5817 standard using a magnetic flaw detector device type REM-230 (ATG, Prague, Czech Republic).

Evaluation of microstructure and mechanical properties:

- Examinations of the microstructure of the samples digested with the Adler reagent using a light microscope (LM),
- Hardness measurement was done focusing on guidelines of the PN-EN ISO 9015-1:2011 [34] and PN-EN ISO 6507-1:2018-05 [35] standards,
- Tensile test was carried out basing on the regimes of the PN-EN ISO 6892-1:2020 standard [36]. The experimental approach was elaborated design to follow the Docol 1200 M weld to Docol 1200 M behavior under static and fatigue loading. Therefore, the U-notched specimen was designed. It was done basing on requirements of the E 468–90 ASTM standard [37] for fatigue tests,

- Fatigue test collecting specimen designing was performed at rules of the ASTM E468-18 standard [38].

3. Results and Discussion

The article presents the possibility of welding a thin-walled structure (3 mm) made of high-strength Docol 1200 M steel. These steels are difficult to weld [39,40], as cracks often occur. The big disadvantage is obtaining much worse strength and elongation of the welded joint in relation to the base material. For this purpose, it was decided to use micro-jet cooling for welding. This section presents selected test results and their analysis.

To further enhance the mechanical properties of joints after MAG welding, an addition of micro-jet cooling technology was selected. For the steel structure welding process, the following micro-jet cooling parameters were implemented:

- quantity of cooling nozzles: 1,
- form of cooling medium: Ar or He,
- pressure of the cooling medium: 0.6−0.7 MPa,
- diameter of micro stream: 60−70 μm,
- distance of the micro-jet nozzle from the welded surface: 20 mm.

After the welding with micro-jet cooling, the following non-destructive tests (NDT) were carried out: visual (VT), magnetic-particle (MT), and radiographic. The gap between the welded elements arm-gap (Figure 2) was checked in a range of 1 to 2 mm, 0.5 mm step by various parameters of micro-jet cooling after MAG welding. A correctly selected gap between elements together with micro-jet cooling parameters affect the cooling conditions of the weld. The results of the welding process are presented in Table 2. The comparison of the results of non-destructive tests (Table 2) confirms that all connections made without micro-jet cooling have welding defects. As it was investigated, the micro-jet cooling has allowed us to manufacture joints that did not show any welding defects or incompatibilities in the tests performed.

Table 2. Results from non-destructive investigations, the joint application-movable platform.

Micro-Jet Stream Diameter (μm)	Micro-Jet Stream Pressure (MPa)	Gap (mm)	Micro-Jet Gas	Symbol of Sample	Observation
without	without	1	without	S1o	Cracks in the weld
60	0.6	1	Ar or He	S1a	Cracks in the weld
60	0.7	1	Ar or He	S1b	Cracks in the weld
70	0.6	1	Ar or He	S1c	Cracks in the weld
70	0.7	1	Ar or He	S1d	Cracks in the weld
without	without	1.5	Ar or He	S1.5o	Cracks in the weld
60	0.6	1.5	Ar or He	S1.5a	No cracks
60	0.7	1.5	Ar or He	S1.5b	No cracks
70	0.6	1.5	Ar or He	S1.5c	No cracks
70	0.7	1.5	Ar	S1.5d	Cracks in the weld
70	0.7	1.5	He	S1.5e	No cracks
without	without	2	without	S2o	Cracks in the weld
60	0.6	2	Ar	S2a	Cracks in the weld
60	0.6	2	He	S2b	No cracks
60	0.7	2	Ar or He	S2c	Cracks in the weld
70	0.6	2	Ar or He	S2d	Cracks in the weld
70	0.7	2	Ar or He	S2e	Cracks in the weld

Results of non-destructive tests exhibited no cracks occurrence in the weld manufactured at the micro-jet cooling technique (Figure 4).

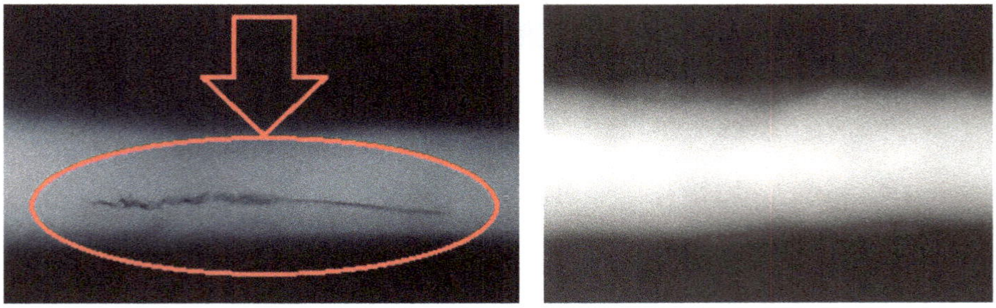

Figure 4. Weld after welding without (**a**) and with (**b**) micro-jet cooling in X-ray inspection.

The table data shows that the gap between elements should be equal to 1.5 mm and micro-jet cooling should not be too intensive (samples S1.5a, S1.5b, S1.5c). For the 2 mm gap, no cracks were observed, only for low-intensity helium micro-jet cooling (sample S2b) (Table 2). It was easy to assume that helium micro-jet cooling gives better results. Further joint hardness distribution was also carried out. Samples with positive results from non-destructive tests (NDT) were only tested (the gap between elements was only 1.5 mm, micro-jet cooling was always not intensive). Two micro-jet gases were used, of which micro-jet cooling helium gave better results (samples S1.5a, S1.5b, S1.5c, S1.5e), which was confirmed by non-destructive tests. Analyzing the data from Table 3, it can be noted that the micro-jet cooling should be used to the welding process. Micro-jet cooling parameters do not have a special effect on the hardness values (Table 3), enabling to create the mechanical parameter of the weld and base metal on the same level.

Table 3. Hardness values in a mixed joint at selected parameters of welding process, HV—Vickers Hardness.

Micro-Jet Stream Diameter (μm)	Micro-Jet Stream Pressure (MPa)	Micro-Jet Gas	Parent Material HV (MPa)	HAZ HV (MPa)	Weld HV (MPa)
60	0.6	Ar	335	359	341
60	0.7	Ar	336	352	343
60	0.6	Ar	334	351	341
	Average value of HV		335	354	342
	Standard deviation of HV		±0.8	±3.6	±0.9
60	0.7	He	333	350	340
70	0.6	He	335	349	341
60	0.6	He	334	347	339
	Average value of HV		334	349	340
	Standard deviation of HV		±0.8	±1.2	±0.8
without	without	without	334	365	351

After the hardness assessment, the strength tests of the welded elements were carried out and the 3369 INSTRON testing machine (INSTRON, High Wycombe, England) was selected for that purpose. Two analyses were performed:

- strength of Docol 1200 M steel with the use of argon micro-jet cooling,
- strength of Docol 1200 M steel with the use of helium micro-jet cooling.

Micro-jet gas pressure (Ar or He) was equal to 0.7 MPa, micro-jet stream diameter was 60 μm. Each test was repeated 3 times (samples 1, 2, 3). Tensile strength and fatigue tests were performed as the main research to check the quality of the joint. The results are presented in Table 4.

Table 4. Tensile tests results of welded Docol 1200 M steel without the use of micro-jet cooling ± measurement uncertainty.

Micro-Jet Gas	Yield Stress, YS (MPa)	Ultimate Tensile Strength, UTS, (MPa)	A_{5min} (%)
without	495 ± 8	723 ± 10	8
Ar	521 ± 12	751 ± 14	10
He	534 ± 10	762 ± 12	11

As captured in the tensile test, the Docol 1200 M steel is a very attractive structural material because of the value of proportional limit taking 771 MPa, ultimate tensile strength exceeding 1200 MPa, and elongation that reaches min. 8% (Figure 5a). These values protect the components made of this kind of steel under operational conditions against deformation and fatigue damages. Taking the fracture zone of the Docol 1200 M steel, the stress state components playing a role in the material cracking under tension can be indicated, i.e., axial and shear. The steel and weld were also examined under the micrographic structure (Figure 6) and tensile test applying hourglass specimen in the geometry and dimensions shown in Figure 7a. It was assessed based on tensile characteristics determined on the flat and hourglass specimens (Figure 7b). The specimen was adopted (Figure 7) to the conducted tests by allocating the weld in the middle of the measuring zone. The comparison of results from the tensile tests indicates an increase of strength parameters at the reduction of elongation as an effect of a stress concentration. In the case of the micro-jet cooling weld of the Docol 1200 M, the same effect was noticed, but the lowering of the elongation value was not significant (Figure 5). The role of the weld joining in the steel behavior under tension is determined by a comparison of the tensile characteristic of the parent material and the weld (Figure 5). It was visible in the 50% lowering of proportional limit, yield stress, and ultimate tensile strength at almost the same elongation.

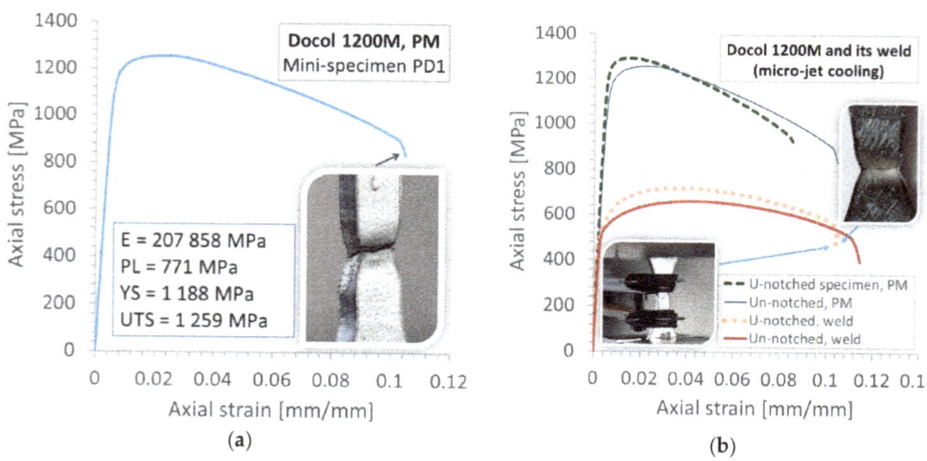

Figure 5. Tensile characteristics of the Docol 1200 M: (**a**) flat (un-notched) mini-specimen (**b**) applying U-notched and un-notched specimens to parent material (PM) and weld (micro-jet), E—Young's modulus, PL—proportional limit, YS—yield stress, UTS—ultimate tensile strength.

Application of micro-jet cooling during the welding process improved YS and UTS values. The average YS value of MAG welded joint without micro-jet cooling was 495 MPa and the average UTS value was 723 MPa. They increased respectively to the values 551 MPa (YS) and 751 MPa (UTS) for argon micro-jet cooling, and respectively to the values 534 MPa (YS) and 762 MPa (UTS) for helium micro-jet cooling. It has been confirmed from NDT observations that helium micro-jet cooling gives better results. Next, the microstructure analysis was carried out. A typical microstructure of weld is presented in Figure 6. This figure illustrates the microstructure of the cross-section of the joint, where the martensitic component is clearly visible. Welds are free from defects and incompatibilities. Finally, fatigue tests were carried out. Only the welding process with helium micro-jet cooling was analyzed. Micro-jet gas pressure (He only) was equal to 0.7 MPa, micro-jet stream diameter was 60 μm.

Figure 6. Microstructure of the joint cross-section without (**a**) and with the use of helium micro-jet cooling (**b**), etched Nital, time 12 s.

Fatigue tests were employed to follow the behavior of the micro-cooling weld joining of the Docol 1200 M under cyclic loading [41,42]. All tests were carried out at room temperature using the 8874 INSTRON servo-hydraulic testing machine (INSTRON, High Wycombe, England) and stress signal having the following parameters: stress ratio = 0, the range of the maximum values of stress from 500 MPa up to 700 MPa, frequency = 5 Hz (Figure 8). The stress signals were in the form of a cyclic function having a sinusoidal shape (Figure 9). The last 50 cycles of the fatigue tests were taken for capturing the weld behavior before fracture appearing. Practically, they were collected in the other files directly related to the analysis taken. The hourglass specimens having the weld in the middle section of the measuring region were used (Figure 7). The specimens were selected with the Docol 1200 M steel in a form of the sheet using a water cutting technology.

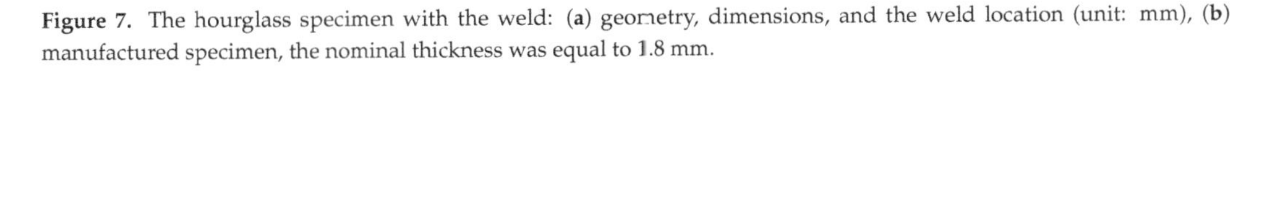

Figure 7. The hourglass specimen with the weld: (**a**) geometry, dimensions, and the weld location (unit: mm), (**b**) manufactured specimen, the nominal thickness was equal to 1.8 mm.

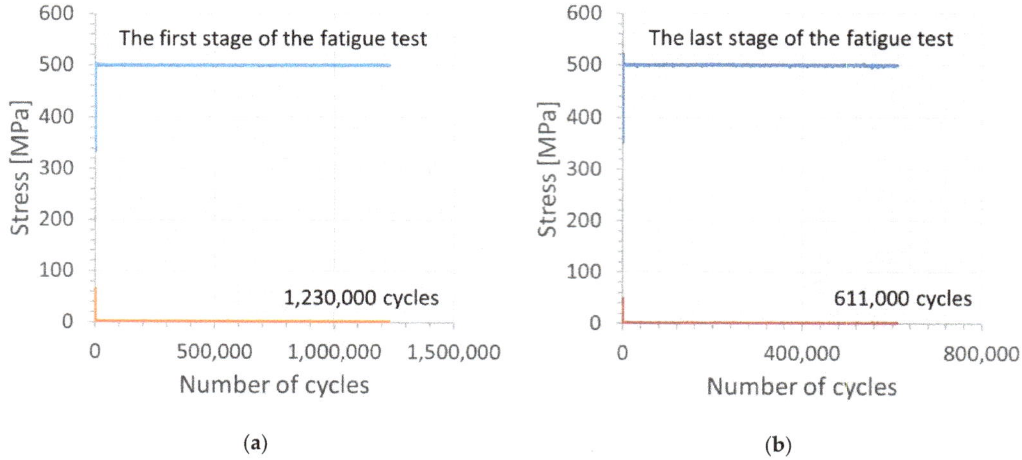

Figure 8. Maximum and minimum values of stress signal for the cycles: (**a**) and (**b**) up to 1,230,000, and 1,841,000 (sum before the material fracture = 1,841,757 cycles), respectively.

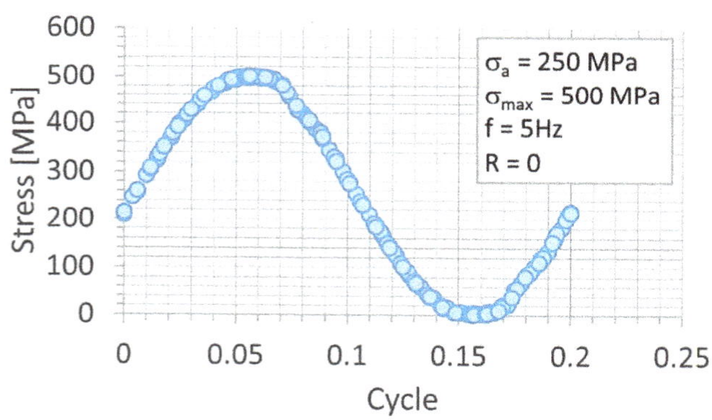

Figure 9. Stress signal at amplitude of 250 MPa and maximum value of 500 MPa, frequency 5 Hz.

The fatigue tests have reflected the weld response on cyclic loading up to fracture (Figure 8 and Figure 11). As can be noticed on the fracture zone from the test at the maximum values of stress of 500 MPa (Figure 10) and 700 MPa (Figure 11), the fatigue damages occurred in the middle of the specimen, following directly the degradation of the micro-jet cooling weld under the cyclic loading. Differences in the weld behavior at various values of stress were expressed by changes in the orientation of the fracture regions, i.e., with an increase of the stress value, the axial stress was more dominant than the shear stress, showing a gradual vanishing of the second-mentioned stress component and it disappearing at the highest value of the stress applied (Figure 11). Moreover, variations in displacement versus a number of cycles, employed to follow the weld response on the cyclic loading, were also clearly visible. They were represented by the reduction of non-linear course of the relationship with the increase of stress value directly before the weld fracture.

Figure 10. The hourglass specimen made of the Docol 1200 M steel with the micro-jet cooling weld after fatigue test at maximum value of stress of 500 MPa.

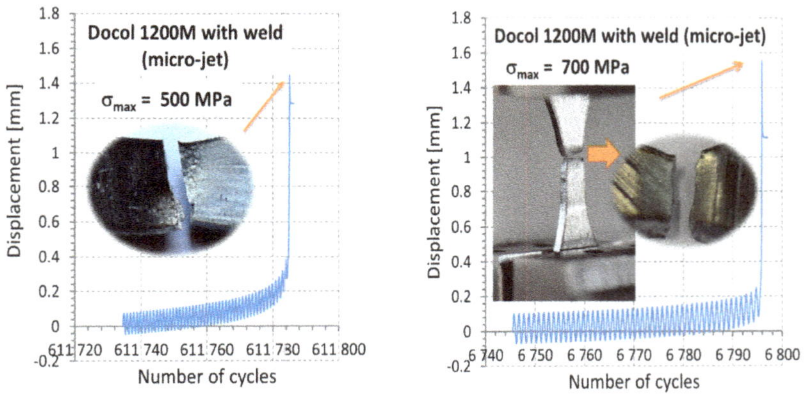

(a) (b)

Figure 11. The last stage of fatigue tests before fracture of the Docol 1200 M steel's micro-jet cooling weld for the following maximum values of stress: (**a**) 500 MPa and (**b**) 700 MPa.

At 500 MPa and 550 MPa, the region degradation occurred at relatively little plastic deformation (Figure 12a,b), compared to the results for the bigger values i.e., 650 MPa and 750 MPa (Figure 12c,d). This kind of data can be very easily connected with results from tensile tests (Figure 5b). As it is possible to notice, the features of the zones captured are dependent on the relationship of values of the cyclic stress to mechanical parameters. If a value of the cyclic stress is close to the ultimate tensile strength, then the plastic deformation is the dominant mechanism in the weld degradation.

Fatigue of the micro-jet cooling weld is also represented by the fracture at the following maximum values of stress: 700 MPa, 650 MPa, 600 MPa, 550 MPa, and 500 MPa, taking the relationship presented in Figure 13. From the engineering point of view, these results have enabled to predict fatigue durability of the weld tested, at cyclic stress within the range from yield stress up to ultimate tensile strength, as well as determine the fatigue limit. It has enabled to indicate the value of stress to operational conditions, which does not lead to fatigue damages initiation. In the case of the weld from the technological process supported by helium micro-jet cooling, the stress reached 480 MPa.

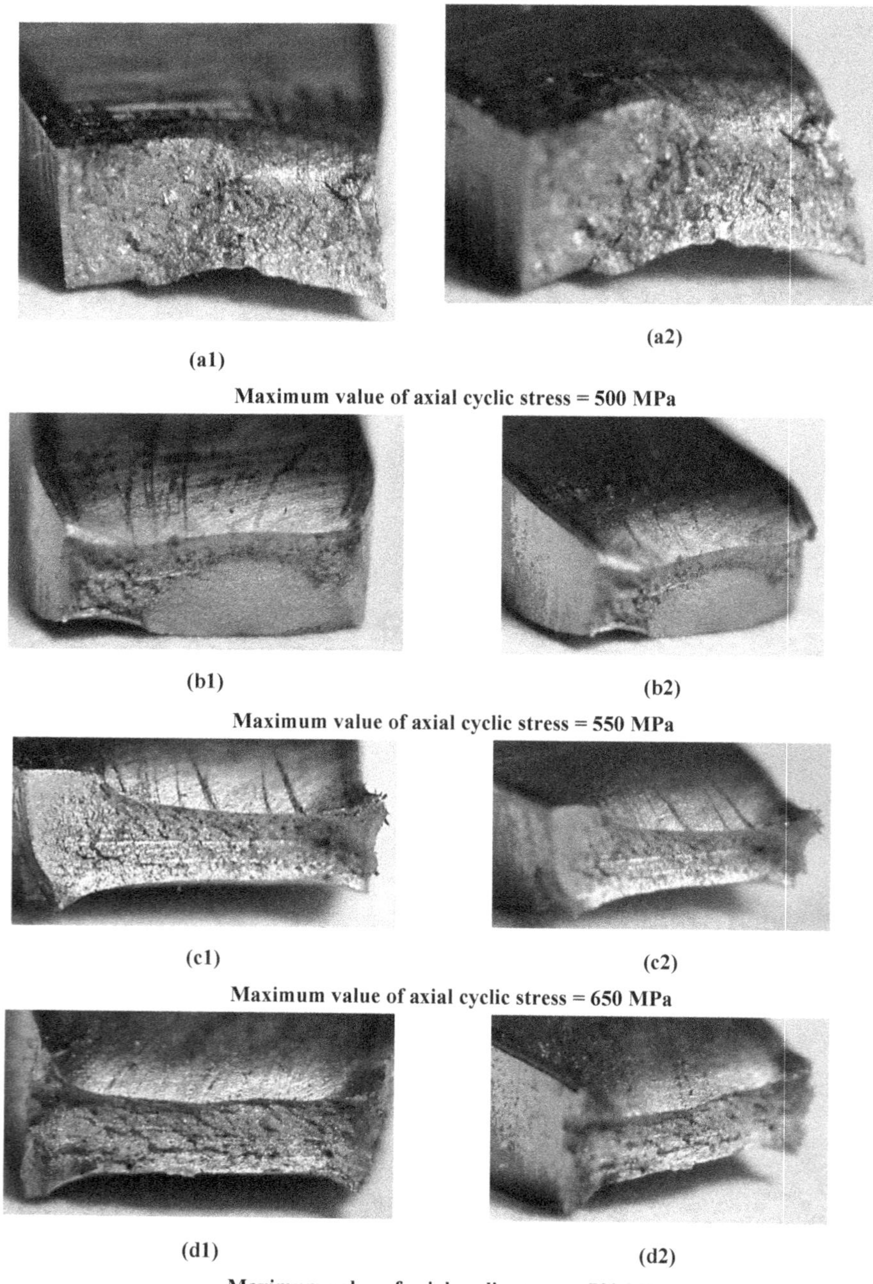

Figure 12. Fracture zones of the Docol 1200 M weld manufactured at the micro-jet cooling technique after fatigue test under the maximum values of stress of 500 MPa to 700 MPa.

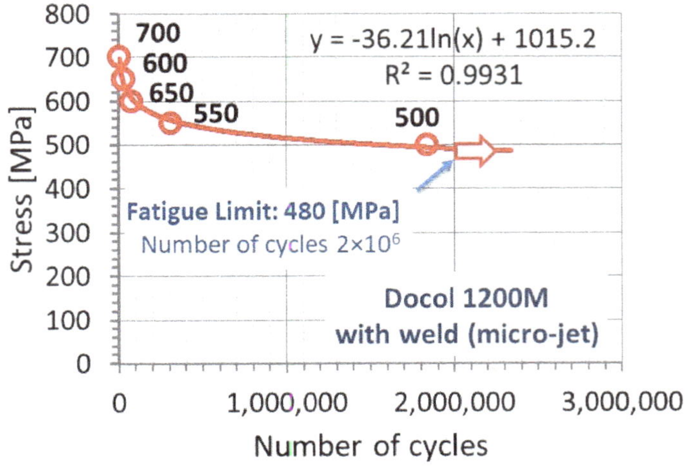

Figure 13. The Wöhler curve of the weld with micro-jet cooling technology on the Docol 1200 M steel.

The obtained results show that:

- Two micro-jet gases were used, of which micro-jet cooling helium gave better results (samples S1.5a, S1.5b, S1.5c, S1.5e), which was confirmed by non-destructive tests (Table 2, Figure 4).
- Tensile strength and fatigue tests were performed as the main research to check the quality of the joint. It was noticed that the joints with better plastic properties were obtained when micro-jet cooling was additionally used for MAG welding. Relative elongation increased from 8% to 11% when micro-jet cooling was used (Table 4, Figure 5).
- The yield stress of the joint without micro-jet cooling was below the required value of 500 MPa, while welding with micro-jet cooling resulted in a yield stress value of 520 MPa both when helium and argon were used in micro-jet cooling. Similarly, a higher tensile strength was obtained thanks to the use of micro-jet cooling.
- Yield stress and tensile strength values were at a similar level of 750 MPa when helium or argon was used for micro-jet cooling. The joint after welding with He micro-jet cooling had slightly higher elongation (11%) in comparison with Ar micro-jet cooling (10%); therefore, helium was selected as the micro-jet gas for fatigue tests. Without micro-jet cooling, tensile strength was obtained at a lower level of 720 MPa (Table 4).

Generally, the fatigue tests confirmed that the joints are of high quality. The Wöhler curve has been established as a key characteristic for determining the quality of the weld examined. The fatigue strength was obtained at a satisfactory level of 480 MPa (Figure 10). Therefore, this kind of data can be used by a lot of engineer and research groups for designing and modeling. It is worth noting that data in the form collected are presented purely indicating an insufficient number of results for designing the high-strength steel behavior. Non-destructive and destructive tests confirmed that the correct joint was obtained from Docol 1200 M steel thanks to the use of micro-jet cooling in MAG welding.

4. Summary

The welded joints were made of the difficult-to-weld material Docol 1200 M. The most favorable welding conditions and parameters were determined. The possibility of welding Docol MAG steel with micro-jet cooling has been tested. The most important parameters of micro-jet cooling have been thoroughly checked. The most suitable micro-jet gas was selected. Non-destructive and destructive tests were carried out to assess the quality of the joint. The microscope (LM) observations, tensile and hardness tests, as well as fatigue approaches were carried out. Fatigue tests were treated as the most important destructive experiment in the assessment of the quality of Docol welds. The results from the tensile test indicated that the Docol 1200M steel is sensitive to the micro-jet welding processes, expressing half-smaller values of mechanical parameters than the parent material, not including elongation. Independently of the material state examined, the value of this parameter was very similar. The weld behavior under fatigue had a range of significant differences in the number of cycles to fracture at the stress value from 700 MPa up to the fatigue limit of 480 MPa. This was also reflected by a gradual fracturing of the steel at the smaller values of stress.

5. Conclusions

Results from non-destructive and destructive tests on the weld joint manufactured at the micro-jet cooling have enabled us to formulate the following concluding remarks:

- the micro-jet cooling technique allows producing the weld for Docol 1200 M steel with a martensitic microstructure similar to the parent material;
- mechanical properties of the weld supported by the micro-jet cooling have expressed 50% lowering and maintaining elongation value in comparison to the steel in the as-received state;
- tensile curve of the weld with respect to the welding process used has received more regular shape, which is requested by engineers for designing and research for modeling;
- fatigue limit of the weld tested was very beneficial because of the 480 MPa that was determined. This value can be directly employed in designing, calculations, and modeling as fundamental data for assessing the technical state of a component;
- the results have confirmed that the elaborated joining technology allows obtaining the assumed parameters of the joint with respect to high plastic properties (elongation = 10%), comparable to the parent material. Publications indicate that, so far, high and repeatable parameters of the joint have not yet been obtained without micro-jet cooling.

Author Contributions: Conceptualization, T.W., T.S. and B.S.-L.; methodology, B.S.-L.; software, T.S.; validation, T.W., T.S. and B.S.-L.; formal analysis, A.J.; investigation, A.J.,T.S.; resources, B.S.-L.; data curation, T.W.; writing—original draft preparation, T.S.; writing—review and editing, B.S.-L.; visualization, T.S.; supervision, A.J.; project administration, T.W.; funding acquisition, B.S.-L. All authors have read and agreed to the published version of the manuscript.

Funding: This research was funded by [Silesian University of Technology] grant number [BK-205/RT1/2020] And The APC was funded by [Authors].

Institutional Review Board Statement: Not applicable.

Informed Consent Statement: Not applicable.

Data Availability Statement: Data sharing is not applicable to this article.

Acknowledgments: The paper is part of the COST project, CA 18223.

Conflicts of Interest: Authors declare no conflict of interest.

References

1. 5 Takeaways from IABC 2019 USA, Automotive Body Congress, Detroit, 2019, 6 pages. Available online: https://www.ssab.com/products/brands/docol/automotive-insights/2019/5-takeaways-from-iabc-2019-usa (accessed on 19 December 2019).
2. Hadryś, D. Impact Load of Welds after Micro-Jet Cooling / Dynamiczne Obciążenie Spoin Chłodzonych Mikrojetowo. *Arch. Met. Mater.* **2015**, *60*, 2525–2528. [CrossRef]
3. Matlock, D.K.; Speer, J.G.; de Moor, E. Recent AHSS developments for automotive applications: Processing, microstructures, and properties. In Proceedings of the Addressing key technology gaps in implementing advanced high-strength steels for automotive light weighting, Southfield, MI, USA, 9–10 February 2012.

4. Szczucka-Lasota, B.; Wegrzyn, T.; Szymczak, T.; Piwnik, J.; Jurek, A.; Wilczyński, K.I. Fatigue of Docol 1200M welds to mobile platform structures. In Proceedings of the XII International Conference Transport Problems, Katowice, Poland, 1 December 2020; pp. 762–769.
5. Matlock, D.K.; Speer, J.G. Processing Opportunities for New Advanced High-Strength Sheet Steels, Mater. *Manuf. Process.* **2010**, *25*, 7–13. [CrossRef]
6. Speer, J.G.; Matlock, D.K.; de Cooman, B.C.; Schroth, J.G. Carbon partitioning into austenite after martensite transformation. *Acta Mater.* **2003**, *51*, 2611–2622. [CrossRef]
7. Varelis, G.E.; Papatheocharis, T.; Karamanos, S.A.; Perdikaris, P.C. Structural behavior and design of high-strength steel welded tubular connections under extreme loading. *Mar. Struct.* **2020**, *71*, 102701. [CrossRef]
8. Crémona, C.; Eichler, B.; Johansson, B.; Larsson, T. Improved Assessment Methods for Static and Fatigue Resistance of Old Metallic Railway Bridges. *J. Bridg. Eng.* **2013**, *18*, 1164–1173. [CrossRef]
9. Pawar, N. *Automotive Advanced High Strength Steel (AHSS) Market Professional Survey Report 2019*; Analytical Research Cognizance (ARCX001); GrandResearchStore: New York, NY, USA, 2019; p. 101.
10. Szczucka-Lasota, B.; Uscilowska, A.; Wegrzyn, T.; Stanik, Z.; Piwnik, J. Implementation of the method of fundamental solutions for the correction of parameters of the thermal HM spraying process. *Comput. Math. Appl. Eng.* **2018**, *98*, 17–26. [CrossRef]
11. Darabi, J. Development of a chip-integrated micro cooling device. *Microelectron. J.* **2003**, *34*, 1067–1074. [CrossRef]
12. Hashimoto, F.; Lahoti, G. Optimization of Set-up Conditions for Stability of The Centerless Grinding Process. *CIRP Ann.* **2004**, *53*, 271–274. [CrossRef]
13. Muszynski, T.; Mikielewicz, D. Structural optimization of microjet array cooling system. *Appl. Therm. Eng.* **2017**, *123*, 103–110. [CrossRef]
14. Barsukov, V.V.; Tarasiuk, V.; Shapovalov, V.M.; Krupicz, B. Express evaluation method of internal friction parameters in molding material briquettes. *J. Frict. Wear* **2017**, *38*, 71–76. [CrossRef]
15. Bleck, W.; Larour, P.; Baeumer, A. High strain tensile testing of modern car body steels. *Mater. Forum* **2005**, *29*, 21–28.
16. Wegrzyn, T.; Szczucka-Lasota, B.; Uscilowska, A.; Stanik, Z.; Piwnik, J. Validation of parameters selection of welding with micro-jet cooling by using method of fundamental solutions. *Eng. Anal. Bound. Elem.* **2019**, *98*, 17–26. [CrossRef]
17. Prijanovič, U.; Tonkovič, M.P.; Trdan, U.; Pleterski, M.; Jezeršek, M.; Klobčar, D. Remote fibre laser welding of advanced high strength martensitic steel. *Metals* **2020**, *10*, 533. [CrossRef]
18. Lahtinen, T.; Vilaça, P.; Peura, P.; Mehtonen, S. MAG Welding Tests of Modern High Strength Steels with Minimum Yield Strength of 700 MPa. *Appl. Sci.* **2019**, *9*, 1031. [CrossRef]
19. Tomków, J.; Haras, J. The influence of welding heat input on the quality and properties of high strength low-alloy dis-similar steel butt joints. *Weld. Tech. Rev.* **2020**, *92*, 15–23. [CrossRef]
20. Mert, T.; Tümer, M.; Kerimak, Z.M. Investigations on Mechanical Strength and Microstructure of Multi-Pass Welded S690QL HSLA Steel Using MAG and FCAW. *Pr. Met.* **2019**, *56*, 634–654. [CrossRef]
21. Szymczak, T.; Makowska, K.; Kowalewski, Z.L. Influence of the Welding Process on the Mechanical Characteristics and Fracture of the S700MC High Strength Steel under Various Types of Loading. *Materials* **2020**, *13*, 5249. [CrossRef] [PubMed]
22. Totten, G.; Howes, M.; Inoue, T. *Handbook of Residual Stress and Deformation of Steel*; ASM International: Ohio, OH, USA, 2002; p. 499.
23. Sága, M.; Blatnická, M.; Blatnický, M.; Dižo, J.; Gerlici, J. Research of the Fatigue Life of Welded Joints of High Strength Steel S960 QL Created Using Laser and Electron Beams. *Materials* **2020**, *13*, 2539. [CrossRef] [PubMed]
24. Yi, H.L.; Bhadeshia, K.H. Stabilisation of ferrite in hot rolled δ-TRIP steel. *Mater. Sci. Technol.* **2011**, *27*, 525–529. [CrossRef]
25. Porter, D.A. Weldable high-strength steels: Challenges and engineering applications. In Proceedings of the IIW International Conference High-Strength Materials-Challenges and Applications, Helsinki, Finland, 2–3 July 2015.
26. Ma, J.-L.; Chan, T.-M.; Young, B. Tests on high-strength steel hollow sections: A review. *Proc. Inst. Civ. Eng. Struct. Build.* **2017**, *170*, 621–630. [CrossRef]
27. Günther, H.-P.; Hildebrand, J.; Rasche, C.; Versch, C.; Wudtke, I.; Kuhlmann, U.; Vormwald, M.; Werner, F. Welded connections of high-strength steels for the building industry. *Riv. Ital. Della Saldatura* **2014**, *66*, 1055–1087. [CrossRef]
28. LaCalle, R.; Alvarez, J.A.; Ferreño, D.; Portilla, J.; Ruiz, E.; Arroyo, B.; Gutiérrez-Solana, F. Influence of the Flame Straightening Process on Microstructural, Mechanical and Fracture Properties of S235 JR, S460 ML and S690 QL Structural Steels. *Exp. Mech.* **2013**, *53*, 893–909. [CrossRef]
29. DOCOL the Automotive Steel. In *The Most Sustainable Solution: Setting the New Standard for the Automotive Industry*; SSAB: Borlänge, Sweden, 2019.
30. Goritskii, V.M.; Shneiderov, G.R.; Guseva, I.A. Effect of Chemical Composition and Structure on Mechanical Properties of High-Strength Welding Steels. *Metals* **2019**, *63*, 21–32. [CrossRef]
31. Chatterjee, S.; Murugananth, M.; Bhadeshia, H.K.D.H. δ-TRIP Steel. *Mater. Sci. Technol.* **2007**, *23*, 819–827. [CrossRef]
32. Mansouri, D.; Sendur, P.; Yapici, G.G. Fatigue characteristics of continuous welded rails and the effect of residual stress on fatigue-ratchetting interaction. *Mech. Adv. Mater. Struct.* **2018**, *27*, 473–480. [CrossRef]
33. Cheng, X. Residual stress modification by post-weld treatment and its beneficial effect on fatigue strength of welded structures. *Int. J. Fatigue* **2003**, *25*, 1259–1269. [CrossRef]

34. *Destructive Tests on Welds in Metallic Materials—Hardness Testing—Part. 1—Hardness Test on Arc Welded Joints*; PN-EN ISO 9015-1:2011; Polish Committee of Standardization: Warsaw, Poland, 2011.
35. *Metallic Materials—Vickers Hardness Test—Part. 1: Test. Method*; PN-EN ISO 6507-1:2018; Polish Committee of Standardization: Warsaw, Poland, 2018.
36. *Metallic Materials—Tensile Testing—Method of Test at Room Temperature*; PN-EN ISO 6892-1:2020; Polish Committee of Standardization: Warsaw, Poland, 2020.
37. *Standard Practice for Presentation of Constant Amplitude Fatigue Test. Results for Metallic Materials*; ASTM E468-90; ASTM International: West Conshohocken, PA, USA, 2004.
38. *Standard Practice for Presentation of Constant Amplitude Fatigue Test. Results for Metallic Materials*; ASTM E468-18; ASTM International: West Conshohocken, PA, USA, 2018.
39. Celin, R.; Burja, J. Effect of cooling rates on the weld heat affected zone coarse grain microstructure. *Met. Mater. Eng.* **2018**, *24*, 37–44. [CrossRef]
40. Jaewson, L.; Kamran, A.; Jwo, P. Modeling of failure mode of laser welds in lap-shear specimens of HSLA steel sheets. *Eng. Fract. Mech.* **2011**, *1*, 347–396.
41. Costanza, G.; Sili, A.; Tata, M.E. Weldability of austenitic stainless steel by metal arc welding with different shielding gas. *Procedia Struct. Integr.* **2016**, *2*, 3508–3514. [CrossRef]
42. Stephens, R.I.; Fatemi, A.; Stephens, R.R.; Fuchs, H.O. *Metal. Fatigue in Engineering*; Wiley Interscience: New York, NY, USA, 2001; p. 496.

materials

MDPI

Article

Role of Bead Sequence in Underwater Welding

Jacek Tomków *, Dariusz Fydrych and Grzegorz Rogalski

Faculty of Mechanical Engineering, Gdańsk University of Technology, G. Narutowicza Street 11/12, 80-233 Gdańsk, Poland; dariusz.fydrych@pg.edu.pl (D.F.); grzegorz.rogalski@pg.edu.pl (G.R.)

* Correspondence: jacek.tomkow@pg.edu.pl; Tel.: +48-58-347-1863

Received: 19 September 2019; Accepted: 14 October 2019; Published: 16 October 2019

Abstract: This paper presents examinations of the role of the bead sequence in underwater welding. Two specimens of wet welded layers made by covered electrodes with the use of normalized S355G10+N steel were welded by a reasonable bead sequence. For each specimen, metallographic macro- and micro-scopic tests were done. Then, Vickers HV10 hardness measurements were conducted for each pad weld in the welded layer. The results show that welding in the water environment carries many problems in the stability of the welding arc, which influences the properties of the welds. The effects of refining and tempering the structure in heat-affected zones of earlier laid beads was observed, which provides a reduction of hardness. The possibility of applying two techniques while welding the layer by the wet method is described. It is stated that a reasonable bead sequence can decrease the hardness in heat-affected zones up to 40 HV10. Tempering by heat from next beads can also change the microstructure in this area by tempering martensite and can decrease susceptibility to cold cracking.

Keywords: underwater welding; covered electrodes; wet welding; cold cracking; pad welding

1. Introduction

The most common method of underwater welding is wet welding. The welder and the welding area are in direct contact with the surrounding environment. The process is often carried out by flux-cored arc welding (FCAW) [1–4]. The most common is welding by covered electrodes, which is cheaper and easier [5,6].

The water environment can generate significant problems during welding. For welders, the biggest problems are instability of the welding arc and limited visibility [7,8], which can result in poor quality of welded joints. From the metallurgical point of view, the high cooling rate, the high hydrogen content in deposited metal, and residual stresses have the greatest influence on the quality of underwater welded joints.

The high cooling rate leads to the formation of hardened structures in the heat-affected zone (HAZ) [9,10]. This is one of the reasons for residual stresses in welded joints after the welding process is completed. The next problem in underwater conditions is high diffusible hydrogen content in deposited metal, which is much higher than welding in air [11]. Three factors are responsible for high susceptibility of underwater welded joints to cold cracking. This type of cracking is located in the HAZ and in welds of joints made in water along the fusion line in the overheated area of the HAZ [6,12]. Sometimes, they can also be found in the welds [13].

Experiments in the field of welding processes in the water environment are developing annually, because underwater processes are used more in engineering [14].

Attempts to reduce the imperfections in welded joints made in the water are made in different ways. Much attention is devoted to improving the stability of the welding arc in underwater conditions and assessing the metal transfer in the welding arc, using the following methods: visual sensing [15], in situ imaging [16], mechanical constraints [17], and ultrasonic waves [3,8]. The other research trend is developing filler materials for welding: modifying the flux coating and wires of coated electrodes [18] and applying a waterproof coating [19]. Several articles present the results of technological methods to improve the weldability of steel: diffusible hydrogen content in deposited metal [11,20], the use of austenitic deposits [9,21], preheating [22], and multilayer welding, including the temper bead welding technique [6,12,13,23]. Controlling the distribution of heat input by controlling the sequence and positioning of the beads is considered to be particularly promising, because it leads to a reduction in the grain size and provides decreased hardness and increased impact strength [24–26]. Different variants of bead sequences are used to weld hard-to-weld metals such as cast iron, nuclear steel, aluminum, quenched and tempered steel, and thermomechanically-treated steel [27–35].

For marine and offshore constructions, high-strength, low-alloy (HSLA) steels are widely used [36]. These constructions may undergo failure during exploitation in marine conditions [37–40]. They often have to be repaired in underwater conditions due to the high cost of transport to air conditions. However, the water environment makes repair welding more difficult, therefore, there is still a search for ways to improve the quality of underwater welds.

The aim of the research was to check the influence of the role of bead sequence on the structure and properties of layers using the wet welding method for specimens welded with covered electrodes.

2. Materials and Methods

2.1. Materials Used

For tests, normalized S355G10 + N steel plates with dimensions of 16 × 100 × 100 mm were chosen as the base metal (BM). This steel is often used as a material for offshore structures that may require repairs in the water environment. The carbon equivalent (Ce_{IIW}) is 0.385 according to the International Institute of Welding (IIW). As a filler material, ISO 2560-A: E 38 0 R 11 [41] rutile electrodes with a diameter of 4.0 mm were used. They produce good plasticity of welds, which may be helpful to avoid cold cracking. The chemical compositions of the materials used are listed in Table 1 and the mechanical properties in Table 2.

Table 1. Chemical compositions of materials used (wt.%).

Material	C	Si	Mn	P	Cr	Mo	Ni	Cu	V	Ce_{IIW}
S355G10 + N according to control analysis	0.11	0.35	1.39	0.01	0.02	0.02	0.25	0.27	0.003	0.385
E 38 9 R 11 electrode deposit according to manufacturer data	0.07	0.44	0.55	0.01	0.04	-	-	0.05	-	-

Ce_{IIW}, carbon equivalent by International Institute of Welding.

Table 2. Mechanical properties of materials used according to manufacturer data.

Material	Re (Mpa)	Rm (Mpa)	A_5 (%)
SS355G10+N	389	521	23.5
E 38 0 R 11 electrode deposit	503	538	26

2.2. Welding Process

For welding, the manual metal arc (MMA) wet welding process was chosen. The schema of this process is presented in Figure 1.

For the tests, 2 specimens were made in the water (0.15 m depth) with 5 beads in each. The time between welding each bead was 120 s. Specimens were welded by different techniques. In sample S1, the beads were laid one after the other from bead 1 to bead 5. In sample S2, the beads were not laid in order. These 2 techniques were chosen because they are representative of welding techniques. In S1, a bead was tempered only by the bead welded later. In S2, beads 1 and 2 were tempered by 2 beads. The use of other techniques, such as 14253, would obtain only one bead (bead 2), which will be tempered by 2 others. That may not have allowed preparation of the planned experiments. The schemas of welding techniques for S1 and S2 specimens are presented in Figure 2. The welding parameters were chosen according to preliminary tests. They provided stability of the welding arc in the experimental conditions and allowed heat input values higher than 0.90 kJ/mm. From the point of view of weldability in underwater conditions, these values may reduce the susceptibility to cold cracking. It has been assumed that the heat input of the tempering bead should be higher than that of the tempered bead. The parameters for both specimens are shown in Table 3. In the case of manual underwater wet welding, the ability to control the value of heat input is very limited. Hence, differences between heat input values of subsequent beads are not significant (0.24 and 0.39 kJ/mm for specimens S1 and S2, respectively) from the point of view of metallurgical transformations.

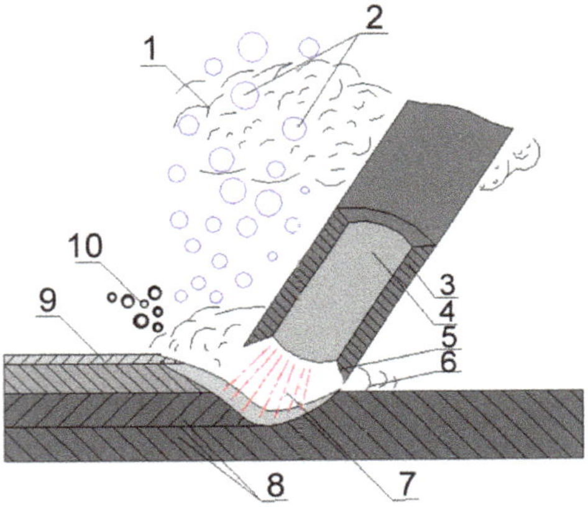

Figure 1. Wet welding by covered electrodes schema. 1: water vapor, 2: gas bubbles, 3: flux coating, 4: wire, 5: melting flux coating, 6: liquid metal, 7: electric arc, 8: base materials, 9: slag, 10: gas from melting flux coating.

Table 3. Welding parameters for specimens S1 and S2.

Bead No.	I (A)	U (V)	Vsp (mm/s)	Ql (kJ/mm)
		Specimen S1		
1	220	21.8	5.95	1.03
2	216	29.5	6.17	1.03
3	224	26.8	5.88	1.02
4	220	29.0	5.38	1.19
5	220	28.8	4.97	1.27
		Specimen S2		
1	220	28.8	6.33	1.00
2	216	30.3	5.85	1.12
3	220	28.8	5.88	1.05
4	216	30.0	5.41	1.20
5	216	29.0	4.58	1.39

(a) (b)

Figure 2. Scheme of welding process in (**a**) specimen S1 and (**b**) specimen S2.

2.3. Examination Procedure

After welding, specimens were cut to cross-sections, and macroscopic and microscopic metallographic examinations were conducted in accordance with the EN ISO 17637:2011 [42] standard, with the use of 4% Nital. Finally, Vickers HV10 hardness measurements were taken in accordance with EN ISO 9015:2011 [43], by using a Sinowon V-10 stand (Sinowon, Dongguan, China), with a measurement error of ± 3 HV10. The investigated S355G10 + N steel is classified as material group 2.1 in accordance with EN ISO 15614-1:2017 [44]. The maximum hardness values of HAZ recommended by this standard cannot exceed 380 HV10.

Macroscopic tests were prepared for cross-sections of both specimens. Microscopic tests were prepared for the base material in welds and in specific areas of the samples. The first areas were located in the HAZ in the axis of each weld. The second areas were located where the HAZ from the previous weld was overlaid by the HAZ from welds deposited later. Schematic views of the areas of microscopic testing are presented in Figure 3. HV10 hardness measurements were prepared for cross-sections of both specimens in the BM, each weld, and the HAZ of each weld in their axis according to the schemas presented in Figure 4.

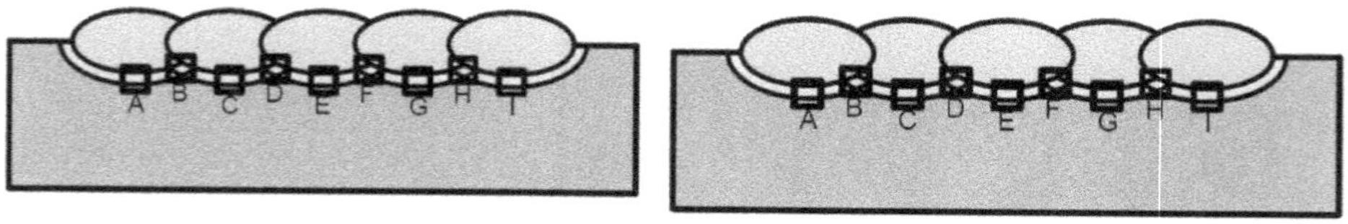

(a) (b)

Figure 3. Schematic view of areas of microscopic testing: (**a**) specimen S1, (**b**) specimen S2.

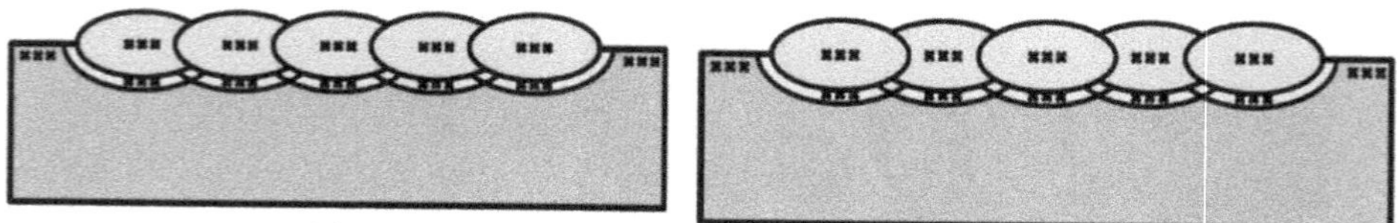

(a) (b)

Figure 4. Hardness distribution points: (**a**) specimen S1, (**b**) specimen S2.

3. Results and Discussion

3.1. Macroscopic Testing

Macroscopic testing showed that both specimens were welded according to the assumed research plan (Figure 2). No imperfections were found in cross-sections of specimens S1 and S2. The characteristic cross-sectional sizes of all beads had similar dimensions, resulting from small differences in the values of heat input (Table 3). A factor that complicates predicting the effects of multi-pass welding is dilution, which affects, among other aspects, hardenability. Dilution values were calculated with the methodology presented in the literature [45,46]. The cross-sectional area of the fusion zone for each pass was used, i.e., D = A_{base}/A, where A is the total molten area and A_{base} is the molten area of the BM [45]. The calculations showed that dilution in S2 is lower than in S1, which may result in the structures and hardness for both specimens [47]. The results of macroscopic testing are presented in Figure 5. The dilution in S1 and S2 is presented in Table 4.

(**a**)

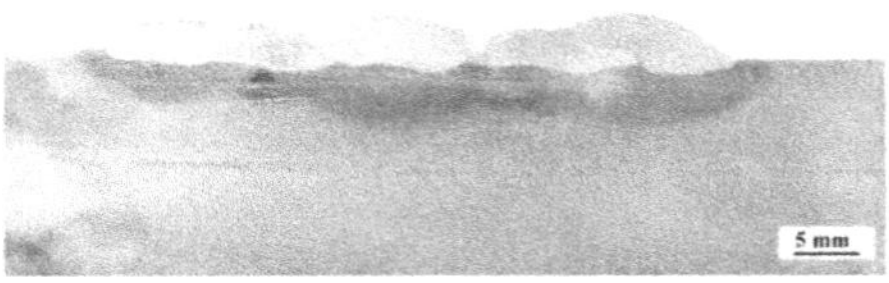

(**b**)

Figure 5. Results of macroscopic testing: (**a**) specimen S1, (**b**) specimen S2.

Table 4. The dilution in the S1 and S2.

S1	D (%)	S2	D (%)
Weld1	0.54	Weld3	0.32
Weld2	0.37	Weld1	0.45
Weld3	0.40	Weld4	0.28
Weld4	0.59	Weld2	0.39
Weld5	0.44	Weld5	0.26

3.2. Microscopic Testing

The results of microscopic testing are presented in Figure 5. Microscopic testing of base material showed that the investigated S355G10 + N steel consisted of fine-grained pearlite and fine-grained ferrite with layers (Figure 6a). The beads consisted of bright fine-grained ferrite arranged in columns, from which grew acicular ferrite at the boundaries of dendrites. Inside them were fine ferrite grains (Figure 6b). These structures are typical for joints made in underwater welding of HSLA steel. In the HAZ of each weld in S1 (regions A, C, E, and G), brittle structures such as bainitic and martensitic were found (Figure 6c). The structures in the HAZ in S2 were the same in areas A, E, and I. In the HAZ of welds 1 and 2 (areas C and G), the structure indicated the presence of refined and tempered low-carbon martensite mixed with normalization structures with fine ferrite and pearlite (Figure 6d), which is a result of the heat influence from welds laid later (Figure 2b). These types of structures have lower susceptibility to cold cracking. The areas where the HAZ of one bead overlapping the HAZ of the bead laid earlier (areas B, D, F, and H) in both specimens are characterized by normalization structures with fine ferrite and pearlite (Figure 6e). While formation and decomposition of austenite occurred twice in regions B, D, F, and H, it happened once in areas A, C, E, and G, followed by in-process tempering with the heat of the next bead. Cracks were found in each HAZ of S1 (Figure 6c), which resulted in brittle structures in this area. In S2, cracks were found only in the HAZ of welds 3, 4, and 5 (Figure 6f) and in area H (Figure 6e). This cracking could have occurred before weld number 5 was laid, which tempered the HAZ of weld 2. Tempering from welds laid later could not repair the microcracks that occurred during the welding of the previous bead and could even propagate these cracks, which is similar to the effect of the temper bead welding technique [12,13]. The location and the length of the cracks in both specimens are summarized in Table 5. The longest cracks were observed in HAZ of welds prepared with the lowest values of heat input.

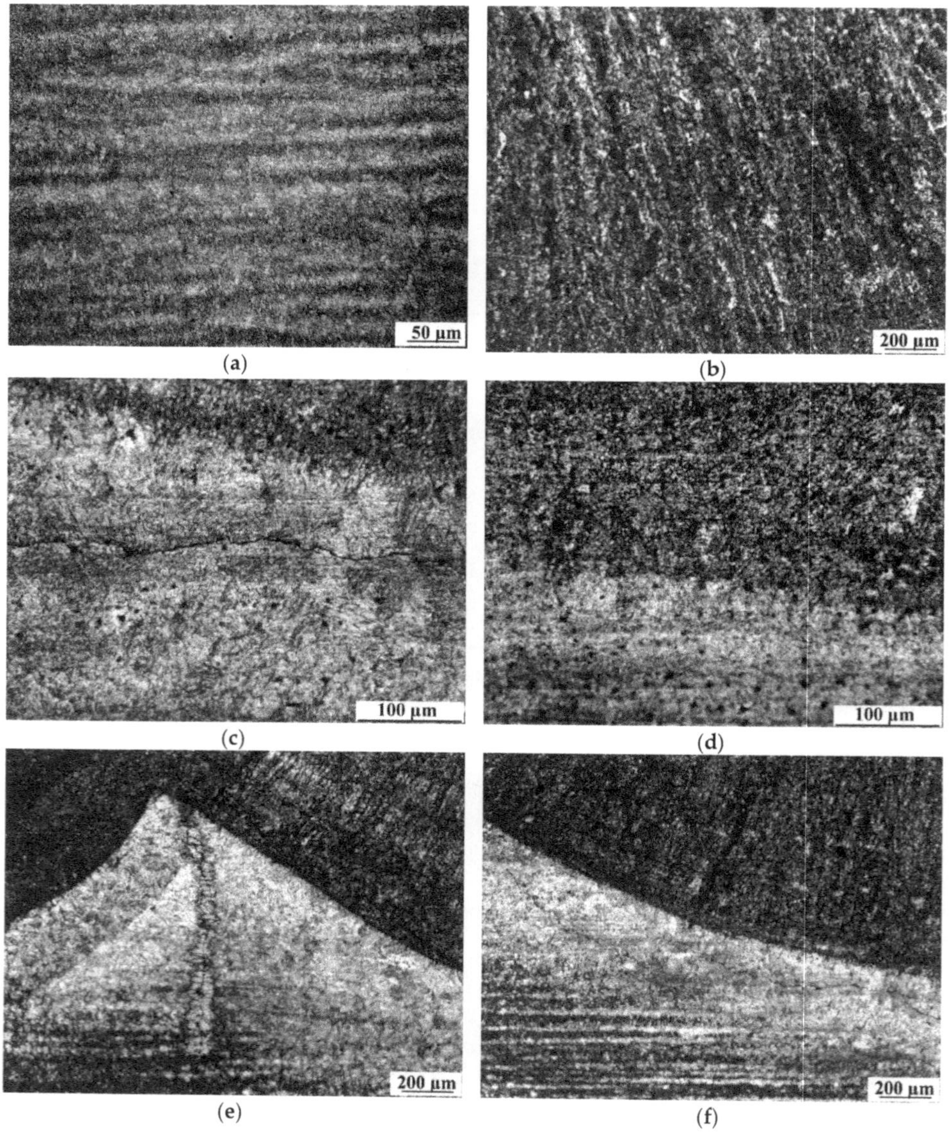

Figure 6. Results of microscopic testing: (**a**) base material, (**b**) weld, (**c**) heat-affected zone (HAZ) in S1, (**d**) HAZ of weld number 1 (area C) in S2, (**e**) overlapping of two HAZs (area H) in S2, (**f**) HAZ of weld 3 (area A) in S2. Etch: 4% Nital.

Table 5. The length of cracks in specimens S1 and S2.

	Area (Figure 3)								
Specimen	A	B	C	D	E	F	G	H	I
S1	550 µm	none	350 µm	100 µm	350 µm	none	250 µm	none	200 µm
S2	350 µm	none	none	none	200 µm	none	none	600 µm	150 µm

3.3. Hardness Measurements

The investigated S355G10 + N steel is classified as material group 2.1 by EN ISO 9015-1:2011. In accordance with the EN-ISO 15614:1 standard, the maximum hardness of the HAZ cannot exceed 380 HV10 for material group 2.1. The hardness in welds for both specimens was lower than that in the HAZ, which is typical for welding HSLA steel in the water environment [5]. Measurements in the HAZ showed significant differences in hardness distribution in the two tested specimens. The HV10 values in the HAZ were higher in S1 than in S2, which is connected with the different dilution values in both specimens. It is stated in the literature [45], that the higher dilution leads, for example, to more martensite in the welded joint, which is a brittle structure. In S1, most of the measured values did not fulfill the criterion of EN ISO 15614-1 standard of 380 HV10. The welding technique used in S2 allows reduced hardness in the HAZ, especially for welds that have been tempered by heat from welds laid later. The reduction of hardness by welding with the technique used in S2 can reduce the probability of cold cracks occurring in the HAZ. The differences of HAZ hardness are associated with the transformation cycle and tempering effect, which is characteristic for multi-pass welds [48].

The hardness distribution in each point is presented in Figure 7. The results shown in this figure are arranged in order of bead appearance (from left to right) in specimens, e.g., "second weld" means this is the second weld from the left. The average results of hardness measurements in each area are presented in Table 6.

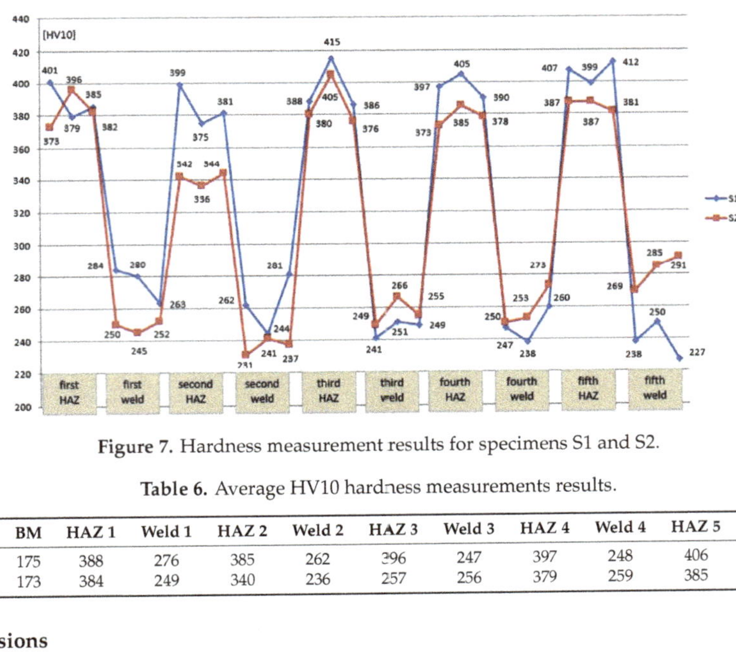

Figure 7. Hardness measurement results for specimens S1 and S2.

Table 6. Average HV10 hardness measurements results.

Specimen	BM	HAZ 1	Weld 1	HAZ 2	Weld 2	HAZ 3	Weld 3	HAZ 4	Weld 4	HAZ 5	Weld 5	BM
S1	175	388	276	385	262	396	247	397	248	406	238	188
S2	173	384	249	340	236	257	256	379	259	385	282	176

4. Conclusions

The results of the prepared experiments show that the technique used has a significant role in the process of repairs in underwater conditions. The technique used for preparation of specimen S1 can cause a need for further repairs, because it allows cold cracks to form.

The prepared experiments allow us to draw the following conclusions:

1. Water as a welding environment causes cold cracks to occur in the HAZ of wet welded HSLA steel. The susceptibility of steel to cold cracking can be reduced by changing the sequence of the beads during the process.

2. The technique used for the preparation of specimen S2 allowed the structure in the HAZ of some welds to be changed. In this area, the structure indicated the presence of refined and tempered low-carbon martensite mixed with normalization structures with fine ferrite and pearlite. In each HAZ in specimen S1, brittle structures such as bainitic and martensitic were found.
3. Welding S355G10 + N steel with the sequence used in specimen S2 (out of order) allowed a reduction of hardness in the HAZ below the critical value of 380 HV10, as stated by the EN-ISO 15614-1 standard. The lowest hardness values were measured in the HAZ in specimen S2, where the heat from beads laid later tempered this area, and in which the dilution is lower than in specimen S1.
4. The technique used in welding specimen S2 should be used during repair procedures that have to be prepared in the water environment.

Author Contributions: Conceptualization, J.T. and G.R.; Methodology, J.T., D.F. and G.R.; Formal Analysis, J.T. and D.F.; Investigation, J.T. and G.R.; Writing—Original Draft Preparation, J.T., D.F. and G.R.; Writing—Review and Editing, J.T. and D.F.

Funding: This research received no external funding.

Conflicts of Interest: The authors declare no conflict of interest.

References

1. Zhang, Y.; Jia, C.; Wang, J.; Zhao, B.; Wu, C. Investigation on the bubble dynamic behaviors and corresponding regulation method in underwater flux-cored arc welding. *Proc. Inst. Mech. Eng. B* **2019**, *223*, 1808–1817. [CrossRef]
2. Sun, K.; Zeng, M.; Shi, Y.; Hu, Y.; Shen, X. Microstructure and corrosion behavior of S32101 stainless steel underwater dry and wet welded joints. *J. Mater. Process. Technol.* **2018**, *256*, 190–201. [CrossRef]
3. Wang, J.; Sun, Q.; Teng, J.; Feng, J. Bubble evolution in ultrasonic wave-assisted underwater wet FCAW. *Weld. J.* **2019**, *98*, 150–163.
4. Guo, N.; Du, Y.; Maksimov, S.; Feng, J.; Yin, Z.; Krazhanovskyi, D.; Fu, Y. Study of metal transfer control in underwater wet FCAW using pulsed wire feed method. *Weld. World* **2018**, *62*, 87–94. [CrossRef]
5. Tomków, J.; Łabanowski, J.; Fydrych, D.; Rogalski, G. Cold cracking of S460N steel in water environment. *Pol. Marit. Res.* **2018**, *25*, 131–136. [CrossRef]
6. Gao, W.B.; Wang, D.; Cheng, F.; Deng, C.; Xu, W. Underwater wet welding for HSLA steels: Chemical composition, defects, microstructure, and mechanical properties. *Acta Metall. Sin. Engl.* **2015**, *28*, 1097–1108. [CrossRef]
7. Fu, Y.; Guo, N.; Du, Y.; Chen, H.; Xu, C.; Feng, J. Effect of metal transfer mode on spatter and arc stability in underwater flux-cored wire wet welding. *J. Manuf. Process.* **2018**, *35*, 161–168. [CrossRef]
8. Wang, J.; Sun, Q.; Ma, J.; Teng, J.; Jin, P.; Feng, J. Investigation of acoustic radiator affecting bubble-acoustic interaction in ultrasonic wave-assisted UWW at shallow water. *J. Manuf. Process.* **2018**, *37*, 563–577. [CrossRef]
9. Li, H.; Liu, D.; Ma, Q.; Guo, N.; Song, X.; Feng, J. Microstructure and mechanical properties of dissimilar welds between 16Mn and 304L in underwater wet welding. *Sci. Technol. Weld. Join.* **2018**, *24*, 1–7. [CrossRef]
10. Tomków, J.; Fydrych, D.; Rogalski, G.; Łabanowski, J. Effect of the welding environment and storage time of electrodes on the diffusible hydrogen content in deposited metal. *Rev. Metall.* **2019**, *55*, e140. [CrossRef]
11. Świerczyńska, A.; Fydrych, D.; Rogalski, G. Diffusible hydrogen management in underwater wet self-shielded flux cored arc welding. *Int. J. Hydrogen Energy* **2017**, *42*, 24532–24540. [CrossRef]
12. Tomków, J.; Rogalski, G.; Fydrych, D.; Łabanowski, J. Improvement of S355G10+N steel -weldability in water environment by Temper Bead Welding. *J. Mater. Process. Technol.* **2018**, *262*, 372–381. [CrossRef]
13. Tomków, J.; Rogalski, G.; Fydrych, D.; Łabanowski, J. Advantages of the application of temper bead welding technique during wet welding. *Materials* **2019**, *12*, 915. [CrossRef] [PubMed]
14. Jia, C.; Zhang, Y.; Wu, J.; Xing, C.; Zhao, B.; Wu, C. Comprehensive analysis of spatter loss in wet FCAW considering interactions of bubbles, droplets and arc–Part 1: Measurement and improvement. *J. Manuf. Process.* **2019**, *40*, 122–127. [CrossRef]

15. Wang, J.; Sun, Q.; Zhang, S.; Wang, C.; Wu, L.; Feng, F. Characterization of the underwater welding arc bubble through a visual sensing method. *J. Mater. Process. Technol.* **2018**, *251*, 95–108. [CrossRef]
16. Chen, H.; Guo, N.; Huang, L.; Zhang, X.; Feng, J.; Wang, G. Effects of arc bubble behaviors and characteristics on droplet transfer in underwater wet welding using in-situ imaging method. *Mater. Des.* **2019**, *170*, 107696. [CrossRef]
17. Wang, J.; Sun, Q.; Pan, Z.; Yang, J.; Feng, J. Effect of welding speed on bubble dynamics and process stability in mechanical constraint-assisted underwater wet welding of steel sheets. *J. Mater. Process. Technol.* **2019**, *264*, 389–401. [CrossRef]
18. Silva, L.F.; Santos, V.R.D.; Paciornik, S.; Rizzo, F.A.; Monteiro, M.J.; Bracarense, A.Q.; Pessoa, E.C.; Vieira, L.A.; Marinho, R.R. Influence of molybdenum in metal weld properties in welding wet with oxy-rutillic electrodes. *Soldag. Inspeção* **2013**, *18*, 102–109. [CrossRef]
19. Winarto, W.; Purnama, D.; Churniawan, I. The effect of different rutile electrodes on mechanical properties of underwater wet welded AH-36 steel plates. *AIP Conf. Proc.* **2018**, *1945*, 020048.
20. Muktepavel, V.; Murzin, V.; Karpov, V.; Kurakin, A. Research on welding and processing behavior of electrodes and features of their application in "wet" underwater arc welding. *Mater. Sci. Forum* **2019**, *946*, 913–920. [CrossRef]
21. Guo, N.; Yang, Z.; Wang, M.; Yuan, X.; Feng, J. Microstructure and mechanical properties of an underwater wet welded dissimilar ferritic/austenitic steel joint. *Strength Mater.* **2015**, *47*, 12–18. [CrossRef]
22. Zhang, H.T.; Dai, X.Y.; Feng, J.C.; Hu, L.L. Preliminary investigation on real-time induction heating-assisted underwater wet welding. *Weld. J.* **2015**, *94*, 8–15.
23. Gao, W.; Wang, D.; Cheng, F.; Di, X.; Deng, C.; Xu, W. Microstructural and mechanical performance of underwater wet welded S355 steel. *J. Mater. Process. Technol.* **2016**, *238*, 333–340. [CrossRef]
24. Yan, Y.; Liu, C.; Wang, C.; Shen, J. Mechanical properties and stress variations in multipass welded joint of low-alloy high-strength steel after layer-by-layer ultrasonic impact treatment. *J. Mater. Eng. Perform.* **2019**, *28*, 2726–2735. [CrossRef]
25. Vargas-Arista, B.; Mendoza-Camargo, O.; Zaragoza-Rivera, I.P.; Medina-Flores, A.; Cuevas-Salgado, A.; Garfias-García, E.; García-Vázquez, F. Influence of heat input on the Charpy ductile fracture behavior of reheated HAZ in GMAW multilayer welded joints on HSLA steel using digital fractographic analysis. *Rev. Metal.* **2019**, *55*, e143. [CrossRef]
26. Winczek, J. Modeling of temperature field during multi-pass GMAW surfacing or rebuilding of steel elements taking into account the heat of the deposit metal. *Appl. Sci.* **2018**, *7*, 6. [CrossRef]
27. Pańcikiewicz, K. Structure and properties of welded joints of 7CrMoVTiB10-10 (T24) steel. *Adv. Mater. Sci.* **2018**, *18*, 37–47. [CrossRef]
28. Yu, J.H.; Choi, Y.S.; Shim, D.S.; Park, S.H. Repairing casting part using laser assisted additive metal-layer deposition and its mechanical properties. *Opt. Laser Technol.* **2018**, *106*, 87–93. [CrossRef]
29. Pandey, C.; Mahapatra, M.; Kumar, P.; Saini, N.; Thakre, J.G.; Vidyarthy, R.S.; Narang, H.K. A brief study on δ-ferrite evolution in dissimilar P91 and P92 steel weld joint and their effect on mechanical properties. *Arch. Civ. Mech. Eng.* **2018**, *18*, 713–722. [CrossRef]
30. Tuz, L. Evaluation of microstructure and selected mechanical properties of laser beam welded S690QL high-strength steel. *Adv. Mater. Sci.* **2018**, *18*, 34–42. [CrossRef]
31. Kurc-Lisiecka, A.; Piwnik, J.; Lisiecki, A. Laser welding of new grade of advanced high strength steel STRENX 1100 MC. *Arch. Metall. Mater.* **2017**, *62*, 1651–1657. [CrossRef]
32. Kik, T.; Górka, J. Numerical simulations of laser and hybrid S700MC T-joint welding. *Materials* **2019**, *12*, 516. [CrossRef]
33. Skowrońska, B.; Chmielewski, T.; Pachla, W.; Kulczyk, M.; Skiba, J.; Presz, W. Friction weldability of UFG 316L stainless steel. *Arch. Metall. Mater.* **2019**, *64*, 1051–1058.
34. Konovalov, S.V.; Kormyashev, V.E.; Gromov, V.E.; Ivanov, Y.F.; Kapralov, E.V.; Semin, A.P. Formation features of structure-phase states of Cr-Nb-C-V containing coatings on martensitic steel. *J. Surf. Investig.* **2016**, *10*, 1119–1124. [CrossRef]
35. Chen, X.; Su, C.; Wang, Y.; Siddiquee, A.N.; Konovalov, S.; Jayalakshami, S.; Singh, R.A. Cold metal transfer (CMT) based wire and arc additive manufacture (WAAM) system. *J. Surf. Investig.* **2018**, *12*, 1278–1284. [CrossRef]

36. Dehghani, A.; Aslani, F. A review on defects in steel offshore structures and developed strengthening techniques. *Structures* **2019**, *20*, 635–657. [CrossRef]
37. Meng, X.; Chen, G.; Zhu, G.; Zhu, Y. Dynamic quantitative risk assessment of accidents inducted by leakage on offshore platforms using DEMATEL-BN. *Int. J. Nav. Archit. Ocean Eng.* **2019**, *11*, 22–32. [CrossRef]
38. Wodtke, M.; Olszewski, A.; Wójcikowski, A. FEM calculations in analysis of steel subsea water injection flowlines designing process. *Pol. Marit. Res.* **2018**, *25*, 84–93. [CrossRef]
39. Kolios, A.; Wang, L.; Mehmanparast, A.; Brennan, F. Determination of stress concentration factors in offshore wind welded structures through a hybrid experimental and numerical approach. *Ocean Eng.* **2019**, *178*, 38–47. [CrossRef]
40. Oh, K.Y.; Nam, W.; Ryu, M.S.; Kim, J.Y.; Epureanu, B.I. A review of foundations of offshore wind energy convertors: Current status and future perspectives. *Renew. Sustain. Energy Rev.* **2018**, *88*, 16–36. [CrossRef]
41. *ISO 2560—A Classification of Coated Rod Electrodes for Arc Welding of Unalloyed Steel and Fine-Grained Steel*; ISO: Geneva, Switzerland, 1908.
42. *EN ISO 17637:2017 Non-Destructive Testing of Welds—Visual Testing of Fusion-Welded Joints*; ISO: Geneva, Switzerland, 2017.
43. *EN ISO 9015-1:2011 Destructive Tests on Welds in Metallic Materials. Hardness Testing. Hardness Test on Arc Welded Joints*; ISO: Geneva, Switzerland, 2011.
44. *EN ISO 15614-1:2017 Specification and Qualification of Welding Procedures for Metallic Materials—Welding Procedure Test—Part 1: Arc and Gas Welding of Steels and Arc Welding of Nickel and Nickel Alloys*; ISO: Geneva, Switzerland, 2017.
45. Sun, Y.L.; Obasi, G.; Hamelin, C.J.; Vasileiou, A.N.; Flint, T.F.; Blakrishna, J.; Smith, M.C.; Francis, J.A. Effects of dilution on alloy content and microstructure in multi-pass steel welds. *J. Mater. Process. Technol.* **2019**, *265*, 71–86. [CrossRef]
46. Sun, Y.L.; Hamelin, C.J.; Flint, T.F.; Vasileiou, A.N.; Francis, A.; Smith, M.C. Prediction of dilution and its impact on the metallurgical and mechanical behavior of a multipass steel weldment. *J. Press. Vessel Technol.* **2019**, *141*, 061405. [CrossRef]
47. Saida, K.; Bunda, K.; Ogiwara, H.; Nishimoto, K. Microcracking susceptibility in dissimilar multipass welds of Ni-base alloy 690 and low-alloy steel. *Weld. Int.* **2015**, *29*, 668–680. [CrossRef]
48. Sun, Y.L.; Obasi, G.; Hamelin, C.J.; Vasileiou, A.N.; Flint, T.F.; Francis, J.A.; Smith, M.C. Characterization and modelling of tempering during multi-pass welding. *J. Mater. Process. Technol.* **2019**, *270*, 118–131. [CrossRef]

materials MDPI

Article

Flame Spraying of Aluminum Coatings Reinforced with Particles of Carbonaceous Materials as an Alternative for Laser Cladding Technologies

Artur Czupryński

Welding Department, Faculty of Mechanical Engineering, Silesian University of Technology, ul. Konarskiego 18A, 44-100 Gliwice, Poland; Artur.Czuprynski@polsl.pl

Received: 20 September 2019; Accepted: 20 October 2019; Published: 23 October 2019

Abstract: The article presents results of the preliminary research of mechanical properties of flame-sprayed aluminum coatings reinforced with carbon materials made on the construction steel S235J0 substrate. For reinforcement the following carbon materials were used: carbon nanotubes Nanocyl NC 7000 (0.5 wt.% and 1 wt.%) and carburite (0.5 wt.%). The properties evaluation was made using metallographic macroscope and microscope, chemical composition, microhardness, abrasion and erosion resistance studies. The obtained results were compared with aluminum powder coatings (EN AW 1000 series). It was proved that the flame spraying of aluminum coatings reinforced with particles of carbonaceous materials can be an effective alternative for laser cladding technology. The preliminary test results will be successively extended by further experiments to contribute in the near future to develop innovative technologies, that can be implemented in the automotive industry for production of components with high strength, wear resistance, good thermal conductivity and low density, such as brake shoes, cylinder liners, piston rings and gears.

Keywords: flame powder spray process; coating; aluminum; carbon nanotubes; carburite; abrasive wear resistance; erosion wear resistance

---

1. Introduction

Modern civilization expects from material engineering scientists to produce lightweight and durable materials that meet the high strength and quality requirements set for innovative constructions made by the automotive and aerospace industries. Under certain structural load conditions, the increase of strength and stiffness of the materials contributes to reducing construction dimensions and consequently also the mass. Because the global oil resources are constantly declining, and renewable energy sources are not effective enough yet, use of lightweight and durable materials becomes a necessity. This type of materials is highly desirable in car, aircraft and space vehicle production because its use has many benefits, such as lower fuel consumption, higher capacity and speed. Insufficient strength and stiffness of the constructions made of metals and alloys led to the development of metal-matrix composites (MMC). The composites of this type achieve high strength and ductility thanks to the metallic matrix, while the stiffness is provided by the reinforcement, which consists of particles-perchance fibers-metallic or ceramic with high stiffness. The microstructure of this materials consists of soft matrix and hard phases which provides increase in abrasion resistance also at high temperatures. Metal-matrix composites can be designed to have specific properties, such as low thermal expansion coefficient and high thermal conductivity so that these materials are suitable for use in applications for installation of electronic microcircuits. Metal-matrix composite materials are widely used in car and air applications nowadays [1–3].

In the 1970s, technologies for producing high-strength carbon fibers were developed. They began to be used for the preparation of advanced composites used for producing rocket engine nozzles, projectile cores, thermal shields, isolators and thermal radiators. Since 1970, carbon fibre reinforced composites have been widely used in the production of aircraft brakes, space constructions, military and commercial airplanes, lithium-ion batteries and sports equipment. Research in the field of carbon materials has been revolutionized by the discovery of carbon nanotubes (CNT) by Sumio Iijima in 1991. The carbon nanotubes (CNTs) have unique mechanical properties compared to carbon fibers, e.g. stiffness to 1000 GPa, strength of 100 GPa and thermal conductivity to 6000 W/(m·K) [4,5]. In recent years, a number of studies have been carried out using CNT carbon nanotubes as reinforcement of various materials: polymers, ceramics and metals, with the majority of research involving polymer composites [6,7], ceramic composites in second place [8,9], and only recently have been published several papers on composites with metallic matrix reinforced with carbon nanotubes (CNT) [10,11]. This is quite surprising considering the fact that most construction materials used in the contemporary world are metals. Publications on this topic concern various aspects such as fabrication [12–15], microstructure [16,17], modelling of mechanical properties and the chemical interaction between carbon nanotubes (CNTs) and metals [4,18–21]

Nanotechnology had a strong influence on the direction of research in the field of surface engineering and related production technology of surface layers and coatings [22,23]. Nowadays, it is possible to use welding methods for producing not only conventional tribological coatings with specific frictional characteristics (high or low coefficient of friction) and resistance to wear, erosion or corrosion but also for producing coatings with unprecedented properties, often intended for special applications and working in difficult conditions, e.g., nanocomposite coatings with high hardness and high resistance to dynamic loads, coatings with frictional characteristics that adapt to changing operating conditions (temperature, humidity), thermal barrier coatings or biocompatible coatings [24,25]. Often, high-quality nanostructure coatings are used on parts of car engines made of aluminum alloys, on copper alloys intended for propellers of vessels, or on heat-resistant intermetals. In surface engineering technology, the implementation of this type of coating is possible by thermal spraying, where the applied metallic layer is bonded to the substrate adhesively or mechanically without melting the base material [24]. The main advantage of thermal spraying technology is a minimal thermal influence on the sprayed materials. Even in the case of laser cladding technologies characterized by the lowest heat input of all the cladding technologies, the substrate material is always partially melted, as well as the additional material, usually in a form of metallic or composite powder. The carbon nanotubes, due to the small dimensions, have very low heat capacity. Additionally, they have high absorption of laser radiation. For this reason, the introduction of carbon nanotubes into the melt pool during laser cladding is basically impossible, because overheating and decomposition of nanotubes [26–33].

Pioneers in the field of thermal spraying processes for composite coatings of aluminum-carbon nanotubes (CNT) were a research group from Florida International University, who successfully deposited carbon nanotubes in the Al-Si matrix in the powder plasma spraying process [34].

S. R. Bakshi and others [10] made multi-layer nanocomposite coatings of aluminum-carbon nanotubes (CNT) in the cold gas spraying process. In order to obtain a good dispersion of carbon nanotubes in Al-Si microparticle eutectic powders, spray drying was used. Spray-dried powders containing 5 wt.% carbon nanotubes (CNTs) were mixed with pure aluminum powder to obtain total nominal carbon nanotube (CNT) compositions in the coating material of 0.5 wt.% and 1 wt.%. As a result of cold spraying, coatings with a thickness of 500 μm were obtained in which the carbon nanotubes were evenly distributed in the matrix. The carbon nanotubes were of shorter length because during the deposition process they fractured as a result of impact and shear between the Al-Si particles and the aluminum matrix.

A. K. Keshri and others [11] compared impact on carbon nanotubes (CNTs) of various heat sources used during thermal spray processes—plasma spraying (PS), high-velocity oxy fuel spraying (HVOF), cold spraying (CS) and plasma spraying of liquid precursor (PSLP). Carbon nanotubes (CNTs) have been successfully preserved as reinforcements in composite metal and ceramic coatings in all thermal spray processes with the exception of PSLP.

There is no data in the literature regarding tribological properties of powder flame-sprayed (PFS) aluminum coatings reinforced with carbon nanotubes (CNT). The purpose of this article is to present the state of knowledge in this area of research and present the possibility of using powder flame spray technology (PFS) for the production of composite coatings with a metallic matrix reinforced with carbon nanotubes (CNT).

2. Materials and Methods

2.1. Aim of Study

The conducted studies were aimed at comparing the structure, chemical composition, hardness and resistance to abrasive and erosive wear of powder aluminum flame-sprayed coatings reinforced with Nanocyl NC 7000 carbon nanotubes in amount of 0.5 wt.%, 1 wt.% and carburite in an amount of 0.5 wt.% with a reference coating made of aluminum powder EN AW 1000 series (Metallisation Ltd., West Midlands, UK) on non-alloy S235J0 steel. Carburite as aluminum matrix reinforcement was used in order to compare tribological properties of this composite coating with coating reinforced with CNTs with equal weight participation of carbon material. The scope of research included:

- Preparation of material for spraying;
- Selection process parameters for each of the coating based on preliminary technological tests;
- Coating manufacturing;
- Examining the structure and tribological properties of aluminum coatings reinforced with carbon nanotubes and carburite;
- Comparison of obtained samples with coatings made of aluminum powder without the addition of carbonaceous materials.

2.2. Materials, Devices and Spraying Parameters

The additional material for flame-spraying was obtained by mixing in the ball mill appropriate proportions of aluminum powder (EN AW 1000 series) with carbon nanotubes and aluminum powder in the form of filter dust carburite (Table 1). Carbon nanotubes that were used in the test are commercially available multi-walled carbon nanotubes MWCNTs, produced in the Catalytic Chemical Vapor Deposition (CCVD) process, NANOCYLTM NC7000 (Belgium Nanocyl SA, Sambreville, Belgium) (Table 2).

Table 1. Specification of aluminum powder EN AW 1000 series.

Specification	Unit	Guaranteed Parameters	Obtained Results
Aluminum content (Al)	%	min 99.7	99.7
Iron content (Fe)	%	max 0.2	0.2
Silicon content (Si)	%	max 0.12	0.12
Copper content (Cu)	%	max 0.004	0.004
Moisture	%	max 0.1	0.1
Bulk density	g/dm^3	min 1000	1050
Granulation above 0.045 mm	%	85.0–100.0	98.0
Granulation above 0.1 mm	%	5.0–30.0	15.3
Granulation above 0.16 mm	%	max 5.0	0.0

Table 2. Structure and specification of Nanocyl NC 7000 carbon nanotubes and of filter dust carburite.

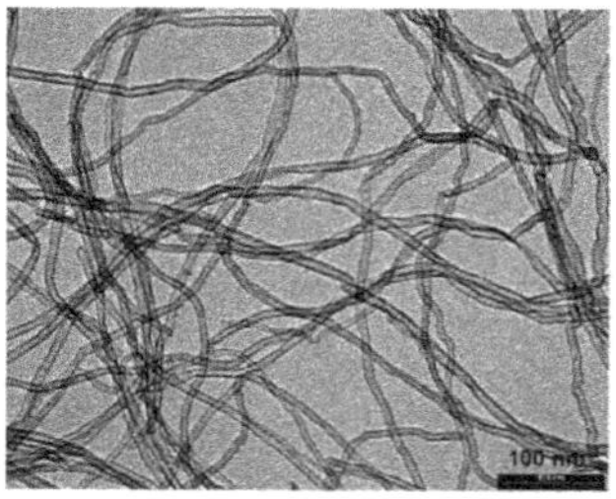

Structure of Nanocyl NC 7000 carbon nanotubes

Properties	Unit	Value	Measurement Method
Average diameter	Nanometer	9.5	TEM
Average length	Micrometer	1.5	TEM
Carbon purity	%	90	TGA
Metal oxide	%	10	TGA
Amorphous carbon	-	1)	HRTEM
Surface area	m^2/g	250–300	BET

Note: 1) pyrolytically deposited carbon on the surface of the NC 7000.

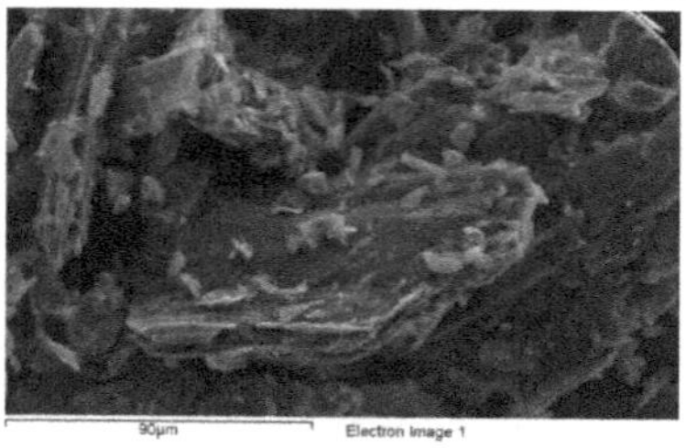

Structure of filter dust carburite

Properties	Unit	Value
Fraction size	Millimeter	>0 up to 1
Dust content	%	5
Moisture content	%	1
Carburite content	%	94
Granulation [mm]	%	>1 mm up to 10%; <0.06 mm up to 70%

The subsonic flame spraying process was carried out cold in accordance with the standard EN 13507:2018 [35] on workstation, equipped with hand-guided modern oxyacetylene system (CastoDyn DS 8000 (Messer Eutectic Castolin, Gliwice, Poland). Final surface preparation was done by shot blasting sheets prior to spraying with sharp-edged cast iron of 0.5–1.5 mm shot grain size in accordance with standard ISO 2063-1:2017 [36]. Final surface roughness of the steel substrate after shot blasting was R_a = 12 μm, R_z = 85 μm. Before the spraying process, steel plates with dimensions of 150 × 150 × 5 mm^3 were preheated with a gas burner up to a temperature of 40 °C (the temperature of preheating was measured using pyrometer). The standard modular nozzles regulating the flame outlet SSM 40 (Messer Eutectic Castolin, Gliwice, Poland) and the neutral flame (ratio O_2/C_2H_2 = 1,2) were used. This allowed to obtain the proper spray jet [37,38]. The flame jet burner was guided in a horizontal position covering the whole surface of the sheet. During the process the spraying direction was changed several times by 90°, until obtained thickness of coating was about 1,0 mm. Distance between the torch and the sprayed surface was 200 mm. The parameters and flame type were constant for each test (Figure 1).

The criterion for visual assessment of the powder coatings' quality, was to make the surface layers characterized by the appropriate thickness, good adhesion to the substrate, low porosity, continuity and uniformity of obtained coatings [39]. Optimal parameters of flame-spraying of aluminum, aluminum with carbon nanotube reinforcement and aluminum with filter dust carburite reinforcement coatings have been determined on the basis of preliminary technological tests (Table 3). The view of representative samples with flame-sprayed coatings on the aluminum matrix are shown in Figure 2.

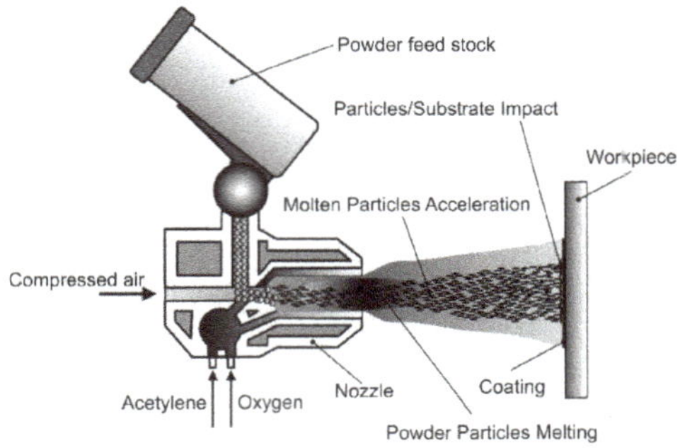

(a)

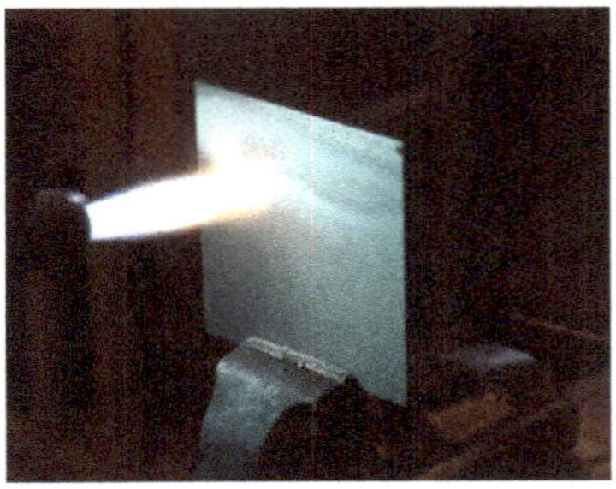

(b)

Figure 1. Flame-spraying process: (**a**) a diagram of handheld flame jet burner; (**b**) a photo from trials of flame-spraying aluminum coatings with CastoDyn DS 8000 burner.

Table 3. Parameters of flame-sprayed aluminum and aluminum coatings reinforced with carbon materials using CastoDyn DS 8000 torch.

Sample Number	Type of Powder	Oxygen Pressure [bar]	Acetylene Pressure [bar]	Assist. Gas (Compressed Air) Pressure [bar]	Number of the Orifice for the Powder	Mass of Used Powder [g]	Powder Yield [%]
1	Al	4	0.7	3	2	93.5	60.3
2	Al + 0.5% CNT [1)]	4	0.7	3	2	97.0	56.2
3	Al + 1% CNT	4	0.7	3	2	100.8	57.0
4	Al + 0.5% C [2)]	4	0.7	3	2	99.7	59.4

Note: [1)] CNT—carbon nanotubes Nanocyl NC 7000 wt.%; [2)] C—carburite wt.%.

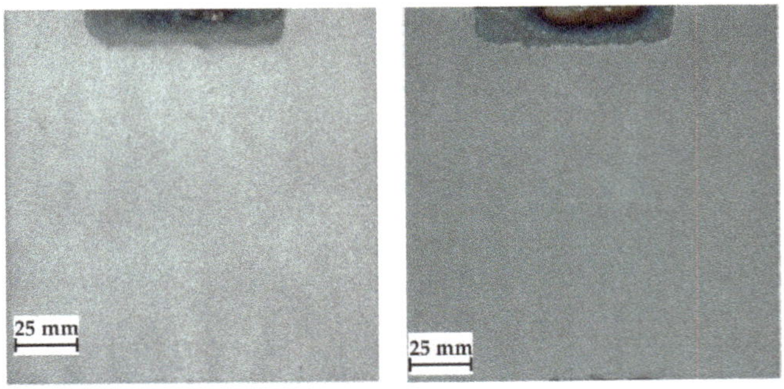

(a) (b)

Figure 2. *Cont.*

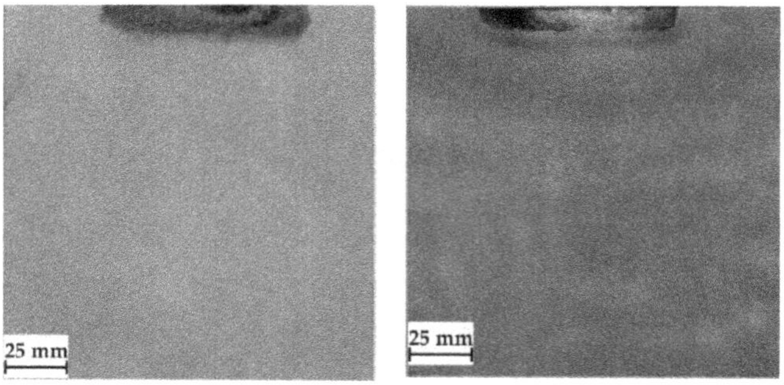

(**c**) (**d**)

Figure 2. View of test samples with flame-sprayed coatings on the aluminum matrix: (**a**) aluminum powder of the EN AW 1000 series; (**b**) aluminum powder EN AW 1000 series with addition of 0.5 wt.% carbon nanotubes (Nanocyl NC 7000); (**c**) aluminum powder of the EN AW 1000 series with the addition of 1 wt.% carbon nanotubes (Nanocyl NC 7000); (**d**) aluminum powder of the EN AW 1000 series with the addition of 1 wt.% carburite.

2.3. *Visual and Metallographic Examination of Coatings*

In each case, the entire surface of the sample was subjected to visual tests to assess the quality and identify any imperfections in the form of cracks, discontinuities, unevenness, porosity or lack of coating adhesion. Macro and microscopic examinations were performed on Olypmus GX 71 optical microscope (Olympus Corporation, Tokyo, Japan). The observations were made on the cross-section of metallographic samples cut from the centre of element. Samples were polished and etched in Aqua Regia. Selected areas of flame-sprayed coatings (aluminum and aluminum with addition of carbon materials) have been subjected to chemical composition analysis on JEOL 5800LV SEM scanning microscope and also EDX (Jeol Ltd., Tokyo, Japan).

2.4. *Hardness Measurements of Coatings*

The coating-hardness measurement was made with the Vickers method using Microhardness Tester 401MVD™ (Wilson Instruments An Instron Company, Norwood, MA, USA). The examinations were carried out in conformity to ISO 6507-1:2018 standard [40]. The load applied during the hardness measurement was 0.98 N. The hardness measurement was made at the polished cross-section of the samples with flame-sprayed coatings. Ten hardness measuring points were made on the cross-section each sprayed coating.

2.5. *Erosive Wear Resistance of Coatings*

The erosive wear tests of flame-sprayed coatings were carried out in accordance with ASTM G76-95 [41], as shown in Figure 3. Aluminum oxide powder (Al_2O_3) with particle diameter of 71 μm was used as the erodent material. Particle velocity was set at 70 ± 2 m/s, the erodent expenditure was 2.0 ± 0.5 g/min and the nozzle distance from the sample surface was 10 mm. The tests were carried out at 90° and 30° erodent impact angle. The average weight loss was determined based on three tests. The erosion rate was determined according to the Equation (1),

$$\text{erosion rate [g/min]} = \text{mass loss of sample [g]:exposure time [min]} \quad (1)$$

However, the erosive wear resistance using Equation (2):

$$\text{erosive wear resistance } [0.001\text{mm}^3/\text{g}] = \text{volume loss of the sample } [\text{mm}^3]\text{: total mass of the erodent used in the test [g]} \quad (2)$$

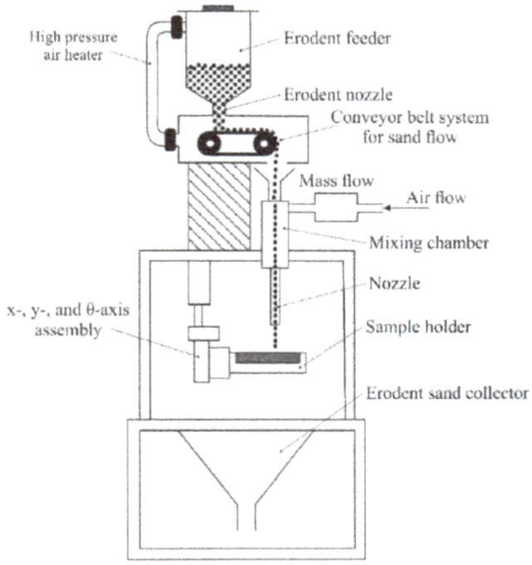

(a) (b)

Figure 3. Erosion resistance testing according to ASTM G 76-95: (**a**) a schematic view, (**b**) the interior view of the erosion measuring chamber.

2.6. Abrasive Wear Resistance of Coatings

Metal-mineral wear resistance tests of aluminum matrix coatings were provided in accordance with ASTM G65-00, Procedure E [42]. During the test, the rubber-wheel made 1000 revolutions and the abrasive flow rate of A.F.S. Testing Sand 50–70 was 335 g/min. The force applied pressing the test coupon against the wheel was TL = 130 N (test load-TL). After the abrasive wear resistance test, the test specimen was weighed at weight sensitivity 0.0001 [g]. Converting mass loss to volume loss was as follows:

$$\text{volume loss } [\text{mm}^3] = \text{mass loss } [\text{g}]\text{:density } [\text{g/cm}^3] \times 1000 \quad (3)$$

The tests were carried out on abrasion tester shown in Figure 4.

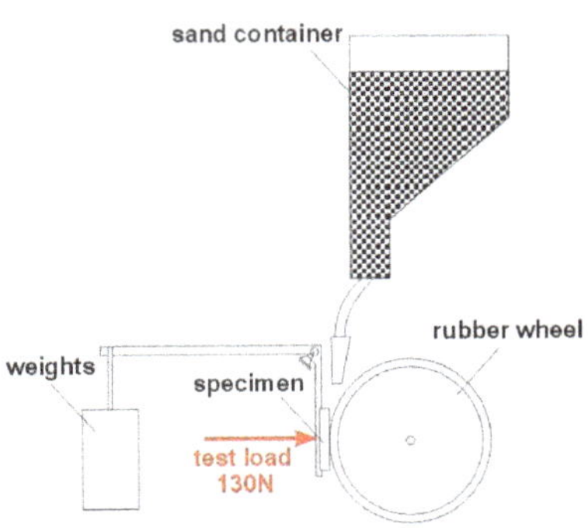

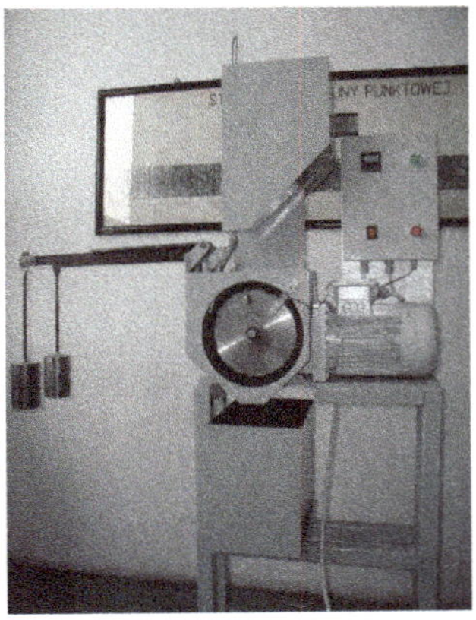

(a) (b)

Figure 4. Metal-mineral wear resistance test stand–according to ASTM G65–00, Procedure E standard (**a**) a schematic view, (**b**) picture of the device used.

3. Results

3.1. Metallographic Test Results

The structure of each tested flame-sprayed coating cross-section is presented in Figure 5. The SEM structures of tested aluminum matrix coatings with chemical composition are presented in Figures 6–9.

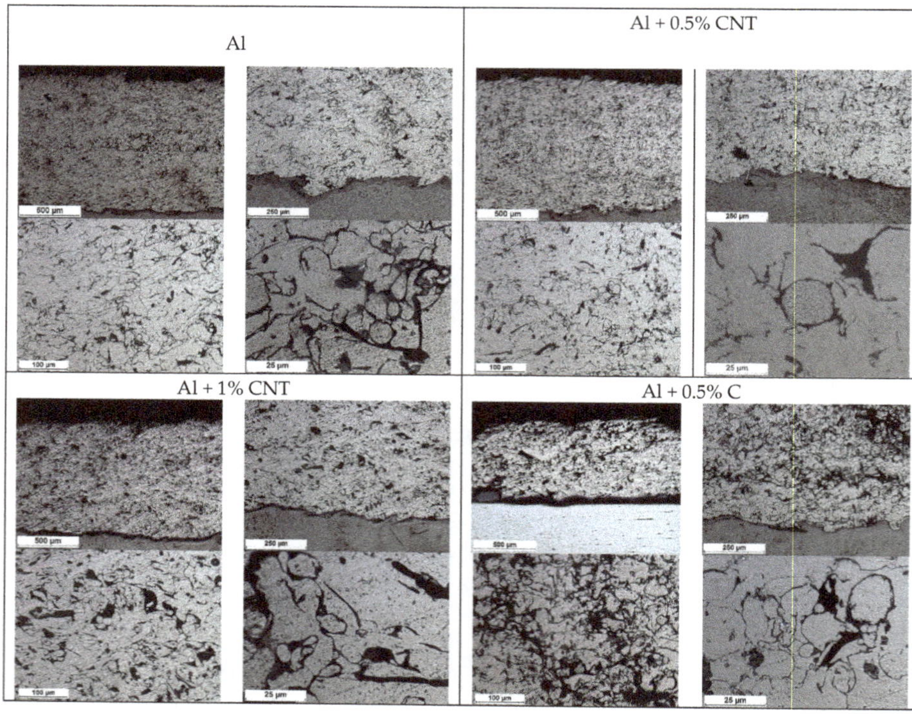

Figure 5. The macro and microstructure of the flame-sprayed pure aluminum and aluminum matrix with carbon nanotubes and carburite reinforcement coatings; etching: $HCl + HNO_3$.

Al			
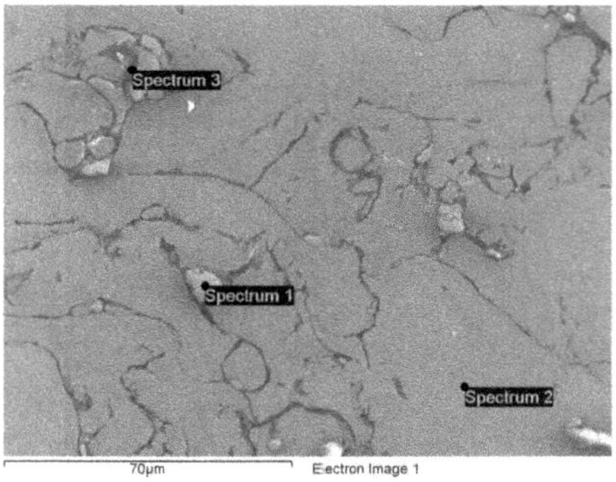			
Measurement area sign	Chemical composition [wt.%]		
	C	O	Al
Spectrum 1	5	2	93
Spectrum 2	6	2	92
Spectrum 3	26	27	47
Standard deviation	12	14	26

Figure 6. The structure of EN AW 1000 aluminum powder flame-sprayed coating with marked chemical composition tested areas on SEM.

Al + 0.5% CNT		
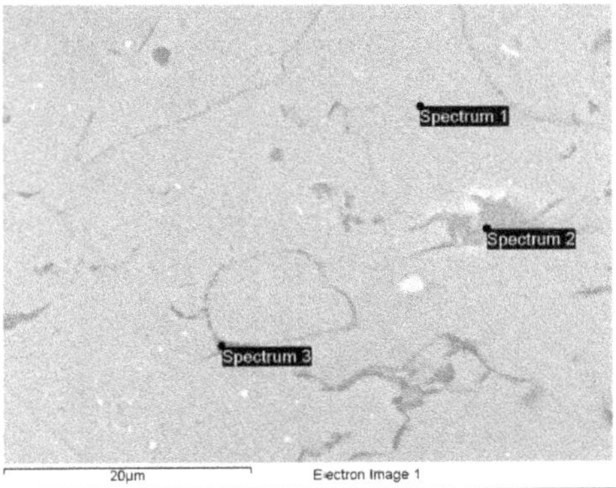		
Measurement area sign	Chemical composition [wt.%]	
	C	Al
Spectrum 1	6	94
Spectrum 2	12	88
Spectrum 3	33	67
Standard deviation	14	14

Figure 7. The structure of aluminum powder with 0.5 wt.% Nanocyl NC 7000 carbon nanotubes flame-sprayed coating with marked chemical composition tested areas on SEM.

Al + 1% CNT

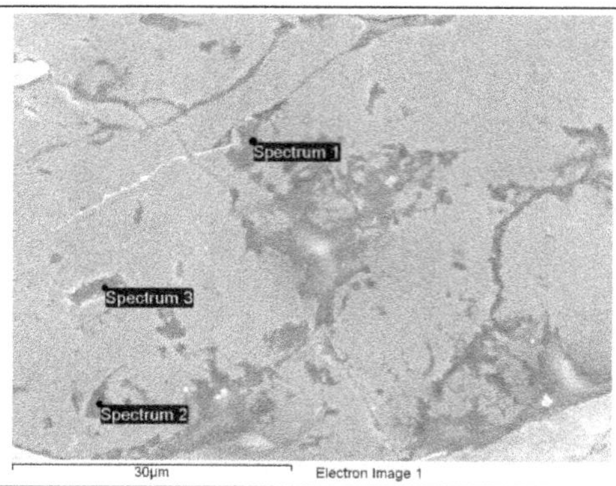

Measurement area sign	Chemical composition [wt.%]	
	C	Al
Spectrum 1	17	83
Spectrum 2	20	80
Spectrum 3	7	93
Standard deviation	7	7

Figure 8. The structure of aluminum powder with 1 wt.% Nanocyl NC 7000 carbon nanotubes flame-sprayed coating with marked chemical composition tested areas on SEM.

Al + 0.5% C

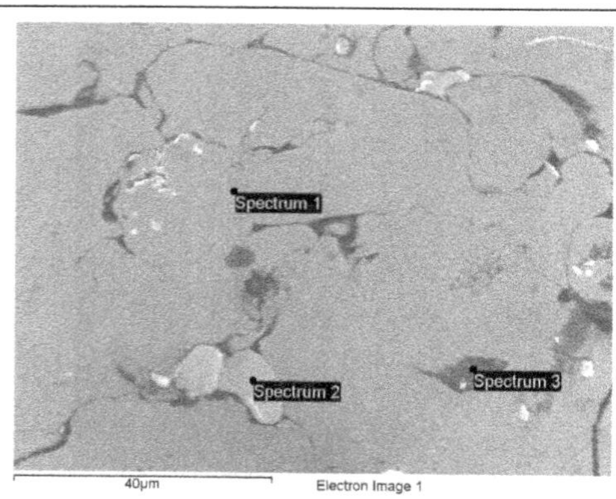

Measurement area sign	Chemical composition [wt.%]	
	C	Al
Spectrum 1	4	96
Spectrum 2	5	95
Spectrum 3	60	40
Standard deviation	32	32

Figure 9. The structure of aluminum powder with 0.5 wt.% carburite carbon nanotubes flame-sprayed coating with marked chemical composition tested areas on SEM.

3.2. Hardness Measurements

The hardness measurements on flame-sprayed aluminum and aluminum matrix reinforced with carbon material coatings, were carried out on the surface at 5 points along one measuring line (Table 4) and on cross-section of the samples (Figures 10 and 11), according to the scheme showed on Figure 10.

Table 4. Surface hardness results of the flame-sprayed aluminum and aluminum reinforced with carbon material coatings.

Specimen Designation	Hardness [HV 0.1]						
	Measuring Point						
	1	2	3	4	5	Average	Standard Deviation
Al	35.2	32.5	34.8	37.2	35.1	34.96	1.67
Al + 0.5% CNT	59.2	62.8	56.5	54.2	55.5	57.64	3.42
Al + 1% CNT	41.2	43.2	37.5	39.2	38.7	39.96	2.25
Al + 0.5% C	30.2	27.3	28.2	27.0	28.0	28.14	1.25

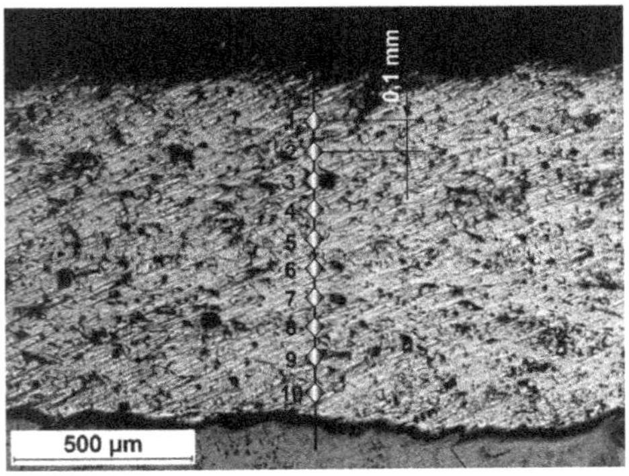

Specimen Designation	Hardness [HV 0.1]											
	Measuring Point											
	1	2	3	4	5	6	7	8	9	10	Average	Standard Deviation
Al	34.9	46.2	31.8	28.3	29.6	41.0	40.5	34.2	30.3	24.1	34.1	6.39
Al + 0.5% CNT	58.8	30.6	39.4	41.2	49.4	33.7	46.1	48.1	38.3	37.3	42.3	7.95
Al + 1% CNT	40.1	54.4	51.0	55.3	42.1	29.2	32.1	34.5	48.1	49.5	43.6	8.95
Al + 0.5% C	27.5	40.1	34.2	41.1	31.6	34.9	31.8	33.7	36.8	41.2	35.3	4.29

Figure 10. Hardness measurements scheme and HV 0.1 results of the flame-sprayed aluminum and aluminum reinforced with carbon materials coatings.

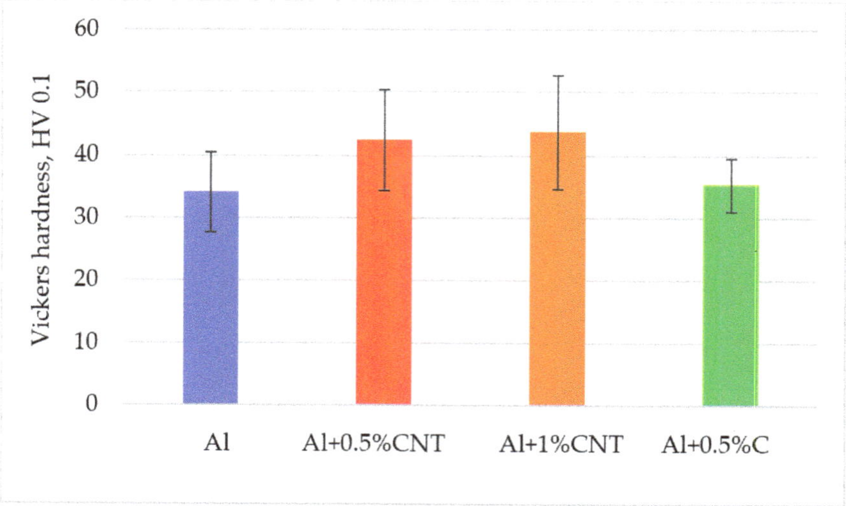

Figure 11. Comparison of average cross-sectional hardness and standard deviation for each of the investigated coatings.

3.3. *Tests Results of the Coatings Erosive Wear Resistance*

The relative erosive wear resistance test results of the flame-sprayed aluminum, aluminum with carbon nanotube reinforcement and aluminum with filter dust carburite reinforcement coatings are presented in Table 5 and Figure 12.

Table 5. Summary of results obtained during the erosion wear test ASTM G76–95 [41].

Erodent Strike Angle [°]	Specimen Designation	Mass Loss [g]	Volume Loss [mm^3]	Erosion Rate [g/min]	Resistance to Erosion as per ASTM G76 [0.001mm^3/g]
90	Al	0.0054	1.985	0.00068	0.12255
	Al + 0.5% CNT	0.0117	4.301	0.00146	0.26552
	Al + 1% CNT	0.0071	2.610	0.00089	0.16113
	Al + 0.5% C	0.0064	2.353	0.00080	0.14524
30	Al	0.0036	1.324	0.00045	0.08170
	Al + 0.5% CNT	0.0066	2.426	0.00083	0.14978
	Al + 1% CNT	0.0045	1.654	0.00056	0.10212
	Al + 0.5% C	0.0039	1.434	0.00049	0.08851

Notes: density of aluminum spray coating 2.72 [g/cm^3], mass of erodent used 16.2 [g], test time 8 [min].

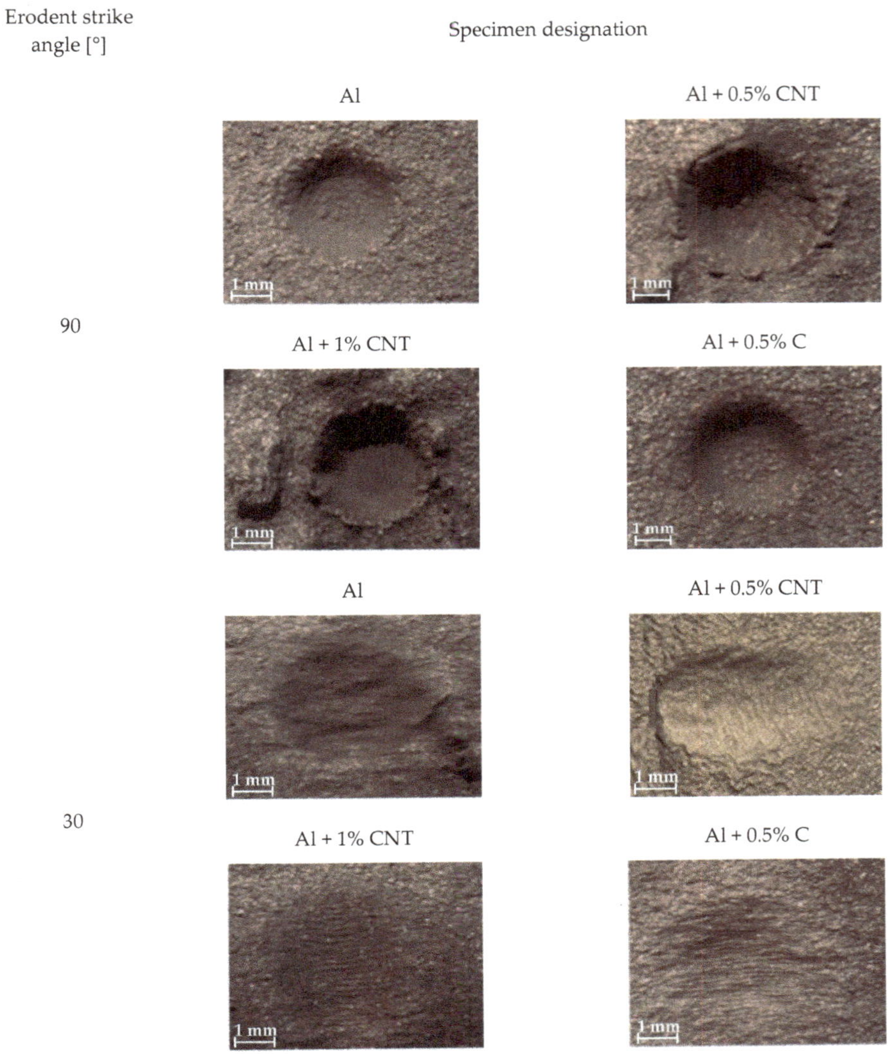

Figure 12. The surfaces of flame-sprayed aluminum and aluminum matrix reinforced with carbon material coatings after erosive wear resistance tests; comparison the erosion effect on samples surfaces for each tested angle of erodent particles incidence.

3.4. *Tests Results of the Coatings' Wear Resistance*

The wear resistance test results of the flame-sprayed aluminum, aluminum with carbon nanotube reinforcement and aluminum with filter dust carburite reinforcement coatings are presented in Table 6 and Figure 13. The metal-mineral type wear resistance of the flame-sprayed aluminum with carbon nanotubes and aluminum with carburite coatings were compared to the flame-sprayed pure aluminum coating.

Table 6. Summary of results obtained during the abrasive wear test ASTM G65 [42].

Specimen Designation	Number of Specimen	Weight Before Test [g]	Weight after Test [g]	Mass Loss [g]	Average Mass Loss [g]	Average Volume Loss [mm^3]	Relative [1)] Abrasion Resistance
Al	S1-1	43.9675	43.8413	0.1262	0.1418	52.1324	1.00
	S1-2	42.3855	42.2281	0.1574			
Al + 0.5% CNT	S2-1	56.5170	56.3924	0.1246	0.1286	47.26103	1.10
	S2-2	53.8604	53.7279	0.1325			
Al + 1% CNT	S3-1	56.9638	56.8322	0.1316	0.1279	47.0221	1.11
	S3-2	57.4587	57.3345	0.1242			
Al + 0.5% C	S4-1	59.4423	59.3199	0.1224	0.1190	43.7500	1.19
	S4-2	61.1152	60.9996	0.1156			

Notes: density of aluminum spray coating 2.72 [g/cm^3]; [1)] relative to sprayed coatings of the aluminum without carbon materials.

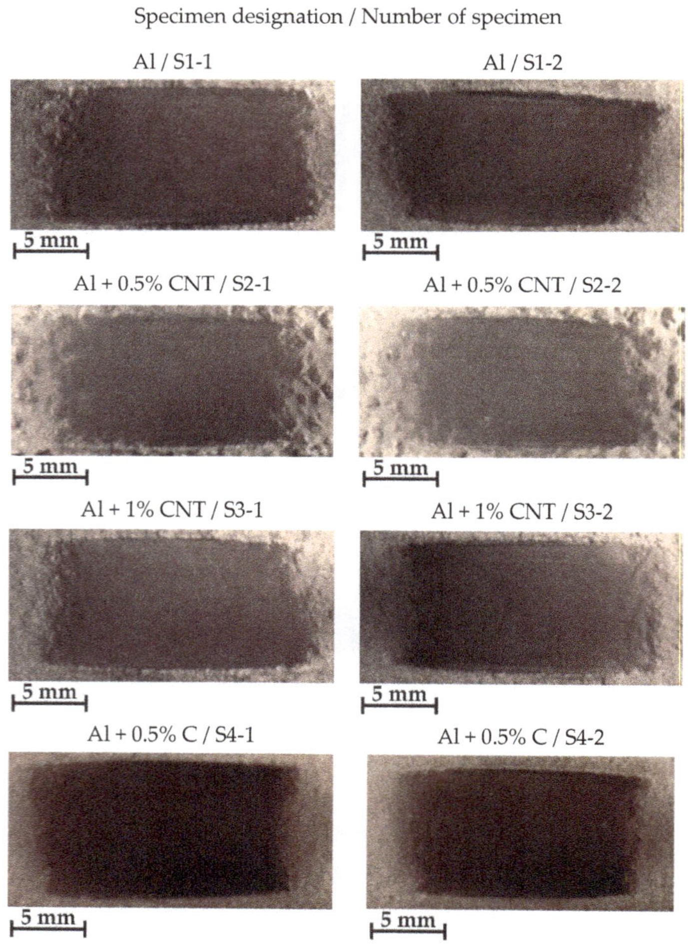

Figure 13. The surfaces of flame-sprayed aluminum and aluminum with carbon material coatings after wear resistance metal-mineral tests.

4. Discussion

Visual and metallographic tests of the flame-sprayed aluminum and aluminum with carbon material reinforcement (0.5 wt.% and 1 wt.% of carbon nanotubes Nanocyl NC 7000 and 0.5 wt.% of carburite) have shown that by using proper parameters of the process it is possible to receive coatings with acceptable quality level, characterized by proper adhesion to the substrate, lack of delamination and even thickness over the entire surface. The outer surface of the coatings was characterized by a slight roughness, lack of porosity and cracks (Figure 2). During the flame-spraying process, carbon material particles added to aluminum powder did not oxidize completely in the oxyacetylene flame. Carbon nanotubes (melting point 4526 °C [43]) and carburite (melting point 3550 °C [44]) in the oxyacetylene flame has formed with aluminum Al-Cx type agglomerates, which due to the large volume and lower heat source temperature than other thermal spraying methods (oxyacetylene flame temperature 3160 °C [45]) migrated in large quantities to the coatings. Partially melted and partially only plasticized in a gas flame, Al-Cx agglomerates collided with the substrate at high speed, (Figure 1b) and in this way formed a fine-grained coating structure. Powder flame spraying process (PFS) in comparison with, for example, plasma spraying, increases the probability of stopping carburite and carbon nanotubes (CNT) in flame sprayed composite coating with aluminum matrix. Presence of carbon materials in aluminum powder flame-sprayed coatings is initially confirmed by metallographic microscopic tests, which revealed areas carburite and carbon nanotubes on specimens (Figure 5, Al + 0.5% CNT). Presence of carbon materials can be observed on the entire cross-section of the coating, also at the outer surface. Inside the Al-Cx composite coatings, no cracks were found, only the presence of individual cavities. The tests made using scanning electron microscope have shown presence of some areas consisting small carbon materials inclusions. These were observed in the all-aluminum coatings with carbon material reinforcement. For the coating with 0.5 wt.% CNT, inclusion areas consisted of 33.05 wt.% C; for coating with 1 wt.% CNT, the carbon content was lower in tested area (20.25 wt.% C), while for the coating with 0.5 wt.% of carburite, carbon content was almost two times higher than in the coating with same content of CNTs and amounted to 59.76 wt.% C. In aluminum coating without carbon material addition, carbon and oxygen were found, (Figure 6). A small amount of carbon in the aluminum coating may be caused by the ease of thermal decomposition of acetylene in the gas flame and the physicochemical properties of unsaturated hydrocarbons [46]. Acetylene is dissociated into active carbon atoms (acetylene black, characterized by high purity) and hydrogen molecule. The oxygen content in the aluminum coating is the result of oxidation of the aluminum particles in the gas flame and the atmosphere. The addition of carbon materials to the aluminum powder causes the carbon to bind oxygen as a strong deoxidizer; that is why its presence was not found in composite coatings with aluminum matrix and carbon material (carburite and carbon nanotubes) reinforcement. These results should still be confirmed using more advanced research methods, e.g. XRD X-ray diffraction or Raman spectroscopy. These studies will be completed and presented in another publication.

The hardness measurements of tested coatings were proceeded using standard ISO 6507 [40]. The measurements were done both on the external surface and the cross-section of the sprayed coatings. These tests showed that using addition of 0.5 wt.% and 1 wt.% carbon nanotubes to aluminum coating (34.1 HV 0.1) caused an increase in its hardness of 8.2 HV 0.1 for the 0.5 wt.% of carbon nanotube reinforcement and by 9.5 HV 0.1 for 1 wt.% of carbon nanotube reinforcement. Addition of carburite to aluminum had not significant influence on the coating hardness (Figure 11).

Erosive wear resistance test results have shown that the addition of carbon materials to aluminum powder does not increase the erosive wear resistance of flame-sprayed coatings. During these tests, the aluminum coatings with carbon nanotubes had worn out by erosion with large and small angles of erodent incidence more than aluminum coatings with carburite and much more than aluminum coatings without carbon materials reinforcement. It was observed that for all tested coatings erosive wear resistance was better during using smaller angle of erodent incidence (Table 4).

The best metal-mineral type wear resistance had the aluminum coating with carburite. The wear resistance of this coating was 19% higher than pure aluminum coating. The aluminum coatings with addition of 0.5 wt.% and 1 wt.% of carbon nanotubes in comparison to pure aluminum coating had better relative wear resistance by 10% and 11% (Table 5). The cause of decreasing wear using aluminum coatings with carbon material addition was increased glide of ceramic abrasive particles on metal.

Based on the conducted study and the obtained results, it can be concluded that it is possible to introduce carbon particles in the form of carbon nanotubes (CNT) and also carburite into the aluminum matrix by means of flame spraying method. The flame spraying is an effective and cheaper alternative to the technology of laser surface treatment of metals. Properly selected parameters of the flame spraying process allow to preserve the properties of particles of carbon materials, their even distribution in the coating, proper bonding with the matrix and prevent the effects of their thermal degradation. The produced composite is characterized by a low friction coefficient. The tribological characteristics of produced test coating of aluminum reinforced by carbon particles in the form of carbon nanotubes and carburite show that the coatings can be classified as sliding materials. Additionally, the coatings are characterized by high wear resistance. The obtained result should be considered as a preliminary information on a new group of materials, which can find application in the automotive industry. They are the basis for the design and optimization of friction materials operated at elevated temperatures (e.g. pistons, engine blocks), systems subject to intensive wear (e.g. brake discs, cylinders), as well as in propulsion systems (e.g. bearings), providing low friction coefficient and also high ability for absorption of vibration. Further research should be focused on the investigation of the effect of doping the aluminum matrix with carbon nanotubes (CNT) on the wear mechanisms, change of microstructure of the counter-specimen and also tests which will allow to determine the tribological characteristics of the materials at elevated temperatures.

5. Conclusions

The analysis carried out comparing properties of flame-sprayed EN AW 1000 aluminum coatings and aluminum matrix coatings with carbon materials reinforcement (0.5 wt.% and 1 wt.% of Nanocyl NC 7000 carbon nanotubes and 0.5 wt.% of carburite) resulted in the following conclusions:

1. Producing aluminum matrix coatings with carbon materials reinforcement with high quality is possible using flame spraying technology.
2. In the aluminum with carburite reinforcement flame-sprayed coating structure, areas were observed with a share of carbon above 61 wt.%. In the aluminum coating with 1 of carbon nanotubes, similar areas were observed with a share of carbon about 33 wt.%.
3. The addition of carbon nanotubes to aluminum powder resulted in increasing the hardness of flame-sprayed coatings by about 10 HV 0.1.
4. The carbon materials reinforced aluminum flame-sprayed coatings have lower erosive wear resistance than pure aluminum coatings with large and small angles of erodent incidence.
5. The metal-mineral type wear resistance of flame-sprayed aluminum coatings reinforced with carbon nanotubes or carburite is 10% to 20% higher in comparison to pure aluminum coating.

Funding: This research was financed from the own resources of the Silesian University of Technology.

Conflicts of Interest: The author declare no conflict of interest.

References

1. Kelly, A. Composite materials after seventy years. *J. Mater. Sci.* **2006**, *41*, 905–912. [CrossRef]
2. Rawal, S. Metal-matrix composites for space applications. *JOM* **2001**, *53*, 14–17. [CrossRef]
3. Shelly, J.S.; LeClaire, R.; Nichols, J. Metal-matrix composites for liquid rocket engines. *JOM* **2001**, *53*, 18–21. [CrossRef]

4. Salvetat, J.P.; Bonard, J.M.; Thomson, N.H.; Kulik, A.J.; Forro, L.; Benoit, W.; Zuppiroli, L. Mechanical properties of carbon nanotubes. *Appl. Phys. A* **1999**, *69*, 255–260. [CrossRef]
5. Coleman, J.N.; Khan, U.; Blau, W.J.; Gun'ko, Y.K. Small but strong: A review of the mechanical properties of carbon nanotube–polymer composites. *Carbon* **2006**, *44*, 1624–1652. [CrossRef]
6. Takashi, I. Overview of trends in advanced composite research and applications in Japan. *Adv. Compos. Mater.* **2006**, *15*, 3–37.
7. Bokobza, L. Multiwall carbon nanotube elastomeric composites. A review. *Polymer* **2007**, *48*, 4907–4920. [CrossRef]
8. Curtin, W.A.; Sheldon, B.W. CNT-reinforced ceramics and metals. *Mater. Today* **2004**, *7*, 44–49. [CrossRef]
9. Saffar, K.P.A.; Najafi, A.R.; Moeinzadeh, M.H.; Sudak, L.J. A Finite Element Study of Crack Behavior for Carbon Nanotube Reinforced Bone Cement. *World J. Mech.* **2013**, *3*, 13–21. [CrossRef]
10. Bakshi, S.R.; Singh, V.; Balani, K.; McCartney, D.G.; Seal, S.; Agarwal, A. Carbon nanotube reinforced aluminum composite coating via cold spraying. *Surf. Coat. Technol.* **2008**, *202*, 5162–5169. [CrossRef]
11. Keshri, A.K.; Balani, K.; Bakshi, S.R.; Singh, V.; Laha, T.; Seal, S.; Agarwal, A. Structural transformations in carbon nanotubes during thermal spray processing. *Surf. Coat. Technol.* **2009**, *203*, 2193–2201. [CrossRef]
12. Wu, Y.; Kim, G. Carbon nanotube reinforced aluminum composite fabricated by semi-solid powder processing. *J. Mater. Process. Technol.* **2011**, *211*, 1341–1347. [CrossRef]
13. Liao, J.; Tan, M.; Ramanujan, R.V.; Shukla, S. Carbon nanotube evolution in aluminum matrix during composite fabrication process. *Mater. Sci. Forum* **2011**, *690*, 294–297. [CrossRef]
14. Feng, Y.; Yuana, H.; Zhanga, M. Fabrication and properties of silver-matrix composites reinforced by carbon nanotubes. *Mater. Charact.* **2005**, *55*, 211–218. [CrossRef]
15. Bakshia, S.R.; Singhb, V.; Seal, S.; Agarwal, A. Aluminum composite reinforced with multiwalled carbon nanotubes from plasma spraying of spray dried powders. *Surf. Coat. Technol.* **2009**, *203*, 1544–1554. [CrossRef]
16. Zeng, X.; Zhou, G.; Xu, Q.; Xiong, Y.; Luo, C.H.; Wu, J. A new technique for dispersion of carbon nanotube in a metal melt. *Mater. Sci. Eng. A* **2010**, *527*, 5335–5340. [CrossRef]
17. Kondoh, K.; Fukuda, H.; Umeda, J.; Imai, H.; Fugetsu, B.; Endo, M. Microstructural and mechanical analysis of carbon nanotube reinforced magnesium alloy powder composites. *Mater. Sci. Eng. A* **2010**, *527*, 4103–4108. [CrossRef]
18. He, X.; Kitipornchai, S.; Liew, K.M. Buckling analysis of multi-walled carbon nanotubes: A continuum model accounting for van der Waals interaction. *J. Mech. Phys. Solids* **2005**, *53*, 303–326. [CrossRef]
19. Tan, H.; Jiang, L.Y.; Huangc, Y.; Liu, B.; Hwang, K.C. The effect of van der Waals-based interface cohesive law on carbon nanotube-reinforced composite materials. *Compos. Sci. Technol.* **2007**, *67*, 2941–2946. [CrossRef]
20. Silva, A.P.; Devezas, T.C.; Segadaes, A.M. Statistical Modelling of the Particle Size Composition of an Alumina Matrix for No-Cement Self-Flowing Refractory Castables. *Mater. Sci. Forum* **2006**, *514–516*, 604–608. [CrossRef]
21. Silva, J.M.A.; Devezas, T.C.; Silva, A.P.; Ferreira, J.A.M. Mechanical Characterization of Composites with Embedded Optical Fibers. *J. Compos. Mater.* **2005**, *39*, 1261–1281. [CrossRef]
22. Czuprynski, A. Properties of Al_2O_3/TiO_2 and ZrO_2/CaO flame-sprayed coatings. *Mater. Tehnol./Mater. Technol.* **2017**, *51*, 205–212. [CrossRef]
23. Czuprynski, A.; Gorka, J.; Adamiak, M.; Tomiczek, B. Testing of flame sprayed Al_2O_3 matrix coatings containing TiO_2. *Arch. Metall. Mater.* **2016**, *61*, 1363–1370. [CrossRef]
24. Czuprynski, A.; Gorka, J.; Adamiak, M. Examining properties of arc sprayed nanostructured coatings. *Metalurgija* **2016**, *55*, 173–176.
25. Adamiak, M.; Czuprynski, A.; Kopysc, A.; Monica, Z.; Olender, M.; Gwiazda, A. The Properties of Arc-Sprayed Aluminum Coatings on Armor-Grade Steel. *Metals* **2018**, *8*, 142. [CrossRef]
26. Lisiecki, A. Mechanisms of hardness increase for composite surface layers during laser gas nitriding of the Ti6A14V alloy. *Mater. Tehnol. Mater. Technol.* **2017**, *51*, 577–583. [CrossRef]
27. Lisiecki, A.; Kurc-Lisiecka, A. Erosion wear resistance of Titanium-Matrix Composite Ti/TiN produced by diode-laser gas nitriding. *Mater. Tehnol. Mater. Technol.* **2017**, *51*, 29–34. [CrossRef]
28. Lisiecki, A.; Piwnik, J. Tribological characteristic of titanium alloy surface layers produced by diode laser gas nitriding. *Arch. Metall. Mater.* **2016**, *61*, 543–552. [CrossRef]

29. Lisiecki, A. Comparison of Titanium Metal Matrix Composite surface layers produced during laser gas nitriding of Ti_6Al_4V alloy by different types of lasers. *Arch. Metall. Mater.* **2016**, *61*, 1777–1783. [CrossRef]
30. Lisiecki, A. Titanium Matrix Composite Ti/TiN Produced by Diode Laser Gas Nitriding. *Metals* **2015**, *5*, 54–69. [CrossRef]
31. Klimpel, A.; Dobrzanski, L.A.; Lisiecki, A.; Janicki, D. The study of properties of Ni-W_2C and Co-W_2C powders thermal sprayed deposits. *J Mater Process Tech.* **2005**, *164*, 1068–1073. [CrossRef]
32. Dobrzanski, L.A.; Klimpel, A.; Bonek, M.; Lisiecki, A. Surface-layer's structure of X40CrMoV5-1 steel remelted and/or WC alloyed with HPDL laser. *Mater. Sci. Forum* **2003**, *437–438*, 69–72. [CrossRef]
33. Klimpel, A.; Dobrzanski, L.A.; Lisiecki, A.; Janicki, D. The study of the technology of laser and plasma surfacing of engine valves face made of X40CrSiMo10-2 steel using cobalt-based powders. *J. Mater. Process. Tech.* **2006**, *175*, 251–256. [CrossRef]
34. Laha, T.; Agarwal, A.; McKechnie, T.; Seal, S. Synthesis and characterization of plasma spray formed carbon nanotube reinforced aluminum composite. *Mater. Sci. Eng. A* **2004**, *381*, 249–258. [CrossRef]
35. Thermal spraying. *Pre-Treatment of Surfaces of Metallic Parts and Components for Thermal Spraying*; EN 13507; CEN: Brussels, Belgium, 2018.
36. *Thermal Spraying—Zinc, Aluminium and Their Alloys—Part 1: Design Considerations and Quality Requirements for Corrosion Protection Systems*; ISO 2063-1; ISO: Geneva, Switzerland, 2017.
37. Zhou, Z.-F.; Hu, M.-Y.; Xin, H.; Chen, B.; Wang, G.-X. Experimental and theoretical studies on the droplet temperature behavior of R407C two-phase flashing spray. *Int. J. Heat Mass Transf.* **2019**, *136*, 664–673. [CrossRef]
38. Zhou, Z.; Wu, W.; Chen, B.; Wang, G.; Guo, L. An experimental study on the spray and thermal characteristics of R134a two-phase flashing spray. *Int. J. Heat Mass Transf.* **2012**, *55*, 4460–4468.
39. *Thermal Spraying—Characterization and Testing of Thermally Sprayed Coatings*; ISO 14923; ISO: Geneva, Switzerland, 2005.
40. *Metallic Materials—Vickers Hardness Test—Part 1: Test Method*; ISO 6507; ISO: Geneva, Switzerland, 2018.
41. *Standard Test Method for Conducting Erosion Tests by Solid Particle Impingement Using Gas Jets*; ASTM G76-95; ASTM: West Conshohocken, PA, USA, 2000.
42. *Standard Test Method for Measuring Abrasion Using the Dry Sand/Rubber Wheel Apparatus*; ASTM G65-00; ASTM: West Conshohocken, PA, USA, 2000.
43. Wei, X.; Wang, M.S.; Bando, Y.; Golberg, D. Thermal stability of carbon nanotubes probed by anchored tungsten nanoparticles. *J. Sci. Technol. Adv. Mater.* **2011**, *12*, 1–6. [CrossRef]
44. Burda, M.; Kik, T.; Koziol, K.; Gruszczyk, A. Development of methods of carbon nanotubes input to weld pool. *Weld. Technol. Rev.* **2011**, *12*, 43–50.
45. Lide, D.R. *CRC Handbook of Chemistry and Physics*, 85th ed.; CRC Press: Boca Raton, FL, USA, 2003; pp. 15–52.
46. Gaskell, D.R. *Introduction to the Thermodynamics of Materials*, 4th ed.; Taylor & Francis: New York, NY, USA, 2017.

materials

MDPI

Article

Comparison of Properties of Hardfaced Layers Made by a Metal-Core-Covered Tubular Electrode with a Special Chemical Composition

Artur Czupryński

Department of Welding Engineering, Faculty of Mechanical Engineering, Silesian University of Technology, Konarskiego 18A, 44-100 Gliwice, Poland; artur.czuprynski@polsl.pl

Received: 27 October 2020; Accepted: 26 November 2020; Published: 29 November 2020

Abstract: In this article, the results of research on the metal-mineral-type abrasive wear of a wear-resistant plate made by a tubular electrode with a metallic core and an innovative chemical composition using the manual metal arc hardfacing process were presented. The properties of the new layer were compared to the results of eleven wear plates manufactured by global suppliers, including flux-cored arc welding gas-shielded (FCAW-GS, Deposition Process Reference Number: 138), flux-cored arc welding self-shielded (FCAW-SS, Deposition Process Reference Number: 114), automated hardfacing, and manual metal arc welding (MMAW, Deposition Process Reference Number: 111) hardfacing T Fe15 and T Fe16 alloys, according to EN 14700:2014. Characterization of the hardfaced layers was achieved by using hardness tests, optical microscopy, confocal microscopy, scanning electron microscopy, and EDS (Energy Dispersive Spectroscopy) and X-ray diffraction analyses. Based on wear resistance tests in laboratory conditions, in accordance with ASTM G65-00: Procedure A, and surface layer hardness tests, in accordance with PN-EN ISO 6508-1, the wear plates most suitable for use in metal-mineral conditions were chosen. The results demonstrated the high metal-mineral abrasive wear resistance of the deposit weld metal produced by the new covered tubular electrode. The tubular electrode demonstrated a high linear correlation between the surface wear resistance and the hardness of the metal matrix of the tested abrasive wear plates. In addition to hardness, size, shape, the dispersion of strengthening phases, and the base metal content, depending on hardfacing technology and technological parameters, impact wear resistance is represented by volumetric loss caused by effect-free or constrained dry abrasive medium contact. The presented results can be used in machine part material selection and wear planning for applications in inspection, conservation, and regeneration interval determination. The obtained results will be applied in a real-time wear rate prediction system based on the measurement of the working parameters.

Keywords: iron-based alloy; hardfacing; surfacing; cladding; tubular hardfacing electrode; FCAW-GS; FCAW-SS; MMAW; wear-resistant steel; wear plate; abrasion; tribology

---

1. Introduction

The intensive wear of machine and apparatus parts in mining, quarrying, petrochemical, metallurgical, cement, construction, and power generation industries, among others, drives increasing demand for wear-resistant plates and liners [1–5].

Wear-resistant materials are dedicated alloys with supreme hardness that are simultaneously weldable and resistant to moderate impact loading. Contemporary wear-resistant steels, due to high metallurgical purity, are characterized by high strength, good weldability, workability, and acceptable machinability. The production costs of these steels, however, remain high [6–11].

Among the alternatives to the above-mentioned cases are prefabricated wear-resistant composite plates manufactured mainly by automated arc hardfacing, vacuum furnace powder melting, or metallurgical bonding of the base carbon steel plate with a wear-resistant surface layer. Hardfaced wear-resistant plates demonstrate significantly better wear resistance compared to wear-resistant hardened plates [12–18]. The gains from hardfaced composite wear plates primarily include increased durability, reduced time and cost of repairs, and increased machine operation safety [15,19,20].

The chemical composition and microstructure of the surface layer are chosen to achieve the highest wear resistance and durability of the finished machine and apparatus components. The impact of alloying elements on the hardness of the surface layer is diverse. For wear-resistant steels, high hardness is achieved by a melting process in suitable metallurgical conditions and the addition of set quantities of carbon, manganese, chromium, niobium, molybdenum, and thermal hardening. For wear-resistant composite plates, surface hardness is achieved by the formation of carbon metal compounds in the form of discrete hard particles, known as carbides [6,21,22].

Iron-chromium-carbon alloys are used in conditions where abrasion resistance is required. Their unique abrasive wear resistance results primarily from high volume and the decomposition of the hard carbide fraction in the weld metal matrix. The study of Fe-Cr-C alloy microstructures has demonstrated that these types of materials contain hypoeutectic, eutectic, and hypereutectic structures. Alloys containing 1.8–3.6 mass% carbon and 11–30 mass% chromium are superior in terms of wear, corrosion, and oxidation resistance, and have been adopted as abrasion resistance materials for wear-resistant parts in the mining industry. Primary carbides form in large amounts at higher carbon concentrations. High chromium and carbon contents in weld metal promote the formation of extreme high-hardness chromium carbide particles (1700 HV), which are embedded in significantly more ductile yet hard metal matrices with an average hardness of approximately 700 HV (60 HRC). Due to the presence of these constituents, a hardness level from 600 HV to 840 HV (55–65 HRC) in the surface layer is achieved [6,18].

The wear resistance achieved by hardfaced layers with carbide compounds (mainly M_7C_3 and M_3C) with hardness in the range of 1500–3000 HV is more than a few times higher than widespread wear-resistant materials [1,23,24].

The formation of alloying carbide elements is not the sole criterion in the consideration for the selection of suitable hardfaced composite wear-resistant plates. The shape and distribution of the carbide are also vital to achieving high wear resistance. The highest wear resistance is achieved when elongated cylindrical chromium carbides are situated perpendicularly to the wear-resistant layer surface. Tightly secured by type and the shape of the metal–ceramic interface, uniformly distributed carbides in a hard metal matrix act as a wear barrier, provided that they are resistant to cracking and pulling from the matrix. The lowest wear resistance is observed when chromium carbides are situated parallel to the layer surface due to the increased chance of carbide cracking and pulling from the metal matrix. The shape and orientation of carbides can be controlled to an extent by cored-wire composition selection and proper hardfacing parameters [25].

The microstructure and hardness of the most important properties of wear-resistant materials contribute to resistance to different types of tribological wear. The hardness of the material is directly dependent on the microstructure and is the easiest material property to measure. However, it is often falsely assumed as the single most important criterion in the assessment of wear-resistant materials. The hardness of the two materials can be identical, whereas their wear resistance can be significantly different due to differences in the microstructure and their impact on the hardness [26]. In the majority of cases, however, the increase in hardness coincides with the increase in wear resistance, especially in hardfaced composite wear plates.

Although there is a wide range of hardfacing electrodes, wires, and filler metals commercially available for protection against abrasion wear, the industry is constantly looking for new material solutions. Chromium-rich hardfacing alloys are commonly used due to their low cost and availability;

however, more expensive tungsten- or vanadium-rich alloys offer improved welding properties due to their suitable combination of hardness and toughness. According to Kim et al. [27], covered tubular electrodes with a metallic and carbides core can also be used, especially when abrasion occurs with other wear mechanisms, such as erosion, corrosion, and moderate shock load. The basic surfacing techniques are oxyacetylene powder surfacing using a modular spray-fuse system, manual metal arc surfacing, and submerged arc surfacing. Other surfacing processes can also be used, ranging from conventional techniques, such as gas flame surfacing, to new and modern processes, such as plasma powder transferred arc surfacing and laser metal deposition techniques. The manual metal arc surfacing technique is frequently used for hardfacing applications due to its adaptability and cost-effectiveness.

The maintenance of wear plates, which includes wear assessment and replacement, is a major component of very high operating costs. Several research centers, mainly from the USA, Australia, and China, are currently conducting research on a wear detection system for wear plates in an operational environment, which would improve efficiency, safety, and profitability in the mining sector. Designing such a system, however, requires obtaining significant quantitative and qualitative data on the phenomenon of abrasive wear, which is to be supported by the presented research.

2. Experimental Procedures

2.1. Aim of the Study

The main purpose of this research was to compare the abrasion resistance of the metal-mineral abrasion of the manual metal arc hardfacing layer of a new self-developed covered tubular electrode with wear-resistant plates from leading manufacturers.

This research obtained selected quantitative and qualitative data to develop an innovative system, enabling the assessment of the wear state of hardfaced wear-resistant plates in industrial conditions through the real-time recording of measurement data, such as the operating temperature and vibration. Real-time wear data processed by a computer system were used to predict maintenance requirements, which would enable industrial operators to properly manage, plan, and reduce the cost of maintenance downtime.

Wear resistance determination was performed according to ASTM G65-00: Procedure A [28]. Hardness measurements were made in accordance with ISO 6508-1:2016 [29]. The wear resistance of selected surface layers was compared to Swedish abrasion-resistant steel with a nominal hardness of 400 HBW (~423 HV). The determination of wear character was made by means of macroscopic, confocal microscopy, and scanning electron microscopy examinations, and the properties of the microstructure were examined by light microscopy and X-ray diffraction.

All of the examined wear-resistant hardfaced composite plates were characterized by:

- Comparable geometric properties, i.e., shape, surface waviness, and surface roughness;
- Lack of unacceptable welding defects;
- Hardfaced layer thickness ≥3 mm;
- Base plate thickness ≥5 mm;
- Suitably low base material content in the hardfaced layer.

The range of performed examinations consisted of:

- Wear test (Procedure A);
- Mass loss during wear test determination;
- Reference sample and the examined surface layers density determination;
- Calculation of relative wear resistance;
- Hardness test of the working surface of plates;
- Comparison of the microstructure of the layers;
- Determination of the linear correlation coefficient between the mean layer hardness and wear resistance of the examined composite plates' surface layers.

2.2. Materials' Characterization

Examinations were performed on samples from 12 wear-resistant hardfaced composite plates (Table 1) produced by different arc hardfacing processes. The selected surfacing technologies differed in the metal deposition rate, penetration depth, surfacing speed, surfacing overlay quality, and the final cost of the product. For a given product type and quality requirements, quality assurance is mainly determined by the coating application process, the technique and technological conditions of coating application, and the type of additional material.

Table 1. Hardfaced composite wear-resistant plates with an arc hardfaced layer by T Fe15 and T Fe16 alloys.

Name of the Hardfacing Wear Plate	Group of Alloy in Accordance with EN 14700 [30] Element Contents Hardfacing Technology Typical Applications and Industry
Hardplate™ 100S	T Fe15 C (5.0%), Cr (27.0%), Mn (1.5%), Si (1.5%), Fe (Balance) Self-shielded flux-cored arc welding (FCAW-SS) Applications: mill shell liners, dump truck bin liners, skid protection on loader buckets, and grinding components Industry: cement, mining, power, ceramic, transport, construction
Alphachrom 7000	T Fe15 C (5.4%), Cr (27.0%), Mn (0.2%), Si (1.3%), Fe (Balance) Self-shielded flux-cored arc welding (FCAW-SS) Applications: chute linings, sieve plates, drum linings, augers, crusher components, mixer linings, and crusher Industry: cement, ceramic, quarries, construction, transport, mining, and power
Kalmetall W 143	T Fe15 C (5.0%), Cr (27.0%), Nb (5.0%), Fe (Balance) Self-shielded flux-cored arc welding (FCAW-SS) Applications: screw conveyors, ventilator housings, cyclones, separators, mixer linings, piping components, screens, troughs, and transport channels Industry: cement, ceramic, quarries, construction, transport, mining, and power
CastoDur Diamond Plate® 1001	T Fe15 C (xx%), Cr (xx%), B (xx%), Fe (Balance) Self-shielded flux-cored arc welding (FCAW-SS) Applications: crusher hammer-breaker bars, secondary crushers, chutes, earthmoving equipment, bucket parts, conveyor chain-crushers, conveyor screw-mill augers, and road repair equipment Industry: cement, disposal, quarries, construction, mining, paper, and power

Table 1. *Cont.*

Name of the Hardfacing Wear Plate	Group of Alloy in Accordance with EN 14700 [30] Element Contents Hardfacing Technology Typical Applications and Industry
Abradur 64 [1]	T Fe15 C (7.0%), Cr (24.0%), Nb (7.0%), Fe (Balance) Manual metal arc welding (MMAW) Applications: cane knives, dozer blades, bucket teeth for excavators, worm wheels, and other mining equipment Industry: cement, quarries, construction, mining, transport, and agriculture
CastoDur Diamond Plate® 4624	T Fe15 C (xx%), Cr (xx%), Mn (xx%), Si (xx%), Fe (Balance) [3] Self-shielded flux-cored arc welding (FCAW-SS) Applications: fan blades and housings, cyclones, pipe, and pipe elbows Industry: cement, steel, utilities, oil and sand mining, waste and recycling, pulp and paper
HCCr	T Fe15 C (4.8%), Cr (32.0%), Mn (1.5%), Si (0.8%), Fe (Balance) Self-shielded flux-cored arc welding (FCAW-SS) Applications: mixer blades and components, pipes, scrapers, and mills Industry: cement, ceramic, power, transport, construction, and agriculture
CastoDur Diamond Plate® 4666	T Fe15 C (4.8%), Cr (32.0%), Mn (1.5%), Si (0.8%), B (0.3%), Fe (Balance) Self-shielded flux-cored arc welding (FCAW-SS) Applications: mill linings, pipelines, coal bunker, chain conveyor, bucket, crusher, tans, cyclones, and blast furnace gas systems Industry: power, cement, mining, steel, paper and pulp
HCNb4B	T Fe15 C (4.5%), Cr (23.0%), Nb (4.5), Mn (1.5%), Si (0.8%), B (0.4%), Fe (Balance) Self-shielded flux-cored arc welding (FCAW-SS) Applications: mixer blades, mills, and chutes Industry: cement, ceramic, power, transport, and construction
CastoDur Diamond Plate® 4695	T Fe16 C (5.0%), Cr (20.0%), Mo (10.0%), Nb (10.0%), W (10.0%), Mn (5.0%), B (5.0%), Si (2.0%), Fe (Balance) Flux-cored arc welding gas-shielded (FCAW-GS) Gas mixture type Ar-CO_2 Applications: screws, rotor and housing, feeding tube, knives and wear pieces, hammers, fan blades, and separators Industry: waste and recycling, cement, ceramic, power, quarries and mining

Table 1. *Cont.*

Name of the Hardfacing Wear Plate	Group of Alloy in Accordance with EN 14700 [30] Element Contents Hardfacing Technology Typical Applications and Industry
Vecalloy 752 Plate®	T Fe16 C (3.5%), Cr (11.5%), Mo (1.2%), Nb (5.0%), B (4.8%), Fe (Balance) Flux-cored arc welding gas-shielded (FCAW-GS) Argon gas Applications: shaker screens, grader blades and other ground-engaging tools, chute blocks, inner pipe walls, mill liners, slurry pipes, shovel wear packages, cutter rings, primary and secondary crusher teeth, grouser risers, and other mining applications Industry: cement, agriculture, quarries, mining, ceramic, and power
PHWPPatented hardfacing wear plate [2]	T Fe16 C (5.0%), Cr (23.0%), Mo (xx%), Nb (xx%), Si (xx%), B (xx%), W (xx%), V (xx%), Fe (Balance) [3] Manual metal arc welding (MMA) with patented cover tubular electrode Potential applications: screeners, blast furnace hoppers, extractor fans Industry: iron and steel, power, quarries, and cement

Remarks: [1] Surface layer manufactured in a three-pass process; [2] surface layer manufactured in a one-pass process; [3] chemistry composition proprietary to its own copyright patent [31].

The advantages of the manual metal arc deposition process are its versatility and the availability of a wide range of hardfacing alloys. The set-up time is short, making the process ideal for small jobs and short production runs. For a working surface, the manual metal arc deposition process has lower deposition rates than the flux-cored arc welding (FCAW) deposition processes, which use a higher current. The process has a low operator duty cycle, with the operator spending significant time changing electrodes and chipping slag. These two factors combine to limit the application of this process, especially if high production rates are required.

The flux-cored arc welding gas-shielded (FCAW-GS) deposition process has the advantage of deeper penetration and a higher charging rate than the manual metal arc process. Thus, the welding process is becoming more economical for jobs in welding workshops. Flux-cored arc welding self-shielded (FCAW-SS) filler metal eliminates the need for external shielding gas and tolerates stronger wind conditions without causing porosity.

The compositions of nine layers contain alloys from group T Fe15, and three layers contain alloys from group T Fe16 (including a hardfaced plate with the name PHWP). The obtained results were compared to the reference abrasion-wear-resistant steel plate (Table 2).

The wear-resistant plate with the name PHWP was obtained by the arc welding process with a tubular electrode 4 mm in diameter. The metallic core of the electrode has a patented metallic composition with a rutile cover. A cross-sectional view of the electrode is presented in Figure 1.

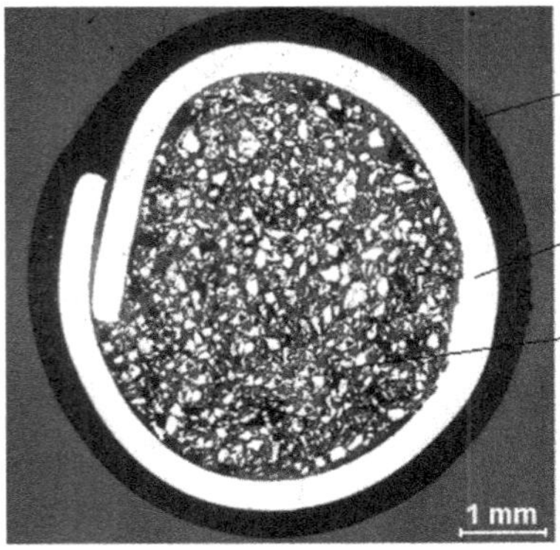

Figure 1. Macrostructure view of the cross-section of the new patented tubular electrode [31].

Table 2. Characteristics of the abrasion-resistant steel.

Name of the Abrasion-Resistant Steel	Steel in Accordance with EN 10029 [32] Element Contents Metal Hardness
Hardox 400 Steel	C (0.32%), Si (0.7%), Mn (1.6%), P (0.025%), S (0.01%), Cr (1.4%), Ni (1.5%), Mo (0.6%), B (0.004%), Fe (Balance) 38–44 HRC (372–435 HV)

The samples were cut from the inner regions of the examined hardfaced composite wear-resistant plates and reference material.

2.3. *Methodology of Research*

For each examined sample, a visual test (VT) on the sample surface was performed to identify potential unacceptable welding defects, such as cracks directed longitudinally to the hardfaced seam, other discontinuities, porosity, irregularity of shape, and lack of fusion. Metal-mineral wear resistance was determined by strictly adhering to the procedure disclosed in the ASTM G65 standard [28]. The determination of surface layer structure and other properties was based on metallographic macro- and microscopic examinations, measuring the base material content in the hardfaced layer and the hardness tests on the working surface.

2.3.1. Nondestructive Testing: Visual Testing

Visual testing of the wear plates was performed in accordance with the procedures, materials, and equipment from the ISO 17637 standard [33]. Visual testing was based on a direct inspection of the antiabrasive wear of the hardfaced layer procedure. Before testing, the surface subjected to observation was cleaned.

2.3.2. Hardness Measurements

The hardness testing of the reference material (Hardox 400 steel) and the hardfaced layer was performed using the Rockwell method Scale C (HRC) on the Nexus 610 RS stationary hardness tester (Innovatest Europe BV, Maastricht, Netherlands). Hardness testing was performed according to ISO 6508 [29]. The test load used was 150 kgf (total test force of 1.471 kN). Hardness tests were performed in five test points on the surface, finished by grinding selected layers manufactured by the arc hardfacing process. The way the measurement points were located on the wear-resistant layer surface is presented in Figure 2.

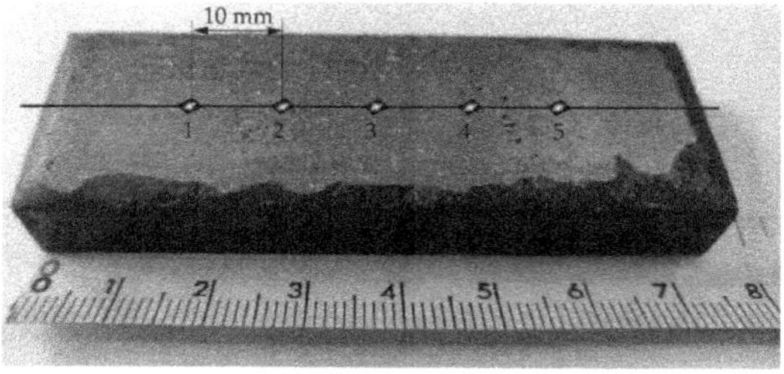

Figure 2. Measurement points location on the wear-resistant layer surface.

The carbide's microhardness measurement was made with the Vickers method at the polished cross-section of the samples. A Sunpoc SMV-1000X (Guizhou Sunpoc Tech Industry Co., Ltd., Guizhou, China) microhardness tester with a diamond pyramidal indenter was used. The load could be varied from 5 to 500 g in fixed steps. The duration was kept at 25 s throughout the study. The examinations were performed in conformity to the ISO 6507-1:2018 standard [34].

2.3.3. Abrasive Wear Test

The metal-mineral wear resistance of all test samples and the reference material (Hardox 400 steel) was carried in accordance with ASTM G 65-00: Standard Test Method for Measuring Abrasion Using the Dry Sand/Rubber Wheel Apparatus, Procedure A (Figure 3) [28].

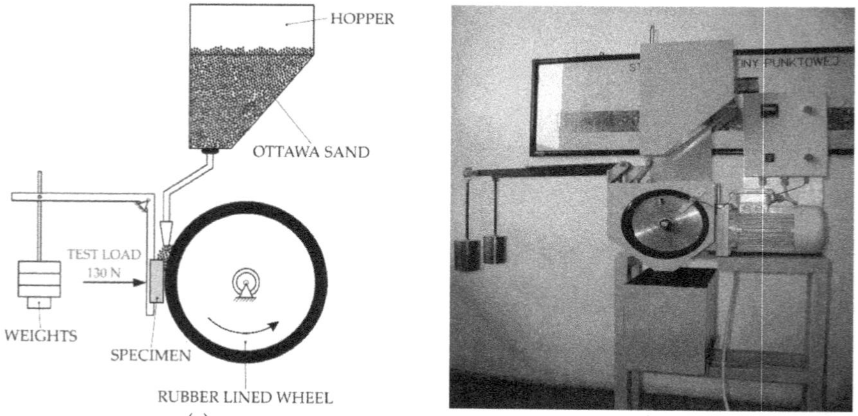

(a) (b)

Figure 3. Schematic diagram of ASTM G65, Procedure A: abrasive wear resistance test (**a**) and apparatus overview (**b**) [35].

The rubber wheel test outlined in ASTM G65, first introduced in 1980, is the most widely applied method in determining material wear resistance. The abrasive used in the test procedure is quartz sand with a grain size of 50–70 mesh (0.300–0.212 mm). The abrasive is fed from a gravity hopper with a 250–350 g/min feed rate. The counterface in the test is rubberized by a hard wheel compound with a total dimension of ϕ 228 mm × 12.7 mm. The specimen under investigation is pressed to the counterface by forces of, depending on the procedure chosen, 45 N or 135 N. The test length is determined by the number of wheel revolutions of the counterface and is contained in the range of 100–6000 revolutions (wheel speed of 200 r/min). The measured results comprised the volume of wear and an examination of the worn surface. According to Hejwowski, under the test condition, the worn material is transferred to the surface of the abrasive grains [36]. In the rubber wheel test, the abrasive grains may rotate in the friction zone or temporarily attach to the rubber material. The movement and interaction of the abrasive material are dependent not only on the test parameters (abrasive flux and type, test force, and counterface revolution speed) but also on the hardness of the sample under investigation. The rolling of abrasive grains is facilitated by a low test force and the limited tested sample hardness. For high test forces and an elevated sample hardness, the sliding movement of the abrasive grains is promoted.

In order to determine the wear resistance of the selected wear-resistant composite plates, two samples (75 mm × 25 mm × 10 mm) were cut from the inner area of the plate. During the test procedure, the counterface wheel made 6000 revolutions, samples were pressed with the force of 130 N, and the abrasive materials' (A. F. S. Testing Sand 50–70 mesh) expense was 335 g/min (Figure 4). The ASTM G65-00 wear test time was 30 min.

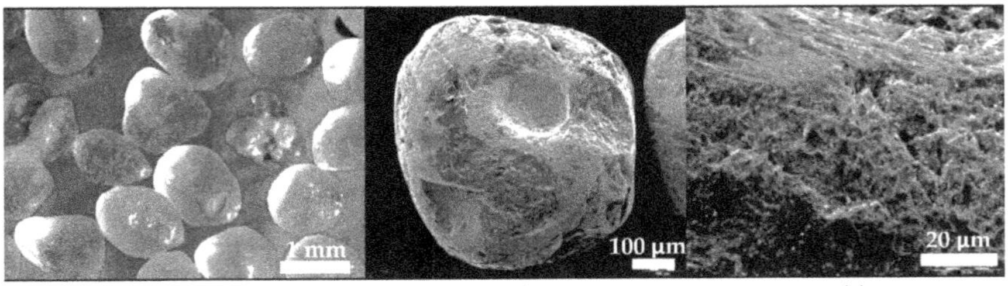

(a) (b) (c)

Figure 4. View of the abrasive material particles (A.F.S Testing Sand 50–70) used in ASTM G65-00, Procedure A: abrasive wear resistance test: (**a**) a quartz sand grain fraction, (**b**) a single grain of quartz sand, (**c**) the surface structure of quartz sand grain.

Before and after the test procedure, the samples subjected to wear testing were weighed on a laboratory scale with an accuracy of 0.0001 g. The mean density of the tested hardfaced layers and reference Hardox steel was determined by measuring the weight of three specimens in air and during submersion in liquid at room temperature (Table 3). The measured mass loss during the test procedure, in conjunction with the determined mean density, was used to calculate volumetric loss according to Formula (1). A similar fraction and distribution of the matrix reinforcing phase over the entire cross-section of the surface layer worn during the 30 min test period were assumed.

$$\text{Volume loss}\left[\text{mm}^3\right] = \frac{\text{mass loss [g]}}{\text{density}\left[\frac{\text{g}}{\text{cm}^3}\right]} \times 1000 \tag{1}$$

The abrasive wear mechanism was assessed in accordance with the criterion, which was the quotient of the cross-sectional area of the sum of the two-sided upsets of the material next to the deepest crack F_1 and crack cavity F_2 [37–39]. The loss of material in the surface layer during abrasive wear was classified as:

- Grooving related to plastic deformation of the contact areas and upset of the material on both sides of the furrow ($F_1/F_2 = 1$);
- Microcutting ($F_1/F_2 = 0$);
- Scratching when the material was partially plastically deformed and partially cut, ($0 \leq F1/F2 \leq 1$), Figure 5.

Table 3. Abrasive wear resistance test results of wear plates hardfaced with alloys from groups T Fe15 and T Fe16 in relation to the Hardox 400 abrasive wear resistance.

Specimen Designation	Spec. Number	Mass Before Test, g	Mass After Test, g	Mass Loss [1], g	Average Mass Loss, g	Clad Layer Density, g/cm^3	Average Volume Loss, mm^3	Relative Abrasive Wear Resistance [2]	Diluted Weld Metal, %
Hardplate P100S	p01	130.6009	130.3432	0.2577	0.2343	7.3229	31.9955	5.94	22.1
	p02	129.6213	129.4118	0.2095					
Alphachrom 7000	p01	157.5045	157.2484	0.2561	0.2328	7.4324	31.3223	6.06	25.4
	p02	157.0426	156.8331	0.2095					
KalmetallW 143	p01	91.8884	91.5677	0.2575	0.2341	7.4943	31.2371	6.08	23.6
	p02	87.1791	86.8916	0.2107					
CDP 1001	p01	173.3469	173.1099	0.2371	0.2155	7.4139	29.0670	6.53	19.1
	p02	178.7772	178.5833	0.1940					
Abradur 64 [3]	p01	136.2893	136.0602	0.22913	0.2083	7.2144	28.8728	6.58	20.7
	p02	139.6675	139.4800	0.18747					
CDP 4624	p01	134.2440	134.0815	0.1625	0.1806	7.4191	24.3426	7.80	21.0
	p02	134.0617	133.8630	0.1987					
HCCr	p01	172.0538	171.8583	0.1955	0.1777	7.4756	23.7707	7.99	22.8
	p02	160.5386	160.3787	0.1599					
CDP 4666	p01	161.9005	161.7996	0.1009	0.1121	7.4894	14.9678	12.69	20.7
	p02	164.5962	164.4729	0.1233					
HCNb4B	p01	174.3838	174.2856	0.0982	0.1091	7.3559	14.8316	12.81	21.3
	p02	176.8729	176.7529	0.1200					
CDP 4695	p01	155.4632	155.3738	0.0894	0.0993	7.7208	12.8614	14.77	22.5
	p02	155.8611	155.7519	0.1092					
Vecalloy 752	p01	167.8436	167.7602	0.0834	0.0758	7.6816	9.8677	19.25	22.7
	p02	168.2761	168.2079	0.0682					
PHWP [4]	p01	161.8431	161.7440	0.0991	0.0999	7.7112	12.9617	14.65	17.2
	p02	163.5409	163.4401	0.1008					
Hardox 400	p01	116.2260	114.7526	1.4734	1.4744	7.7620	189.9510	1.00	-
	p02	116.7526	115.2773	1.4753					

Remarks: [1] Mass loss in 30 min; [2] relative abrasive wear resistance to Hardox 400 steel; [3] surface layer manufactured in a three-pass process; [4] surface layer manufactured in a one-pass process.

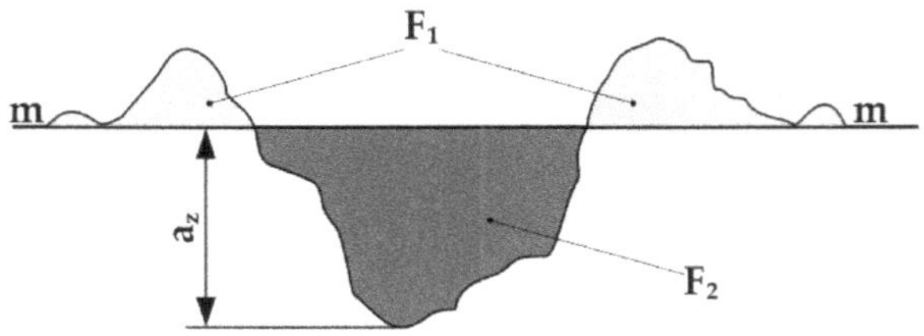

Figure 5. The abrasive wear mechanism criterion: a_Z—groove depth; m-m—reference line.

2.3.4. Metallographic Examination and X-ray Diffraction Analysis

Microscopic examinations were performed on standard metallographic specimens. The etchant chemical composition and etching parameters were determined individually for each hardfaced layer. The observation and acquisition of the macro- and microstructure of the specimens were performed using the Olympus SZX7 (Olympus Corporation, Tokyo, Japan) stereoscopic microscope, the Olympus GX 71 inverse metallographic microscope (Olympus Corporation, Tokyo, Japan), the Zeiss Smartproof 5 confocal microscope (Carl Zeiss AG, Oberkochen, Germany), and the Zeiss Supra 25 scanning microscope (Carl Zeiss AG, Oberkochen, Germany). Precise determination (surface and volumetric) of the surfaced layer chemical composition was performed by means of energy-dispersive spectroscopy (EDS).

The obtained macroscopic images enabled the determination of surface layer thickness as well as the base material content in the hardfaced layer. The dilution of weld metal, U, was calculated according to Formula (2), as the ratio between F_{BM}, the area of the fusion-based metal and a sum of F_{BM} and F_R, the area of reinforcement, Figure 6.

$$U = \frac{F_{BM}}{F_R + F_{BM}} \times 100\% \tag{2}$$

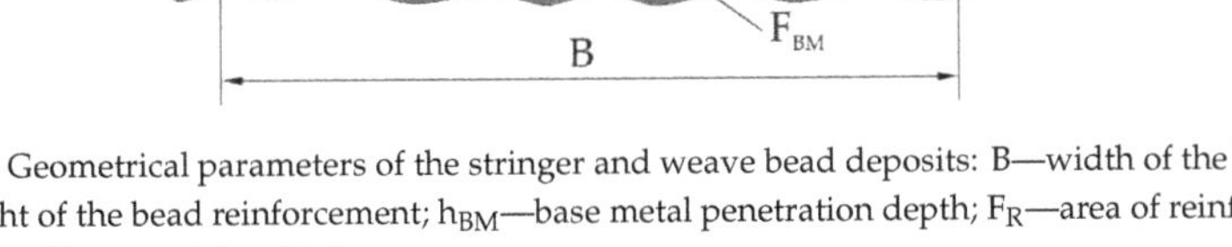

Figure 6. Geometrical parameters of the stringer and weave bead deposits: B—width of the bead face; h_R—height of the bead reinforcement; h_{BM}—base metal penetration depth; F_R—area of reinforcement; F_{BM}—area of base metal melted.

The X-ray diffraction phase examinations were carried on X'Pert Pro PANalytical (Malvern Panalytical Ltd., Malvern, UK) diffractometer with Cu lamp (λ = 1.54056 nm). The samples were examined in a Bragg–Brentano geometry.

3. Results and Discussion

3.1. Nondestructive Testing: Visual Testing Results

During the visual tests of selected hardfaced composite wear-resistant plates produced by the automated FCAW and MMA methods, imperfections of cracks (100) perpendicular to the hardfacing direction and surface spatter (602) were found, as presented in Figure 7.

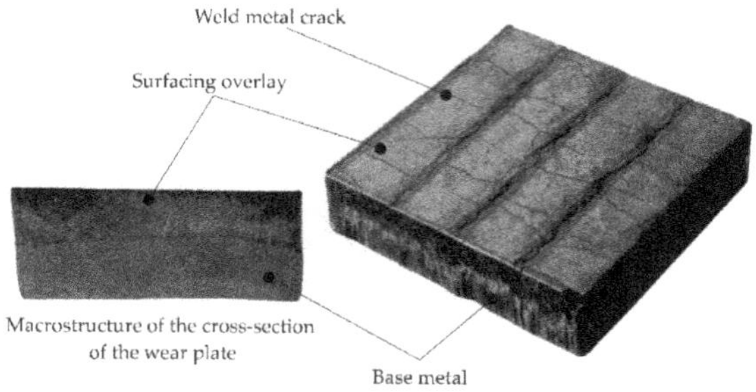

Figure 7. Overview and cross-section of the hardfaced composite wear-resistant plate produced by the automated flux-cored arc welding self-shielded (FCAW-SS) alloy.

The hardfacing process is an important process associated with welding; therefore, it is necessary for the verification of the applied hardfacing technology according to accepted standards. For hardfaced composite wear-resistant plates, the ISO 15614-7 [40] standard is suitable. However, applying this standard precisely can be difficult. In acceptance tests, wear-resistant hardfaced layers are often unable to conform to acceptable quality levels and cannot be accepted without remarks. Regular transverse cracks in hardfaced layers contribute to the reduction of stress levels in hardfaced elements and can act as a lubricant reservoir. In the context of product application, transverse cracks in the hardfaced layer can be deemed acceptable.

3.2. Hardness Measurements Test Results

The results of the mean Rockwell hardness test on the working surface of wear-resistant plates, arc hardfaced with alloys from groups T Fe15 and T Fe16 and subjected to examinations, were significantly higher compared to the Hardox 400 reference material. The results from the five measurement points for each specimen are presented in Figure 8. The hardness measurements placed the hardness of examined plates in the range of 60–70 HRC (from approximately 700 HV to over 900 HV), which is mostly consistent with additional material characteristics and published data in scientific articles [1,41–44]. Only the hardness of the Alphachrom 700 wear-resistant plate's working surface was lower than that declared by the manufacturer, which could be caused by high dilution of the weld metal. The average value of the hardness test for the PHWP wear-resistant plate was the lowest among the T Fe16 alloys. The reason for this phenomenon is the high plasticity of the metallic matrix. No major dispersion in the hardness results was observed.

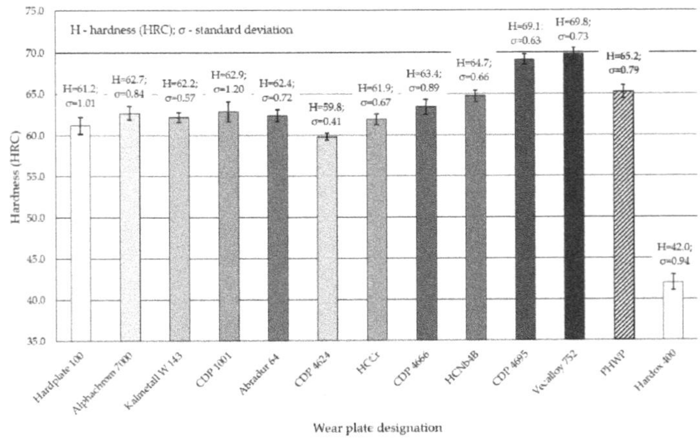

Figure 8. Comparison of the working surface hardness arc hardfaced with alloys from groups T Fe15 and T Fe16: wear-resistant plates and the Hardox 400 reference material.

3.3. Abrasive Wear Test Results

The results obtained in the metal-mineral wear resistance test of selected wear plates hardfaced with alloys from groups T Fe15 and T Fe16 (Table 3) were compared to the wear resistance of a Hardox 400 steel plate, resulting in a relative abrasive wear resistance number. The surface view of the samples after the metal-mineral abrasive wear resistance test performed according to ASTM G65-00: Procedure A is presented in Figure 9. Selected surface areas of the representative samples after the metal-mineral abrasive wear resistance test observed under a confocal microscope and scanning microscope are presented in Figures 10 and 11. Among the composite hardfaced wear-resistant plates examined, two produced by the automated FCAW-GS method, with trade names CastoDur Diamond Plate® 4695 and Vecalloy 752 Plate®, deserve special attention. The corresponding relative abrasive wear resistance was 15 and 19 times higher than the Hardox 400 reference material. The average results of mass loss in the ASTM 65 test of the CastoDur Diamond Plate® 4695 and Vecalloy 752 Plate® plates were presented by Górka et al. [45] and Gucwa et al. [46]. Under the conditions of the experiment performed, the lowest relative metal-mineral abrasive wear resistance was achieved by the hardfaced composite wear-resistant plate under the trade name Hardplate™ 100S (Figure 12). The average value of the metal-mineral wear resistance of the PHWP plate was higher than the result obtained for the CastoDur Diamond Plate® 4695 plate made by NanoAlloy using the FCAW-SS automated hardfacing process.

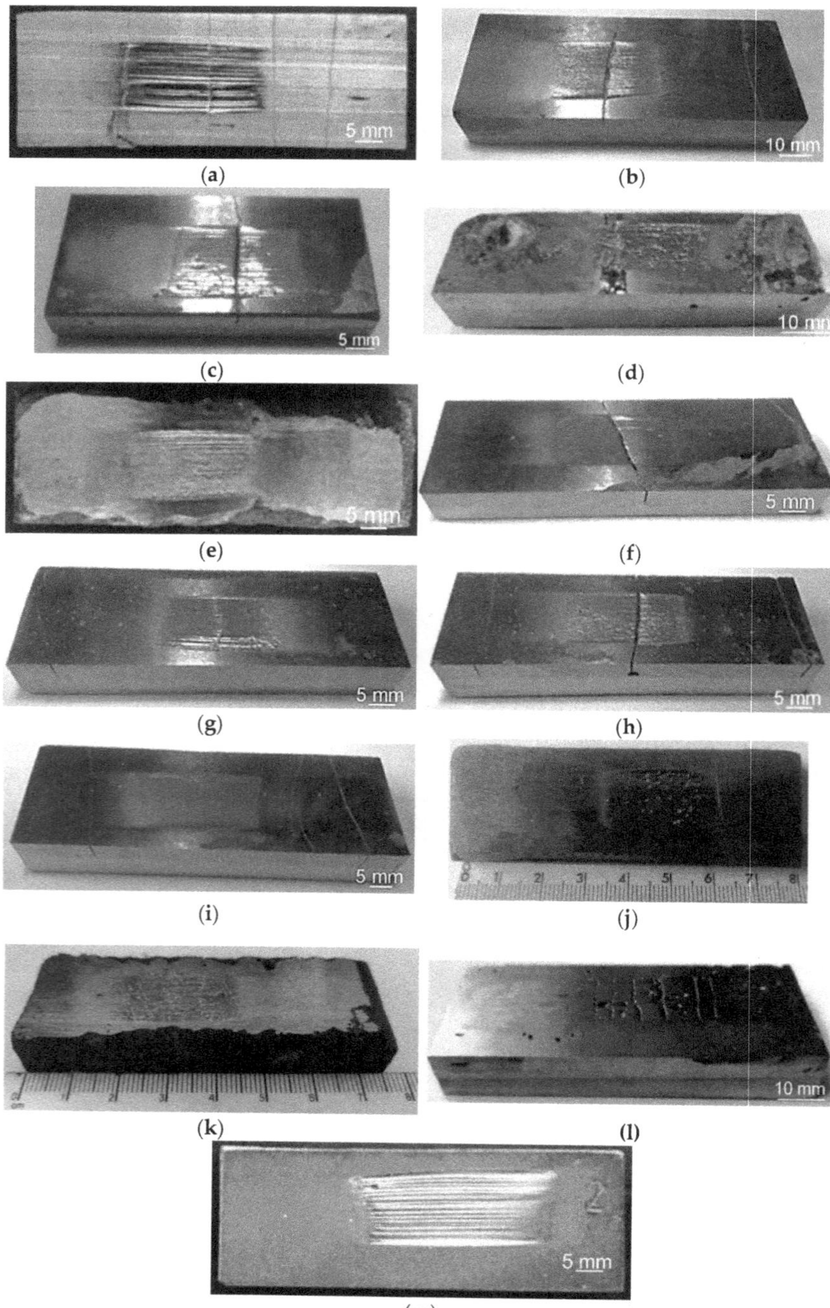

Figure 9. Surface view of the representative sample after the metal-mineral abrasive wear resistance results performed according to ASTM G65-00: Procedure A. Samples from the: (**a**) Hardplate™ 100S wear plate; (**b**) Alphachrom 7000 wear plate; (**c**) Kalmetall W 143 wear plate; (**d**) CDP® 1001 wear plate; (**e**) Abradur 64 wear plate; (**f**) CDP® 4624 wear plate; (**g**) HCCr wear plate; (**h**) CDP® 4666 wear plate; (**i**) HCNb4B wear plate; (**j**) CDP® 4695 wear plate; (**k**) Vecalloy 752 Plate wear plate; (**l**) PHWP wear plate; (**m**) Hardox 400 abrasion-resistant steel.

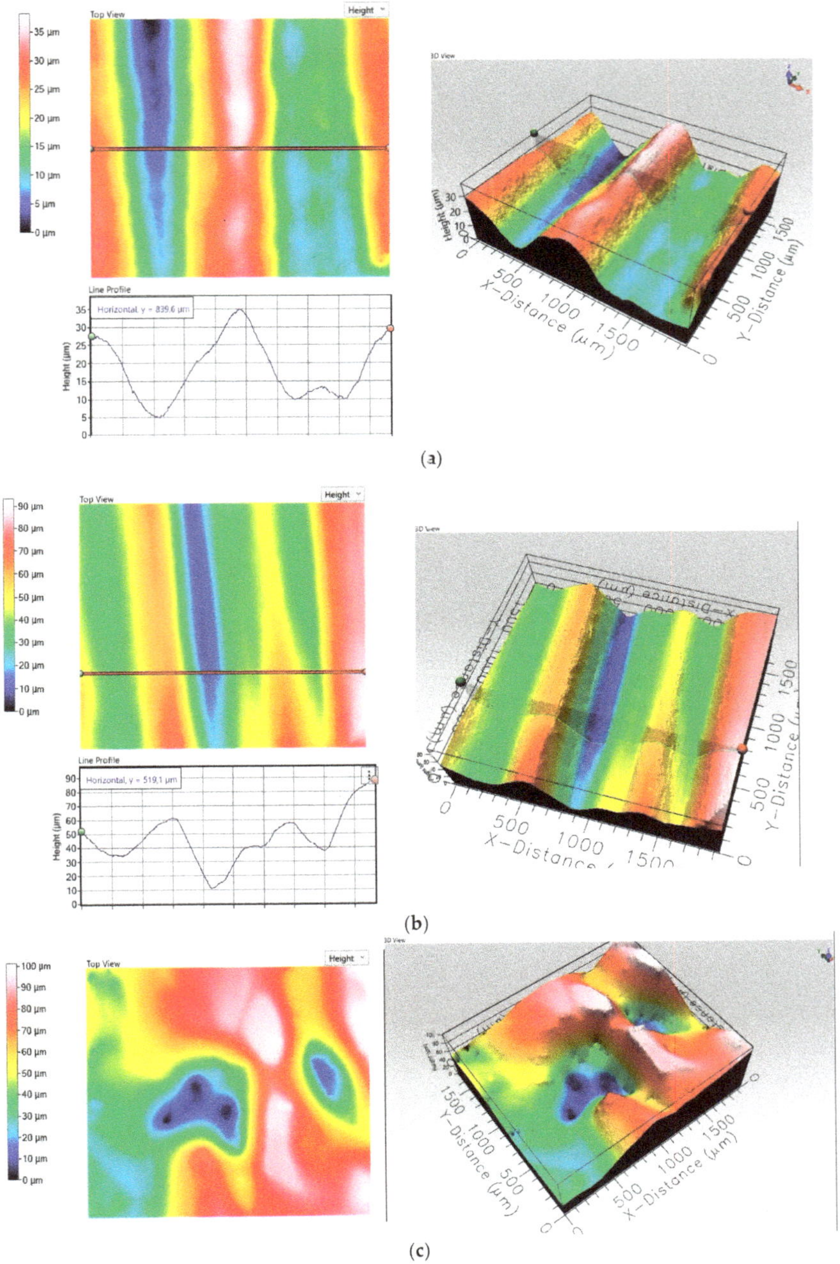

Figure 10. View of the surface of the representative sample after the metal-mineral abrasive wear resistance test observed under a confocal microscope: (**a**) HCCr wear plate; (**b**) CDP® 4666 wear plate; (**c**) PHWP wear plate.

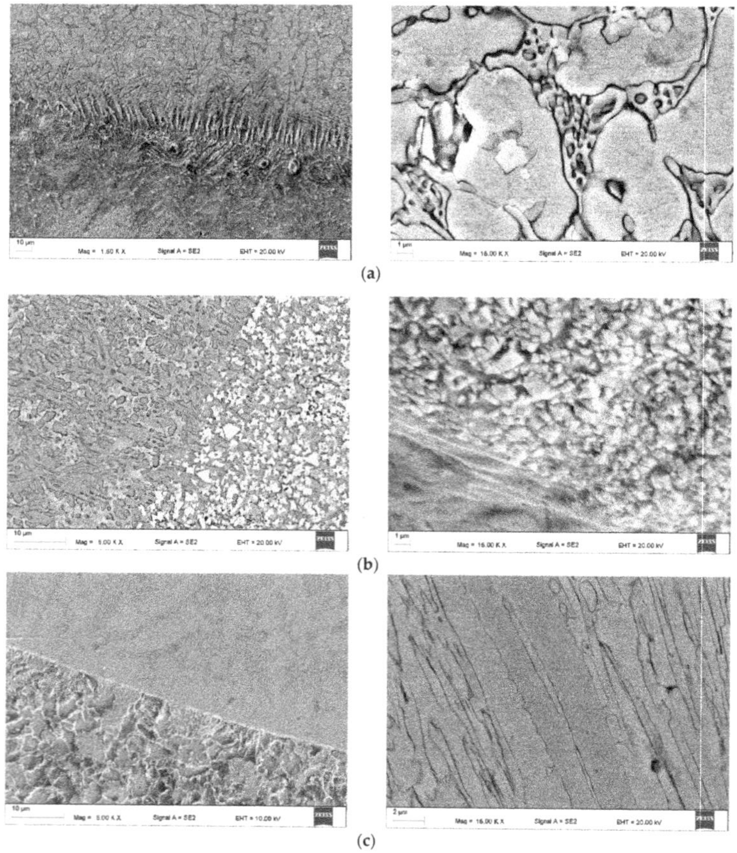

Figure 11. View of the surface of the representative sample after the metal-mineral abrasive wear resistance test observed under a scanning electron microscope: (**a**) HCCr wear plate; (**b**) CDP® 4666 wear plate; (**c**) PHWP wear plate.

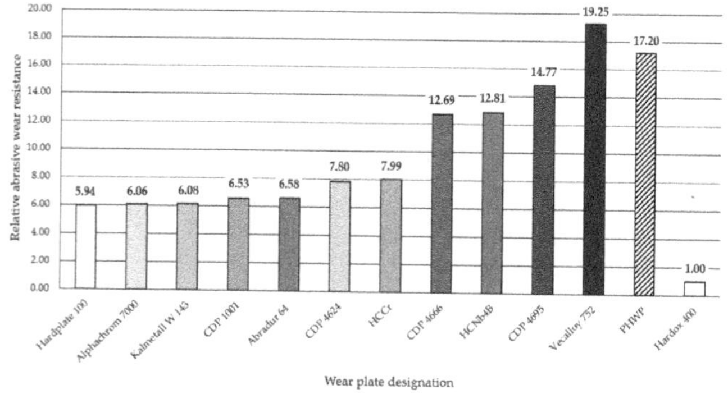

Figure 12. Relative metal-mineral abrasive wear resistance (ASTM 65-00, procedure A) of wear plates hardfaced with alloys from groups T Fe15 and T Fe16 in relation to Hardox 400 abrasive wear resistance.

3.4. Metallographic Test Results and Results of the X-ray Diffraction Analysis

The microscopic metallographic examinations enable the determination of the microstructures of the plated layer of wear-resistant plates hardfaced with alloys from groups T Fe15 and T Fe16 and the Hardox 400 steel reference material (Figure 13). Moreover, X-ray diffraction analysis allowed determining the phase composition. The selected diffractograms are presented in Figure 14. The results of X-ray diffraction analysis were confirmed by means of energy-dispersive spectroscopy (EDS) (Figure 15).

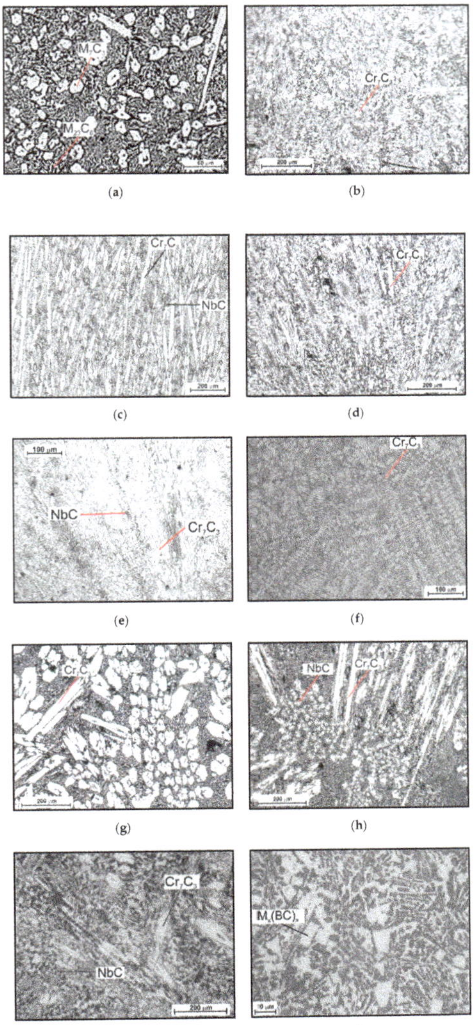

Figure 13. *Cont.*

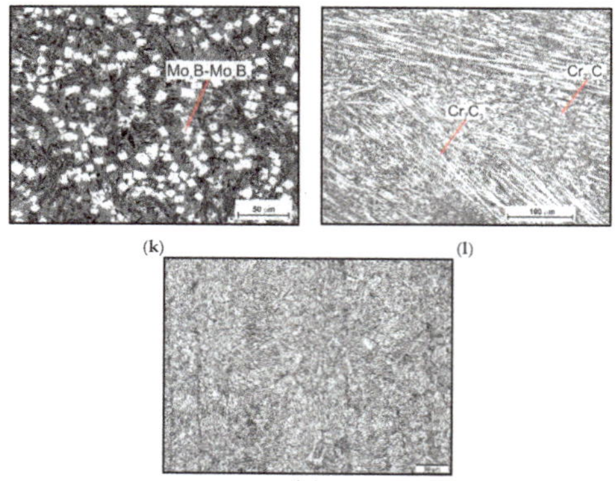

(k) (l)

(m)

Figure 13. Microstructure of wear-resistant material of wear plates hardfaced with alloys from groups T Fe15, T Fe16, and the reference material: (**a**) Hardplate™ 100S wear plate; (**b**) Alphachrom 7000 wear plate; (**c**) Kalmetall W 143 wear plate; (**d**) CDP® 1001 wear plate; (**e**) Abradur 64 wear plate; (**f**) CDP® 4624 wear plate; (**g**) HCCr wear plate; (**h**) CDP® 4666 wear plate; (**i**) HCNb4B wear plate; (**j**) CDP® 4695 wear plate; (**k**) Vecalloy 752 Plate wear plate; (**l**) wear plate PHWP; (**m**) Hardox 400 abrasion-resistant steel.

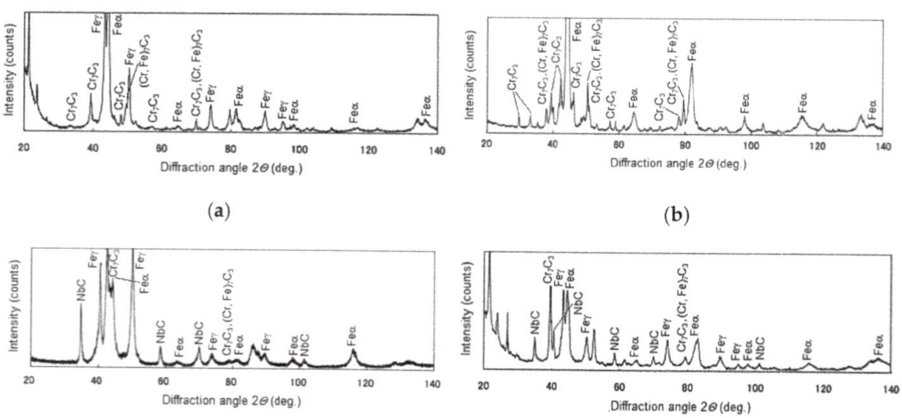

(**a**) (**b**)

(**c**) (**d**)

Figure 14. *Cont.*

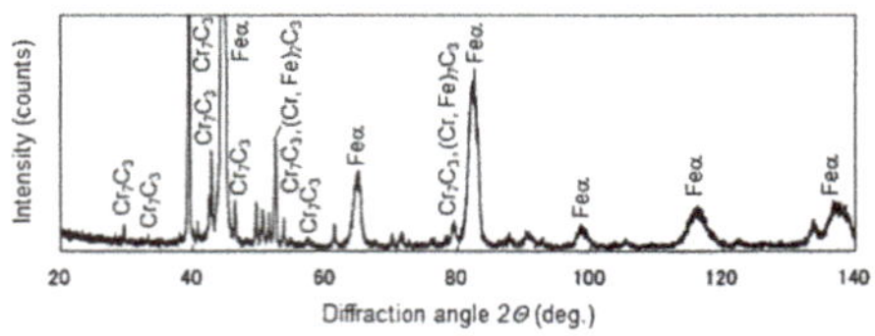

(**e**)

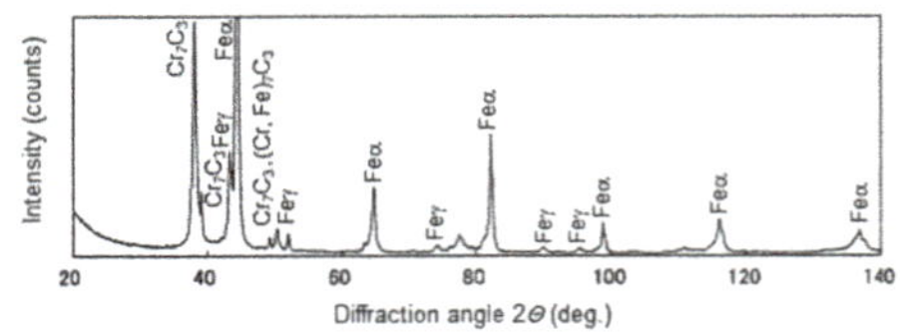

(**f**)

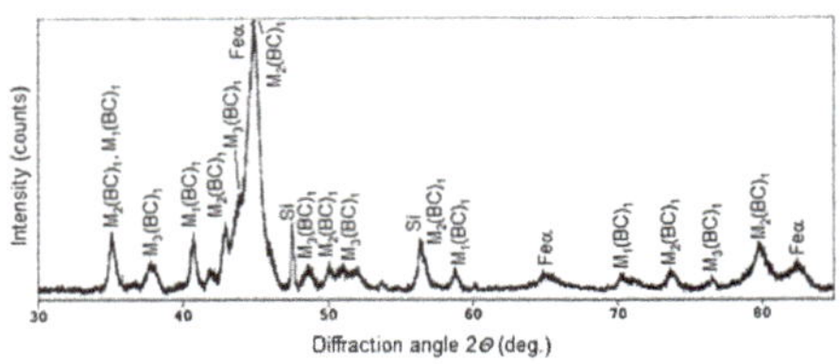

(**g**)

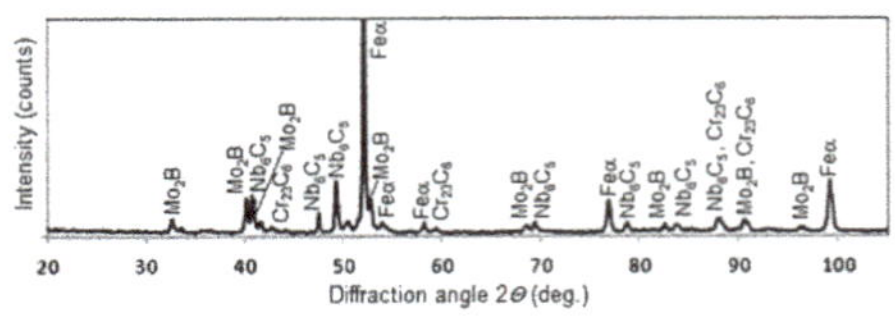

(**h**)

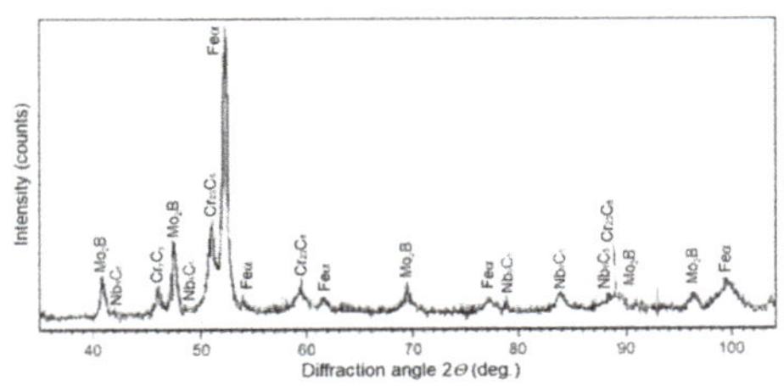

(**i**)

Figure 14. Selected diffractograms of the wear-resistant material of wear plates hardfaced with alloys from groups T Fe15 and T Fe16: (**a**) Hardplate™ 100S wear plate; (**b**) Alphachrom 7000; (**c**) Kalmetall W 145 wear plate; (**d**) Abradur 64; (**e**) CDP® 4624 wear plate; (**f**) HCCr wear plate; (**g**) CDP® 4695 wear plate; (**h**) Vecalloy 752 Plate® wear plate; (**i**) PHWP wear plate.

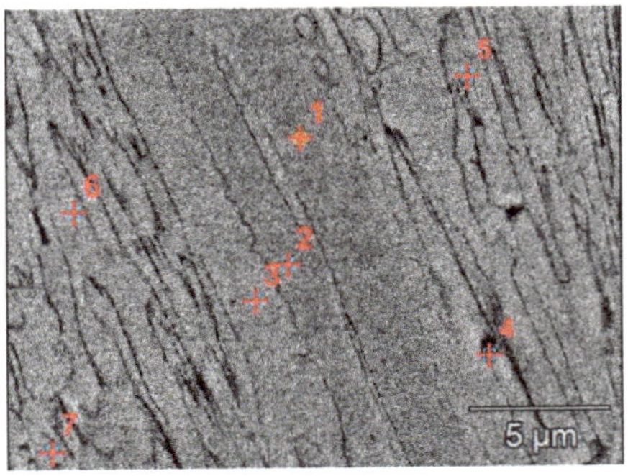

(**a**)

Figure 15. *Cont.*

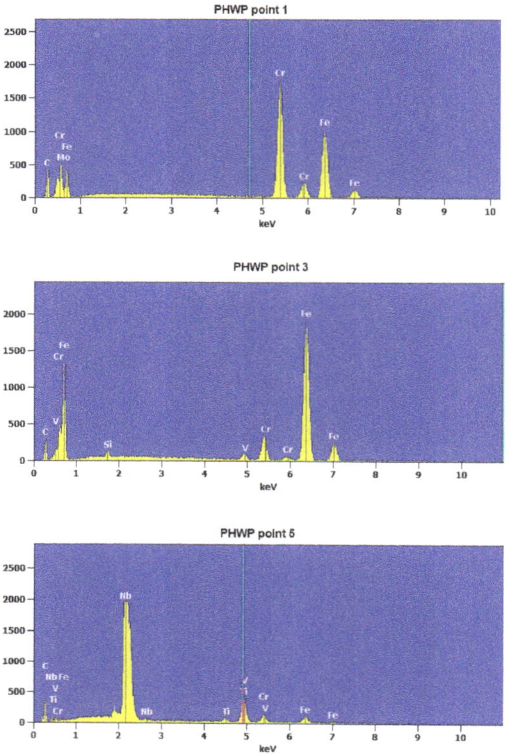

Weight %	C	Ti	Si	Cr	Nb	V	Fe	Mo
PHWP Point 1	8.15	-	-	25.96	-	-	65.73	0.15
PHWP Point 3	7.03	-	1.78	22.98	-	0.97	67.23	-
PHWP Point 5	5.51	1.02		20.60	7.03	0.86	64.97	-

(**b**)

Figure 15. Sample BSE (Back Scattered Electrons) image of the surface layer microstructure with the results of EDS (Energy Dispersive Spectroscopy) point microanalysis, mag., 15,000×, high tension, 20 kV (PHWP sample): (**a**) view of the carbide structure; (**b**) point chemical analysis of the study area (measurement points: 1, 3, 5).

The metallographic analysis of the wear-resistant plates hardfaced with an abrasion-resistant layer of group T Fe 15 alloys revealed the existence of the martensitic Fe-Cr-C ternary system phase in all of the samples and an austenitic Fe-Cr-C ternary system phase in some of the samples. These two phases formed a metal matrix. For the hardfaced layers with a high chromium cast alloy composition, it is presumed that a primary austenitic phase was partially transformed into a martensitic phase during the hardfacing thermal cycle. The degree of transformation varied among the examined wear-resistant plates. Several metal carbides can be observed in the X-ray diffractogram patterns. The main identified metal carbide in all the layers produced by hardfacing with alloys from group T Fe15 is the primary Cr_7C_3 carbide, and $Cr_{23}C_6$ eutectic carbides can sometimes be observed. Moreover, in the abrasion-resistant layers of Kalmetall W 143, Abradur 64, CastoDur Diamond Plate® 4666, and HCNb4B plates, indices of NbC phases were found. The microstructures of abrasion-resistant layers hardfaced with alloys from group T Fe16 demonstrated a more complex composition characterized by the presence of borocarbides and molybdenum borides.

The hardfaced layer of the Hardplate™ 100S wear-resistant plate (Figure 13a) was characterized by a chromium cast iron structure with a fine Cr_7C_3 carbide precipitation of approximately 20 μm and a microhardness in the range of 950–1450 HV, uniformly distributed in the austenitic matrix (Figure 14a). The weld metal dilution was around 22%.

A similar microstructure was observed in the Alphachrom 7000 wear-resistant plate (Figure 13b). The hardfaced layer was characterized by an austenitic structure with a precipitation of primary Cr_7C_3 chromium carbides oriented perpendicularly to the surface (Figure 14b). The weld metal dilution was the highest and exceeded 25%.

In contrast, the hardfaced layer of the Kalmetall W 145 wear-resistant plate (Figure 13c) was identified as a chromium and carbon-rich iron alloy with a hard martensitic matrix structure, with 30 a vol % of primary Cr_7C_3 chromium carbides, a microhardness around 2200 HV, and niobium carbides NbC with a microhardness around 2400 HV (Figure 14c). The carbides were oriented parallel to the working surface, which increased abrasive wear resistance. The calculated weld metal dilution was under 24%.

The microstructure of the hardfaced layer of the CastoDur Diamond Plate® 1001 wear-resistant plate (Figure 13d) was composed of primary Cr_7C_3 chromium carbides with a microhardness in the range of 1500–2200 HV and M_2B metal borides with a microhardness of 1800 HV, regularly and densely situated in a plastic metal matrix. The base metal content was low, which promotes uniform and high abrasive wear resistance on the whole cross-section [26,47]. The hardfaced layer was distinguished by the lowest weld metal dilution rate of just over 19%. Abrasive wear of the surface layer of the wear plate resulted from microcutting. In the area of the abrasion, material discontinuities were present, caused by shearing unevenness of the surface with abrasive. The hardfaced layer of the Abradur 64 wear-resistant plate was characterized by a hypereutectic microstructure with chromium and niobium carbide precipitations (Figure 13e). The metallographic examination revealed the existence of a buffer layer with a high chromium, nickel, and boron content, presumed to act as a barrier for crack development. The MMA hardfaced wear-resistant layer of the Abradaur 64 plate, similar to the CastoDur Diamond Plate® 1001 plate, was characterized by a uniform distribution of chromium and niobium carbides on the whole cross-section (Figure 14d). The weld metal dilution under 21%.

The hardfaced layer of the CastoDur Diamond Plate® 4624 wear-resistant plate (Figure 13f) was characterized by a high chromium cast alloy microstructure of approximately 30 vol % of primary Cr_7C_3 chromium carbides (Figure 14e) 1800–2000 HV in hardness, uniformly distributed in the matrix. The carbide volume and distribution positively impacted the abrasive wear resistance of the plate. The weld metal dilution was slightly above 21%.

In contrast, the matrix of the hardfaced layer of the HCCr wear-resistant plate (Figure 13g) was austenitic in structure, with a high volume (over 50 vol %) of precipitations, mainly Cr_7C_3 chromium carbides (Figure 14f). Due to the high chromium and carbon content in the weld, a metal hardfaced layer with high abrasion resistance was achieved. The weld metal dilution was under 23%.

The above-characterized wear-resistant layers were subjected to intensive abrasive wear. The volumetric material loss was in the range of 23–32 mm^3. The wear mechanism was deeply grooved, with plastic deformation of the contact area and raising of both groove edges (Figure 9a,b) or microcutting (Figure 9c–g). The wear was accompanied by the formation of surface defects, e.g., scuff marks or microcraters.

The hardfaced layer of the CastoDur Diamond Plate® 4666 wear-resistant plate (Figure 13h) was characterized by a supereutectic ferrous alloy microstructure with a high chromium content. The microstructure was composed of a high fraction (over 50 vol %) of hard (2000–2100 HV) Cr_7C_3 chromium carbides, niobium carbides NbC, and chromium borides Cr_2B, as well as other metal compounds distributed in a hard–austenitic matrix. The calculated weld metal dilution was under 21%.

The hardfaced layer of the HCNb4B plate (Figure 13i) was characterized by an alloyed austenitic microstructure with optimally oriented Cr_7C_3 chromium carbide precipitations and hard niobium carbides NbC and a hardness of approximately 1800 HV. The weld metal dilution was over 21%.

The structure of overlays produced on the CastoDur Diamond Plate® 4695 wear plate (single weld layer on a carbon steel substrate) is presented with eight micron-scale images in Figure 13j. The high volume fraction of the ultrahard complex borocarbides ($M_{23}(BC)_6$), metal carbides (MC), and metal borides (M_2B) is finely distributed within a mesomorphous α–Fe alloy matrix (Figure 14g). The abrasion-resistant layer possessed very high abrasive wear resistance due to the in situ formation of a high fraction of the complex borocarbide phase (−70 vol %). The weld metal dilution was over 22%.

The FCAW-GS hardfaced layer of the Vecalloy 752 Plate® wear-resistant plate was unique in the microstructure of the fine martensitic ferrous alloy (Figure 13k). The microstructure was similar to composite layers reinforced with tungsten carbide (WC). The microstructure was composed of cubic molybdenum borides (33–50 vol %), with a hardness on the level of tungsten carbide, embedded in a hard martensitic matrix with a high fraction of primary and eutectic Cr_7C_3 and $Cr_{23}C_6$ chromium carbides, niobium carbides, and metal borides (Figure 14h). The size of the complex molybdenum borides was around 10 µm, with spacing lower than 50 µm. The weld metal dilution was around 23%. The slightly lower abrasion wear resistance of the wear plate CDP 4695 compared to the Vecalloy 752 wear plate can be explained by the lesser volume fraction of stable borocarbides, carbides, and borides in the deposit.

The hardfaced layer of the PHWP wear-resistant plate (Figure 13l) was characterized by a highly alloyed chromium carbide microstructure with a high concentration of complex carbides. The microstructure was composed of a high eutectic fraction of hard (1000–1600 HV) $Cr_{23}C_6$ chromium carbides; niobium carbides NbC, with a hardness of approximately 1800 HV and molybdenum borides Mo_2B; and other metal compounds distributed in a plastic austenitic–martensitic matrix (Figure 14i).

Advantageous properties under the conditions of abrasion by mineral grains and erosion were also revealed by padding welds made of iron-based materials in the Fe-Cr-C-Mo-B alloy. The addition of boron increases the flowability and wettability of the metal and lowers the melting point. The abrasive wear resistance of borides was much higher than that of carbides. According to Hejowski [36], an increase in the boron content with constant chromium content increases the volume fraction and hardness of the primary phases. The calculated weld metal dilution was the smallest and amounted to 17%.

The wear mechanism of the wear-resistant layers presented in Figure 9h–l consisted only of the microcutting of the contact surface by abrasive grains. A volumetric loss of 10 to 15 mm^3 was measured. As a result of the wear process, strengthening phases embedded in the ductile matrix were revealed.

3.5. Linear Regression Model

In the majority of cases, the increase in hardness coincides with the increase in wear resistance, especially in hardfaced wear plates. Research conducted by Marulanda-Arévalo et al. [1] and Ban et al. [4] demonstrated a strong linear relationship between the hardness and the wear resistance of the hardfaced wear plates. ASTM G 64 tests were performed on samples that had undergone a grinding treatment. The surface condition of the wear plate was taken into account for the purpose of determining wear resistance.

The obtained abrasive wear resistance test results of arc-hardfaced alloys from groups T Fe15 and T Fe16 were used to obtain a mathematical linear regression model describing the dependence between metal-mineral wear resistance and the hardness of hardfaced layers. A linear predictor function with an equation in the form $y = a + bx$, where x is the explanatory variable (surface hardness) and y is the dependent variable (abrasive wear resistance), was used (Figure 16).

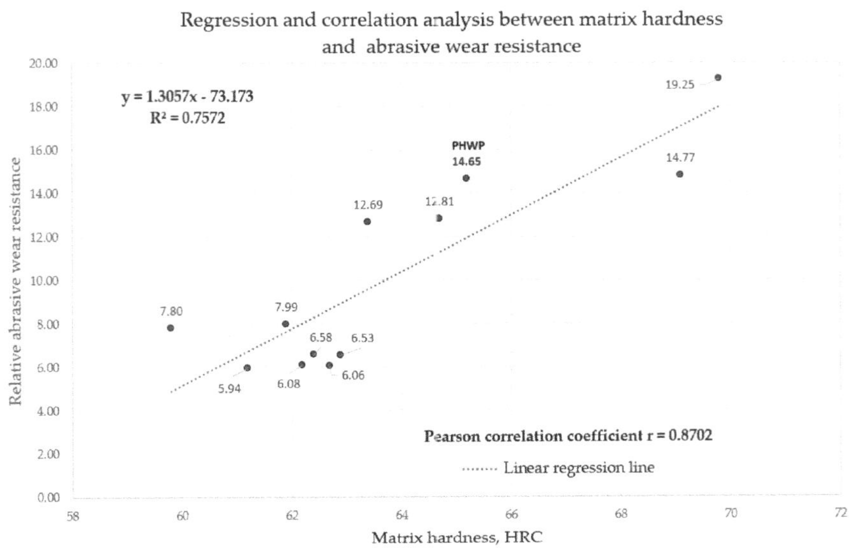

Figure 16. Regression and correlation analysis between matrix hardness and abrasive wear resistance.

Pearson's r correlation coefficient was calculated for the proposed model and was equal to 0.8702, which indicates the high accuracy of the description of the linear model between the hardness of the layer and its resistance to abrasive wear. The calculated coefficient of determination $R^2 = 0.7572$ proves the accuracy of the model. The proposed mathematical linear regression model describes the variability of the dependent variable (abrasive wear resistance) and the independent variable (surface layer hardness) by nearly 80%.

4. Conclusions

The aim of this research was to compare abrasive wear resistance and determine the linear correlation between surface hardness and the metal-mineral wear resistance of 11 commercially produced and industrially applied wear-resistant composite plates and hardfaced layers produced by a patented covered tubular electrode with a special chemical composition of the metallic core. The comparative analysis allowed for the formulation of the following conclusions:

1. The layers of commercially produced, arc-hardfaced, additional modern materials from alloy group T Fe16 composite wear-resistant plates possess higher hardness and abrasive wear resistance compared to conventional chromium carbide overlay (CCO) wear plates with layers hardfaced with alloys from the T Fe15 group. The higher wear resistance of the CastoDur Diamond Plate® 4695, Vecalloy 752 Plate®, and PHWP plates results from the higher volume fraction and dispersion of more uniform, finer, and harder precipitations in the matrix. For chromium carbide overlay (CCO) wear plates with layers hardfaced with alloys from the T Fe15 group, the more malleable metal matrix was subjected to more intensive wear from abrasive interaction. This revealed brittle carbides, causing cracking and ripping of the carbides, promoting further degradation of the matrix. The type, size, and dispersion of strengthening phases further impact abrasive wear resistance.
2. The metallographic examination and X-ray diffraction analysis indicated that layers hardfaced with alloys from the T Fe15 group have microstructures composed mostly of primary M_7C_3 carbides uniformly distributed in the Fe-Cr-C-type matrix, with occasional harder NbC-type carbides. The microstructure of the layers hardfaced with alloys from the T Fe16 group was more complex and consisted of ultrahard complex borocarbides $M_{23}(BC)_6$, MC metal carbides,

M_2B metal borides, or nearly cubic complex Mo_2B molybdenum borides, Cr_7C_3 primary chromium carbides, $Cr_{23}C_6$ eutectic chromium carbides, and Nb_6C_5 niobium carbides densely distributed in a martensitic matrix.

3. The metal-mineral abrasion resistance of layers made by the covered tubular electrode with an innovative chemical composition of metal core is equal to the plate made in the FCAW-SS process with the cored electrode wire EnDOtec DO*395N, which, according to the manufacturer, allows obtaining a nanostructured weld metal.
4. The obtained results suggest the high linear relationship between an increase in surface hardness increase and an increase in the metal-mineral abrasive resistance of the wear-resistant plates' hardfaced layers. The developed linear regression model will be applied in real-time wear prediction systems for wear plates and liners. It is worth noting that no factors, such as the type and chemical composition of the additional material, distribution, and geometric properties of the strengthening phase, were considered.

In the future, the results of laser and plasma layers hardfaced with alloys from the Ni20 group and metal-mineral abrasive wear resistance will be published by the author.

Funding: The research was founded by the Silesian University of Technology Rector's proquality grant and the Faculty of Mechanical Engineering Silesian University of Technology Dean's proquality grant.

Conflicts of Interest: The author declares no conflict of interest.

References

1. Marulanda-Arévalo, J.L.; Cañas-Mendoza, L.A.; Barón-Jaimez, J.A. Abrasive wear in wear plates and hard coatings applied by welding with shielded electrode. *Rev. Fac. Ing.* **2017**, *26*, 105–112. [CrossRef]
2. Lisiecki, A. Tribology and surface engineering. *Coatings* **2019**, *9*, 663. [CrossRef]
3. Jankauskas, V.; Kreivaitis, R.; Milcius, D.; Baltusnikas, A. Analysis of abrasive wear performance of arc welded hard layers. *Wear* **2008**, *265*, 1626–1632. [CrossRef]
4. Ban, M.; Hasegawa, N.; Ueno, Y.; Shinozaki, H.; Aoki, T.; Fukumoto, H. Wear resistance property of hardfacing weld overlays containing metal carbides. *Tribol. Online* **2012**, *7*, 207–212. [CrossRef]
5. Lisiecki, A. Mechanisms of hardness increase for composite surface layers during laser gas nitriding of the Ti_6A1_4V alloy. *Mater. Technol.* **2017**, *51*, 577–583.
6. Janicki, D. Fabrication of high chromium white iron surface layers on ductile cast iron substrate by laser surface alloying. *Stroj. Vestn. J. Mech. Eng.* **2017**, *63*, 363–372.
7. Klimpel, A.; Dobrzanski, L.A.; Lisiecki, A.; Janicki, D. The study of the technology of laser and plasma hardfacing of engine valves face made of X40CrSiMo10-2 steel using cobalt-based powders. *J. Mater. Process. Technol.* **2006**, *175*, 251–256. [CrossRef]
8. Czupryński, A.; Wyględacz, B. Comparative analysis of laser and plasma surfacing by nickel-based superalloy of heat resistant steel. *Materials* **2020**, *13*, 2367. [CrossRef]
9. Lisiecki, A. Titanium matrix composite Ti/TiN produced by diode laser gas nitriding. *Metals* **2015**, *5*, 54–69. [CrossRef]
10. Woldman, M.; van der Heide, E.; Schipper, D.J.; Tinga, T.; Masen, M.A. Investigating the influence of sand particle properties on abrasive wear behavior. *Wear* **2012**, *294*, 419–426. [CrossRef]
11. Bremerstein, T.; Potthoff, A.; Michaelis, A.; Schmiedel, C.; Uhlmann, E.; Blug, B.; Amann, T. Wear of abrasive media and its effect on abrasive flow machining results. *Wear* **2015**, *342*, 44–51. [CrossRef]
12. Lisiecki, A.; Piwnik, J. Tribological characteristic of titanium alloy surface layers produced by diode laser gas nitriding. *Arch. Metall. Mater.* **2016**, *61*, 543–552. [CrossRef]
13. Singh, K.; Khatirkar, R.K.; Sapate, S.G. Microstructure evolution and abrasive wear behavior of D2 steel. *Wear* **2015**, *328*, 206–216. [CrossRef]
14. Lisiecki, A.; Ślizak, D. Hybrid laser deposition of Fe-based metallic powder under cryogenic conditions. *Metals* **2020**, *10*, 190. [CrossRef]

15. Ojala, N.; Valtonen, K.; Heino, V.; Kallio, M.; Aaltonen, J.; Siitonen, P.; Kuokkala, V.T. Effects of composition and microstructure on the abrasive wear performance of quenched wear resistant steels. *Wear* **2014**, *317*, 225–232. [CrossRef]
16. Zahiri, R.; Sundaramoorthy, R.; Lysz, P.; Subramanian, C. Hardfacing using ferro-alloy powder mixtures by submerged arc welding. *Surf. Coat. Technol.* **2014**, *260*, 220–229. [CrossRef]
17. Lisiecki, A.; Kurc-Lisiecka, A. Erosion wear resistance of titanium-matrix composite Ti/TiN produced by diode-laser gas nitriding. *Mater. Technol.* **2017**, *51*, 29–34. [CrossRef]
18. Buchanan, V.E.; McCartney, D.G.; Shipway, P.H. A comparison of the abrasive wear behaviour of iron-chromium based hardfaced coatings deposited by SMAW and electric arc spraying. *Wear* **2008**, *264*, 542–549. [CrossRef]
19. Cora, O.N.; Ağcayazı, A.; Namiki, K.; Sofuoğlu, H.; Koç, M. Die wear in stamping of advanced high strength steels—Investigations on the effects of substrate material and hard-coatings. *Tribol. Int.* **2012**, *52*, 50–60. [CrossRef]
20. Dobrzanski, L.A.; Klimpel, A.; Bonek, M.; Lisiecki, A. Surface-layer's structure of X40CrMoV5-1 steel remelted and/or WC alloyed with HPDL laser. *Mater. Sci. Forum.* **2003**, *437*, 69–72. [CrossRef]
21. Lisiecki, A. Comparison of titanium metal matrix composite surface layers produced during laser gas nitriding of Ti_6Al_4V alloy by different types of lasers. *Arch. Metall. Mater.* **2016**, *61*, 1777–1784. [CrossRef]
22. Tabrett, C.P.; Sare, I.R.; Ghomashchi, M.R. Microstructure-property relationships in high chromium white iron alloys. *Inter. Mater. Rev.* **1996**, *41*, 59–82. [CrossRef]
23. Zum Gahr, K.H.; Eldis, G.T. Abrasive wear of white cast irons. *Wear* **1980**, *64*, 175–194. [CrossRef]
24. Buchely, M.F.; Gutierrez, J.C.; Leon, L.M.; Toro, A. The effect of microstructure on abrasive wear of hardfacing alloys. *Wear* **2005**, *259*, 52–61. [CrossRef]
25. Miller, R.F. *Cladded Wear Plates Find Mining Applications*; Clad Technologies Inc.: Moncton, NB, Canada, 1985; 1-800-978-9780.
26. Fulcher, J.K.; Kosel, T.H.; Fiore, N.F. The effect of carbide volume fraction on the low stress abrasion resistance of high Cr-Mo white cast irons. *Wear* **1983**, *84*, 313–325. [CrossRef]
27. Kim, C.K.; Lee, S.; Jung, J.Y.; Ahn, S. Effects of complex carbide fraction on high-temperature wear properties of hardfacing alloys reinforced with complex carbides. *Mater. Sci. Eng. A* **2003**, *349*, 1–11. [CrossRef]
28. ASTM G65-00. *Standard Test Method for Measuring Abrasion Using the Dry Sand/Rubber Wheel Apparatus*; American Society for Testing and Materials: West Conshohocken, PA, USA, 2015.
29. ISO 6508. *Metallic Materials—Rockwell Hardness Test—Part 1: Test Method*; ISO: Geneva, Switzerland, 2016.
30. EN 14700. *Welding Consumables. Welding Consumables for Hard-Facing*; CEN: Brussels, Belgium, 2014.
31. Czupryński, A.; Stawarz, M. Cast Iron Powder Core, Especially for the Production of Corrosion and Wear Resistant Layers. Poland Patent 234,142, 26 January 2018.
32. EN 10029. *Hot-Rolled Steel Plates 3 mm Thick or Above—Tolerances on Dimensions and Shape*; CEN: Brussels, Belgium, 2010.
33. ISO 17637. *Non-Destructive Testing of Welds—Visual Testing of Fusion-Welded Joints*; ISO: Geneva, Switzerland, 2016.
34. ISO 6507. *Metallic Materials—Vickers Hardness Test—Part 1: Test Method*; ISO: Geneva, Switzerland, 2018.
35. Tomków, J.; Czupryński, A.; Fydrych, D. The abrasive wear resistance of coatings manufactured on high-strength low-alloy (HSLA) offshore steel in wet welding conditions. *Coatings* **2020**, *10*, 219. [CrossRef]
36. Hejwowski, T. Erosive and abrasive wear resistance of overlay coatings. *Vacuum* **2009**, *83*, 166–170. [CrossRef]
37. Mulhearn, T.O.; Samuels, L.E. The abrasion of metals a model of the processes. *Wear* **1962**, *5*, 478–487. [CrossRef]
38. Archard, J.F.; Hirst, W. The wear of metals under unlubricated conditions. *Proc. R. Soc.* **1956**, *236*, 397–405.
39. Buttery, T.C.; Archard, J.F. Grinding and abrasive wear. *Proc. Inst. Mech. Eng.* **1971**, *188*, 43–51. [CrossRef]
40. ISO 15614. *Specification and Qualification of Welding Procedures for Metallic Materials—Welding Procedure Test—Part 7: Overlay Welding*; ISO: Geneva, Switzerland, 2016.
41. Górka, J.; Czupryński, A.; Żuk, M.; Adamiak, M.; Kopyść, A. Properties and structure of deposited nanocrystalline coatings in relation to selected construction materials resistant to abrasive wear. *Materials* **2018**, *11*, 1184. [CrossRef]
42. Klimpel, A.; Górka, J.; Czupryński, A. Comparison of chromium cast iron deposits of wear plates. *J. Achiev. Mater. Manuf. Eng.* **2006**, *18*, 387–390.

43. Klimpel, A.; Dobrzański, L.A.; Janicki, D. A study of worn wear plates of fan blades of steel mill fumes suction system. *J. Mater. Process. Technol.* **2005**, *164–165*, 1062–1067. [CrossRef]
44. Adamiak, M.; Górka, J.; Kik, T. Comparison of abrasion resistance of selected constructional materials. *J. Achiev. Mater. Manuf. Eng.* **2009**, *37*, 375–380.
45. Górka, J.; Czupryński, A.; Adamiak, M.; Kopyść, A. Evaluation of the properties and structure of nanocrystalline surface layers in relation to selected constructional materials resistant to abrasive wear. *Weld. Technol. Rev.* **2018**, *90*. [CrossRef]
46. Gucwa, M.; Winczek, J.; Mičian, M. The effect of the preheating on to properties of the wear resistant welds. *Weld. Technol. Rev.* **2020**, *92*. [CrossRef]
47. Czupryński, A.; Adamiak, M.; Bayraktar, E.; Wyględacz, B. Comparison of tribological properties and structure of coatings produced in powder flame spraying process on grey cast iron. *Weld. Technol. Rev.* **2020**, *92*, 7–21. [CrossRef]

materials

MDPI

Article

Comparative Analysis of Laser and Plasma Surfacing by Nickel-Based Superalloy of Heat Resistant Steel

Artur Czupryński and **Bernard Wyględacz** *

Department of Welding Engineering, Faculty of Mechanical Engineering, Silesian University of Technology, Konarskiego 18A, 44-100 Gliwice, Poland; artur.czuprynski@polsl.pl

* Correspondence: bernard.wygledacz@polsl.pl

Received: 28 April 2020; Accepted: 14 May 2020; Published: 21 May 2020

Abstract: In this article, the results of surfacing technology development, and structural, and mechanical properties examinations of 16Mo3 steel pipes with an outside coating of Inconel 625 deposited by automated plasma powder transferred arc (PPTA) and automated high power direct diode laser (HPDDL) surfacing were presented. Based on the results of non-destructive, metallographical macro- and microscopic, chemical composition, and thickness and hardness examinations optimal technology for use in high temperature energy or chemical industry applications was selected. The examinations conducted for each of the aforementioned technologies revealed the proper structure and high quality of coating. Dendritic structure with primary crystals growing in the direction of heat dissipation was revealed. No defects such as cracks, lack of fusion or porosity were found. Iron content in the most outer area of the layer made by PPTA with a heat input of 277–514 J/mm, thickness from 1.2 to 1.7 mm, between 4% and 5.5% was observed. Iron content in the most outer area of the layer made by HPDDL surfacing with output power of 1000–1600 W and scanning speed 3.3–4.7mm/s, from 0.6 to 1.3 mm in thickness, between 5.1% and 7.5% was observed. In coated pipes made by either technology high quality of surfaced layers, conforming to requirements posed on protective layers manufactured for prolonged exploitation in temperatures up to 625 °C, were observed. High temperature resistance examinations are the focus of further, yet unpublished, research. The obtained results point to slight differences in the parameters and properties of nickel-based superalloy layers surfaced on 16Mo3 pipes based on the technologies used. However, the process parameters optimization in the case of PPTA was simpler compared to HPDDL surfacing.

Keywords: nickel-based superalloy; heat resistant steel; surfacing; PPTA; HPDDL

1. Introduction

The devolvement of the energy industry is determined by techno-economical, ecological, legal, but most of all material-technology factors [1–6]. Achieving energetic security and profitability is based on the correct design and application of proper high temperature resistant materials, the application of new and advanced manufacturing technologies, as well as the addition of capabilities for burning alternative fuels such as biomass, non-recyclable waste. In the manufacturing of machines and plants used in energy production, conversion, distribution, fuel processing, and storage high temperature and creep resistant steels are used, as well as more and more often other high temperature resistant metallic materials [2]. Parameters, operational period, and the need for periodic maintenance of machines and elements in the energy industry are based not only on technological and design parameters but mainly on the combined impact of temperature, external forces, mechanical and thermal cycling and chemical composition of the medium on metallic materials.

The use of alternative fuels as an energy resource is a result of the international effort to reduce CO_2 production. However, introducing organic matter and waste to fuel mix results in the production of highly reactive chlorides and fluorides, drastically increasing the rate of power plant elements corrosion [5,7]. The combined application of high temperature and corrosive environments induces a need for a change of used materials of application of protective layers in the power industry. Structural degradation of steel or layer, applied by means of surfacing or thermal spraying, on power industry elements operated in creep range is described by evolving with time set of structural and physicochemical properties depending on applied stress and exploitation temperature. Exploitation conditions of the designed element can lead to the formation and further development of gas pores, voids, and microcracks, leading to premature failure [4]. Structural instability of temperature resistant steels and nickel alloys is influenced by: level of substructural changes i.e., dislocation density evolution, recrystallization and recovery, intermetallic phase and carbide formation and dissolution, change of phase morphology (distribution, shape, size), decomposition of perlite, bainite and martensite, impoverishment of metal matrix in Cr, Mo, W. Aforementioned factors regardless of chromium content in steel or protective layer impact corrosion resistance including adhesion of passive layer, as well as, reduce material strength and resistance to cracking. The stability of this process is based on the chemical composition of the alloy, metal structure in the base state, and decides on the usage of high temperature steel or preventative surfaced/thermal sprayed layer in a given temperature, stress, and other environmental conditions [1,2]. The heat exchanger pipes in waste disposal plants are most susceptible to the corrosive environment. The fact results in a continuous decrease in parameters impacting the operational period and the occurrence of premature failures.

Pressure vessel steel 16Mo3 (1.5415) is a low alloyed chromium-molybdenum steel designed for application in elevated temperature, exhibiting high plasticity and ductility. Molybdenum, as a ferrite promoting element, has a high affinity to carbon. In pressure vessel steels with molybdenum content (e.g., 15Mo3, 16Mo3) there is a possibility of alloying elements carbides formation. Molybdenum content in steel reduces the temperature of perlitic transformation and has nearly no impact on bainitic transformation temperature. Molybdenum alloyed steels posses ferritic-perlitic, ferritic-bainitic, or bainitic microstructure [1,2]. Molybdenum addition in a range of 0.25–1% increases steel creep resistance, hardness, yield strength, and ultimate tensile strength, simultaneously decreasing elongation. Mo as an alloying element has a positive impact on brittle cracking resistance, impact toughness, decreases the critical cooling rate, and increases wear resistance. Moreover, Mo addition slightly decreases plastic workability and machinability. Chromium content in 16Mo3 steel increases corrosion resistance in steam environment. These grades of steels can be successfully applied in energy and chemical industries up to the working temperature of 530 °C (e.g., boilers, pressure vessels, hot medium pipelines) [4,5].

One of the methods used to increase the operational period of boiler elements in high temperature and corrosive environments is the usage of complex nickel-based alloy (superalloys) coatings. Nickel-based superalloys are characterized by very high resistance to the above-mentioned conditions compared to typical high temperature resistant steels. This metallic material group is characterized by the following properties: working temperature up to 1250 °C, limited susceptibility to cyclic and dynamic loading, resistance to nitrogen, sulfur, and carbon compounds [5,7–10]. A widespread group of nickel superalloys is Inconel. Inconel alloys were developed in the 1940s by a research group of Wiggin Alloy (Great Britan) and are still used in power, aero, and space industries among others [11–16].

Applied on the industrial scale mechanized, automated and robotized surfacing technologies still require complex studies, which lead to achieving high quality protective layers on pressure vessel elements posed by the evolving needs of the power industry. The most widespread fabrication method of Inconel-based protective layers is a consumable electrode in inert gas shielding surfacing—131 (metal inert gas—MIG) [2,10]. However, achieving high quality requires no surface or inner surfacing defects, low iron content in the surfaced layer, and the low heat affected zone (HAZ) depth is difficult or impossible in case of MIG surfacing. This fact induces the need for the application of modern and advanced coating technologies. The selection of proper surfacing, surfacing or thermal spraying technology for manufacturing nickel-based superalloy layers on pressure vessel steel is multivariable and requires extended studies [17–40].

For this reason, the present work is a study about the possibility of high power density methods application in manufacturing on Inconel 625 protective layers.

2. Experimental

2.1. The Aim of the Study

The aim of the study was to conduct a comparative analysis of basic technological parameters of automated PPTA surfacing and HPDDL surfacing on structure, chemical composition, geometry, base material content and mechanical parameters of nickel-based superalloy Inconel 625—NiCr 22M09Nb (2.4856) acc. EN 10095:2002 [41] surfaced layers on the outer surface of 16Mo3 (1.5415) acc. EN 10216–2 [42] heat resistant steel pipes.

Both assessed technologies were performed under a wide range of technological parameters to establish technical guidelines for producing surfaced layers conforming to the following criteria:

- Lack of surfacing defects,
- The thickness of a single surfaced layer in range 1–1.5 mm,
- Depth of HAZ ≤ 1.5 mm,
- Iron content in surfaced layer ≤ 7% [2,10],
- Relatively high process productivity in comparison with conventional surfacing methods.

The study covered among others:

- Determination of optimal process parameters range of surfacing based on preliminary technological trials,
- Assessment of geometry change and base material content in the surfaced layer depending on process parameters,
- Comparison of the microstructure of surfaced layer, heat affected zone and base material using metallography,
- Determination of atomic and weight composition of surfaced layers,
- Hardness testing of surfaced layer, heat affected zone and base material,
- Carrying and evaluation out of scratch testing.

2.2. Materials

The examinations were performed on a segment of 16Mo3 pressure vessel steel pipe 180 mm in length, 51 mm in outer diameter and the wall thickness of 5 mm surfaced with a singular layer of nickel-based superalloy Inconel 625 through robotized plasma transfer arc and laser surfacing. In both surfacing methods, additional material in the form of metallic powder 50–150 µm in diameter was used. The surfaced layer was not subjected to post surfacing heat treatment. Chemical composition of base and additional material, acc to the manufacturer and obtained by spectrometric analysis, are presented in Tables 1–4. Mechanical properties of nickel-based superalloy weld metal are presented in Table 5.

Table 1. Chemical composition of 16Mo3 steel according to manufacturer data (Margo Ltd., Stalowa Wola, Poland).

Chemical Composition, wg.%											
C	Si	Mn	P	S	N	Cr	Cu	Mo	Ni	Fe	CE [1)]
0.12–0.2	≤0.35	0.4–0.9	≤0.025	≤0.01	≤0.012	≤0.3	≤0.3	0.25–0.35	≤0.3	Bal.	0.52

Notes: [1)] carbon equivalent calculated according to International Institute of Surfacing (IIW) guidelines.

Table 2. Chemical composition of 16Mo3 steel according to spectrometric analysis.

Chemical Composition, wg.%										
C	Si	Mn	P	S	N	Cr	Cu	Mo	Ni	Fe
0.17	0.34	0.58	0.022	0.008	0.01	0.26	0.19	0.32	0.28	Bal.

Notes: In the table mean values of 5 measurements were presented.

Table 3. Chemical composition of Inconel 625 superalloy; Böhler L625 (EN NiCr22Mo9Nb) powder according to manufacturer data (voestalpine Böhler Surfacing Germany GmbH, Hamm, Germany).

Chemical Composition, wg.%												
C	Si	Mn	P	S	Cr	Mo	Co	Ti	Al	Nb+Ta	Fe	Ni
≤0.03	≤0.4	≤0.5	≤0.01	≤0.01	21–23	8–10	≤1	≤0.4	≤0.4	3.2–3.8	≤5	Bal.

Table 4. Chemical composition of Inconel 625 superalloy, Böhler L625 (EN NiCr22Mo9Nb) powder deposit weld according to spectrometric analysis.

Chemical Composition, wg.%												
C	Si	Mn	P	S	Cr	Mo	Ni	Co	Ti	Al	Nb+Ta [1)]	Fe
0.03	0.26	0.38	0.006	0.004	20.98	8.46	63.37	0.74	0.32	0.17	3.42	1.86

Notes: In the table mean values of 5 measurements were presented, [1)] no Tantalum was identified.

Table 5. Mechanical properties of Böhler L625 (EN NiCr22Mo9Nb) powder deposit weld.

Mechanical Properties of Deposit Weld [1)]							
Hardness, HB 30	Yield Strength $R_{p0,2}$, MPa	Ultimate Tensile Strength R_m, MPa	Elongation A_5, %	Maximum Operation Temperature T, °C	Young Modulus E, GPa	Impact Energy ISO-V KV, J	
210 (≤240)	540 (≥460)	800 (≥760)	38 (≥35)	1000	200 (≤209)	+20 °C 160	−196 °C 130 (≥32)

Notes: [1)] metal in as welded condition (without PWHT), test temperature 20 °C, unless otherwise specified.

Prior to surfacing, according to the manufacturer recommendations, the powder was subjected to drying, by baking at a temperature of 200 °C for 10 min and mixed in a laboratory planetary stirrer. After preparation, the powder was placed in feeding hopper of a used powder surfacing station. The base material surface was prepared by cleaning the outer pipe surface from rust, mill scale, and grease. The base metal surface preparation consisted of abrasive blasting in cabinet sandblaster followed by mechanized brushing to remove any remaining electro corundum from surface and chemical degreasing. The used abrasive material was electro corundum normal brown with particle diameter 850–1000 µm (F22 acc. Federation of European Producers of Abrasives). The chemical agent used for degreasing was Tetrachloroethylene. Surface roughness parameters measured after preparation were Ra = 12 µm, Rz = 85 µm. The prepared pipes were mounted on an automated surfacing station. The station consisted of a horizontal pipe rotator (with self-centering chucks for pipe mounting) and pipe cooling equipment (with liquid coolant contact on the inner pipe surface). Additional station equipment changed based on surfacing technology applied.

2.3. Plasma Processing

PPTA was carried out with surfacing machine Durweld 300T PTA with machine powder plasma surfacing torch PT 300AUT (Durum Verschleiss-Schutz GmbH, Willich, Germany), mounted on industrial robot Fanuc R-2000iB (FANUC Ltd., Oshino-mura, Japan) arm.

Based on a preliminary surfacing trial, nine parameter sets were used for the fabrication of samples. The initial evaluation of samples has shown the fabrication of surfaced layers of acceptable quality. Optimal parameters of automated PPTA enabling the formation of nickel-based superalloy Inconel 625 surfaced layers on Surface of pressure vessel steel 16Mo3 pipe of sufficient quality are presented in Table 6.

Table 6. Processing parameters of robotic plasma powder transferred arc surfacing of Inconel 625 superalloy layer on the outer surface of the 16Mo3 steel pipe.

Parameters	Sample Designation								
	P1	**P2**	**P3**	**P4**	**P5**	**P6**	**P7**	**P8**	**P9**
Surfacing current I, (A)	160	160	160	170	170	170	190	190	190
Arc voltage U, (V)	20	20	20	20	20	20	20	20	20
Surfacing velocity v [1)], (mm/s)	5.2	6.5	7.9	7.9	7.9	9.2	11.8	14.4	14.4
Powder feed rate q, (g/min)	17	19	17	15	21	21	21	25	27
Plasma gas flow rate Q_p [2)], (l/min)	1.5	1.5	1.5	1.5	1.5	1.5	1.5	1.5	1.5
Shielding gas flow rate Q_o [2)], (l/min)	12	12	12	12	12	12	12	12	12
Transport gas flow rate Q_s [2)], (l/min)	4	4	4	4	4	4	4	4	4
Nozzle-workpiece distance l, (mm)	5	5	5	5	5	5	5	5	5
Overlap ratio, (%)	33	33	33	33	33	33	33	33	33
Heat input E_u [3)], (J/mm)	642	514	428	455	455	390	339	277	277

Notes: [1)] defined as the resultant velocity of rotational pipe movement and linear industrial robot manipulator movement parallel to pipe rotation axis, [2)] Argon 5.0 (99.999%) acc. ISO 14175—I1: 2009 [43] was used as plasma, shielding and transport gas, [3)] calculated acc. to the formula: E_u = k·(U × I)/v The thermal efficiency coefficient for plasma transferred arc k = 0.6 was used.

2.4. *Laser Processing*

The powder laser surfacing process was carried out on the robotized station, equipped with a modern laser system for surfacing mounted on a six-axis robot system, ABB IRB 2600 (Asea Brown Boveri, Zurich, Switzerland). A 2 kW 808 ± 5 nm wave length high power direct diode laser Rofin DL020 (ROFIN-SINAR Technologies Inc., Hamburg, Germany) with a rectangular beam with the top-hat intensity distribution in the slow-axis direction and a near Gaussian in the fast-axis direction was used in this study. The laser spot size in the focal plane, measured by the Prometec Laserscope UFF100, was approximately 1.8 mm × 6.8 mm. The fast-axis of the beam spot was set parallel to the traverse direction and the focal plane of the beam was positioned at the surface of the substrate material. The powder was injected directly into the molten pool by an off-axis powder injection system. To ensure a uniform powder distribution on the surface of the molten pool, the geometry of the powder injection nozzle has been fitted to the laser beam spot. Details of the used powder injection system are available elsewhere [12,16]. Argon was used as a shielding gas. In an effort to establish the range of optimal surfacing parameters, a series of single-pass clads have been made at laser powers of 1000, 1200, 1400, and 1600 W with traverse speeds ranging from 2.6 to 4.7 mm/s. The powder feed rate was in the range of 5 to 15 g/min. The optimal surfacing parameters (Table 7) were determined as the parameters providing the single-pass overlapping clad with a uniform and low fusion penetration into the base material.

The schematic of outer 16Mo3 steel pipe surface surfacing with nickel-based superalloy Inconel 625 by powder plasma transferred arc surfacing and powder laser beam surfacing is presented in Figure 1.

Table 7. Processing parameters of robotic laser powder surfacing of Inconel 625 superalloy layer on the outer surface of the 16Mo3 steel pipe.

Parameters	Oznaczenie Próbki								
	L1	L2	L3	L4	L5	L6	L7	L8	L9
Laser power, (W)	1200	1400	1600	1600	1000	1000	1000	1000	1200
Surfacing velocity [1], (mm/s)	3.9	3.9	3.9	4.7	3.9	3.3	3.3	2.6	2.6
Powder feed rate, (g/min)	10	10	10	15	10	10	7.5	5	5
Shielding gas flow rate Q_o [2], (l/min)	12	12	12	12	12	12	12	12	12
Transport gas flow rate Q_s [2], (l/min)	2	2	2	2	2	2	2	2	2
Overlap ratio, (%)	33	33	33	33	33	33	33	33	33
Heat input [3], (J/mm)	308	359	410	340	256	303	303	385	462

Notes: [1] defined as the resultant velocity of rotational pipe movement and linear industrial robot manipulator movement parallel to pipe rotation axis, [2] Argon 5.0 (99.999%) acc. ISO 14175—I1: 2009 [43] was used as shielding and transport gas, [3] defined as the laser power divided by the traverse speed.

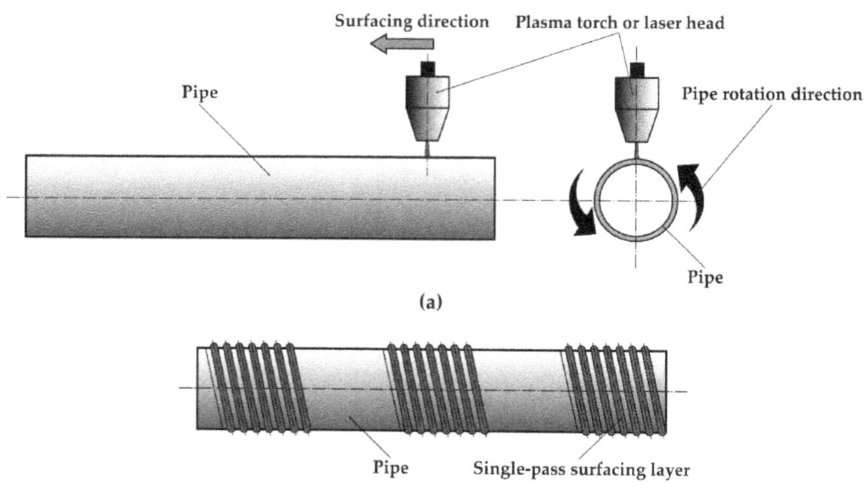

Figure 1. Schematic of outer 16Mo3 steel pipe surface surfacing with nickel-based superalloy Inconel 625: (**a**) movement of plasma/laser head and pipe rotation, (**b**) arrangement of surfacing passes on the pipe surface.

2.5. Methodology of Research

To establish quality and reveal surfacing defects such as cracks, discontinuities, gas pores, non-uniform geometry or lack of surfaced layer adhesion visual (VT) and penetrant-dye (PT) testing were carried out on each sample. Surfaced layers properties evaluation was based on macro and microscopic metallographic examinations, surfaced layer thickness measurements, HAZ measurements, determination of base material content in surfaced layer, chemical composition analysis, and X-ray diffraction of the most outer area of surfaced layer, microhardness testing and surfaced layer scratch resistance testing.

2.5.1. Non-Destructive Testing

Visual and penetrant-dye testing was carried out in accordance with procedures, material, and equipment from normative ISO 17637 [44] and ISO 3452–2 [45]. Visual testing was based on a direct inspection with the naked eye of the surfaced layer surface procedure. Before the testing surface subjected to observation were cleaned and dried. Penetrant-dye inspection was carried out with the use of color contrast penetrant System Designation Type II, Sensitivity 2 (PT ISO 3452-2 II Cd-2).

2.5.2. Metallographic Examination

Microscopic examinations were carried out on a cross-section subjected to typical metallographic specimen preparation. The samples were etched in two stages: steel pipe base material structure was revealed by etching in FeCl3Et (Mi19Fe) solution, nickel-based superalloy structure was revealed by electrochemical etching in reagent: 20 cm^3 HCl, 10g $FeCl_3$, 30 cm^3 distilled water. Etching parameters were selected by trial. Observation and recording of macro- and microstructure was carried out on Olympus SZX7 stereoscopic microscope (Olympus Corporation, Tokyo, Japan) and Olympus GX 71 inverted metallographic microscope (Olympus Corporation, Tokyo, Japan). Obtained macroscopic images enabled the determination of surfaced layer thickness, HAZ depth and base material content in the surfaced layer. Chemical composition examinations, including iron content determination, on the surface area of the nickel-based superalloy surfaced layer were done on XRF X-MET8000 Expert mobile spectrometer (Hitachi High-Technologies Corporation, Tokyo, Japan). Precise determination (surface and volumetric) of surfaced layer chemical composition were carried out on ZEISS SUPRA 25 scanning electron microscope (Carl Zeiss AG, Oberkochen, Germany) by means of Energy Dispersive Spectroscopy (EDS) method. High tension of 15 kV and probe current of 5 nA were used. X-ray diffraction testing, enabling phase content determination in surfaced layer, were carried out on X'Pert Pro PANalytical diffractometer (Malvern Panalytical Ltd., Malvern, UK) Cu (λ = 1.54056) lamp. X-ray diffraction was done in Bragg-Brentano geometry.

2.5.3. Hardness Measurements

Surfaced layer hardness testing was done using Vickers method on Nexus 423D stationary hardness tester (Innovatest Europe BV, Maastricht, Nederland). Hardness testing was carried out acc. to ISO 6507 [46]. The test load used was 300 gf (2.942 N). Hardness tests were performed in 9 test points on polished metallographic transversal crosssection of chosen surfaced layers manufactured by powder plasma surfacing and powder laser surfacing.

2.5.4. Scratch Test

To establish scratch resistance and kinetic friction coefficient of surfaced layers, macroscale scratch tests—Revetest (RST) on Revetest Xpress Scratch Tester machine (Anton Paar Instruments, Graz, Austria) were performed. Scratch tests were carried out with the use of Rockwell diamond indenter of radius R = 200 μm. Constant Load Scratch Test (CLST) mode of scratch testing was used. The indenter travel direction was parallel to the pipe axis and velocity was v = 0.3 mm/min on test length l = 5 mm with constant load P = 100 N. Schematic diagram of the scratch test is presented in Figure 2.

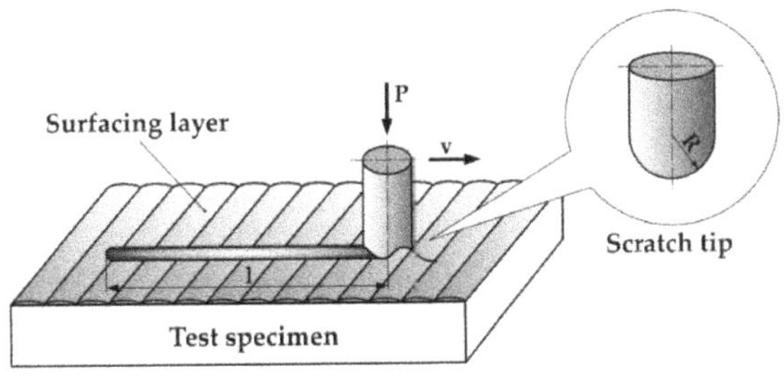

Figure 2. Schematic diagram of Inconel 625 nickel-base superalloy surfaced layer scratch test.

3. Results and Discussion

3.1. Non-Destructive Testing Results

During surfaced layers manufactured by powder plasma surfacing and powder laser surfacing no surfacing defects of type: cracks (100), surface pores (2017), excessive weld metal (502), incorrect weld toe (505), spatter (602) and other were found. Manufactured surfaced layers had high surface smoothness and symmetry of surfaced seam overlaps (Figure 3). Both powder plasma surfacing and powder laser surfacing enabled the formation of surfaced layers with quality level B. According to ISO 5817 [47] norm level B corresponds to the highest quality of manufactured layers.

(**a**) (**b**)

Figure 3. Photograph of Inconel 625 nickel-based superalloy surfaced layer on 16Mo3 steel pipe: (**a**) after surfacing, (**b**) after dye-penetrant application, removal and developer application (visible pipe length L = 180 mm, diameter D = 51 mm and thickness t = 5 mm).

3.2. Metallographic Test Results

Carried out macroscopic examinations of Inconel 625 surfaced layer and 16Mo3 steel pipe base material did not reveal any defect such as: cracks, lack of penetration, gas pores, and other types of discontinuities in fusion line area both powder plasma surfacing and powder laser surfacing. Lack of defects presence in the fusion line area indicates correct parameter selection and sufficient preparation of the pipe surface. Sample macrostructures of the fusion line between nickel-based superalloy surfaced layer and steel pipe base material are shown in Figure 4. The measured thickness of nickel-based superalloy layer manufactured by powder plasma surfacing was 1.2–1.7 mm (Table 8). The measured thickness of nickel-based superalloy layer manufactured by powder plasma surfacing was 0.6–1.3 mm (Table 9). The measured values are lower than recommended 2.5 mm [8,9]; however, taking into consideration the high price of nickel alloy additional material and the difference in heat transfer coefficient between nickel- and iron-based alloys surfaced layers can be considered to be conforming to criteria for exploitation and optimal [35,36]. Surfaced layers were manufactured as a single layer surfaced without post-weld heat treatment.

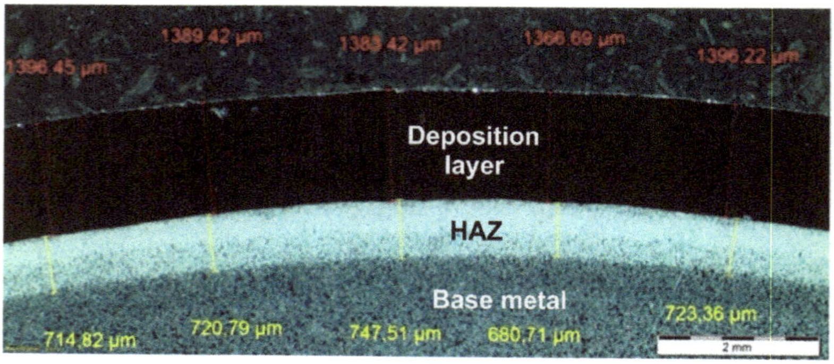

(**a**)

Figure 4. *Cont.*

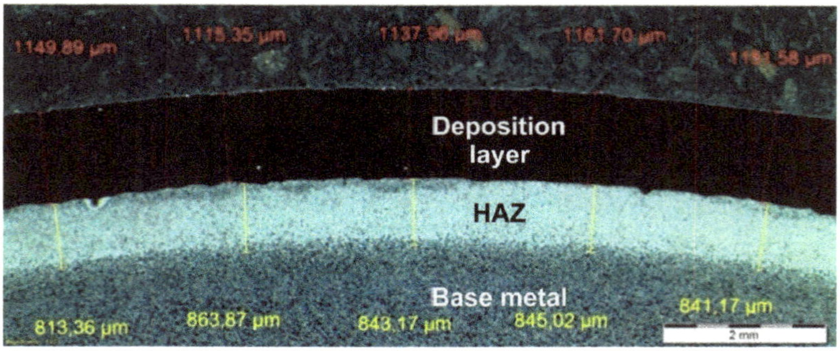

(b)

Figure 4. Sample macrostructures of the fusion line between nickel-based superalloy surfaced layer and steel pipe base material (**a**) powder plasma surfacing (sample P2), (**b**) powder laser surfacing (sample L4).

Table 8. The average thickness of the surfaced layers, iron content in the coating layers and HAZ width for samples obtained in the process of plasma powder transferred arc surfacing (PPTA), Table 6.

Sample Designation	P1 1)	P2	P3	P4	P5	P6	P7	P8	P9
Average surfacing thickness *g*, (µm)	1275	1339	1269	1132	1416	1193	1292	1392	1646
Average visible HAZ depth *s*, (µm)	1252	719	743	795	831	742	715	633	622
Surfaced layer base metal content 2), *U* (%)	10.7	5.0	3.2	4.2	3.7	5.1	4.1	4,0	3.8
Iron content in surfaced layer 3), *Fe* (wt. %)	11.8	5.2	4.0	4.8	4.2	5.5	4.5	4.3	4.3

Notes: 1) Sample was not considered consecutive examinations as iron content was to high, 2) base metal content in surfaced layer was calculated by formula U = P/(S + P) where P—fused material area on cross-section S—added material area on cross-section P, 3) mean value of five measurements.

Table 9. The average thickness of the coating layers, iron content in the coating layers and HAZ width for samples obtained in the process of laser surfacing, Table 7.

Sample Designation	L1 1)	L2 1)	L3 1)	L4	L5	L6	L7 1)	L8 1)	L9 1)
Average surfacing thickness *g*, (µm)	610	810	770	1143	650	1000	180	260	320
Average visible HAZ depth *s*, (µm)	1030	1128	1330	841	1070	1180	1120	1350	1340
Surfaced layer base metal content 2), *U* (%)	11.5	13.6	14.0	3.9	4.8	6.9	12.4	20.6	29.8
Iron content in surfaced layer 3), *Fe* (wt. %)	10.4	14.9	15.1	5.1	6.8	7.5	8.9	19.1	28.7

Notes: 1) Sample was not considered consecutive examinations as iron content was to high, 2) base metal content in surfaced layer was calculated by formula U = P/(S + P) where P—fused material area on cross-section S—added material area on cross-section P, 3) mean value of five measurements.

Microscopic metallographic examinations revealed the ferritic-perlitic microstructure of 16Mo3 steel base material (Figures 5d and 6d). In the case of both surfacing technologies in HAZ diversified microstructure was observed from martensitic, through martensitic-bainitic to ferritic-bainitic. Moreover, high heating and cooling rates of the base material in the heat affected zone, due to higher achievable surfacing velocity, have lowered grain growth rates observed in the case of powder plasma surfacing (Figures 5c and 6c) [37–39]. In both powder plasma surfacing and powder laser surfacing depth of HAZ was low, a probable cause was liquid cooling of pipe inner surface during the process. Visible HAZ depth evolved with the change of heat input into the material. With the increase of surfacing linear energy depth of the base material layer subdued to microstructural changes. In the range of examined parameters slightly higher depth of HAZ occurred as a result of powder laser surfacing, but did not exceed 1350 µm (Table 9).

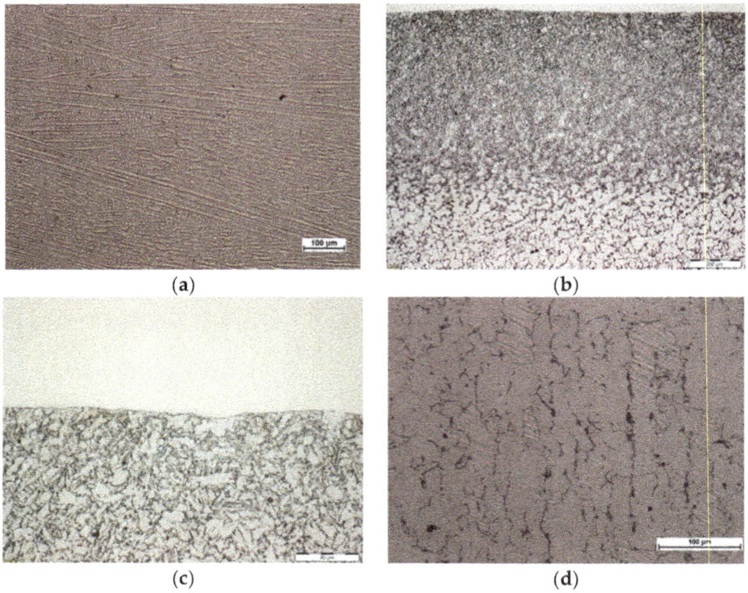

Figure 5. View of the microstructure of Inconel 625 superalloy coating layer obtained in the process of plasma powder transferred arc surfacing on the outer surface of 16Mo3 steel pipe (sample P2): (**a**) coating layer area, mag. 100×, (**b**) general view HAZ area mag. 100×, (**c**) HAZ area at the fusion line, mag. 500×, (**d**) base material, mag. 200×.

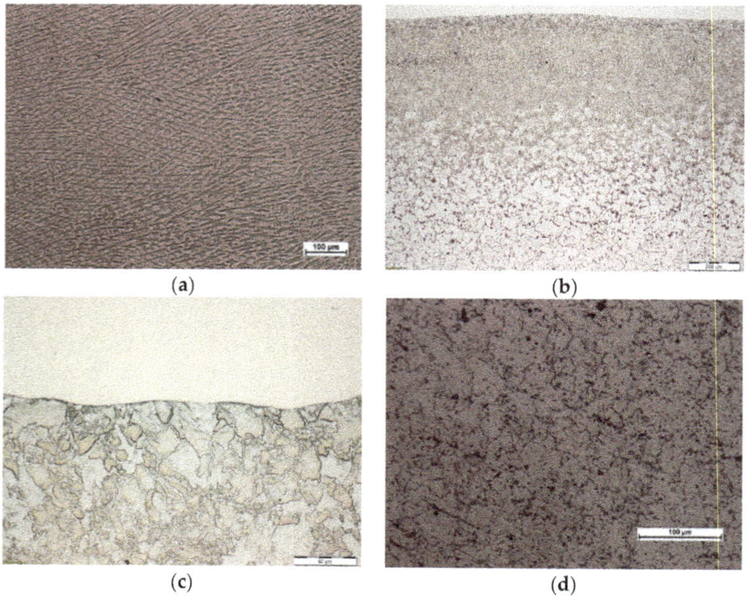

Figure 6. View of the microstructure of Inconel 625 superalloy coating layer obtained in the process of powder laser surfacing on the outer surface of 16Mo3 steel pipe (sample L4): (**a**) coating layer area, mag. 100×, (**b**) general view HAZ area, mag. 100×, (**c**) HAZ area at the fusion line, mag. 500×, (**d**) base material, mag. 200×.

Surfaced layers in each case were composed of organized, primary, fine-grained microstructure of dendrites and precipitations in interdendritic space—which is typical for nickel-based superalloy Inconel 625. Packets were orthogonal, dendritic, and oriented in the heat dissipation direction (Figures 5a and 6a). The structure was highly uniform and possessed no gas pores or cracks on the microscopic level. Slight grain refinement on the boundary layer between surfaced material and base material was observed. (Figures 5b and 6b). For each layer, slight fusion into the base material (enabling monolithic bonding of the surfaced layer with the pipe outer surface) was observed. One of the base criteria for the outer surface layer of heat exchangers pipe exploitation in biomass and waste fueled furnaces is the assessment of iron content in the surface layer. The guidelines of 7% maximal wg. content of Iron in case of automatic surfacing and 10% maximal wg. content of Iron in case of manual surfacing should not be exceeded [2,10]. Exciding the maximal iron content in the surfaced surface layer cand lead to iron oxide (Fe_2O_3) formation, which due to discontinuous and lamellar structure are prone to flaking during exploitation [2,10]. Measured iron content in the case of powder plasma surfaced layers, with the exclusion of sample P1, was in the range of 4.0–5.5 wg.% (Table 8), whereas in the case of powder laser surfaced layer only samples L4, L5, L6 had around 7 wg.% of iron (Table 9). EDS chemical composition microanalysis (Figure 7) revealed that the interdendritic area is rich in Niobium content. Iron and Chromium are equally present intra- and interdendritically. Iron presence in the surface layer is due to the mixing of base and additional metal. Chemical composition of interdendritic precipitations suggests the presence of γ/NbC eutectics, probably derivative of Ni2Nb Laves phase.

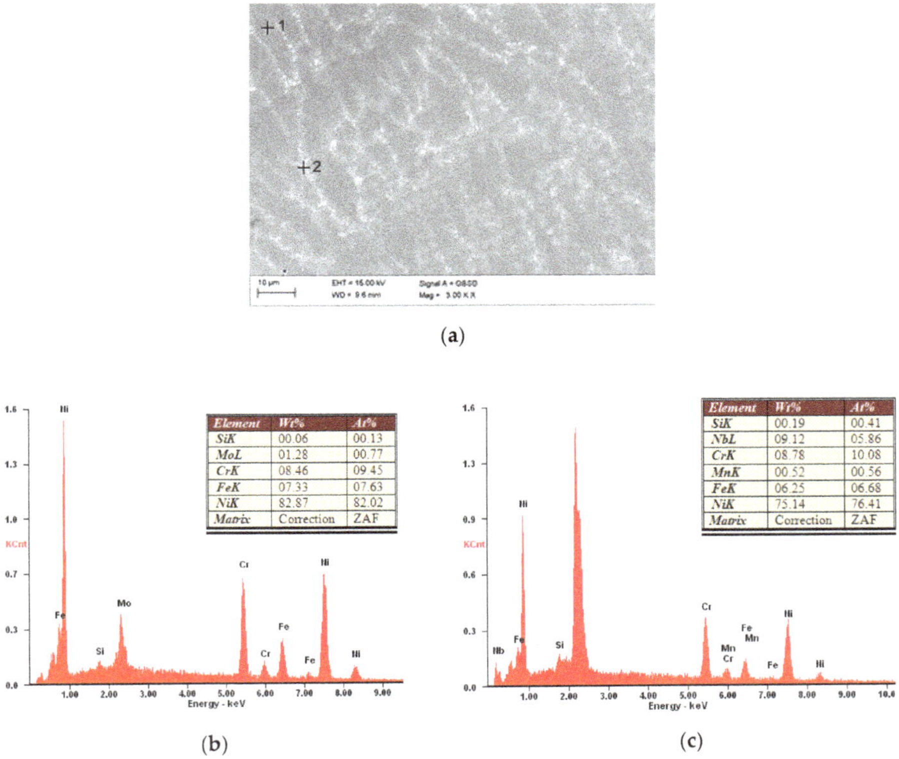

Figure 7. Sample BSE image of surfaced layer microstructure with results of EDS point microanalysis, mag. 3000×, high tension 15 kV (sample L4): (**a**) a view of dendritic structure, (**b**) chemical analysis of intradendritic area (measurement point 1), (**c**) chemical analysis of interdendritic area (measurement point 2), higher content of Nb compared to Inconel 625 powder chemical composition is notable.

One of the main factors determining iron content in the surfaced layer is surfacing linear heat input. According to [7,11,15] the linear heat input should no exceed 300 J/mm. In the case of PPTA with intensive liquid cooling of the inner pipe surface, modified heat dissipation mode enabled obtaining of sufficiently low iron content in the surfaced layer when linear surfacing energy was kept under 500 J/mm. Maximum linear heat input into the material in case of powder laser surfacing, with other technology parameters unchanged, which resulted in the sufficiently low iron content was 340 J/mm.

3.3. Results of the XRD Analysis

X-ray diffractogram for singular surfaced layer of nickel-based superalloy Inconel 625 manufactured by powder plasma surfacing (sample P2) was presented in Figure 8. Similarly in powder laser surfaced layer and powder plasm surfaced layers examinations main counts peaks from nickel are observed on diffractogram with angles: $2\theta = 43.66°, 50.85°, 75.34°, 90.83°$, and $96.74°$. However, main count peaks for pure nickel crystalline structure from JCPDS-ICDD database angles are: $2\theta = 44.51°, 51.85°, 76.37°, 92.94°$, and $98.45°$. In the case of analyzed sample counts peaks from crystallographic planes γ (111), γ (200), γ (220), and γ (311) were observed with slightly higher angle 2θ. This difference can be caused by the change of lattice parameters due to the solid solution of Inconel 625 alloying elements in Ni lattice and strengthening phases precipitations. In the surfaced layer Ni-Si phase with lattice Miller indices (101), (111), (120), (121), (301), and (310) was found. Moreover, in the surfaced layers manufactured by powder plasma surfacing or powder laser surfacing no other phases with parameters comparable to pure Ni were found.

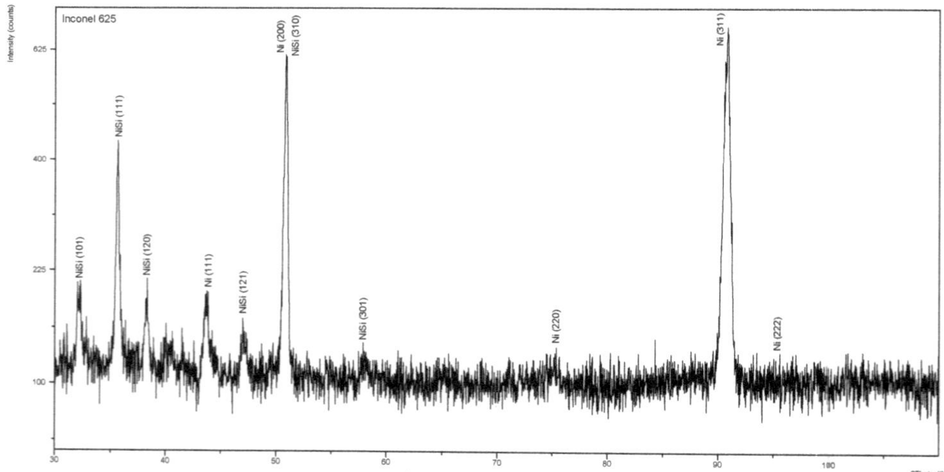

Figure 8. Examples of X-ray spectrum of overlay weld in the surface of Inconel 625 superalloy layer obtained in the process of plasma powder transferred arc surfacing (sample P2).

3.4. Hardness Measurements Test Results

Measured in the course of macroscopic metallographic examinations of powder plasma surfaced pipes and powder laser surfaces pipes, the depth of HAZ (Figure 5) was confirmed in microhardness tests. Mean microhardness HV0.3 results tested on cross-section on 16Mo3 steel plates surfaced with nickel-base superalloy Inconel 625 manufactured by surfacing were presented in Tables 10 and 11. The presented hardness results are mean from five measurements done at 0.1 mm intervals in the base material, HAZ, and surfaced layer. The base material was characterized by hardness in range 160–168 HV0.3. In the range of used surfacing parameters, the surfacing thermal cycle increased the hardness of HAZ to maximally 210 HV0.3 in case of plasma surfacing and 260 HV0.3 in case of laser surfacing. The HAZ microstructure evolved with the increasing distance from the fusion line. Near the fusion line bainitic-ferritic microstructure with a hardness of around 200–210 HV0.3, as a result of normalization, was present. In the distance of around 1.5 mm from the fusion line, microstructure changed to ferritic-pelitic-bainitic with hardness in the range 170–180 HV0.3. In the distance of 2.0 mm and more from the fusion line, the perlitic-ferritic microstructure of base material was present. The surfaced layer manufactured by powder plasma surfacing was characterized by hardness of over 240 HV0.3, while the hardness of the powder laser surfaced layer decreased sharply, from 261 HV0.3 to 187 HV0.3, with the increase of base material content in the surfaced layer (Table 11).

Table 10. Average microhardness HV0.3 measured on the cross-section of 16Mo3 steel pipes plasma clad with Inconel 625 superalloy powder (Table 6).

Microhardness Test Area	Sample Designation								
	P1	P2	P3	P4	P5	P6	P7	P8	P9
	Mean Microhardness HV 0.3								
Base material (16Mo3)	161.5	165.3	163.8	162.7	162.9	163.1	162.9	161.4	160.7
Heat affected zone	209.8	193.4	196.4	199.2	190.5	186.1	182.2	180.3	178.8
Surfaced layer (Inconel 625)	242.8	245.4	241.0	246.2	248.7	249.6	242.2	245.1	242.9

Table 11. Average microhardness HV0.3 measured on the cross-section of 16Mo3 steel pipes laser clad with Inconel 625 superalloy powder (Table 7).

Microhardness Test Area	Sample Designation								
	L1	L2	L3	L4	L5	L6	L7	L8	L9
	Mean Microhardness HV0.3								
Base material (16Mo3)	163.5	166.3	168.2	161.8	160.3	164.0	163.7	160.6	163.2
Heat affected zone	259.5	249.7	240.5	251.2	253.7	253.0	250.5	250.7	249.9
Surfaced layer	234.2	231.4	229.0	265.9	254.3	252.0	245.2	204.1	186.7

3.5. Scratch Test Results

Scratch tests enabled the determination of kinetic friction coefficient μ, scratch depth h, and friction force T on the outer surface of the surfaced layer. For scratch testing, one sample from powder plasma surfaced and powder laser surfaced pipes were chosen. The surfaced layers subjected to scratch testing had a comparable iron content of under 7%. Scratch test was performed on slightly leveled by grinding surface to exclude the impact of layers convexity on the results. As a result, powder laser surfaced layer (sample L4), with mean iron weight content of 5.1%, has slightly lower (0.38) mean outer surface kinetic coefficient of friction, than powder plasma surfaced layer (sample P2, Fe—5.2 wg.%), for which mean outer surface kinetic coefficient of friction was 0.39. Obtained results correlate with a higher mean microhardness of the sample L4 surfaced layer (Tables 10 and 11). In cases of both P2 and L4 scratch test results regions with a lower and higher coefficient of friction are visible. These results can be attributed to changes in surface scratch resistance due to partial heat treating of the previous surfacing pass by the next pass. The scratch test results are presented in Figures 9 and 10.

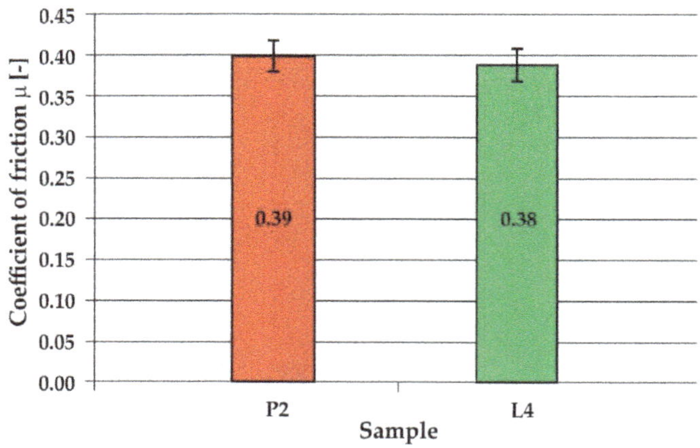

Figure 9. Coefficient of friction for Inconel 625 superalloy layer obtained in the process of plasma powder transferred arc surfacing (sample P2) and of laser surfacing (sample L4).

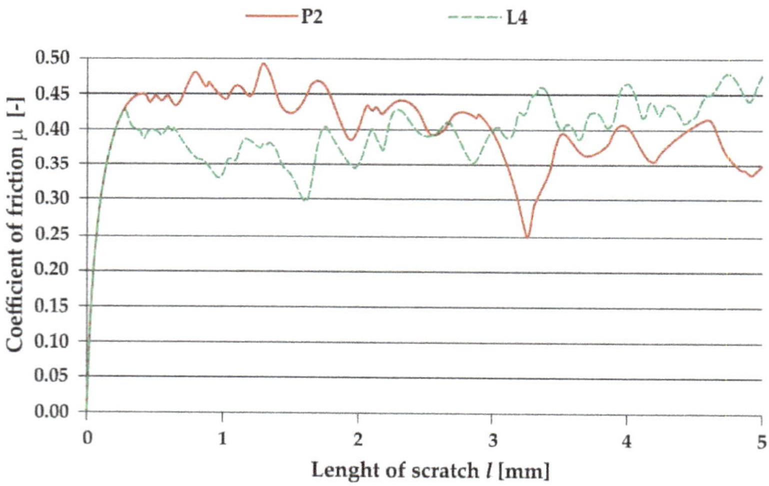

Figure 10. Coefficient of friction as a function of scratch length for Inconel 625 superalloy layers obtained in the process of plasma powder transferred arc surfacing (sample P2) and of laser surfacing (sample L4).

4. Conclusions

The aim of this study was to compare robotized powder plasma transferred arc surfacing and powder high power diode laser surfacing parameters on structure, chemical composition, geometry, base material content and mechanical properties of nickel-based superalloy Inconel 625 layers surfaced on 16Mo3 pressure vessel grade steel pipe. The carried comparative analysis enabled the formation of the following conclusions:

1. In the case of both powder plasma surfacing and powder laser surfacing narrow parameter range enabling the formation of nickel-based superalloy Inconel 625 layer with dependable fusion into 16Mo3 steel pipe base material, minimal content of base material, iron content on the outer surfaced layer surface under 7 wg.%, and low HAZ depth is present.

2. Achieving low base material content in the surfaced layer and reducing detrimental microstructural changes in the base material is possible, even when high surfacing linear energy is used, by application of intensive liquid cooling of inner pipe surface during the process.
3. Singular layer depth in case of powder plasma and laser surfacing was two times lower than in the case of MIG surfacing [2,10], moreover finishing machining overmeasure can be reduced in the case of PPTA and HPDDL surfacing. As a result, additional material usage can be reduced in the case of tested technologies.
4. In the powder plasma surfaced and powder laser surfaced layers no additional phases, aside from Ni-Si, with lattice parameters close to pure Nickel were found.
5. The nickel-based superalloy Inconel 625 surfaced layers manufactured by powder laser surfacing exhibit higher hardness compared to powder plasma surfaced layers.

In the future, results of holding temperature impact on nickel-based superalloy Inconel 625 layer will be presented.

Author Contributions: Conceptualization, A.C.; Data curation, A.C.; Formal analysis, A.C. and B.W.; Funding acquisition, A.C.; Investigation, A.C.; Methodology, A.C.; Project administration, B.W.; Resources, A.C.; Supervision, A.C.; Validation, B.W.; Visualization, B.W.; Writing-original draft, A.C.; Writing-review & editing, B.W. All authors have read and agreed to the published version of the manuscript.

Funding: This research was financed from the own resources of the Silesian University of Technology.

Conflicts of Interest: The authors declare no conflict of interest.

References

1. Dobrzański, J. Materials science interpretation of the life of steels for power plants. Second edition. *Open Access Libr.* **2016**, *3*, 1–228.
2. Golański, G.; Lachowicz, M.; Słania, J.; Jasak, J.; Marszałek, P. Research on 16Mo3 (16M) steel pipes overlaid with Haynes NiCr625 alloy using MIG (131) method. *Arch. Metal. Mater.* **2015**, *60*, 2521–2524. [CrossRef]
3. Węgrzyn, T.; Piwnik, J. Low alloy surfacing with micro-jet cooling. *Arch. Metal. Mater.* **2012**, *57*, 540–543. [CrossRef]
4. Zieliński, A.; Dobrzański, J. Characteristics of changes in properties and structure in X10CrMoVNb9-1 steel due to long-term impact temperatures and stress. *Arch. Mater. Sci. Eng.* **2013**, *60*, 72–81.
5. Uussitalo, M.A.; Vuoristo, P.M.J.; Mantyla, T.A. High temperature corrosion of coatings and boiler steels in reducing chlorine—Containing atmosphere. *Surf. Coat. Technol.* **2002**, *161*, 275–285. [CrossRef]
6. Chmielewski, T.; Golański, D. New method of in-situ fabrication of protective coatings based on Fe-Al intermetallic compounds. *Proc. Inst. Mech. Eng. Part B J. Eng. Manuf.* **2011**, *225*, 611–616. [CrossRef]
7. Adamiec, J. High temperature corrosion of power boiler components surfaced with nickel alloys. *Mater. Charact.* **2009**, *60*, 1093–1099. [CrossRef]
8. Frei, J.; Alexandrov, B.T.; Rethmeier, M. Low heat input gas metal arc surfacing for dissimilar metal weld overlays part I: The heat-affected zone. *Surf. World* **2016**, *60*, 459–473.
9. Frei, J.; Alexandrov, B.T.; Rethmeier, M. Low heat input gas metal arc surfacing for dissimilar metal weld overlays part II: The transition zone. *Surf. World* **2018**, *62*, 317–324.
10. Rozmus-Górnikowska, M.; Blicharski, M.; Kusiński, J. Influence of weld overlaying methods on microstructure and chemical composition of Inconel 625 boiler pipe coatings. *Kov. Mater.* **2014**, *52*, 1–7. [CrossRef]
11. Lippold, J.C.; Kiser, S.D.; DuPont, J.N. *Welding Metallurgy and Weldability of Nickel-Base Alloys*, 1st ed.; Wiley: Hoboken, NJ, USA, 2011.
12. Janicki, D. Laser surfacing of Inconel 625-based composite coatings reinforced by porous chromium carbide particles. *Opt. Laser Technol.* **2017**, *94*, 6–14. [CrossRef]
13. Węgrzyn, T.; Miros, M.; Hadryś, D.; Silva, A. Truck frame surfacing reparation by steel covered electrodes with varied amount of Ni and Mo. *Transp. Probl.* **2010**, *5*, 87–94.
14. Lisiecki, A. Tribology and Surface Engineering. *Coatings* **2019**, *9*, 663. [CrossRef]

15. Rajkumar, V.; Arjunan, T.V.; Rajesh Kannan, A. Metallurgical and mechanical investigations of Inconel 625 overlay welds produced by GMAW-hardfacing process on AISI 347 pipes. *Mater. Res. Express* **2019**, *6*, 076534. [CrossRef]
16. Janicki, D.; Musztyfaga-Staszuk, M. Direct Diode Laser Cladding of Inconel 625/WC Composite Coatings. *Stroj. Vestn. J. Mech. Eng.* **2016**, *62*, 363–372. [CrossRef]
17. Tang, B.; Tan, Y.; Zhang, Z.; Xu, T.; Sun, Z.; Li, X. Effects of Process Parameters on Geometrical Characteristics, Microstructure and Tribological Properties of TiB_2 Reinforced Inconel 718 Alloy Composite Coatings by Laser Surfacing. *Coatings* **2020**, *10*, 76. [CrossRef]
18. Tomków, J.; Rogalski, G.; Fydrych, D.; Łabanowski, J. Advantages of the application of the temper bead surfacing technique during wet surfacing. *Materials* **2019**, *12*, 915. [CrossRef]
19. Tello, L.; Castejon, L.; Malon, H.; Valladares, D.; Luque, P.; Mantaras, D.A.; Ranz, D.; Cuartero, J. Development of a fatigue life prediction methodology for welded steel semi-trailer components based on a new criterion. *Eng. Fail. Anal.* **2020**, *108*, 104268. [CrossRef]
20. Bayraktar, E.; Jansen, J.P.; Hugele, D.; Kaplan, D. Evaluation of pipeline laser girth weld properties by Charpy (V) toughness and impact tensile tests. *J. Mater. Process. Technol.* **2004**, *147*, 155–162. [CrossRef]
21. Cardoso, J.V.; Gamboa, P.V.; Silva, A. Effect of surface pre-treatment on the behaviour of adhesively-bonded CFRP T-joints. *Eng. Fail. Anal.* **2019**, *104*, 1188–1202. [CrossRef]
22. Janicki, D. Fabrication of high chromium white iron surface layers on ductile cast iron substrate by laser surface alloying. *Stroj. Vestn. J. Mech. Eng.* **2017**, *63*, 363–372.
23. Wang, S.-H.; Yang, C.-W.; Lee, T.-M. Evaluation of Microstructural Features and in Vitro Biocompatibility of Hydrothermally Coated Fluorohydroxyapatite on AZ80 Mg Alloy. *Ind. Eng. Chem. Res.* **2016**, *55*, 5207–5215. [CrossRef]
24. Miraoui, I.; Elimi, E.; Boujelbene, M.; Bayraktar, E. Analysis of Roughness and Microstructure for High-Power Laser Cutting of Stainless Steel. *Adv. Sci. Lett.* **2013**, *19*, 483–486. [CrossRef]
25. Roskosz, S.; Adamiec, J. Methodology of quantitative evaluation of porosity, dendrite arm spacing and grain size in directionally solidified blades made of CMSX-6 nickel alloy. *Mater. Charact.* **2009**, *60*, 1120–1126. [CrossRef]
26. Lisiecki, A. Mechanisms of hardness increase for composite surface layers during laser gas nitriding of the Ti6A14V alloy. *Mater. Technol.* **2017**, *51*, 577–583.
27. Zhou, Z.; Hu, M.-Y.; Xin, H.; Chen, B.; Wang, G.-X. Experimental and theoretical studies on the droplet temperature behavior of R407C two-phase flashing spray. *Int. J. Heat Mass Transf.* **2019**, *136*, 664–673. [CrossRef]
28. Lisiecki, A.; Kurc-Lisiecka, A. Erosion wear resistance of Titanium-Matrix Composite Ti/TiN produced by diode-laser gas nitriding. *Mater. Technol.* **2017**, *51*, 29–34. [CrossRef]
29. Lisiecki, A.; Piwnik, J. Tribological characteristic of titanium alloy surface layers produced by diode laser gas nitriding. *Arch. Metall. Mater.* **2016**, *61*, 543–552. [CrossRef]
30. Zhou, Z.; Wu, W.; Chen, B.; Wang, G.; Guo, L. An experimental study on the spray and thermal characteristics of R134a two-phase flashing spray. *Int. J. Heat Mass Transf.* **2012**, *55*, 4460–4468.
31. Lisiecki, A. Comparison of Titanium Metal Matrix Composite surface layers produced during laser gas nitriding of Ti_6Al_4V alloy by different types of lasers. *Arch. Metall. Mater.* **2016**, *61*, 1777–1783. [CrossRef]
32. Lisiecki, A. Titanium Matrix Composite Ti/TiN Produced by Diode Laser Gas Nitriding. *Metals* **2015**, *5*, 54–69. [CrossRef]
33. Lisiecki, A. Study of Optical Properties of Surface Layers Produced by Laser Surface Melting and Laser Surface Nitriding of Titanium Alloy. *Materials* **2019**, *12*, 3112. [CrossRef] [PubMed]
34. Lisiecki, A.; Ślizak, D. Hybrid Laser Deposition of Fe-Based Metallic Powder under Cryogenic Conditions. *Metals* **2020**, *10*, 190. [CrossRef]
35. Moskal, G.; Grabowski, A.; Lisiecki, A. Laser remelting of silicide coatings on Mo and TZM alloy. *Solid State Phenom.* **2015**, *226*, 121. [CrossRef]
36. Mikułowski, B. *Heat Resistant and High Temperature Creep Resistant Alloys—Superalloys*; Publishing House of the AGH University of Technology: Krakow, Poland, 1997. (In Polish)
37. Adamiec, J.; Kierzak, A. Padding of the components of waste combustion boilers with the use nickel alloys. *Mater. Eng.* **2008**, *4*, 380–385.

38. Cieślak, M.J.; Headley, T.J.; Kollie, T.; Roming, A.D., Jr. A melting and solidification study of Alloy 625. *Metall. Trans. A* **1988**, *19*, 2319–2331. [CrossRef]
39. Cieślak, M.J. The welding and solidification metallurgy of Alloy 625. *Weld. J.* **1991**, *70*, 49–56.
40. Sridhar, R.; Devendranath, R.K.; Arivazhagan, N. Characterization of microstructure, strength and toughness of dissimilar weldments of Inconel 625 and duplex stainless steel SAF 2205. *Acta Metall. Sin. (Engl. Lett.)* **2014**, *27*, 1018–1030. [CrossRef]
41. CEN. *Heat Resisting Steels and Nickel Alloys*; EN 10095:1999; CEN: Brussels, Belgium, 1999.
42. CEN. *Seamless Steel Tubes for Pressure Purposes—Technical Delivery Conditions—Part 2: Non-Alloy and Alloy Steel Tubes with Specified Elevated Temperature Properties*; EN 10216-2; CEN: Brussels, Belgium, 2004.
43. ISO. *Surfacing Consumables—Gases and Gas Mixtures for Fusion Surfacing and Allied Processes*; ISO 14175; ISO: Geneva, Switzerland, 2008.
44. ISO. *Non-Destructive Testing of Welds—Visual Testing of Fusion-Welded Joints*; ISO 17637; ISO: Geneva, Switzerland, 2016.
45. ISO. *Non-Destructive Testing—Penetrant Testing—Part 2: Testing of Penetrant Materials*; ISO 3452; ISO: Geneva, Switzerland, 2013.
46. ISO. *Metallic Materials—Vickers Hardness Test—Part 1: Test Method*; ISO 6507; ISO: Geneva, Switzerland, 2018.
47. Welding—Fusion-Welded Joints in Steel, Nickel, Titanium and Their Alloys (Beam Welding Excluded)—Quality Levels for Imperfections. Available online: https://www.iso.org/standard/54952.html (accessed on 19 May 2020).

MDPI
St. Alban-Anlage 66
4052 Basel
Switzerland
Tel. +41 61 683 77 34
Fax +41 61 302 89 18
www.mdpi.com

Materials Editorial Office
E-mail: materials@mdpi.com
www.mdpi.com/journal/materials

www.ingramcontent.com/pod-product-compliance
Lightning Source LLC
Chambersburg PA
CBHW041938260926
51850CB00018B/246

9783036567013